Managing Plant Viral Diseases With Advanced Nanoparticles

Javed Iqbal
https://orcid.org/0000-0002-9032-2622
Bacha Khan University, Pakistan

Muhammad Anas
https://orcid.org/0000-0001-5127-8043
Selçuk University, Turkey

Ayush Madan
https://orcid.org/0000-0001-5089-1794
People's University, India

Ghulam Murtaza
Yunnan University, China

IGI Global
Scientific Publishing
Publishing Tomorrow's Research Today

Vice President of Editorial Melissa Wagner
Director of Acquisitions Mikaela Felty
Director of Book Development Jocelynn Hessler
Production Manager Mike Brehm
Cover Design Jose Rosado

Published in the United States of America by
IGI Global Scientific Publishing
701 East Chocolate Avenue
Hershey, PA, 17033, USA
Tel: 717-533-8845 | Fax: 717-533-7115
Website: https://www.igi-global.com E-mail: cust@igi-global.com

Library of Congress Cataloging-in-Publication Data

Names: Iqbal, Javed (Professor of botany)- editor | Anas, Muhammad, 1997- editor | Madan, Ayush, 1994- editor | Murtaza, Ghulam, (Agricultural scientist), editor
Title: Managing plant viral diseases with advanced nanoparticles / edited by Javed Iqbal, Muhammad Anas, Ayush Madan, Ghulam Murtaza.
Description: Hershey, PA : IGI Global Scientific Publishing, 2026. | Includes bibliographical references and index. | Summary: "This book explores how nanotechnology is revolutionizing plant virology through advanced diagnostics, antiviral delivery systems, and sustainable crop protection strategies. By integrating insights from plant pathology, nanoscience, and biotechnology, it offers innovative solutions to combat viral diseases and safeguard global agriculture against evolving threats"-- Provided by publisher.
Identifiers: LCCN 2025043921 (print) | LCCN 2025043922 (ebook) | ISBN 9798337349138 hardcover | ISBN 9798337349145 paperback | ISBN 9798337349152 ebook
Subjects: LCSH: Virus diseases of plants | Nanoparticles | Phytopathogenic microorganisms--Control | Plant biotechnology
Classification: LCC SB736 .M36 2025 (print) | LCC SB736 (ebook)
LC record available at https://lccn.loc.gov/2025043921
LC ebook record available at https://lccn.loc.gov/2025043922

British Cataloguing in Publication Data
A Cataloguing in Publication record for this book is available from the British Library.

Table of Contents

Detailed Table of Contents

Plant viral diseases are a major challenge to global agriculture, causing yield losses, poor crop quality, and food supply disruptions. Their management is difficult since viruses lack independent metabolism, limiting control to preventive measures. Traditional strategies—resistant cultivars, cultural practices, and vector control—often fail due to viral diversity, rapid evolution, and climate-driven spread. Globalization further accelerates cross-border transmission. The economic toll includes reduced productivity, trade restrictions, and heavy burdens on smallholder farmers. Innovative solutions such as molecular breeding, genome editing, RNAi, integrated vector control, and nanotechnology offer promise, though they need validation, regulation, and accessibility. Future progress depends on interdisciplinary collaboration to build sustainable, resilient, and equitable food systems against viral threats.

This chapter provides an in-depth primer on nanotechnology, with a focus on its application in plant virology. It explores the fundamental principles behind nanotechnology, highlighting the unique properties of nanoparticles (NPs) due to their size (1–100 nm). The chapter outlines the various types of nanoparticles, including metallic, carbon-based, organic, and inorganic, and examines their physicochemical properties such as size, shape, surface charge, and stability. The size-dependent reactivity and surface area of nanoparticles make them highly effective in plant disease diagnostics, antiviral treatments, and real-time monitoring of plant health. The chapter further discusses the methods for nanoparticle synthesis and characterization and investigates their interactions with plant systems and viral pathogens. By integrating interdisciplinary knowledge from nanoscience, plant pathology, and biotechnology, this chapter serves

Plant viral diseases severely impact yields and food security, while conventional controls remain inefficient. Metal oxide and metallic nanoparticles (NPs) such as Ag, Cu, ZnO, TiO2, and MgO offer promise through viral inactivation, replication interference, ROS generation, and defense activation. NPs also act as nanocarriers for RNAi-based antivirals. Their efficacy depends on size, charge, solubility, application method, plant anatomy, and environment. Studies show enhanced resistance, growth, and physiology in crops. Yet risks like phytotoxicity, accumulation, non-target effects, and food chain transfer require assessment. Green synthesis and surface functionalization improve safety and sustainability. This chapter summarizes NP-mediated antiviral mechanisms, uptake, applications, ecological impacts, and their role in advancing sustainable plant protection, particularly in South Asia and Pakistan.

Plant viral diseases are a serious threat to global food security, causing yield losses and showing the limits of resistant cultivars, chemical antivirals, and vector control. This chapter examines nanotechnology as a sustainable solution, focusing on lipidic and polymeric nanoparticles for targeted virus control. Lipid-based carriers such as liposomes, solid lipid nanoparticles, and nanostructured lipid carriers protect and deliver nucleic acids for RNA interference approaches. Polymeric systems, including natural polymers like chitosan, alginate, and starch and synthetic types like PLGA and dendrimers, provide stability, biodegradability, and immune priming through systemic acquired resistance. Comparative strengths, delivery routes including foliar sprays, root absorption, seed coatings, and post-harvest treatments, plus safety and regulatory issues, are assessed. Advances such as stimuli-responsive carriers, CRISPR/Cas delivery, and smart agriculture integration position these nanocarriers as promising tools for climate-smart, sustainable crop protection.

Plant viral diseases pose a critical threat to global food security, causing severe yield and economic losses. Conventional control measures such as resistant cultivars, crop rotation, and vector management are often ineffective due to the rapid evolution of viral pathogens. Green-synthesized nanoparticles (NPs) have emerged as sustainable alternatives, exhibiting multifunctional antiviral roles including direct virion inactivation, activation of host defence responses, and targeted delivery of antiviral agents. Biogenic NPs derived from plants, microbes, fungi, algae, and biopolymers are inherently stable, biocompatible, and environmentally safe. Metal-based (Ag, ZnO, Cu), biopolymeric, and carbon-based nanomaterials have shown efficacy against major viruses such as TMV, CMV, TYLCV, and PRSV. However, large-scale synthesis, formulation stability, biosafety, and regulatory gaps remain key challenges. Integration with precision agriculture offers a strategic route to sustainable viral disease management.

Plant viruses significantly constrain global agriculture by reducing crop yield and quality, with no direct chemical control options available. Conventional detection methods, including ELISA, RT-PCR, and next-generation sequencing, are limited by low sensitivity, infrastructure requirements, and inability to detect novel viral strains. Nanobiosensors integrating nanomaterials such as gold nanoparticles, graphene, carbon nanotubes, and quantum dots with antibodies, aptamers, nucleic acid probes, or CRISPR systems offer rapid, ultra-sensitive, and portable diagnostics. These platforms enable early detection, multiplexing, and field deployment, and their integration with smartphone-based readouts and IoT applications promises real-time surveillance and precision agriculture. By bridging laboratory assays and field-ready tools, nanobiosensors hold transformative potential for sustainable crop protection and food security.

Nanoparticle-based immunoassays and rapid diagnostic platforms are revolutionizing diagnostics by overcoming low sensitivity, long turnaround times, and dependence on centralized laboratories. Functionalized nanomaterials—gold nanoparticles, quantum dots, magnetic particles, carbon structures, and polymers—enable strong signal amplification, lower detection limits, and portable, field-ready use. Their integration into point-of-care devices, lateral flow assays, microfluidics, smartphone-based tools, and biosensors has enhanced detection of pathogens, cancer biomarkers, foodborne contaminants, and toxins with high precision. Applications extend beyond medicine to agriculture, veterinary health, food safety, and environmental monitoring. Remaining hurdles include reproducibility, scalable production, regulatory approval, and field stability. Emerging solutions such as multiplexed assays, CRISPR–nanoparticle hybrids, and AI-driven interpretation mark a paradigm shift toward accessible, precise, and sustainable global diagnostics.

Plant viruses cause crop losses and over half of emerging plant diseases are viral, yet effective cures are scarce. Nanoformulations offer a targeted strategy to deliver antivirals or boost plant defenses—an emerging field called nanophytovirology. Nanoparticles can inactivate virions, enhance innate immunity, or carry genetic and chemical agents into cells. Classes include metal/metal-oxide NPs (Ag, Au, Cu, ZnO), carbon materials, silica/nano-clays, polymeric and lipid carriers (chitosan, dendrimers, liposomes), and biogenic NPs. This chapter reviews virus classes and impacts; core mechanisms (virion inactivation, RNA interference, immune priming) and delivery routes (foliar spray, root uptake, vascular injection, seed/tissue application). We integrate laboratory and field results, highlight environmental and regulatory issues (toxicity, safety, scalability), and map future directions—NP-mediated CRISPR delivery, smart carriers, and sustainable nano-agriculture. The goal is to turn lab advances into practical tools that strengthen virus surveillance and plant-health security.

Plant diseases caused by viruses pose a significant threat to food production, resulting in an annual loss of approximately $ 30 billion USD globally. RNA interference (RNAi) and nanocarrier-based delivery systems have emerged as transformative technologies for the sustainable suppression of plant viruses.

The expanding use of silver (AgNPs) and gold nanoparticles (AuNPs) in plant health nanotechnology necessitates a clear understanding of their safety profiles in human cells. This chapter examines the cytotoxic and genotoxic consequences of AgNPs and AuNPs by integrating evidence on oxidative stress, mitochondrial impairment, DNA damage, and long-term genomic instability. Emphasis is placed on how nanoparticle size, surface chemistry, and biological interactions govern toxicity, with comparative insights into AgNP-induced ionic effects and the relative stability of AuNPs. The chapter further links human cell toxicity to broader agricultural safety, highlighting exposure risks and the need for safe-by-design nanoformulations. These insights aim to guide the responsible and sustainable application of nanotechnology in plant health.

Currently, there is an extensive demand for the application of nanotechnology in sustainable agricultural systems, nutrient delivery, pest control and addressing resistance of plant diseases. This quick implementation of nanoparticles-based agricultural products has outperformed the development of comprehensible regulatory frameworks. In this chapter the current global rules for the nanoparticles-based products used in plant health management are comprehensively reviewed. Scientific assessment as well as the harmonization efforts led by international frames such as FAO/WHO are also highlighted in this chapter. Reviewing the current guidelines reveals the absence of standardized definitions, limitation of risk assessment protocols and contradictory labelling requirements. Addressing these issues and adoption of rational harmonized evidence-based regulatory strategy will ensure environmental safety and biological systems health while fostering innovation in nano-based agricultural technologies.

Preface

INTRODUCTION

As editors of *Managing Plant Viral Diseases With Advanced Nanoparticles*, we are pleased to present this comprehensive volume at a pivotal moment in global agriculture. Plant viral diseases continue to intensify in scope and complexity, posing immense challenges to crop productivity, economic stability, and food security worldwide. Traditional management strategies—though foundational—are increasingly strained by the rapid evolution of viral pathogens and the limitations of existing detection and control measures. These realities underscore the pressing need for innovative, sustainable approaches to safeguarding plant health.

Nanotechnology has emerged as one of the most promising frontiers in this endeavor. Its unique capabilities—ranging from precision diagnostics to targeted delivery of antiviral agents—have begun to reshape how we understand, detect, and manage plant viruses. In recent years, the convergence of plant pathology, nanoscience, molecular biology, and materials engineering has generated unprecedented opportunities to address viral threats with greater accuracy, efficiency, and environmental responsibility. Advanced nanoparticles hold particular promise in enabling early detection, enhancing host resistance, and minimizing reliance on chemical interventions, thereby advancing more resilient agricultural systems.

This edited volume brings together cutting-edge research, technological advances, and critical analyses from leading experts across diverse scientific disciplines. The chapters explore a wide spectrum of topics, including nanoparticle synthesis, mechanisms of antiviral activity, nano-enabled diagnostic platforms, RNAi and CRISPR-based delivery systems, biosafety considerations, and emerging regulatory frameworks. Collectively, these contributions offer an integrated perspective on how nanotechnology can be effectively harnessed to detect, manage, and mitigate plant viral diseases. Our aim is to provide readers with both foundational knowledge and applied strategies that reflect the rapidly evolving landscape of nano-enabled plant protection.

In curating this book, we sought to fill an important gap in the existing literature by offering a cohesive and up-to-date resource that bridges laboratory innovation with practical agricultural applications. We believe that this volume will support researchers, academicians, and industry professionals in advancing novel solutions for viral disease management, while also informing policymakers tasked with guiding the responsible development and deployment of nanotechnologies in agriculture.

This book is intended for a broad scientific audience, including researchers and students in plant virology, plant pathology, nanotechnology, agricultural biotechnology, and crop protection. Professionals working in agri-tech industries, developers of nanoformulations and diagnostic tools, and policymakers shaping sustainable agriculture initiatives will also find valuable insights within these pages. Our multidisciplinary approach ensures that the book serves both as a rigorous reference and as a catalyst for new lines of inquiry in the development of eco-friendly, nano-enabled crop protection strategies.

We extend our gratitude to all contributing authors for their expertise and dedication, and to the broader scientific community whose ongoing innovations continue to inspire progress in this vital field. It is our hope that this volume will not only enrich current understanding but also encourage further research and collaboration aimed at securing plant health and global food resilience through the transformative power of nanotechnology.

ORGANIZATION OF THE BOOK

Chapter 1: Plant Viral Diseases—Global Challenges and Economic Implications

In this opening chapter, the contributors illuminate the profound and escalating challenges posed by plant viral diseases to global agriculture. They describe how the unique biology of plant viruses—lacking independent metabolism and relying entirely on host machinery—renders conventional control measures only partially effective. The chapter underscores the mounting pressures of viral diversity, rapid evolutionary rates, climate-driven shifts in vector ecology, and the accelerating global movement of plant materials, all of which magnify the risk and reach of viral outbreaks. Importantly, the authors quantify the economic toll, highlighting how yield losses, trade barriers, and disproportionate impacts on smallholder farming communities destabilize food systems. By introducing emerging solutions such as molecular breeding, genome editing, RNAi, integrated vector management, and nanotechnology, the chapter sets the stage for the innovations examined throughout this volume and calls for interdisciplinary collaboration to build resilient, equitable agricultural futures.

Chapter 2: Nanotechnology Primer—Principles, Types, and Physicochemical Properties of Nanoparticles

This foundational chapter provides readers with the scientific grounding necessary to understand nanotechnology's role in plant virology. It unpacks the defining properties of nanoparticles—particularly their size-dependent reactivity, surface area, charge, and shape—which underlie their exceptional performance in diagnostics, antiviral interventions, and plant health monitoring. The authors compare major classes of nanoparticles, from metallic and metal oxides to carbon-based, organic, and inorganic materials, and discuss synthesis and characterization techniques that shape functionality and stability. The chapter also examines how nanoparticles interact with plant tissues and viral pathogens, offering critical insights into uptake, transport, and molecular interfaces. By merging perspectives from nanoscience, biotechnology, and plant pathology, this chapter provides a conceptual scaffold for the advanced applications that follow.

Chapter 3: Nanoparticle–Plant Interactions—Uptake, Mobility, and Biocompatibility

Chapter 3 explores the complex behavior of nanoparticles once they encounter plant systems—a central consideration for designing effective antiviral strategies. The authors detail the multiple pathways of nanoparticle entry through leaves, roots, and seeds, showing how physicochemical traits such as size, surface charge, coatings, and shape dictate uptake efficiency and intracellular movement. They describe nanoparticle trafficking through apoplastic, symplastic, and vascular routes and demonstrate how environmental variables—including soil chemistry, moisture, and microbial activity—modulate nanoparticle fate. The chapter also addresses biocompatibility, emphasizing dose-dependent outcomes ranging from enhanced immunity and viral suppression to oxidative stress and phytotoxic effects. Through case studies featuring silver, zinc oxide, silica, and chitosan nanoparticles across major crop systems, the authors provide practical guidance for achieving safe, effective nanoparticle use in sustainable agriculture.

Chapter 4: Mechanisms of Nanoparticle-Induced Antiviral Activity in Plants

Building upon earlier discussions of nanoparticle–plant interactions, Chapter 4 delves into the mechanistic underpinnings of nanoparticle-mediated antiviral activity. The contributors outline both direct antiviral actions—such as inhibition of viral entry, destabilization of capsid structures, and induction of reactive oxygen species—and indirect effects involving the stimulation of plant immune pathways and antioxidant defenses. While these mechanisms reveal tremendous promise, the authors also address key concerns related to phytotoxicity, environmental persistence, and the scalability of nanoparticle-based solutions. They highlight the emerging role of green-synthesized and biodegradable nanoparticles, as well as the potential synergy between nanotechnology and precision agriculture, offering a balanced and forward-looking assessment of this rapidly advancing field.

Chapter 5: Metallic and Metal Oxide Nanoparticles in Plant Virus Suppression

Chapter 5 focuses on metallic and metal oxide nanoparticles—such as Ag, Cu, ZnO, TiO_2, and MgO—which have gained significant attention for their potent antiviral properties. The authors examine how these materials inactivate virions, perturb viral replication, generate ROS, and prime plant defense responses. They also review nanoparticle-mediated delivery of RNAi-based antiviral agents, expanding their utility beyond direct antiviral action. Extensive experimental evidence is presented to illustrate improvements in plant health and viral resistance across diverse crop systems. At the same time, the chapter critically examines risks associated with nanoparticle accumulation, food-chain transfer, and ecological disturbance. Discussions of green synthesis, surface functionalization, and region-specific applications—including insights from South Asia and Pakistan—underscore the importance of tailoring nanoparticle-based strategies for safety, sustainability, and local agricultural needs.

Chapter 6: Lipidic and Polymeric Nanoparticles for Viral Control in Plant Protection

This chapter shifts attention to lipidic and polymeric nanocarriers—platforms that offer significant promise for targeted, controlled delivery of antiviral agents. The contributors detail how lipid-based systems, including liposomes and solid lipid nanoparticles, protect nucleic acids for RNAi-mediated silencing of viral genomes, while polymer-based carriers such as chitosan, alginate, PLGA, and dendrimers provide stability, controlled release, and immune-priming benefits. The authors compare delivery modalities—foliar sprays, root absorption, seed coatings, and postharvest treatments—and assess safety and regulatory considerations tied to these emerging technologies. With discussions of stimuli-responsive materials, CRISPR/Cas delivery, and integration with smart agriculture systems, the chapter positions these nanocarriers as essential tools for advancing climate-smart, sustainable plant health management.

Chapter 7: Green-Synthesized Nanoparticles for Eco-Friendly Plant Virus Management

Chapter 7 centers on green-synthesized nanoparticles, highlighting their ecological advantages and expanding role in sustainable virus management. Biogenic nanoparticles derived from plants, microbes, fungi, algae, and biopolymers are presented as inherently stable, biocompatible, and environmentally conscious alternatives to chemically synthesized nanomaterials. The authors review antiviral mechanisms of green-synthesized metallic, biopolymeric, and carbon-based nanoparticles, citing efficacy against major viral pathogens such as TMV, CMV, TYLCV, and PRSV. Despite their potential, challenges in large-scale production, formulation stability, and regulatory oversight are acknowledged. The chapter concludes by emphasizing opportunities to integrate green nanomaterials with precision agriculture and digital disease surveillance for next-generation viral management.

Chapter 8: Nanobiosensors for Early and Accurate Detection of Plant Viruses

In this chapter, the authors highlight the transformative potential of nanobiosensors in overcoming long-standing limitations of conventional viral diagnostics. They describe how nanomaterials—including gold nanoparticles, graphene, CNTs, and quantum dots—enable ultra-sensitive, portable, and multiplexed detection when paired with antibodies, aptamers, nucleic acid probes, or CRISPR systems. Emphasis is placed on the value of point-of-care platforms, smartphone-linked devices, and IoT-integrated sensors for real-time field detection and precision agriculture. By bridging laboratory capabilities with field-ready practicality, this chapter demonstrates how nanobiosensors can substantially strengthen global plant virus surveillance.

Chapter 9: Nanoparticle-Based Immunoassays and Rapid Field Diagnostic Tools

Chapter 9 expands the diagnostics landscape by exploring nanoparticle-enhanced immunoassays and rapid field tools that offer high precision with minimal infrastructure. The authors illustrate how functionalized nanomaterials drive strong signal amplification and low detection limits across technologies such as lateral flow assays, microfluidic devices, magnetic separation platforms, and smartphone-based readers. While highlighting successes in detecting agricultural pathogens, contaminants, and toxins, the chapter also addresses remaining challenges involving reproducibility, large-scale manufacturing, regulatory approval, and outdoor functionality. Emerging innovations—including multiplex diagnostics, CRISPR–nanoparticle platforms, and AI-assisted readouts—point toward increasingly accessible and scalable diagnostic solutions.

Chapter 10: Nanoformulations for Targeted Antiviral Delivery in Plants

Chapter 10 synthesizes emerging research on nanoformulations designed to deliver antiviral compounds and immune-priming agents directly to plant tissues. The contributors survey multiple nanoparticle classes—metal-based, carbon-based, silica, polymeric, lipidic, and biogenic—highlighting their capacity to inactivate virions, transport RNAi molecules, and activate plant defense networks. Delivery routes such as foliar application, root uptake, vascular injection, and seed treatment are examined in detail. The authors also address ecological and regulatory considerations, including toxicity, biosafety, and manufacturing scalability, and outline future opportunities for NP-mediated CRISPR delivery, intelligent nanocarriers, and integrated nano-agricultural systems.

Chapter 11: Nanocarriers for RNAi-Mediated Plant Virus Suppression

This chapter focuses on RNA interference as a precision, non-GMO approach for sustainable virus suppression and examines how nanocarriers have revolutionized RNAi delivery in real-world crop environments. The authors detail how various nanomaterials protect RNA molecules from degradation, enable controlled release, and facilitate transport across biological barriers. They discuss recent innovations in plant-compatible nanocarrier engineering and present case studies demonstrating field-relevant performance. By integrating concepts from plant molecular biology, virology, and nanotechnology, the chapter establishes RNAi-nanocarrier systems as a cornerstone for next-generation antiviral strategies.

Chapter 12: Field Application Approaches—Foliar Sprays, Soil Additives, and Seed Treatments

Chapter 12 translates laboratory advances into practical application strategies, assessing how nanoparticles can be deployed effectively across agricultural production systems. The authors analyze foliar spraying as a highly accessible method for achieving immediate plant–nanoparticle contact, while soil incorporation strategies are presented as longer-term approaches for sustained nanoparticle availability. Seed-based treatments, including nanopriming, are highlighted for their capacity to protect seedlings from early viral attack and enhance initial vigor. The chapter stresses the importance of understanding nanoparticle persistence, mobility, and interaction with traditional management practices, and it calls for further research into customizable delivery systems tailored to diverse agroecosystems.

Chapter 13: Phytotoxicity and Environmental Impact of Nanoparticles in Agriculture

With safety as a central concern in agricultural nanotechnology, this chapter examines the potential risks associated with nanoparticle use in farming systems. The authors detail how nanoparticle size, concentration, and surface chemistry influence phytotoxicity and how interactions with soil microbes, nutrient cycles, and trophic networks can shape ecological outcomes. While acknowledging their role in improving nutrient efficiency and pest control, the chapter urges careful evaluation of bioaccumulation, leaching, and long-term environmental implications. Strategies such as green synthesis, biodegradable carriers, and functional surface modifications are proposed to balance technological innovation with responsible environmental stewardship.

Chapter 14: Cytotoxic and Genotoxic Effects of Silver and Gold Nanoparticles in Human Cells: Implications for Safe Applications in Plant Health Nanotechnology

Chapter 14 shifts attention to human health considerations, exploring the cytotoxic and genotoxic effects of AgNPs and AuNPs to guide safe applications in plant health nanotechnology. The authors synthesize evidence on oxidative damage, mitochondrial dysfunction, DNA instability, and other cellular responses, noting how nanoparticle size, coating, and chemical transformations shape toxicity. Differences between ionic effects associated with AgNPs and the relative biostability of AuNPs are discussed. By linking human cellular toxicity findings to agricultural exposure scenarios, the chapter emphasizes the need for safe-by-design nanoparticle formulations that protect human and environmental health.

Chapter 15: Global Regulatory Guidelines for Nanoparticle Use in Plant Health Management

This chapter provides a comprehensive review of regulatory frameworks governing agricultural nanotechnology worldwide. The authors highlight inconsistencies in definitions, gaps in risk assessment protocols, and divergent labeling requirements that complicate product approval and international trade. They examine ongoing harmonization efforts led by global organizations such as FAO and WHO and emphasize the importance of evidence-based, transparent, and globally aligned regulatory strategies. The chapter advocates for policies that simultaneously safeguard environmental and human health while enabling innovation in nano-enabled agricultural technologies.

Chapter 16: Future Directions in Nano-Virology—From Lab to Field Deployment

The concluding chapter looks ahead to emerging horizons in nano-virology, showcasing how lab-on-a-chip diagnostics, nanosensors, nano-enabled antivirals, and next-generation vaccine technologies may transform global plant-virus surveillance and response. The authors describe how bioengineered nanomaterials are poised to enhance detection accuracy, therapeutic targeting, and resistance management. They also address challenges related to scalability, biocompatibility, regulatory readiness, and environmental considerations that must be addressed for real-world implementation. The chapter closes by outlining a roadmap for translating laboratory innovations into field-deployable technologies capable of strengthening global agricultural resilience and biosecurity.

CONCLUSION

As editors of *Managing Plant Viral Diseases With Advanced Nanoparticles*, we conclude this preface with a deep appreciation for the breadth, rigor, and forward-looking vision represented by the chapters that follow. Together, they reflect a field in rapid transformation—one in which plant virology, nanoscience, biotechnology, and environmental stewardship converge to address some of the most urgent challenges facing global agriculture. While plant viral diseases continue to impose significant economic, ecological, and social burdens, the collective research presented in this volume demonstrates that innovative, sustainable solutions are within reach.

Across these chapters, our contributors illuminate how nanoparticles can reshape every dimension of plant virus management—from early detection and diagnostic precision to targeted antiviral delivery, immune priming, and environmentally conscious crop protection. They also confront the complexities that accompany the adoption of emerging technologies, acknowledging the critical importance of biosafety, phytotoxicity assessment, regulatory harmonization, and ethical deployment. These discussions are essential for guiding the responsible integration of nanotechnology into agricultural systems worldwide.

This volume aims not only to consolidate current knowledge but also to chart a course for future discovery. We hope it inspires new collaborations, informs evidence-based policy, and catalyzes innovations that translate scientific advances into practical benefits for farmers, industry professionals, and communities dependent on secure food systems. By bridging fundamental science with applied solutions, and laboratory insights with field realities, this book underscores a central message: the intelligent and ethical application of nanotechnology holds transformative potential for managing plant viral diseases and strengthening global agricultural resilience.

We invite readers to engage deeply with the perspectives offered here and to join us in advancing a future where cutting-edge science supports sustainable, equitable, and resilient plant health management.

Chapter 1
Plant Viral Diseases:
Global Challenges and Economic Implications

Muhammad Anas
https://orcid.org/0000-0001-5127-8043
Selcuk University, Turkey

Vildan Altun
Necmettin Erbakan University, Turkey

Javed Iqbal
https://orcid.org/0000-0002-9032-2622
Bacha Khan University, Pakistan

Aqsa Shabbir
https://orcid.org/0009-0005-9800-8768
Ordu University, Turkey

ABSTRACT

Plant viral diseases are a major challenge to global agriculture, causing yield losses, poor crop quality, and food supply disruptions. Their management is difficult since viruses lack independent metabolism, limiting control to preventive measures. Traditional strategies—resistant cultivars, cultural practices, and vector control—often fail due to viral diversity, rapid evolution, and climate-driven spread. Globalization further accelerates cross-border transmission. The economic toll includes reduced productivity, trade restrictions, and heavy burdens on smallholder farmers. Innovative solutions such as molecular breeding, genome editing, RNAi, integrated vector control, and nanotechnology offer promise, though they need validation, regulation, and accessibility. Future progress depends on interdisciplinary collaboration to build sustainable, resilient, and equitable food systems against viral threats.

DOI: 10.4018/979-8-3373-4913-8.ch001

1. INTRODUCTION

Plant viral diseases are one of the most problematic and widespread issues related to global agriculture, which causes a crucial challenge to food production, stability in the ecosystem, and economic sustainability (R. A. Jones & Naidu, 2019). In comparison to bacterial or fungal pathogens, which frequently produce conspicuous symptoms and may be treated with chemical interventions, viruses are obligate intracellular pathogens that rely on host cells on the metabolic activities of their plant hosts to replicate. The resulting inherent dependence complicates their control significantly, and there are no direct antiviral agents that can be used safely in the field without attacking the physiology of the host itself. Plant viruses are extraordinarily diverse in the context of genome organization, modes of transmission, and host, and over 1000 species of plant viruses have already been discovered, and new variants continue to emerge as agricultural systems change and transform (Lefeuvre et al., 2019). These infect almost all cultivable plants, including cereals, legumes, fruits, as well as vegetables and ornamental plants, and they cause symptoms in the form of mosaics on leaves or chlorosis, necrosis, dwarfing, curling of leaves, malformation of fruits, and their reduced vigor (Anas et al., 2025).

These infections undermine both yield and quality, resulting not only in reduced quantity of harvest but also in refusal of produce on the international market under the strict quarantine conditions. Plant viral diseases have been estimated to cause losses of up to 10–15% per year in high-volume crops, with some local outbreaks destroying whole crop harvests. Examples include rice tungro disease in Asia, cassava mosaic disease in Africa, and tomato yellow leaf curl virus in Mediterranean areas, which have all shown that epidemics by viruses can destroy staple foods and destabilize food security in an area. Plant viral diseases are added by the rapid adaptation of genetic mutation and recombination, high association of insect vectors like whiteflies, aphids, leafhoppers, and thrips, and the increased intensity of modern agriculture that supports viral transmission (Krishnareddy, 2013).

1.1 Overview of Plant Viral Diseases

Plant viral disease management has, traditionally, been based on a three-pronged tool of resistant cultivars, cultural practices, and whole vector management techniques. Developing resistant types has been one of the building blocks, and with the speed of viral evolution and resistance, new versions and strains are breaking through genetic constraints (Gomez et al., 2009; Meshram, 2025). The inoculum pressure can be mitigated using cultural practices, which include crop rotation, sanitation, and elimination of alternate hosts, but such types of practices are labor-intensive and are not necessarily practical on a large scale. Insecticides have proven to be effective in the control of vectors, but excessive innovation in chemicals has contributed to the pollution of the environment, the emergence of resistance in insect populations, and the misbalancing of the natural populations of the insect community, which is injurious. Subsequently, viral diseases still serve as significant challenges towards the attainment of stable and sustainable crop production (Wilson et al., 2020). The intricate relations among the host plants, the viruses, vectors, and the environment make the development of a universal solution difficult. Further, trade globalization, climate change, and enhanced agricultural activities have also contributed to the extent to which and how intensely the viral outbreaks have been increasing (Baker et al., 2022). It is in this depressed scenario that the proposal of innovative, combined, and enduring managerial measures has never been more significant, and it is in this regard that scientific endeavors to comprehend the biology, epidemiology, and management of plant viruses acquire renewed significance (Tatineni and Hein, 2023).

1.2 Historical Perspective and Emerging Trends

The historical standpoint reveals that plant viral diseases have been considered since the end of the 19th century, when tobacco mosaic disease was among the initial plant pathologies, which were advocated scientifically (Bos, 1999). The mosaic symptoms in leaves of tobacco were described in 1886 by Adolf Mayer, who proposed that this was a result of an infectious agent. In later parts of the 1892 case of Dmitri Ivanovsky and 1898 of Martinus Beijerinck, it was shown that the causal agent was a smaller filterable, non-bacterial agent that led to the idea of a virus. The unrestricted finding of the Tobacco mosaic virus (TMV) not only became the dawn of virology but also demonstrated the importance of viral pathogens in plants long before it was discovered in animals and humans. During the first parts of the 20th century, more plant viruses infecting economically valuable crops like potatoes, sugar beets, and cereals were identified, and, in most cases, these viral infections were associated with insect vectors (R. Jones, 2014). Plant virologists started to view the aphids, whiteflies, and leafhoppers with increased respect as a means of the viral diseases epidemiology, which provided the basis of integrated control mechanisms currently (Sarwar et al., 2020). With the progress in the field of microscopy and molecular biology, the research on plant viruses has gained a new dimension. In the middle of the 20th century, electron microscopy enabled researchers to see viral particles, as well as describe their morphology and sizes, which prompted them to classify viruses into morphologically and genome-type families (Wick, 2023). Subsequently, the diagnostics would be revolutionized in later years as molecular methods like nucleic acid hybridization, polymerase chain reaction (PCR), and sequencing of the nucleic acid facilitated the ability to identify viral species and strains with speed and precision. (Niraula & Fondong, 2021). Nevertheless, it was the inoculation that required a very short resistance to break, and the overall argument of genetically modified organisms (GMOs) in societies limited the number of common strategies implemented.

During the last several decades, the globalization process and environmental alteration have influenced plant virology. The viral pathogen transmissions have added fresh threats to regions never affected before due to the accelerated international trade (R. A. Jones & Naidu, 2019). As an example, the transportation of contaminated planting substances has been a significant motivation in how the Banana bunchy top virus, Cassava mosaic begomoviruses, and Citrus tristeza virus spread in the world (R. Jones, 2021). Simultaneously, climate change has changed the population and patterns of the disease, and an increase in temperatures, changing rain patterns, and other growing seasons provide favorable conditions that lead to the spread of the virus. The modern agricultural systems are at risk of being overcome by a viral epidemic reflected by emerging viral diseases, including Maize lethal necrosis in East Africa and the rebirth of Tomato brown rugose fruit virus (R. A. Jones, 2020).

The growing prominence of the usage of sophisticated technologies in plant virology is considered among the most noticeable tendencies in the contemporary context. An unprecedented wealth of genomes of viruses is presented by high-throughput sequencing (HTS), which has led to the detection of known and undiscovered viruses at a level unacknowledged before (Ibaba & Gubba, 2020). This has not only deepened our knowledge of the evolution of a virus, its recombination, as well as its interaction with a host or potential host, but it also provides difficulty in defining pathogenic viruses and viruses living in symbiosis commonly found in plants. The other emerging tools that could be used to work out innovative solutions in combating plant virus diseases include nanotechnology, synthetic biology, and CRISPR-based genome editing. Its application has been investigated to interfere with the replication and transmission of viruses using nanoparticles and to integrate into different crops to produce virus-resistant crops using CRISPR systems (Anas et al., 2024). These radical technologies are only in an experimental phase, but there could be a time when controlling plant viruses will pass beyond a state of reactive control and be initiated through proactive and specific means.

Historical patterns in plant virology show that the relationship between science, need, and necessity has been a constant interaction between science and agriculture, and society (Zuo et al., 2024). Nevertheless, the difficulties are overwhelming: due to the huge assortment of plant viruses, their flexibility and integration into sophisticated agro-ecosystems, the solution will always be different. Rather, there should be combined solutions of resistant varieties, agricultural methods, insect management, biotechnological solutions, and policy measures in taking care of the world's agriculture (Anderson et al., 2019). Taking the historical reflection approach, we can realize how much we have evolved to grasp plant virology and its emerging future, taking into consideration the identified trends can hint towards the creation of resilient food systems with a sustainable future. Since the 21st century has been characterized by constant development, the investigation of plant viral diseases will continue to take center stage in ensuring food production, a result of the unabated viral threat and the resultant improvement in the set of scientific tools against Southeastern plagues (Kreuze et al., 2023; Nigam, 2024).

2. GLOBAL BURDEN OF PLANT VIRAL DISEASES

All plant viral diseases hold a huge burden on the global agricultural sector, making crop yields small, the quality of produce declines, and provoking a significant number of economic losses experienced by farmers, industries, and countries. Working against the formidable decisions of the other pathogens, viruses emerge at a fast rate in the numbers of vectors, vegetative propagation, and trade networks, and they are harder to control once in the vicinity. It is estimated that the cumulative annual yield losses caused by plant viruses are in billions of dollars, be it due to the high dependency of developing countries on staple crops as a source of subsistence and, secondly, as a source of income. In addition to this, viruses are often neglected in contrast with bacterial and fungal infections, partly because their effects can overlap with those of a deficiency of nutrients or other stressors and partly because of the shortage of diagnostic capabilities in resource-limited areas. This underestimation fails to help in diffusing the problem that plagues the world virtually (Vurro et al., 2010): viral diseases are spreading in areas and finding hosts without detection until outbreaks devastate world economies and cause irreparable effects (**Figure 1**).

2.1 Major Viral Diseases Affecting Staple Crops

Food systems that include rice, wheat, maize, cassava, potatoes, and bananas are the staple crops worldwide, and infections among these crops due to viruses have a significant impact globally (Oraon et al., 2024). Being one of the main staples, with more than half of the world's population consuming rice, it is a very dangerous product exposed to viral pathogens. The virus that remains to be discussed is rice tungro disease, a complex of two viruses transmitted by leafhoppers, overwhelming in South and Southeast Asia, which can reduce the productivity of the affected fields by up to 70% (Dey et al., 2024). Equally, the Rice grassy stunt virus and Rice ragged stunt virus have led to repeated epidemics, especially where monoculture systems have thrived abundantly, with the various areas serving as vector propagation zones. Wheat is the mainstay of diets in the temperate areas, but suffers from Wheat streak mosaic virus (WSMV), which is transmitted through a wheat curl mite, or Barley yellow dwarf virus (BYDV), which is spread by aphids (Flint, 2014). Besides a decrease in grain yields, these pathogens compromise the quality of the grain as they are a constant menace to the North American continent, Europe, and Asia. Maize is another world crop that has been badly affected by Maize streak virus in sub-Saharan Africa, as well as recent Maize lethal necrosis (MLN), a disease-causing agent generated as a result of synergistic interaction between Maize chlorotic mottle virus and potyviruses (Redinbaugh & Stewart, 2018). MLN has rapidly propagated in East Africa, as an outbreak was reported in Kenya in 2011 and which led to complete failure of crops in some areas and posed a risk of food insecurity affecting millions of families. There is a chronic impact of effects on cassava, a vital subsistence crop within both Africa and parts of Asia, by Cassava mosaic begomoviruses (CMVs) and Cassava brown streak virus (CBSV) (Chikoti & Tembo, 2022). A combination of these viruses is found to be the largest limitation to cassava production in Africa, where the crop forms an important source of calories for about 800 million people. The epidemics of cassava mosaic disease have a long history of leading to starvation in East Africa, and the more recently surfaced brown streak disease is spreading and developing (Tomlinson et al., 2018).

A critical crop in the temperate and tropical highlands, potatoes are exceptionally vulnerable to virus attacks due to production using tubers as vectors (which can carry the virus through different systemic plants); moreover, tubers can accumulate the virus over time whilst cultivating different crops (Kreuze et al., 2020). The Potato virus Y (PVY) is one of the most destructive and decreasing the yield and quality of the tuber, whereas the Potato leaf roll virus (PLRV) causes rolling, chlorosis, and high rates of disease yield. These viruses are spread with the assistance of aphid vectors and the use of infected seed tubers, especially in smallholder farms, which have limited access to virus-free planting materials (Fingu-Mabola & Francis, 2021). The banana bunchy top virus (BBTV), which is transmitted by the banana aphid, affects bananas and plantains, another critical staple in Africa, Asia, and Latin America. BBTV results in serious stunting, distortion of the leaf, and the end of functioning, which in effect kills the plant. Due to the clonal nature of bananas, as soon as the fruit is infected, it spreads within the plantations and further with the spread of the disease to other regions due to the movement of the infected suckers (Drenth & Kema, 2021). Tomatoes and citrate are globally important, but not typically staple crops, as in addition to these, they demonstrate the disease complication that is fronted by plant viral diseases. The white fly virus that causes tomato yellow leaf curl (TYLCV) is one of the most destructive viruses in the world, especially in warm weather, which kills both the small holder and commercial tomato production. Aphids, the carrier of citrus tristeza virus (CTV) and infected propagation material, have killed millions of citrus trees worldwide the resulting in changes to the industries of whole regions in Latin America and Asia (Figure 1). This set of examples helps to conclude that plant viruses are not a national matter but a worldwide opponent, which mostly makes use of networks of ecology, agriculture, and trade that can usually make themselves available worldwide (Table 1) (R. A. Jones and Naidu, 2019).

Table 1. Major viral diseases affecting global staple crops (rice, wheat, maize, cassava, potatoes, and bananas), along with their causal agents, vectors, impacts, and regions of occurrence. These viruses represent significant threats to food security due to their widespread distribution, severe yield losses, and reliance on insect vectors or infected planting materials for transmission

Crop	Viral Disease	Causal Virus/Complex	Vector/Transmission	Major Impact	Regions Affected
Rice	Rice Tungro Disease	Rice tungro bacilliform virus (RTBV) + Rice tungro spherical virus (RTSV)	Green leafhoppers (*Nephotettix* spp.)	Yield losses up to 70%, severe stunting	South & Southeast Asia
	Rice Grassy Stunt	Rice grassy stunt virus (RGSV)	Brown planthopper (*Nilaparvata lugens*)	Epidemics in monoculture fields, severe stunting	Asia
	Rice Ragged Stunt	Rice ragged stunt virus (RRSV)	Brown planthopper	Ragged leaves, reduced tillering, epidemics	Asia
Wheat	Wheat Streak Mosaic	Wheat streak mosaic virus (WSMV)	Wheat curl mite (*Aceria tosichella*)	Reduced yield, grain quality loss	North America, Europe, Asia
	Barley Yellow Dwarf	Barley yellow dwarf virus (BYDV)	Aphids	Stunted growth, yellowing, and yield decline	Global (esp. temperate regions)
Maize	Maize Streak Disease	Maize streak virus (MSV, *Mastrevirus*)	Leafhoppers (*Cicadulina* spp.)	Chlorotic streaks, major yield loss	Sub-Saharan Africa
	Maize Lethal Necrosis (MLN)	Maize chlorotic mottle virus (MCMV) + Potyviruses (e.g., Sugarcane mosaic virus)	Thrips, aphids, and seed transmission	Complete crop failure, food insecurity	East Africa (Kenya, Tanzania, Uganda, Rwanda)
Cassava	Cassava Mosaic Disease (CMD)	Cassava mosaic begomoviruses (CMVs)	Whiteflies (*Bemisia tabaci*), cuttings	Severe leaf mosaic, yield losses up to 90%	Africa, Asia
	Cassava Brown Streak Disease (CBSD)	Cassava brown streak virus (CBSV) & Ugandan CBSV	Whiteflies, infected cuttings	Root necrosis, famine threats	East & Central Africa
Potato	Potato Virus Y (PVY)	PVY (*Potyvirus*)	Aphids, infected seed tubers	Yield reduction, tuber quality loss	Global
	Potato Leaf Roll Disease	Potato leaf roll virus (PLRV)	Aphids, infected tubers	Leaf rolling, chlorosis, severe yield loss	Global
Banana	Banana Bunchy Top Disease	Banana bunchy top virus (BBTV)	Banana aphid (*Pentalonia nigronervosa*), infected suckers	Stunting, leaf distortion, and plant death	Africa, Asia, Latin America
Others	Tomato Yellow Leaf Curl	Tomato yellow leaf curl virus (TYLCV)	Whiteflies (*Bemisia tabaci*)	Severe yield losses in tomato production	Global (warm regions)
	Citrus Tristeza Disease	Citrus tristeza virus (CTV)	Aphids (e.g., *Toxoptera citricida*), infected grafts	Millions of citrus trees lost, industry collapse	Latin America, Asia

2.2 Impact on Food Security

The impact of viral diseases is much more than short-term losses of crops. Food security has four aspects, which include the availability dimension, access dimension, usage dimension, and stability, but all of them are adversely impacted by plant viral infections (Savary et al., 2017). At the most basic level, viruses lower the amount of food by diminishing the harvests of food staple crops. In other areas where the daily diet is rice-based, maize, cassava, and bananas can directly result in hunger and malnutrition as a direct consequence. Cassava mosaic disease epidemics in East Africa throughout the 1990s triggered an acute deficit of food provisions, which helped to make the rural settings famine-stricken. Similarly, the outbreaks of MLN in Kenya and the adjacent nations have weakened the production of maize, which directly brought down the number of grains to be sold and consumed as household foods and in the local markets (Mahuku et al., 2015). There is also a factor of food access, where the viral diseases interfere with the income-generating channels of farming households. Smallholder farmers depend on cash crops, in most cases, tomatoes, potatoes, and citrus, to earn money.

Viruses destroy these crops, limiting farmers to sales of the crops as well as having financial powers to buy other foods, thus exposing them to increased vulnerability. Viral outbreaks in commercial agriculture may result in heavy panhandling of prices, especially at the regional and even continental level for the consumers. The problem caused mass losses in terms of oranges produced in Brazil, which affected the world price of juices and price stability in the market through CTV outbreaks. Viral diseases will tend to diminish the nutritional quality of crops on the dimension of utilization. Potatoes have viruses like the PVY, which lead to poor fruit development and stunted growth, leading to reduced size of produce and, consequently, less nutrition (Figure 1). Tomatoes have viruses like the TYLCV that cause poor fruit growth and stunted development, hence reduced-sized produce that is less nutritious. Cassava Brown streak disease causes non-edible cassava roots and therefore robs the population of an important carbohydrate source (Robson et al., 2024).

This health effect is extremely great, especially in low-income areas where there is a lack of nutritional variety, and the staple food supplies the majority of the required nutrients. Lastly, viral illnesses raise the instability of food systems because of their recurrent and unpredictable losses in production (Béné, 2020). Viral illnesses tend to be incurable in a developing cycle, whether by vegetative spread, by seed contamination, or by host vectors, unlike pests or lack of nutrients, which at times can be managed within a growing season. The outcome is that this exposes a permanent strain on farming systems, with each season having the threat of new or recurring epidemics. Climate change exacerbates this instability by expanding vibrant habitats of vectors, swelling breeding proceedings, and extending the conveying seasons (Yadav & Upadhyay, 2023).

The successful migration of the whiteflies harboring TYLCV to the north or the general increase of aphids infected by BYDV shows that ecological alterations due to the climate are redefining the geography of viral infections. To be more specific, the effects of plant viral diseases on the food security of people are not just justified on the local level but also of national and even international significance. The export of crops with high infection rates in the case of quarantine and trade limitations during an outbreak can be an economic loss to the producing nations and a disruption of the supply chain in the case of importing countries. Tombco and tomato booths in the UAE, where greater tomatoes can be sold by virtue of possessing TYLCV, or banana booths in the UAE, where greater bananas can be sold by virtue of possession of BBTV, directly impact the international markets and consumer prices. These precautions underscore the global intercropping of plant health and food systems with viral diseases in one area potentially cracking all the continents (R. A. Jones and Naidu, 2019; Ristaino et al., 2021).

Plant viral disease burden is therefore interchangeable across dimensions, including agricultural production, economic stability, and human welfare around the world. Although the breeding of plants, methods of vector control, and the introduction of molecular diagnostics have enhanced how far it is possible to go in combating these pathogens, the adaptability and survival of viruses make them a formidable opponent (Mehetre et al., 2021). The proposed challenges will need joint efforts that not only consider the pathogens and vectors of the plant viral diseases, but also build agricultural resilience, availability of planting materials that are free of the virus, and for countries to develop a capacity to work together in the area of surveillance and control of pathogens. Unless this is undertaken, the major impact of the viral diseases on the staple crops will continue to jeopardize the food security of millions of people around the world as well as their livelihoods.

Figure 1. Global burden of plant viral diseases. Viruses cause billions in crop losses annually, spreading rapidly through vectors, trade, and propagation, while often being misdiagnosed or underestimated. Major staples—rice, wheat, maize, cassava, potato, and banana—are severely impacted by viral infections, with additional threats in tomato and citrus. These diseases undermine food security by reducing availability, limiting farmer access to income, lowering nutritional quality, and destabilizing production under recurrent epidemics and climate change. The global consequences include trade restrictions, market price shocks, and long-term agricultural burdens.

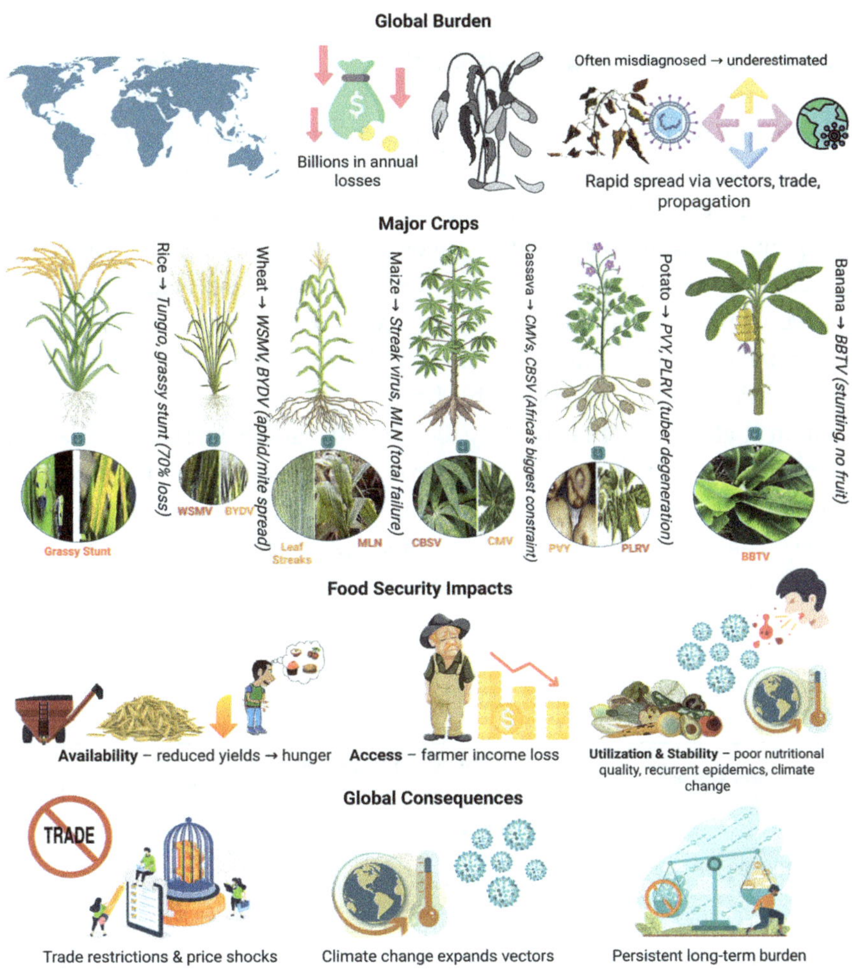

3. ECONOMIC IMPLICATIONS

Viral plant diseases not only undermine the health and crop yield of secondary host plants, but also have expansive economic effects at both micro and macro scales (R. A. Jones and Naidu, 2019). The effects are not only limited to the loss in the short term, but they challenge the dynamics of the market, global trade, rural livelihoods, and long-term sustainability of agriculture. The economic cost of a virus may be impressive, with certain estimates saying that, collectively, plant viruses cost billions of dollars a year for crop losses the world over. The distribution of the losses is skewed, whereby migratory spaces where agriculture is the primary source of livelihoods are attracted by the losses. The magnitude of attack has factors related to the affected crop, disease's geographic location, access to markets, and provision of effective control strategies (Kamal et al., 2025). It is important to know the economic dimensions of plant virus diseases to make an investment in diagnostic measures, sustainable management, and effective surveillance systems across the world.

3.1 Yield Losses and Crop Quality Decline

The most apparent manifestation of the economic impact of viral infection is yield loss. Viruses interfere with physiological activities like photosynthesis, nutrient uptake, and water use efficiency, which results in the retardation of plant growth and decreases the harvest number (Ray, 2024). Likewise, Sub-Saharan Africa suffers a maximum loss of 90 percent in affected cultivars due to cassava mosaic virus and rice tungro virus destruction by millions of tons of lost grain every year in Asia. Other than cutting back on volumes, viruses have a devastating impact on the quality of the buffer production. Fruits and vegetables present in the fresh produce markets might experience chlorosis or mottling and malformations due to viral infections, thus rendering them unmarketable because these aspects influence customer preference in the market (Gautam et al., 2015). In relatively the same fashion, sugarcane mosaic virus reduces the level of sucrose content, resulting in less industrial processing value. Viral diseases in some crops, such as grapes, are known to not only reduce production but also the flavor and aroma compounds, causing a direct effect on the business of the wine industry. This type of qualitative loss enhances the economic losses further because infected produce has lower prices or may not sell at all (Table 2).

Table 2. Highlighting the economic impact of viral infections on crops, based on yield loss, physiological effects, and market value reduction

Crop Affected	Viral Disease	Key Findings on Economic Impact	Region/Scale	Study/Reference
Rice	Rice tungro virus	Yield losses up to millions of tons annually; disruption of photosynthesis and nutrient uptake.	Asia	Ray (2024)
Vegetables (tomato, cucurbits)	Various mosaic & leaf curl viruses	Chlorosis, mottling, and fruit malformations reduced marketability	Global fresh produce markets	Gautam et al. (2015)
Rice	Rice tungro disease (RTBV + RTSV)	Productivity loss up to 70%; severe outbreaks linked to food insecurity	South & Southeast Asia	Dey et al. (2024)
Wheat	Wheat streak mosaic virus (WSMV), Barley yellow dwarf virus (BYDV)	Grain yield and quality reduction: a persistent economic threat	North America, Europe, Asia	Flint (2014)
Maize	Maize lethal necrosis (MLN)	Complete crop failures; food insecurity for millions	East Africa	Redinbaugh & Stewart (2018)

continued on following page

Table 2. Continued

Crop Affected	Viral Disease	Key Findings on Economic Impact	Region/Scale	Study/Reference
Cassava	Cassava mosaic begomoviruses (CMVs)	Up to 90% yield loss in susceptible cultivars; famine risk	Sub-Saharan Africa	Chikoti & Tembo (2022)
Cassava	Cassava brown streak virus (CBSV)	Root necrosis leads to loss of edible yield, a major economic limitation	East Africa	Tomlinson et al. (2018)
Potato	Potato virus Y (PVY)	Reduced tuber yield & quality; cumulative loss across generations of tubers	Global	Kreuze et al. (2020)
Potato	Potato leaf roll virus (PLRV)	Leaf rolling, chlorosis, high yield loss, and lower tuber quality	Global smallholder farms	Fingu-Mabola & Francis (2021)
Banana	Banana bunchy top virus (BBTV)	Plantation collapse due to clonal spread; economic threat to the banana industry	Africa, Asia, Latin America	Drenth & Kema (2021)

3.2 Market Disruptions and Trade Restrictions

The occurrence of viral diseases in the area tends to implement severe quarantine measures and trade bans that lead to indirect economic costs, which, at times, could be more than losses on the yield. Phytosanitary certifications are also important in the international trade of agricultural commodities, and an outbreak of viruses may result in delivery cancellation, export bans, or expensive delays (Mabunda et al., 2025). To illustrate, an outbreak of tomato brown rugose fruit virus (ToBRFV) has resulted in more inspections and trade restrictions by the various regions in Europe, North America, and the Middle East, which has greatly increased the costs of production and compliance. On the same note, the banana bunchy top virus has minimized the exportation of bananas in countries with the disease, which has minimized specific foreign exchange incomes. Consumers' lack of confidence in producing safety and quality is are cause of market disruptions due to low demand and causes price shock. The frequent epidemics of viruses may cause farmers to give up profitable crops, which will also lead to destabilizing the supply chains at a local and international scale (Figure 2). The result of this format of an interrupted market, a decrease in profitability, and the rise in compliance costs is a demanding economic climate for both growers and exporters (Roberts, 2025).

3.3 Socioeconomic Consequences for Smallholder Farmers

Smallholder farmers, especially in the developing world, who rely mainly on subsistence farming, are the worst affected by the socioeconomic effects of the viral diseases that affect plants. Smallholders may also not have access to certified planting material free of virus, effective diagnostic facilities, and resistant cultivars, making them extremely susceptible to a disease epidemic. It is found that the loss of yields involves a direct agreement between the low availability of household food, low-income level, and heightened poverty opportunities (Hertel et al., 2010). Cassava Mosaic disease and cassava brown streak virus in East Africa have the potential to disrupt the major source of food and income of millions of rural households. Besides, frequent viral outbreaks can propel farmers to divert their attention from staple or cash crops, and this can lead to instability in nutrition and economic stability (Figure 2).

Smallholders are also marginalized because they cannot compete in export markets due to viral pollution, which increases the distinction between the industrialized systems of agriculture and resource-limited systems of farming (Amelework et al., 2021). The effects of the program on poor people are long-term, as they result in rural indebtedness, migration, and underinvestment in agro-technology circles of the poverty trap and the susceptibility cycle (Figure 2).

4. CHALLENGES IN MANAGEMENT OF PLANT VIRAL DISEASES

The control of plant viral diseases remains a daunting challenge despite the tremendous breakthroughs in the study of plant pathology and crop protection. As opposed to fungal or bacterial pathogens, no machinery can directly fight the viruses via a metabolic pathway, and thus, controlling them becomes difficult. The mitigation concepts about the effects of viruses are mainly based on the aspects of prevention, resistant crops, and control of the vectors. Nevertheless, these measures can be compromised due to the dynamic character of viruses, ecological intricacies, as well as technological restrictions. It is also vital that the issue with managing plant viral diseases is understood to come up with adaptation and sustainable methods (Zhan et al., 2014).

Figure 2. Economic implications of plant viral diseases. Losses include billions in annual yield reductions and poor crop quality. Market disruptions arise from quarantine measures, export losses, and price shocks. Smallholder farmers face severe food insecurity and poverty, with long-term impacts such as debt, migration, and underinvestment in agriculture.

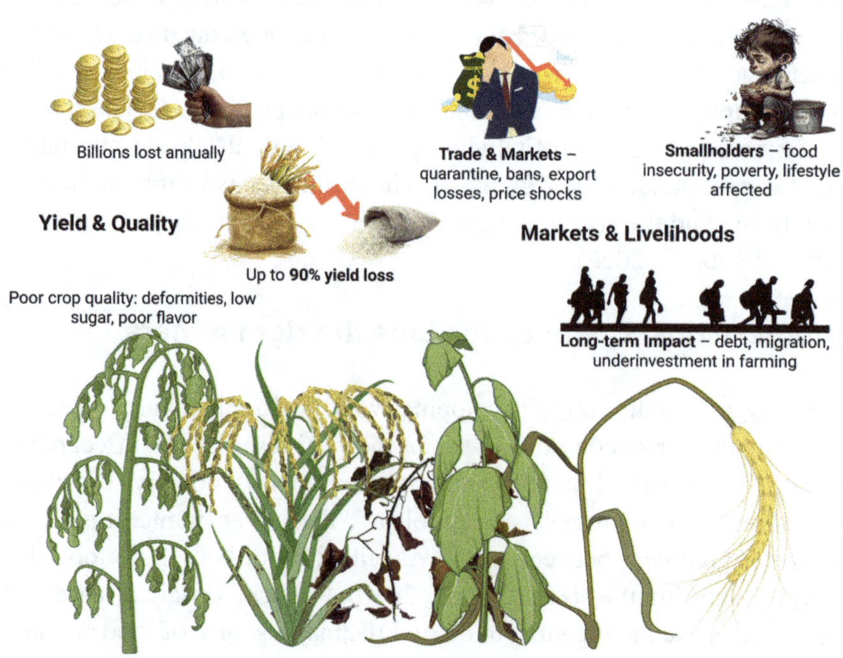

4.1 Rapid Viral Evolution and Genetic Diversity

Viruses are also the most adaptable, hence they might contain a large wealth of mutation rate, frequently combine, and stand a chance of recombining whenever present with segmented genomes. Genetic plasticity helps begomoviruses, which are spread by whiteflies, to overcome resistance genes in plants in a remarkably short time, and new strains that can overcome new host resistance emerge frequently, again by choosing in favor of reduced C4 (4 CAM) (Vaulreal et al., 2019). On the same note, the rice tungro virus complex has a significant genome variation level and thus cannot be easily developed to formulate long-lasting control strategies. Not only does the rapid evolution of viruses make it more difficult to produce long-term resistance, but it is also difficult to diagnose because the diagnostic tools should not stagnate, since the variants they can detect keep evolving. Such developments in evolution enable conditions favor virus spread of ecological alterations and enter new areas where they are a constant threat to the global agricultural world (Figure 3) (McDonald & Stukenbrock, 2016).

Figure 3. Viral adaptability in crops driven by mutation rate, genetic plasticity, diagnostic challenges, and ecological exploitation, enabling rapid evolution and persistent threats to agriculture

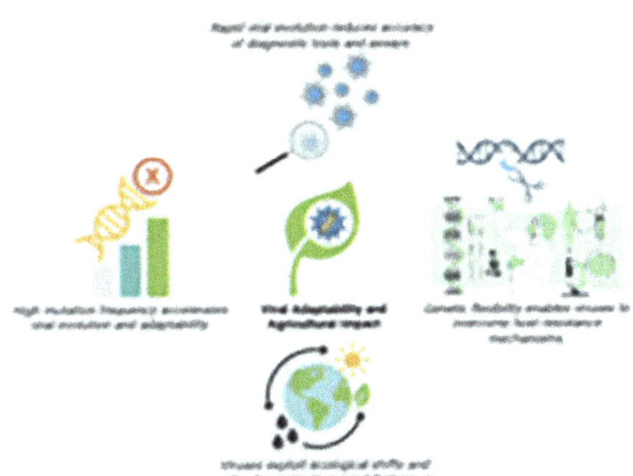

4.2 Limitations of Resistant Cultivars

Resistant cultivars serve as a central theme in the measure of viral diseases, but they do not last long when such viruses develop resistance to preserving their invulnerability. Breeding of resistance is an exercise that consumes a lot of time and resources, as it normally takes years to prepare a new variety, by which time, viruses have advanced to resistance-breaking strains. An example of this is when the genes used to protect against potato virus-Y (PVY) were used in solanaceous crops; the virulent types of viral strains break through the protocols (Torrance & Talianksy, 2020). More so, in most cases, resistance is specific to crops and regions, which limits its wider application. The situation with smallholder farmers is another issue, since the resistant cultivars do not always fit in with the preferences of the locality when it comes to taste, cooking quality, and requirements of the market. The other disadvantage is that in most cases, resistance is partial, thus minimizing but not eliminating the viral incidences, hence requiring supplementary management mechanisms. Such limitations pinpoint the urgent need to combine resistance breeding with the molecular-based techniques of Clustered Regularly Interspaced Short Palindromic Repeats (CRISPR)-based systems of genome editing and RNA interference (Wang et al., 2022).

4.3 Vector Transmission and Environmental Influences

The majority of plant viruses are carried by insect vectors, like aphids, whiteflies, leaf hoppers, nematodes, and fungi (Butter, 2018). These vectors are difficult to control due to their high reproduction ability, locomotion, and development of resistance to the use of insecticides. Bemisia tabaci, a key driver of begomoviruses, has also emerged as one of the greatest depressed scenario because it has been established to dwell on a wide range of environmental elements and is developing resistance to chemical suppression rather quickly. Climatic factors further dictate the population of vectors, and as temperatures rise, the rainfall cycle changes, and there are other extreme weather conditions that increase the range of activities of vectors (de Souza & Weaver, 2024). This increases the uncertainty and inability to contain viral epidemics. This non-persistent and semi-persistent transmission of most viruses further makes controlling vectors such a challenge, because during a short interval, such as insect feeding, the virus can be passed, which leads to insecticides acting before the enemy. Besides, environmental sustainability ensures restricting the use of chemical pesticides in the long term, which requires alternative and environmentally friendly methods. (Singh, 2024).

4.4 Shortcomings of Current Control Strategies

The new approaches to the treatment of plant viral diseases remain largely preventive, given their nature, and can also be reactive and not very curative. They include phytosanitary rules, planting vegetable material based on certification of virus-free certification, rotation of crops, culture, the application of vectors, and resistant varieties. Nevertheless, there are serious malfunctions with these measures. In areas having loose seed assembly or loose laws, phytosanitary measures are hard to implement (Kuhlmann & Dey, 2021). Smallholder farmers do not have or cannot afford certified virus-free growing materials. Chemical insecticide-based strategies of controlling vectors are unsustainable through environmental factors, human resistance, and undesired effects on the non-target organisms, including pollinators and predatory factors. Besides, most of the cultural practices, although beneficial to use, are inadequate when implemented independently. Combinations are frequently suggested, but their implementation is still rather rare because of a lack of knowledge, limitations of resources, and untrained farmers (Bottrell & Schoenly, 2018). Moreover, diagnostic tools that are fast and applicable in the field are absent in most regions of the world, making it hard and slow to detect and respond to cases of viruses. Such weaknesses put forward the challenge of the need to develop novel and multi-prong approaches in integrating molecular biology, nanotechnology, precision agriculture, and educating farmers to achieve effective and sustainable control.

5. ROLE OF CLIMATE CHANGE AND GLOBALIZATION

5.1 Influence of Climate Variability on Disease Incidence

The variable climatic changes also have a major impact on plant viral diseases as they affect their occurrence, spread, and severity. Global warming can induce pest vectors to multiply faster and reproduce faster, e.g., aphids, whiteflies, leafhoppers, resulting in the enhancement of virus spread. To give one example, increased temperatures have been indicated to increase the dispersal of barley yellow dwarf virus through an increase in the season of activity of its vector aphids. Likewise, the unpredictability of the rainfall and drought stressors causes plants to be weak and, therefore, the hosts are predisposed to infection by the viruses. The spread of climate change also allows the relocation of the population of vectors to high latitudes and altitudes, putting viruses in areas that were previously not conducive to their survival because of climate fluctuations (Ramos et al., 2019). Different drastic weather conditions like storms and floods may again contribute to the expansion of the viruses, either physically through the dispersion of the vectors or through the encouragement of continuous congestion patterns of crop troughs. All these forces combined put climate change as an effective force behind the rise in viral diseases and their maintenance.

5.2 Spread Through International Trade and Travel

Globalization has transformed trade practices in agricultural activities, but a negative consequence is that globalization has contributed to the cause and spread of plant viruses faster across borders than before. The viruses and their vectors have found an avenue to pass through natural geographic barriers through the ever-shifting of seeds, planting material, fruits, and ornamentals worldwide. As an example, the international trade of tomato seedlings closely corresponded to the dissemination of tomato brown rugose fruit virus (ToBRFV) in the whole world, whose outbreak occurred within a surprisingly short period in Europe, the Middle East, and North America. On the same note, the banana bunchy top virus has already been distributed through the transport of infected planting materials, restricting the exportability of bananas in most countries. Ornamental crops, which are typically not strictly controlled compared to the crops that are the staples, fall victim to new forms of viruses, which subsequently end up spreading into the farms. It is also human travel, which augments risks, also because infested plant materials or contaminated tools may release viruses into new territory. Introduction does not have very high levels of prevention due to loopholes in phytosanitary or other regulations, the absence of quick diagnostic tools in ports, and hair of many introductions of the plants without showing symptoms in the planting materials. (Ambe, 2015).

5.3 Emerging Threats to Non-Endemic Regions

The appearance of viral diseases in non-endemic areas is one of the most frightening outcomes of climate change and globalization. Introduction of new viruses or vectors in agro environments that are naive can cause severe epidemics since there is no co-evolved host defense and management experience. To illustrate, the introduction of cassava brown streak virus to the uninfected regions in East Africa has affected the food security of the region with disastrous impacts. In the same vein, the citrus tristeza virus that was previously found hewn to the South American continent has found its way to some areas of Asia and Africa, which poses a threat to the citrus industry of the world (Sun et al., 2024). In most instances, these emerging threats are realized once they are well established, since the systems of surveillance and diagnostic capabilities are poor in most of the developing regions. Such introductions are projected to increase with further occurrences of climate change, which would ensure favorable ecological niches of vectors in inhospitable places. The threats are also not limited to the staple crops; high-value horticultural and industrial crops are also facing threats and are becoming susceptible to posing a threat to both the food and economic security of the susceptible regions (Figure 4) (Faour-Klingbeil & Todd, 2018).

Figure 4. Emerging threats of plant viral diseases, including introductions of new viruses/vectors, severe epidemics, impacts on food security, poor surveillance systems, and increasing risks to staple, horticultural, and industrial crops under climate change

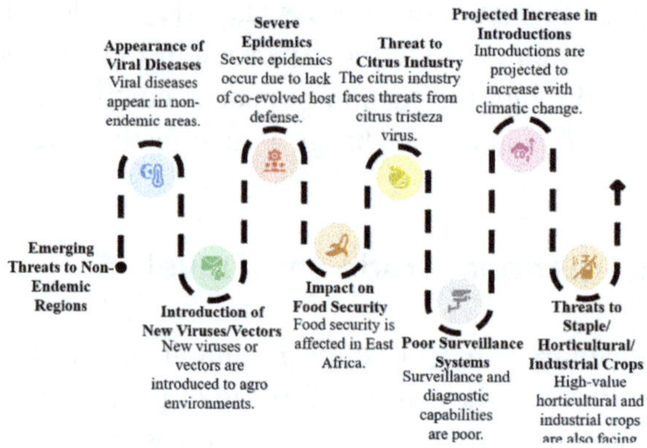

6. INNOVATIVE APPROACHES IN DISEASE MANAGEMENT

Plant viral diseases can no longer be managed using traditional methods, which pose drawbacks in terms of their versatility, finesse, and the amount of information provided. The conventional methods like cultural management, resistant cultivars, and chemical-controlled vectors have not been completely or temporarily effective, mainly due to the adaptability of the viruses, complex conjectures related to the vectors, and environmental sustainability issues. New developments in compound breeding, biotechnology, integrated pest and vector control, as well as nanotechnology, can be adopted as alternatives. Such new developments are not only more effective in the treatment of diseases, but are also in line with the sustainable approach of development, where there will be little ecological impact and food security.

6.1 Advances in Molecular Breeding and Biotechnology

Biotechnology and molecular breeding have changed the ability to breed resistant crops and enhance the defense mechanism of plants against viruses. WA All carbon dioxide seeping through the soil pollutes the soil with the annual and chemical web (Mayer 2016). Demonstratively, molecular markers have been successfully applied in tracking resistance to rice tungro and cassava mosaic viruses. In the later stages, genome editing technology has also been introduced in the form of CRISPR-Cas systems that can be utilized as powerful tools for developing virus-resistant crops. CRISPR provides the ability to make specific alterations in host susceptibility genes or the whole genome sequences of viruses, and is a useful force in providing a stable defense against an infection that is very specific infection. RNA Interference (RNAi) has also been successful in using RNAi to silence viral genes, and, therefore, this reduces the severity of infection in crops such as papaya, banana, and cassava. Biotechnology has been instrumental in dealing with viral diseases, with transgenic methods of dealing with them included in these practices, including transgenic papaya in resistance to papaya ringspot virus development. Nevertheless, regulation issues, societal acceptance, and biosafety are major concerns that should be overcome in order to maximize the utilization of these innovations.

6.2 Integrated Pest and Vector Management

Evidence has shown that most plant viruses depend on vectors as a method of transmission, and therefore, effective management of vectors is at the core of controlling the disease. I am talking of Integrated Pest and Vector Management (IPVM), which is a convergence of biological, cultural, mechanical, and chemical modes of attaining sustainable suppression of the vector populations. Biological control agents have been used more effectively to control aphids, whiteflies, and leafhoppers — the main vectors of most viral diseases — using parasitoids, entomopathogenic fungi, as well as predatory insects. To minimize the appeal of fields to the vectors, cultural practices such as intercropping, crop rotation, and vegetation control are used in the management of habitats that destroy the viral vectors' reproduction. Mechanical obstacles are beneficial, such as steering off the landing of vectors with reflective mulches, with precision pesticide spraying so that there is minimal impact on the environment, and resistant insects would be postponed. Increased use of remote sensing and digital agriculture can also promote the timeliness and accuracy of intervention in affecting vector populations by conducting real-time tracking. The use of the IPVM methodology, which is holistic and ecologically sensitive, is a substitute for high dependency on the use of chemical insecticides, as it helps to build resilience in dealing with viral epidemics.

6.3 Nanotechnology-Based Solutions

Nanotechnology provides an opportunity for the control of viral diseases in plants, using the ability of nanoparticles (NPs), which uniquely possess physicochemical characteristics (Abbasi et al., 2025). Direct antiviral action against viral capsids, viral replication, or release of reactive oxygen species (ROS), killing viral infectivity. Metallic nanoparticles, which include silver (AgNPs), zinc oxide (ZnONPs), or copper oxide (CuONPs), have direct antiviral effects (Tortella et al., 2022). Indirectly, nanoparticles can condition or program the plant's immune system through the activation of systemic acquired resistance (SAR), induction of antioxidant enzymes by activating increased antioxidant enzyme activity, which will help reduce the fixed impacts of a virus on the physiology of a host. Silver nanoparticles are effective in the case of tobacco mosaic virus (TMV), whereas silica nanoparticles have been possible to enhance the host defense using stirs. Also, formulations based on nanoparticles can be used along with traditional agrochemicals or biological agents to increase the delivery and decrease the amount of chemicals. The previously mentioned technology is also more sustainable because the issue of toxicity and environmental retention of nanoparticles produced via plant extracts and microbial systems is solved through green synthesis (Khan et al., 2025). Although it is showing promising laboratory and greenhouse outcomes, a scaling exercise in the field, phytotoxicity testing, and regulatory guidelines have been identified to be the major stumbling blocks and impediments to one-day widespread adoption.

7. FUTURE PERSPECTIVES

In the future, the control of plant viral diseases needs progressive mindsets beyond the conventional discipline to consider the interrelated issues of agriculture, the environment, and the food security of the world. It is important that interdisciplinary approaches are needed because any single strategy that is applied will not be enough to counter the dynamic nature of the plant viruses, the complexity of their vectors, or the vulnerabilities of the socio-economic factors that they take advantage of. The implementation of plant pathology, molecular biology, ecology, nanotechnology, bioinformatics, as well as social sciences, will develop an integrated framework of a holistic approach into disease management, joined by innovative science and farmer-based solutions. As it is described, collaborative systems which would combine genomic information with climate prediction and modeling and vector ecology can prove useful in anticipating possible occurrences and developing proactive activities. On the same note, sustainable and earth-friendly approaches to management should be put forward as more and more people worry about their environment, the existence of pesticide-resistant insects, and the disappearance of other biodiversity, posed by the conventional approaches. Green-synthesized nanoparticles, CRISPR-based precision breeding, RNA interference, and biologically developed pesticides are likely suppliers of an approach that will yield control of the viruses but with minimal environmental effects. These must be combined with agroecological techniques such as crop diversification, intercropping, as well as conserving natural enemies that integrate resilience into agricultural systems and smaller chances of epidemics. Moreover, the current challenges facing plant viral insecurity in assuring global food security also require that these technologies and solutions should be affordable, easily accessible, and dynamic for the needs of smallholder farmers, who are the most exposed to viral epidemics. This does not need just technological advances, but also well-established policy structures, extension services, and cross-national collaboration in order to have a fair dispensation of resistant varieties, virus-cleaning planting materials, as well as diagnostic equipment. The smart farming system with the integration of precision, artificial intelligence, and remote sensing will be central in early disease detection, monitoring air vectors, and interventions at the location, thus making it efficient and sustainable in resource utilization. Nevertheless, the future of managing plant viral diseases will be a convergence of state-of-the-art science, sustainable agricultural causes, and arrangements of governance that are mindful of being not only productive but also environmentally caring. The global community can find new solutions to reduce the effects of viral diseases and avoid environmental negativity through an interdisciplinary approach to this issue by investing in green and scale technologies, equipping farmers with expertise and resources, and retaliating against other motives to protect food availability and food security for generations to come.

CONCLUSION

Plant viral diseases have been among the most pressing issues facing agricultural activity in the world, including developed and developing countries, as they threaten crop productivity, food quality, and even the economies of countries. The viruses are extremely complicated and more of a preventive process compared to an elimination process. Unlike other pathogens, they cannot be attacked specifically with chemicals. As this chapter has described, the world provides a comprehensive list of plant viral disease burdens, which cannot be underestimated due to their severe effects on nature and the economy as a whole, and smallholder farmers in particular. The old-fashioned approaches, like using resistant cultivars, culture, and taking control of the vectors, have been partially helpful, but are mostly subdued by the breakneck development of the viruses, genetic diversity, and the impact of climate change and globalization. Meanwhile, new prospects are provided by innovative approaches. The future of managing plant viral diseases is changing with the changes in both molecular breeding, genome editing, RNA interference, integrated pest and vector management, as well as nanotechnology. These plans have the prospects of offering a long-lasting, environmental-friendly, and green solution, particularly in combination with precision farming and digital technology in the processes of initial identification and surveillance. Its extensive use must overcome severe issues to consider such aspects as environmental protection, government policy and regulations, affordability, and availability to resource-deprived agriculture models. In the future, the interdisciplinary interactions that span across plant science, biotechnology, ecology, nanotechnology, and socio-economic studies will enable the successful management of plant viral diseases. The key elements should lie in sustainable and environmentally friendly management practices that will safeguard the biodiversity and decrease the use of chemical interventions. It is also essential that the innovative solutions are scalable and available to all farmers across the globe, hence creating sustainable food and nutritional security in the long run. Through the adoption of integrative and sustainable practices, the world's above-ground agricultural fraternity is set to be a step closer to resilience to food systems that can stand the current and manifesting menace of transforming viral diseases in plants.

REFERENCES

Abbasi, A. J., Anas, M., Elahi, M., Khan, A., Khattak, W. A., Saleem, M. H., Fahad, S., Elansary, H. O., Mahmoud, E. A., & Ahmed, T. (2025). Restoring wheat productivity and nutrient balance under cadmium stress through reducing toxicity, metal uptake, and oxidative damage using selenium nanoparticles. *Journal of Trace Elements in Medicine and Biology*, *89*, 127644. PMID: 40233590

Ambe, S. (2015). *Invasive plant pests detection difficulties at the port of entry: A phenomenological study of the lived experiences of agriculture specialists.*

Amelework, A. B., Bairu, M. W., Maema, O., Venter, S. L., & Laing, M. (2021). Adoption and promotion of resilient crops for climate risk mitigation and import substitution: A case analysis of cassava for South African agriculture. *Frontiers in Sustainable Food Systems*, *5*, 617783.

Anas, M., Falak, A., Hassan, S., Khattak, W. A., Saleem, M. H., Khan, K. A., Khalid, A., & Fahad, S. (2025). Microbial Interactions and Bacterial Responses to Metal Stress in Plants: Mechanisms, Adaptations, and Applications for Sustainable Agriculture. *Journal of Crop Health*, *77*(1), 36.

Anas, M., Khattak, W. A., Fahad, S., Alrawiq, N., Alrawiq, H. S., Abdelsalam, N. R., Jaremko, M., & Quraishi, U. M. (2024). Mitigating heavy metal pollution in agriculture: A multi-omics and nanotechnology approach to safeguard global wheat production. *Journal of Hazardous Materials Advances*, 100584.

Anderson, J. A., Ellsworth, P. C., Faria, J. C., Head, G. P., Owen, M. D., Pilcher, C. D., Shelton, A. M., & Meissle, M. (2019). Genetically engineered crops: Importance of diversified integrated pest management for agricultural sustainability. *Frontiers in Bioengineering and Biotechnology*, *7*, 24. PMID: 30842944

Baker, R. E., Mahmud, A. S., Miller, I. F., Rajeev, M., Rasambainarivo, F., Rice, B. L., Takahashi, S., Tatem, A. J., Wagner, C. E., & Wang, L.-F. (2022). Infectious disease in an era of global change. *Nature Reviews. Microbiology*, *20*(4), 193–205. PMID: 34646006

Béné, C. (2020). Resilience of local food systems and links to food security–A review of some important concepts in the context of COVID-19 and other shocks. *Food Security*, *12*(4), 805–822. PMID: 32837646

Bos, L. (1999). Beijerinck's work on tobacco mosaic virus: Historical context and legacy. *Philosophical Transactions of the Royal Society of London. Series B, Biological Sciences*, *354*(1383), 675–685. PMID: 10212948

Bottrell, D., & Schoenly, K. (2018). Integrated pest management for resource-limited farmers: Challenges for achieving ecological, social, and economic sustainability. *Journal of Agricultural Science*, *156*(3), 408–426.

Butter, N. S. (2018). *Insect vectors and plant pathogens*. CRC Press.

Chikoti, P. C., & Tembo, M. (2022). Expansion and impact of cassava brown streak and cassava mosaic diseases in Africa: A review. *Frontiers in Sustainable Food Systems*, *6*, 1076364.

de Souza, W. M., & Weaver, S. C. (2024). Effects of climate change and human activities on vector-borne diseases. *Nature Reviews. Microbiology*, *22*(8), 476–491. PMID: 38486116

Dey, S. R., Das, R., & De, M. (2024). A brief review on the present status of rice tungro disease: Types of viruses, vectors, occurrence, symptoms, control, and resistant rice varieties. *Int. J. Adv. Life Sci. Res*, *7*, 15–23.

Drenth, A., & Kema, G. (2021). The vulnerability of bananas to globally emerging disease threats. *Phytopathology*, *111*(12), 2146–2161. PMID: 34231377

Faour-Klingbeil, D., & Todd, E. C. (2018). A review on the rising prevalence of international standards: Threats or opportunities for the agri-food produce sector in developing countries, with a focus on examples from the MENA region. *Foods*, *7*(3), 33. PMID: 29510498

Fingu-Mabola, J. C., & Francis, F. (2021). Aphid–plant–phytovirus pathosystems: Influencing factors from vector behaviour to virus spread. *Agriculture*, *11*(6), 502.

Flint, L. J. (2014). *Identification, prevalence, and impacts of viral diseases of UK winter wheat.*

Gautam, S., Variyar, P. S., & Sharma, A. K. (2015). Quality and safety aspects of vegetables. In *Handbook of Vegetable Preservation and Processing* (pp. 925–954). CRC Press.

Gómez, P., Rodríguez-Hernández, A., Moury, B., & Aranda, M. (2009). Genetic resistance for the sustainable control of plant virus diseases: Breeding, mechanisms, and durability. *European Journal of Plant Pathology*, *125*(1), 1–22.

Hertel, T. W., Burke, M. B., & Lobell, D. B. (2010). The poverty implications of climate-induced crop yield changes by 2030. *Global Environmental Change*, *20*(4), 577–585.

Ibaba, J. D., & Gubba, A. (2020). High-throughput sequencing application in the diagnosis and discovery of plant-infecting viruses in Africa, a decade later. *Plants*, *9*(10), 1376. PMID: 33081084

Idrees, M. A., Abbas, A., Saddam, B., Bashir, M. H., Naveed, H., Khan, A. K., & Dara, M. Z. N. (2024). A comprehensive review: Persistence, circulative transmission of begomovirus by whitefly vectors. *International Journal of Tropical Insect Science*, *44*(2), 405–417.

Jones, R. (2014). Plant virus ecology and epidemiology: Historical perspectives, recent progress and prospects. *Annals of Applied Biology*, *164*(3), 320–347.

Jones, R. (2021). Global plant virus disease pandemics and epidemics. *Plants*, *10*, 233. PMID: 33504044

Jones, R. A. (2020). Disease pandemics and major epidemics arise from new encounters between indigenous viruses and introduced crops. *Viruses*, *12*(12), 1388. PMID: 33291635

Jones, R. A., & Naidu, R. A. (2019). Global dimensions of plant virus diseases: Current status and future perspectives. *Annual Review of Virology*, *6*(1), 387–409. PMID: 31283443

Khan, A., Anas, M., Ullah, N., Khattak, W. A., Ali, M., Maaza, M., & Shinwari, Z. K. (2025). Harnessing green nanotechnology: Exploring the potential of silver, copper, and their nanocomposites for biomedical and agricultural applications. *Pakistan Journal of Botany*, *57*, 6.

Kreuze, J. F., Cuellar, W. J., Kumar, P. L., Boddupalli, P., & Omondi, A. B. (2023). New technologies provide innovative opportunities to enhance understanding of major viral diseases threatening global food security. *Phytopathology*, *113*(9), 1622–1629. PMID: 37311729

Kreuze, J. F., Souza-Dias, J., Jeevalatha, A., Figueira, A., Valkonen, J., & Jones, R. (2020). Viral diseases in potato. In *The potato crop: Its agricultural, nutritional and social contribution to humankind* (pp. 389–430). Springer International Publishing Cham.

Krishnareddy, M. (2013). Impact of climate change on insect vectors and vector-borne plant viruses and phytoplasma. In *Climate-resilient horticulture: Adaptation and mitigation strategies* (pp. 255–277). Springer.

Kuhlmann, K., & Dey, B. (2021). Using regulatory flexibility to address market informality in seed systems: A global study. *Agronomy (Basel)*, *11*(2), 377.

Lefeuvre, P., Martin, D. P., Elena, S. F., Shepherd, D. N., Roumagnac, P., & Varsani, A. (2019). Evolution and ecology of plant viruses. *Nature Reviews. Microbiology*, *17*(10), 632–644. PMID: 31312033

Mabunda, G. P., Nemukondeni, N., & Selaledi, L. (2025). Sanitary and phytosanitary (SPS) measures and their implications for international agricultural trade: Challenges and opportunities; comprehensive review. *Discover Agriculture*, *3*(1), 1–35.

McDonald, B. A., & Stukenbrock, E. H. (2016). Rapid emergence of pathogens in agro-ecosystems: Global threats to agricultural sustainability and food security. *Philosophical Transactions of the Royal Society B: Biological Sciences, 371*(1709), 20160026.

Mehetre, G. T., Leo, V. V., Singh, G., Sorokan, A., Maksimov, I., Yadav, M. K., Upadhyaya, K., Hashem, A., Alsaleh, A. N., & Dawoud, T. M. (2021). Current developments and challenges in plant viral diagnostics: A systematic review. *Viruses*, *13*(3), 412. PMID: 33807625

Meshram, P. (2025). *Plant Breeding for Resistance to Pests and Diseases*. Academic Guru Publishing House.

Nigam, R. (2024). *Plant diseases and food security in the 21st Century*. Academic Guru Publishing House.

Niraula, P. M., & Fondong, V. N. (2021). Development and adoption of genetically engineered plants for virus resistance: Advances, opportunities, and challenges. *Plants*, *10*(11), 2339. PMID: 34834702

Oraon, S., Padamini, R., Shahni, Y. S., Das, N., Sinha, D., Sujatha, G., Singh, O. B., & Karanwal, R. (2024). Impact of emerging pathogens on crop production. *Microbiology Research Journal International*, *34*(7), 80–92.

Ramos, R. S., Kumar, L., Shabani, F., & Picanço, M. C. (2019). Risk of spread of tomato yellow leaf curl virus (TYLCV) in tomato crops under various climate change scenarios. *Agricultural Systems*, *173*, 524–535.

Ray, R. V. (2024). Effects of pathogens and disease on plant physiology. In *Agrios' Plant Pathology* (pp. 63–92). Elsevier.

Redinbaugh, M. G., & Stewart, L. R. (2018). Maize lethal necrosis: An emerging, synergistic viral disease. *Annual Review of Virology*, *5*(1), 301–322. PMID: 30059641

Ristaino, J. B., Anderson, P. K., Bebber, D. P., Brauman, K. A., Cunniffe, N. J., Fedoroff, N. V., Finegold, C., Garrett, K. A., Gilligan, C. A., & Jones, C. M. (2021). The persistent threat of emerging plant disease pandemics to global food security. *Proceedings of the National Academy of Sciences of the United States of America*, *118*(23), e2022239118. PMID: 34021073

Roberts, O. (2025). Assessing the Impact of Economic Sanctions on Small Business Sustainability and Growth. *Available at SSRN* 5129887.

Robson, F., Hird, D. L., & Boa, E. (2024). Cassava brown streak: A deadly virus on the move. *Plant Pathology*, *73*(2), 221–241.

Sarwar, M., Shad, N. A., & Batool, R. (2020). Integrated management of vectored viral diseases of plants. In *Applied Plant Virology* (pp. 707–724). Elsevier.

Savary, S., Bregaglio, S., Willocquet, L., Gustafson, D., Mason D'Croz, D., Sparks, A., Castilla, N., Djurle, A., Allinne, C., & Sharma, M. (2017). Crop health and its global impacts on the components of food security. *Food Security*, *9*(2), 311–327.

Singh, M. (2024). Pesticide predicament: Exploring environmental and health impacts, and possible eco-friendly solutions. *Int. J. Environ. Health Sci*, *6*, 38–47.

Sun, Y., Yokomi, R. K., & Folimonova, S. Y. (2024). Citrus tristeza virus: A century-long challenge for the world's citrus industries. *Annals of Applied Biology*, *185*(3), 304–322.

Tatineni, S., & Hein, G. L. (2023). Plant viruses of agricultural importance: Current and future perspectives of virus disease management strategies. *Phytopathology*, *113*(2), 117–141. PMID: 36095333

Tomlinson, K. R., Bailey, A. M., Alicai, T., Seal, S., & Foster, G. D. (2018). Cassava brown streak disease: Historical timeline, current knowledge and prospects. *Molecular Plant Pathology*, *19*(5), 1282–1294. PMID: 28887856

Torrance, L., & Talianksy, M. E. (2020). Potato Virus Y emerged and evolved from the Andes of South America to become a major destructive pathogen of potato and other solanaceous crops worldwide. *Viruses*, *12*(12), 1430. PMID: 33322703

Tortella, G., Pieretti, J., Rubilar, O., Fernández-Baldo, M., Benavides-Mendoza, A., Diez, M., & Seabra, A. (2022). Silver, copper, and copper oxide nanoparticles in the fight against human viruses: Progress and perspectives. *Critical Reviews in Biotechnology*, *42*(3), 431–449. PMID: 34233551

Vurro, M., Bonciani, B., & Vannacci, G. (2010). Emerging infectious diseases of crop plants in developing countries: Impact on agriculture and socio-economic consequences. *Food Security*, *2*(2), 113–132.

Wang, Y., Zafar, N., Ali, Q., Manghwar, H., Wang, G., Yu, L., Ding, X., Ding, F., Hong, N., & Wang, G. (2022). CRISPR/Cas genome editing technologies for plant improvement against biotic and abiotic stresses: Advances, limitations, and future perspectives. *Cells*, *11*(23), 3928. PMID: 36497186

Wick, C. H. (2023). *Virus Detection*. CRC Press.

Wilson, A. L., Courtenay, O., Kelly-Hope, L. A., Scott, T. W., Takken, W., Torr, S. J., & Lindsay, S. W. (2020). The importance of vector control for the control and elimination of vector-borne diseases. *PLoS Neglected Tropical Diseases*, *14*(1), e0007831. PMID: 31945061

Yadav, N., & Upadhyay, R. K. (2023). Global effect of climate change on seasonal cycles, vector population, and rising challenges of communicable diseases: A review. *Journal of Atmospheric Science Research*, *6*(1).

Zhan, J., Thrall, P. H., & Burdon, J. J. (2014). Achieving sustainable plant disease management through evolutionary principles. *Trends in Plant Science*, *19*(9), 570–575. PMID: 24853471

Zuo, K., Gao, W., Wu, Z., Zhang, L., Wang, J., Yuan, X., Li, C., Xiang, Q., Lu, L., & Liu, H. (2024). Evolution of virology: Science history through milestones and technological advancements. *Viruses*, *16*(3), 374. PMID: 38543740

Chapter 2
Nanotechnology Primer: Principles, Types, and Physicochemical Properties of Nanoparticles

Wajiha Zaka Ansari
University of Queensland, Australia

Tauqeer Ahmed Qadri
COMSATS University Islamabad, Pakistan

Amjid Khan
University of South Africa, South Africa

Zeeshan Ahmad
COMSATS University Islamabad, Pakistan

Malik Maaza
University of South Africa, South Africa

Haleema Tariq
McGill University, Canada

Khaoula Hkiri
University of Cape Town, South Africa

Hamza Mohamed
https://orcid.org/0000-0003-0550-8607
IMMM - Institute of Molecules and Materials, Le Mans University, France

ABSTRACT

This chapter provides an in-depth primer on nanotechnology, with a focus on its application in plant virology. It explores the fundamental principles behind nanotechnology, highlighting the unique properties of nanoparticles (NPs) due to their size (1–100 nm). The chapter outlines the various types of nanoparticles, including metallic, carbon-based, organic, and inorganic, and examines their physicochemical properties such as size, shape, surface charge, and stability. The size-dependent reactivity and surface area of nanoparticles make them highly effective in plant disease diagnostics, antiviral treatments, and real-time monitoring of plant health. The chapter further discusses the methods for nanoparticle synthesis and characterization and investigates their interactions with plant systems and viral pathogens. By integrating interdisciplinary knowledge from nanoscience, plant pathology, and biotechnology, this chapter serves as a foundation for the use of nanoparticles in managing plant viral diseases and offers insights into their future applications in sustainable agriculture.

DOI: 10.4018/979-8-3373-4913-8.ch002

1. INTRODUCTION

Nanotechnology is one of the most vibrant and disruptive knowledge areas of the recent past; it lies across physics, chemistry, materials science, biology, and engineering (Hassan et al., 2023; Khan, Ashiq, et al., 2025). Nanotechnology is fundamentally about these processes of designing, manipulating, and using matter on the nanoscale; that is, a scale defined as being in a broad sense of 1-100 nanometers (Bibi et al., 2024; Khan, Anas, Bibi, et al., 2025). Products in the range would be treated as nano objects as defined by ISO and the ASTM corresponding to nanoparticles, nanofibers, and nanoplates, where at least one dimension is of the nanoscale. This dimensional limit is not random but is the realm where the properties of matter are essentially different from their bulk neighbors. Even materials that seem to be the same as each other at a very large scale can, when forced below the nanoscale, rearrange themselves in such a way that they have distinct optical, electronic, catalytic, and mechanical properties (Kalaoglu-Altan et al., 2025; Sani et al., 2025).

Intellectually, the roots of nanotechnology can be traced to 1959, when in his classic lecture entitled "*There's Plenty of Room at the Bottom*" Richard Feynman envisaged the possibility of constructing functional systems by playing with individual atoms and molecules. At the time of his remarks, his speculations were subsequently proven, but his remarks catalyzed a paradigm shift that inspired many generations of scientists (Sachdeva et al., 2021). The following decades were marked by a swift growth in the synthesis, characterization, and use of nanomaterials. One of the first breakthroughs that made that dream possible was in the 1980s, when scanning tunneling microscopy (STM) was developed and allowed direct observation and manipulation of atoms. Concurrent developments in bottom-up chemical synthesis, top-down lithographic processes, and bio-scaffolded methods increased the fold of nanostructures that could be fabricated predictably. The resultant events put nanotechnology on the map, as it were, instead of it standing as a hypothetical interest among scholars (Chaikittisilp et al., 2022).

The problem with nanoscale processes is that they are the consequence of the interplay between surface interactions and quantum. The surface volume ratios become extremely large at sizes less than 100 nm, and therefore a sizeable fraction of atoms or molecules would reside in the interfacial or "wetting" regions. Such a structural defect can enhance the reactivity, modify catalysis efficiency, and strength (Zhao et al., 2025). For example, bulk gold is inert physically but not chemically, whereas gold nanoprecipitates are chemically active and can act as catalysts, resulting from a pack of highly reactive surface atoms. In the same way, titanium dioxide nanoparticles have better photocatalytic power than bulk TiO_2, and this has made them very relevant in cases of environmental remediation and self-cleaning coatings (Mahdi et al., 2023). There is another complication due to the quantum confinement that we need to consider besides the surface effect. Quasi-spherical semiconducting nanoparticles, also referred to as quantum dots (QDs), have size-dependent band gaps and thus provide tunable fluorescence over the entire visible spectrum. Single-domain magnetic nanoparticles, such as FeOF, exhibit large biomedical utility because they are superparamagnetic when these particles are at the nanoscale, whereas in bulk forms, they lose that property (Lyle, 2022).

Nanotechnology is a science of contradiction that must be controlled in perspective of its huge potential. In contrast, including nanoscience in the curriculum would provide several alternative materials and devices that can be used to address next-generation challenges. Nanoparticles as dose carriers for the specific drug, nanosensors for prompt diagnosis, and nano nanoscale catalyst to fasten the reaction, which is part of green chemistry with effectiveness. These traits, on the other hand, which would be beneficial for function, i.e., decreased volume, increased reactivity, and altered surface, may also be detrimental.

TNs are known to cross biological barriers and engage with biomolecules in unusual manners, as well as bioaccumulate in the ecosystem (Chaikittisilp et al., 2022; Khan, Anas, Qadri, et al., 2025).

Nanotechnology has emerged as a transformative approach in advancing climate-smart agriculture, which seeks to enhance productivity, build resilience against climate change, and promote environmental sustainability. The integration of nanoparticles in farming systems enables efficient nutrient delivery, soil health improvement, and reduced agrochemical losses, thereby supporting sustainable resource management (Abbasi et al., 2025; Khan, Anas, Ullah, et al., 2025). Nano-fertilizers and nano-pesticides, for instance, can release active compounds in a controlled manner, minimizing leaching and volatilization while improving crop uptake efficiency (Ullah et al., 2025). In addition, nanotechnology contributes to soil and water remediation, stress-tolerant crop development, and precision monitoring of environmental conditions through nanosensors (Anas et al., 2024; Ehsan et al., 2025). Such applications align closely with global efforts to ensure food security under changing climatic conditions. By merging fundamental nanoscience with agricultural innovation, researchers are paving the way for sustainable, efficient, and environmentally responsible farming practices. This chapter explores the fundamental principles of nanoscience, focusing on the physicochemical properties, synthesis, and characterization of nanoparticles. It highlights how size, shape, surface chemistry, and crystallinity impact performance at the nano-bio interface, with an emphasis on practical applications. The importance of reporting standards, reproducibility, and eco-safe design for sustainable nanotechnology applications in climate-smart agriculture is also discussed.

2. FUNDAMENTAL PRINCIPLES OF NANOSCIENCE

Nanoscience is based on the number of physical and chemical properties that will manifest themselves upon formatting the materials into the nanoscale (1-100 nm). In these dimensions, the classical state of bulk materials transforms to the new behaviors of surface, interface, and quantum effects, with capabilities unavailable to macroscopically sized systems (Anas et al., 2025; Maaza et al., 2022). It is imperative to understand how these principles can be utilized to rationally design and create nanomaterials and the basis of their application in energy, medicine, catalysis, and environmental remediation. All the fundamental principles of nanoscience are illustrated in Figure 1.

Nanoscaling down the dimensions changes the dynamics and the electronic structure of the substance. Such a transition to quantization may be illustrated by quantum confinement in semiconductor nanocrystals or quantum dots: the ability to tune the optical emittance by size, leading to quantum-confined optoelectronics and bioimaging (Udoisoh et al., 2025). In analogy, noble metal nanoparticles like gold and silver are also characterized by localized surface plasmon resonance (LSPR) wherein resonant absorption and scattering are observed due to the collective oscillation of the conduction electrons under excitation by light. The ability of this resonance to be very sensitive to both the size and the shape of particles well as to the dielectric surrounding, makes it possible to use it in sensing, imaging, and even photothermal therapy (Shipunova et al., 2022). The stability of thermodynamic systems at the nanoscale is mostly dominated by the contributions of the surface. Nanoparticles have a large surface area to mass ratio, which means they have a high surface energy, which in many cases causes them to be metastable over the bulk phase (Khan, Younis, et al., 2025; Sani et al., 2024).

This is crucial since the largest proportion of functional nanoparticles is used in colloidal dispersion. The DLVO theory of colloidal stability relates the forces acting between colloidal particles to determine whether they are in stability or not to a balance between the van der Waals force of attraction on the one hand and the electrostatic repulsion between charged colloids, the other theory (Krucker-Velasquez et al., 2025). Besides electrostatics, steric stabilization is achieved due to the presence of adsorbed or grafted polymer molecules that produce entropic and enthalpic repulsion and therefore prevent the formation of aggregates. Surface potential is empirically employed as a good measure of colloidal stability; a value greater than +/-30 mV is representative of a well-dispersed system. These principles, when mastered, ensure that the nanoparticles can be stable in terms of uniformity, bioavailability, and reactivity, especially where biomedical research or catalytic performance is subject to aggregation (Roucoux & Philippot, 2021).

The strong interface between structure and functionality characterizes nanoscience. The size, shape, and aspect ratio all cause a direct effect on the optical, electronic, mechanical, and catalytic characteristics (Hassan et al., 2024). As another example, longitudinal plasmon bands of gold nanorods are tunable by aspect ratio, which is widely used in plasmonic sensing and photothermal conversion. Carbon nanotubes (CNTs) are an example of structure-property optimization, with their tube structure, giving rise to very strong tensile strength, high thermal conductivity, and electronic tunability (Li et al., 2025). These applications demonstrate how one can design and control nanostructures to impart them with application-specific capabilities, which holds a deeper significance for the premise that structure dictates function at the nanoscale, in ways that it does not in bulk matter.

Figure 1. Some fundamental principles of nanoscience (a) Quantum confinement: size-dependent band gap tuning in quantum dots. (b) Localized surface plasmon resonance (LSPR) in metal nanoparticles. (c) Superparamagnetism in magnetic nanoparticles with loss of remanence. (d) Nano thermodynamics and Ostwald ripening showing growth of larger particles at the expense of smaller ones. (e) DLVO theory illustrates the balance of attractive and repulsive forces in colloidal stability. (f) Structure-property relationships exemplified by gold nanorods (aspect ratio-dependent plasmon bands) and carbon nanotubes (high strength and conductivity)

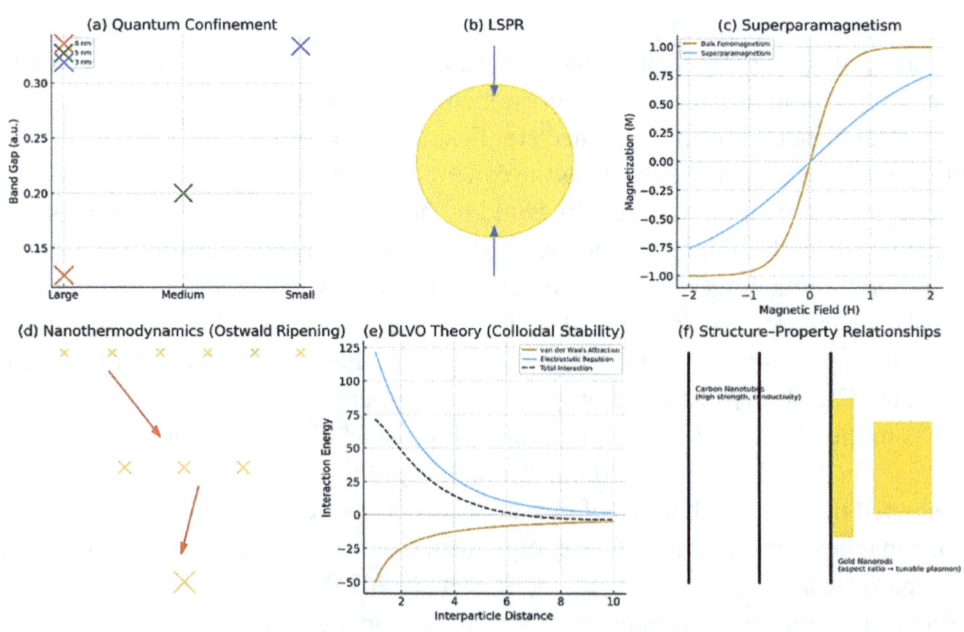

3. PHYSICOCHEMICAL PROPERTIES OF NANOPARTICLES

The physicochemical characteristics of nanoparticles (NPs) determine their specific behavior that forms the basis of multifaceted usage in catalysis, biomedicine, agriculture, energy, and environmental sciences. In contrast to bulk materials, NPs have particular characteristics because of their downsizing (1-100 nm), quantum size impact, and augmented surface area relationship (Anas et al., 2025; Hassan & Sani, 2025). It is crucial to know these properties not only to take advantage of them but also to understand possible biosafety and environmental risks. The essential details regarding the characterization, critical physicochemical properties, and functional significance of nanoparticles are discussed in the following sections.

3.1. Size and Distribution

The key characteristic of nanoparticles that determines their physical, chemical, and biological properties is the size of the particles. As particle size is reduced, an increasing fraction of particles are at the surface and therefore are more reactive to the surface, or more catalytically active, or more reactive with biological systems (Falak et al., 2025). As an example, nanoparticles that are less than 10 nm may have quantum confinement in their electronic structure, where their electronic states become discrete, dramatically changing their optical and electronic properties. In biomedical applications, size can cause differences in biodistribution, cellular uptake, and rates of clearance; particulate sizes below 100 nm may exploit the enhanced permeability and retention (EPR) effect to target tumors In biomedical applications, size can cause differences in biodistribution, cellular uptake, and rates of clearance; particulate sizes below 100 nm may exploit the enhanced permeability and retention (EPR) effect to target tumors (De et al., 2025).

Transmission Electron Microscopy (TEM) can be used to obtain an accurate analysis of nanoparticle size and distribution due to a direct image of particle morphology at the nanoscale (Wilson & Prud'homme, 2021). Dynamic Light Scattering (DLS) is used to determine the hydrodynamic size in suspension, and Nanoparticle tracking Analysis (NTA) allows dynamic visualization of particle size and decomposition rate while it is in suspension. Narrow size distribution is preferred in most applications, reproducibility, and similar performance across applications, including drug delivery or catalysis (Rodriguez-Loya et al., 2023).

3.2. Shape and Morphology

The geometry of the nanoparticles has a profound influence on their surface energy, reactivity, and light and biomolecule interaction. Although the most prevalent nanoparticles are spherical because of their thermodynamic stability, non-spherical nanoparticle shapes, including nanorods, nanocubes, nanostars, and nanosheets, can demonstrate some advantages (Joudeh & Linke, 2022). As an example, gold nanorods have been shown to display anisotropic plasmon resonance, which has resulted in high efficiency in such fields as photothermal therapy and biosensing. As would be expected, the nanocubes of cerium oxide are also catalytically more efficient because of high-energy facial exposure (Badir et al., 2025). In order to characterize the morphology, the instrumentation used is normally SEM, TEM, and AFM. The now accepted design variable to modify nanoparticle shape synthetically has come to be viewed as an important design parameter in nanotechnology, since it is possible to optimize a specific property of a nanoparticle for a specific application (Jagadeesh et al., 2024).

3.3. Crystal Structure Phase

The electronic, catalytic, and mechanical properties of nanoparticles depend on their crystal lattice structures and phases. The properties of material also vary from one form to another, with titanium dioxide (TiO_2) being an example of a material that has differing properties in its anatase and rutile forms; the former, becoming highly photocatalytically active, the latter becomes more stable under high-temperature conditions (Sukarman et al., 2025). According to established practice, X-ray Diffraction (XRD) and Raman Spectroscopy often find use in crystal structure determination, crystallinity, and phase composition. These have the effect of allowing material properties to be tuned or stabilized in specific phase forms that would otherwise be impossible at bulk levels, thus broadening the technological interest in them (Bulavchenko & Vinokurov, 2023).

3.4. Surface Chemistry and Functionalization

The surface chemistry of nanoparticles determines the properties of the interaction of these nanoobjects and liquid environments, biomolecules, and other substances. Surfactant functional groups may determine binding affinity, dispersity, and solubility toward a target (Ahmad et al., 2022). Conjugation or functionalization methods (e.g., with polymers such as polyethylene glycol, with small organic molecules, or with biomolecules such as peptides, antibodies) may improve biocompatibility and targeting specificity, along with the circulation half-life within the body (Kaymaz et al., 2023). Fourier Transform Infrared Spectroscopy (FTIR) and X-ray Photoelectron Spectroscopy (XPS) are used in the characterization of the surface chemistry, to identify functional groups and elements, respectively, coupled with providing element and chemical state information. Surface modification not only enhances stability in suspension, but also broadens the spectrum of applications, including targeted drug delivery and biosensing, and remediation (Pasieczna-Patkowska et al., 2025a).

3.5. Surface Charge and Zeta Potential

Nanoparticle surfaces are effectively charged, a fact that is an important determinant in colloidal stability, aggregation behavior, and interactions with biological membranes. The zeta-potential, characterized by electrophoretic light scattering, is the potential difference at the slipping plane of dispersed particles. Large absolute zeta potential values (>30 mV) are usually taken to suggest that electrostatic repulsion is sufficient to ensure dispersion stability, whilst lower values predispose to aggregation (Rodriguez-Loya et al., 2023). Another important phenomenon is the affecting force in biological interactions, namely the surface charge. Negatively charged nanoparticles are expected to have less cellular uptake as there is no electrostatic interaction between nanoparticles and the cell membrane, but can have lower cytotoxicity. In contrast, more biocompatibility and enhanced circulation are dependent on negatively charged or neutral nanoparticles (Zhang et al., 2022).

3.6. Surface Area and Porosity

Nanoparticles have a remarkably large surface-to-volume ratio that leads directly to increased reactivity, catalysis, and adsorption properties. The benefit is further intensified by porous nanoparticles (i.e., mesoporous silica or zeolitic materials) due to high internal surface area and adjustable pore sizes. Such property is of special significance in such applications as heterogeneous catalysis, gas adsorption, and controlled drug release (Joudeh & Linke, 2022; Rizzi et al., 2021). There are a few established methods to quantify specific surface area and porosity, but the most commonly used is the Brunauer-Emmett-Teller (BET) analysis technique. Regulation of pore size and surface area can be used to design nanoparticles with high-capacity drug delivery, selective sieving of certain molecules, and enhanced energy storage materials (López-Pérez et al., 2021; Mane et al., 2024).

3.7. Magnetic Properties

Sub-micron-sized magnetic NPs (MNPs), e.g., Fe_3O_4 and Co-based NPs, exhibit size-dependent magnetic properties best known as superparamagnetism. In nanoscale thickness, the magnetic moments are randomized by the thermal fluctuations, and the remanence and coercivity are lost, leaving large magnetization in the external magnetic field. This has the advantage of making MNPs suitable to be used in biomedical applications such as magnetic resonance imaging (MRI), targeted drug delivery, and heated therapies (Díez et al., 2022; Joudeh & Linke, 2022). Magnetic properties are generally measured by means of methods such as Vibrating Sample Magnetometry VSM) or Superconducting Quantum Interference Device (SQUID) magnetometry. By optimizing particle size, particle composition, and surface coatings, magnetic properties can be fine-tuned to a desired degree, making them of greater use in the medical community, as well as industrial (Stiufiuc & Stiufiuc, 2024).

3.8. Interplay of Physicochemical Properties

It should also be stressed that the physical/chemical characteristics of nanoparticles do not always have an isolated effect; they are characterized by the synergistic effect. As an example, circulation and biodistribution profiles will depend on size and surface charge in combination, whereas catalytic activity will be determined by both crystal structure and surface chemistry (Ewii et al., 2025). In the same manner, optical and electronic characteristics are closely related to the size, 3D structure, and surface functionalization. The general knowledge of these interconnected parameters is, therefore, of critical significance to rational nanoparticle model development specific to a given application (Mekuye & Abera, 2023).

4. CLASSIFICATION AND TYPES OF NANOPARTICLES

Nanoparticles are a heterogeneous group of materials that display peculiar physicochemical properties due to their nanodimensional size. They are broadly divided according to chemical makeup, structural composition, and usage fields of application (Al-Fakeh & Alsaedi, 2021). A systematic knowledge of the quality of nanoparticle types is important to connect the inherent physical and chemical characteristics to performance on biomedical, environmental, and agricultural platforms, and industrial platforms

(Manzoor et al., 2022). In the following section, nanoparticles will be divided into five broad categories: inorganic, carbon-based, polymeric, lipid-based, and hybrid/composite nanoparticle systems

4.1. Inorganic Nanoparticles

One of the most represented and widely used ones is inorganic nanoparticles. They are highly stable, their physicochemical properties can be tuned, and they can comprise a wide spectrum of functionalities, which engineering researchers need (Zhou & Dong, 2025)

4.1.1. Metals (Ag, Au, Pt)

The maximum optical absorption and scattering in the light are characterized by strong SPR of metallic nanoparticles. Gold (Au) nanoparticles have been used significantly in plasmonics, biosensing, photothermal therapy, and imaging because they are highly compatible, and a wide variety of functionalization is possible (Mohseni & Bahram, 2024). Silver (Ag) nanoparticle has a strong antimicrobial effect that destroys microbial membranes, leaks reactive oxygen species (ROS), and affects DNA replication. Platinum (Pt) nanoparticles have high source activity and are eminent in the catalytic reaction, which includes fuel cells and hydrogenation (Rozhin et al., 2021)

4.1.2. Metal Oxides (TiO_2, ZnO, Fe_3O_4)

Metal Oxides (TiO_2, ZnO, Fe_3O_4), such as TiO_2, nanoparticles, find extensive application in the areas of photocatalysis, self-cleaning paints, and solar cells, due to their large bandgap and photo reactivity (Garjaei et al., 2024). Zinc oxide (ZnO) nanoparticles have UV absorption, antimicrobial, and photocatalytic degradation properties, thus relevant in remedying the environment and personal care products. Fe_3O_4 (magnetite) magnetic nanoparticles have found use in magnetic resonance imaging (MRI), drug targeting, and hyperthermia treatment, based on their superparamagnetic nature at the nanoscale (Dehghani et al., 2024).

4.1.3. Quantum Dots (QDs)

Quantum Dots (QDs): Semiconductor quantum dots (CdSe, InP) are nanocrystals that demonstrate size-dependent bandgap tuning because of quantum confinement. Their distinct photoluminescence features can be used to perform high-resolution bioimaging, optoelectronic applications, and light-emitting devices. Improvements in cadmium-free quantum dots (i.e., ZnS, carbon dots) are also enhancing safer uses in biomedical and environmental applications (Ham et al., 2022)

4.2. Carbon-Based Nanostructures

Carbon nanomaterials constitute another high-profile type of nanoparticle that takes advantage of the ample bonding capability of carbon atoms (sp2/sp3 hybridization) to yield structures with superior mechanical/electrical/chemical characteristics (Ezenkwa et al., 2022) Fullerene (C60) molecules, commonly known as buckyballs, are spherical carbon cages with electron-taking ability and, as such can be used as components in photovoltaics, radical scavenging, and drug delivery systems

(Gorelik et al., 2022) Carbon Nanotubes (CNTs) are formed in tubular shapes as a single-walled shape (SWCNTs) or multi-walled (MWCNTs) and possess excellent tensile strength with great conductivity. They are also applied in the semiconductor industry as composites in structural reinforcement, energy storage apparatus, and nanoelectronics (Tsukamoto, 2023)

Graphene is a two-dimensional lattice or sheet containing single atoms of carbon with electrically remarkable conductivity, high surface area, and flexible characteristics. It has seen applications in energy devices, biosensors, and environmental remediation (Ambade et al., 2023). Nanodiamonds can be synthesized by means of detonation synthesis, and have presented biocompatibility, high hardness, and the ease with which the surface can be functionalized. They are developed with potential drug delivery and image applications, tissue engineering (Shen et al., 2021). As a group the carbon nanostructures should have considerable utility in biomedical and environmental technologies, because of their mechanical strength, electrical conductivity, and variability. Their biopersistence and cytotoxicity potential, however, raise concerns, which is why further evaluation is necessary (Shahazi et al., 2023).

4.3. Polymeric Nanoparticles

Polymeric nanoparticles are formed by synthetic or natural polymers that have been dispersed or entrapped by a variety of methods. Capsules consist of a core that contains the therapeutic agent and the surrounding, which is the shell. The vehicles include proteins, DNA, synthetic polymers, natural polymers (e.g., albumin), antigen/epitope-PE-complex, and fluorophores (Shinde, 2023). Nanoparticles composed of polymers are biodegradable, biocompatible, and versatile materials, and hence are highly applicable in drug delivery and engineering ventures, especially those in the biomedical field (Khuddus et al., 2024). Biodegradable Polymers like Poly (lactic-co-glycolic acid) (PLGA) nanoparticles will have controlled and sustained drug release as they are degraded in the body into lactic and glycolic acids, which are naturally metabolic in the body (Blagojevic & Kamaly, 2023). Polymers Polyethylene glycol (PEG) is commonly used to modify the surfaces to have better circulation (PEGylation) and minimize immunogenicity. Chitosan is a cationic natural polysaccharide that is broadly explored as a delivery carrier of genes, as an antimicrobial agent, and as a cohesion vehicle (Al-Anil et al., 2021).

4.4. Lipid Nanoparticles (LNPs)

Lipid-based nanoparticles are an important element in the current pharmaceutics because they are biocompatible and can encapsulate nucleic acids, hydrophobic drugs, and peptides. The benefits of Solid Lipid Nanoparticles (SLN) and Nanostructured Lipid Carriers (NLC) include drug release control and greater stability, whereas there is a limited drug-loading capacity. To overcome this, NLCs use solid and liquid lipids to make drug encapsulation more effective and cut down on polymorphic transitions (Agosti et al., 2024) Ionizable LNPs in Vaccines (mRNA COVID-19), the worldwide administration of mRNA COVID-19 vaccines attracted attention to the innovative power of ionizable lipid nanoparticles. Their capacity to pack negatively charged mRNA and allow endosomal escape led to successful translation in vivo and the induction of potent immune responses, and transformed vaccinology (Borrajo et al., 2024)

4.5. Hybrid and Composite Nanoparticles

The hybrid and composite nanoparticles are designed to consist of more than one material and, therefore, incorporate complementary features and provide multifunction. Core and shell nanoparticles are fabricated to see improved stability and synergy. Such as Au@SiO_2 particles combine the plasmonic performance of gold with the protective shield of silica to create stable biosensing platforms. Fe_3O_4/Au constructs are magnetic targeting/optical imaging constructs (Liang et al., 2021). Metal-Organic frameworks (MOFs) have very large surface areas, and they can be tuned to have variable pore sizes, formed by metal ions and organic bridging ligands. They are used as gas storage, catalysis, and vehicles for therapeutic Biomolecules (Batisai & Mbonzhe, 2023). Dendrimer-Metal Hybrids are used as support to increase the solubility and stability, as well as enhance the performance of metal nanoparticles (Maumela & Bingwa, 2023). These hybrids find application in catalysis and imaging, and drug delivery HNPs can combine various functionalities into a single application simultaneously to diagnosis, treatment and environment.

Table 1. Classification of nanoparticles with composition, key properties, synthesis routes, and hallmark applications

Nanoparticle Class	Composition	Key Properties	Common Synthesis Routes	Hallmark Applications	References
Metal Nanoparticles	Au, Ag, Pt, Pd, Cu, etc.	Localized surface plasmon resonance (LSPR), high conductivity, catalytic activity, tunable optical absorption	Chemical reduction (e.g., citrate, borohydride), electrochemical deposition, green synthesis (plant extracts, microbes)	Biosensing, photothermal therapy, antimicrobial coatings, catalysis, nanoelectronics	(Abbas et al., 2024; Serna-Gallén & Mužina, 2024)
Metal Oxide Nanoparticles	TiO_2, ZnO, Fe_3O_4, CeO_2, CuO	Photocatalytic activity, magnetic/ electronic properties, high stability, redox activity	Sol–gel, hydrothermal, co-precipitation, combustion synthesis, green methods	Photocatalysis (wastewater treatment), UV protection, MRI contrast agents, drug delivery, fertilizers	(Belew & Assege, 2025; Shah et al., 2025)
Carbon-Based Nanomaterials	Fullerenes, CNTs, graphene, carbon quantum dots	High surface area, mechanical strength, electrical conductivity, fluorescence (CQDs)	Arc discharge, chemical vapor deposition (CVD), laser ablation, hydrothermal/ solvothermal	Supercapacitors, biosensors, drug/ gene delivery, pollutant adsorption, structural composites	(Díez-Pascual, 2021; Sheikhzadeh et al., 2021)
Polymeric Nanoparticles	Natural (chitosan, alginate) or synthetic (PLA, PLGA, PCL) polymers	Biodegradability, tunable size, controlled drug release, biocompatibility	Emulsion–solvent evaporation, nanoprecipitation, self-assembly, microfluidics	Drug delivery systems, vaccine carriers, tissue engineering, and controlled agrochemical release	(Geszke-Moritz & Moritz, 2024; A. Kumari et al., 2010; Parveen et al., 2023)

continued on following page

Table 1. Continued

Nanoparticle Class	Composition	Key Properties	Common Synthesis Routes	Hallmark Applications	References
Lipid-Based Nanoparticles	Liposomes, solid lipid nanoparticles (SLNs), nanostructured lipid carriers (NLCs)	High biocompatibility, ability to encapsulate hydrophobic drugs, and membrane fusion	High-pressure homogenization, ultrasonication, microfluidics, thin-film hydration	mRNA vaccines (COVID-19), gene therapy, nutraceuticals, transdermal delivery	(M. Mehta et al., 2023; Nguyen et al., 2025; Zeng et al., 2023)
Quantum Dots (QDs)	CdSe, CdTe, InP, PbS, carbon QDs	Size-tunable fluorescence, high quantum yield, photostability	Colloidal synthesis, hydrothermal, sol-gel, green synthesis (carbon QDs)	Bioimaging, diagnostics, photovoltaics, LED displays, sensors	(Aftab et al., 2021; Hassan Ahmed & Soylak, 2024; Moradialvand et al., 2025)
Silica Nanoparticles	Mesoporous silica, silica spheres	High porosity, surface functionalization, stability, and large loading capacity	Stöber method, sol–gel, microemulsion	Drug/gene delivery, chromatography, biosensors, imaging agents	(Fatima et al., 2025; Gonçalves, 2018; Gupta et al., 2023)
Dendrimers	Branched synthetic polymers (e.g., PAMAM)	Monodispersity, high surface functionality, nanoscale precision	Divergent and convergent synthesis approaches	Gene/drug delivery, imaging probes, nanocatalysis	(Kianfar, 2025; Mittal et al., 2021)
Hybrid Nanoparticles	Organic–inorganic composites, core–shell (e.g., Au@SiO_2, polymer-coated metal NPs)	Multifunctionality, enhanced stability, synergistic properties	Layer-by-layer assembly, sol-gel encapsulation, in-situ hybridization	Theranostics, smart drug delivery, multifunctional sensors, energy storage	(Khan et al., 2023; Sanchis-Gual et al., 2023)

5. NANOPARTICLE SYNTHESIS: ROUTES AND STRATEGIES

Nanoparticle synthesis is also a fundamental technology of nanotechnology since particle size, morphology, crystallinity, surface chemistry, and functional performance are all direct reflections of the type of synthesis performed (S. Kumari et al., 2023). The great diversity of the physical, chemical, and biological pathways has been established, and they have different limitations and advantages. These approaches can be broadly classified into top-down methods that involve the disintegration of bulk materials and bottom-up methods that involve the assembly of nanoparticles using atoms or molecules. Green synthesis and improved strategies have become more recent and sustainable, and precise alternatives. It is important to have comparative knowledge on these strategies to make a guided nanoparticle modification to fit the biomedical, agricultural, electronic, and environmental purposes (Nyabadza et al., 2023).

5.1. Top-Down Approaches

Top-down techniques are based on reduced sizes by mechanical fragmentation of bulk materials to the nanoscale. Intensive applications are mechanical milling, ion-beam, and high-tech lithography. To make nanocrystalline powders, this can be accomplished using ball milling as a technique whereby nanocrystalline powders can be formed due to the formation via repetitive fracture, and subsequent cold welding, but custom manufacturing difficulties due to contamination, overall distribution particle size, and the morphology are disadvantages concerning nanomaterials of high purity (Saleh & Hassan, 2023). Lithographic techniques such as photolithography, electron-beam lithography, and nanoimprint lithography provide exceptional levels of precision and patterning control, but are limited in use by their high costs and low scalability, and clean-room production. Such practices are especially useful in semiconductor businesses where nanometers of precision hold such importance, yet their shortcomings make it imperative to have alternative, scalable measures (Kim et al., 2024).

5.2. Bottom-Up Approaches

Bottom-up approaches involve chemical or physicochemical tools to fabricate nanoparticles starting with atoms, ions, or molecules. One of the most versatile methods is the so-called sol-gel process, which yields oxide nanoparticles and thin films with a controllable porosity and crystallinity. Hydrothermal and solvothermal synthesis employs precisely controlled temperature and pressure, in closed reactors to produce highly crystalline nano materials with well-controlled particle size distributions (Taniguchi et al., 2022). Precipitation and co-precipitation methods can also be performed on a large scale relatively easily, although it is difficult to control the particle size and to avoid secondary aggregation. Recent methods, employing microemulsions in which droplets have nanoscale dimensions, are used as confined reactors, and allow precise control of particle shape (Vinukonda et al., 2025). Generally, bottom-up methods provide superior tunability, homogeneity, and reproducibility compared to top-down methods, but issues around scalability and solvent-related environmental concerns remain a limiting factor.

5.3. Biological and Green Synthesis

The increasing concern about the environmental impact of conventional synthesis has catalyzed the development of eco-friendly methods. Nanoparticle synthesis can be realized using biological systems, including plants, bacteria, fungi, and algae (Figure 2), which include natural reducing and capping agents, obviating the need for hazardous chemicals (Jangid et al., 2025). Plants containing abundant polyphenols, terpenoids, and flavonoids were effectively exploited to synthesize silver, gold, and zinc oxide nanoparticles, which were biocompatible in nature. By the same token, microbial synthesis utilizes the enzyme composition of bacteria and fungi in nanoparticle production, with the overall benefits of sustainability and mild reaction conditions (Zuhrotun et al., 2023). Large molecules with green solvents (e.g., ionic liquids, deep eutectic solvents) and green-derived starting materials provide further expansion of an already extensive set of tools useful to green synthesis. Such techniques are not only compatible with the principles of green chemistry but also produce nanoparticles with specific surface functionality desirable in biomedical and agricultural applications (Padinhattath et al., 2025).

Figure 2. Green synthesis of nanoparticles

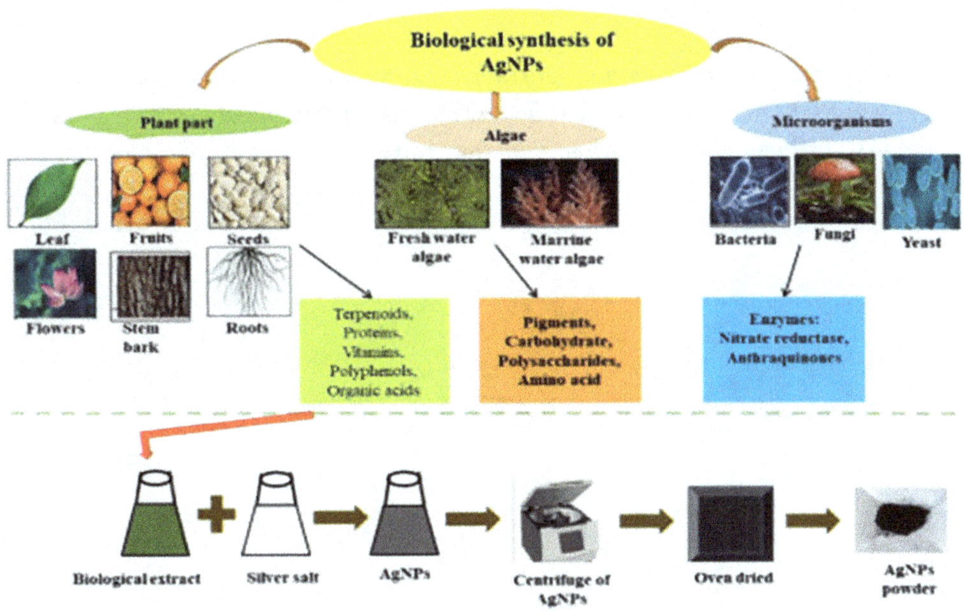

Adapted from (Githala & Trivedi, 2023)

5.4. Emerging and Advanced Methods

New technologies of nanoparticle manufacturing are concerned with precision, scalability, and compatibility with other technologies of the contemporary world. Continuous-flow synthesis platforms, such as microfluidic platforms, allow conditions to be highly controlled, and the variability of results between different batches is minimized and reproducible (Bi et al., 2025). Biomimetic approaches include designs inspired by nature, hoping to recreate the way the natural world self-assembles, as in biomineralization, by finding another route to functional nanostructures. Artificial intelligence (AI) and machine learning are being used to optimize synthesis conditions and make predictions on nanoparticle properties, as well as to increase the speed of discovery activities (Qin et al., 2024). The progressive approaches promise to address the drawbacks of the traditional approaches and enable the production of nanomaterials in the industrial and clinical sectors.

6. NANO-BIO INTERFACE AND PROTEIN CORONA

The nano bio interface is a dynamic boundary between engineered nanoparticles and biology, comprising cells, tissues, fluids, and ecological systems. When exposed to a biological medium, e.g., blood plasma, interstitial fluid, or even environmental matrices, bio-compatible nanoparticles are soon coated with biomolecules, especially proteins (Kumar et al., 2025). This results in the existence of a layer of proteins or "protein corona," which dramatically changes the physicochemical properties of the nanoparticle and, as such, its biological identity (Figure 3). An insight into the complexities of such an interface is essential in predicting the behavior of these NPs in biological systems, the optimum use of their biomedical exploitation, and their environmental safety assessment (Sun et al., 2024).

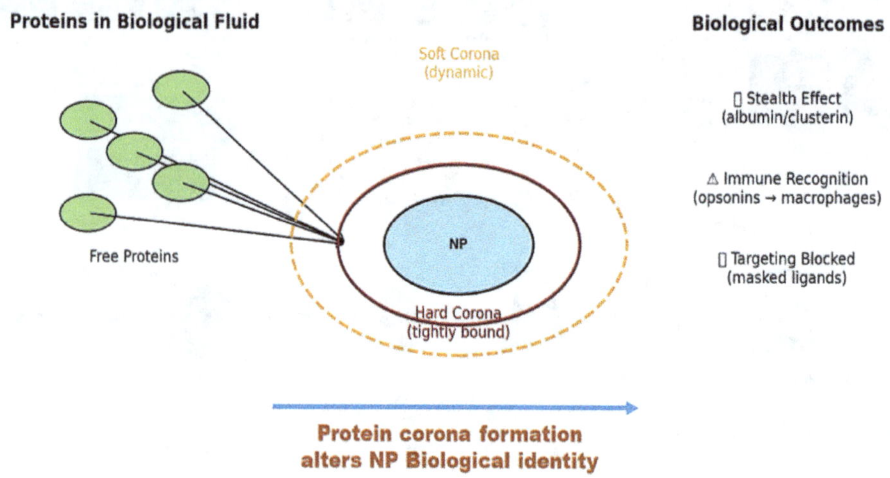

Figure 3. Protein corona formation on nanoparticles. Free proteins in biological fluids adsorb onto nanoparticles, forming a hard corona (stable) and a soft corona (dynamic). This new protein layer alters nanoparticle identity, influencing circulation, immune recognition, and targeting ability

6.1. Concept of the Protein Corona

On entering biological settings, NPs are readily exposed to a heterogeneous ensemble of biomolecules, most notably proteins, that adsorb to the NPs surface and create a dynamic layer, referred to as the protein corona. This corona is dynamic in nature and continuously exchanges based on the scale of the physicochemical characteristics of the NP, the concentration of the local proteins, and the environmental factors (Sun et al., 2024). The protein corona is divided into two classes: hard corona and soft corona. The hard corona is the protein that binds with high affinity and is retained in the protein corona over longer times, whereas the soft corona is those with low affinity, rapidly exchanging the protein corona. This involves a dual-layer arrangement of phenomenally new life and biological identity, often radically different than the synthetic or pristine version, and the determination of how cells, tissues, and organisms perceive, recognize, and respond to the NP (Vilanova et al., 2024).

6.2. Impact on Biological Identity

The protein corona has profound implications beyond the biological effects of nanoparticles. Firstly, it is applied to regulate cellular uptake, by conformational masking or exposure of ligand moieties that modify receptor-mediated endocytosis or non-specific internalization (Bresinsky & Goepferich, 2025). For instance, opsonized (Ig's complement proteins), coronas may favor recognition and clearance by macrophages, which seems to be the limiting factor of the therapy, although immune activation concerns will spike. On the other hand, dysopsonin-rich coronas (built using either albumin or clusterin) can facilitate circulation to some extent, so that a stealth effect would be possible (Panico et al., 2022). Second, the immune reaction may be strongly altered by protein corona formation. Hard corona proteins may trigger the inflammatory response, a complement or even cytoxicin release, which would be avoided with an engineered soft corona (Lee et al., 2014). In addition, protein coronas affect the targeting specificity. This renders the nanocarrier corona composition a dominant factor regarding whether the process of the nanocarriers meets their intended purpose, either theranostics or diagnostics (Chou & Lin, 2023).

6.3. Analytical Methods to Characterize Protein Corona

The features of the protein corona require thorough characterization using high-resolution analytical techniques that permit qualitative and quantitative aspects to be addressed. The application of mass spectrometry in proteomics strategies provides a comprehensive profiling of the bound proteins, allowing the creation of a map of corona complexity, variation over time, and changes (Al-Amrani et al., 2021). CD, FTIR, and SPR spectroscopic techniques for investigating protein structural changes upon adsorption as well as determining the kinetics of binding. Transmission electron and atomic force microscopy, as well as small-angle x-ray studies, are again the imaging tools that reveal the coronal structures and interaction of NPs with proteins (Sun et al., 2024). Moreover, for size distribution and surface potential change after dynamic formation of the corona, dynamic light scattering (DLS) and zeta potential measurements are standard techniques. Using the benefits of the synergistic interaction of these complementary methods, they can give us a multi-dimensional description of how coronas influence biodistribution, clearance, and bioactivity (Rodriguez-Loya et al., 2023).

6.4. Design Strategies to Modulate the Corona

Due to its vital role in defining any biological interactions, there are multiple strategies devised to control or regulate protein corona development. Another common method is the use of surface coatings, e.g., polyethylene glycol (PEG), to provide a hydrophilic and steric barrier to reduce nonspecific protein adsorption (Zheng et al., 2025). PEGylation has had significant application in clinically approved nanomedicines to increase half-life in circulation and reduce immune clearance. Other alternatives now emerging include PEG, coatings with zwitterionic functionalities and hydrophilic polymers, including poly(carboxybetaine) and poly(sulfobetaine). Biomimetic functionalization is also an option: natural membranes surrounding red blood cells or platelets, or cancer membranes can be used to protect nanoparticles and ensure desired biodistribution (Zheng et al., 2025). More recently, it became possible to engineer nanoparticles to selectively bind with certain ligands, which could facilitate the preferential adsorptions of dysopsonins, directing the composition of the coronas towards more favorable profiles. The intended outcome of these design strategies is to optimize and/or control the protein corona effect and provide nanomaterials with predictable, safe, and effective biological behaviors (Chehelgerdi et al., 2023).

6.5. Environmental Protein Corona: An Eco-Nanotoxicology Perspective

Beyond biomedical, protein coronas are also an important part of environmental systems where nanoparticles interact with proteins, humic compounds, and natural organic matter. These kinds of environmental coronas change the stability of nanoparticles, their transport in the soil, aquatic, and atmospheric environments, and their activity (Mahmoudi et al., 2023). As an example, coronas generated in wastewater or in natural waters may affect the aggregation of nanoparticles, their sedimentation, and their uptake by microorganisms and, thus, may have an impact on the ecological health. These eco-coronas too can alter the bioavailability of the toxic metals, or organic pollutants that are adsorbed on the nanoparticle surface, increasing the environmental risks or diminishing them (Doskocz et al., 2025). Noteworthy, the composition of environmental coronas is dissimilar compared to the coronas found in biological fluids, because often they are mainly composed of extracellular polymeric substances (EPS) of microbial communities, plant-derived biomolecules, or humic acids. Determining these activities is crucial to the further development of eco-nanotoxicology and in the safe and sustainable implementation of nanomaterials in agriculture, industry, and in environmental restoration (Junaid & Wang, 2021).

7. CHARACTERIZATION TECHNIQUES: TOOLBOX FOR NANOSCIENCE

The characterization of nanoparticles should comprise a full ensemble of physicochemical characterization tools. The different methods offer distinct information on morphology, size distribution, surface chemistry, crystallinity of charge and porosity, and stability to allow the researcher to determine the relationship between morphology and performance (Nogueira et al., 2024). Nanoparticles are dynamic in their interaction with biological and environmental systems; complete and multi-dimensional characterization is key to the further development of nanoscience applications in the medical, agricultural, catalytic, and environmental remediation fields. The major instrumentation in the nanoparticle characterization toolbox is outlined in the sections that follow (Altammar, 2023).

7.1. Microscopy

The TEM refers to a vital technique of nanoscale visualization. Because TEM can transmit an electron beam through an ultrathin specimen, it can give high-resolution detail sufficient to resolve features below 1 nm. It gives important details about the size of the particles, morphology, lattice fringe, and crystallinity, to which Selected Area Electron Diffraction (SAED) may be utilized further to obtain information on the structure (Mahanta et al., 2023). In comparison to SEM, however, Scanning Electron Microscopy only depends on the electron scattering against the surface of a sample, giving it three dimensions worth of topographical information. Although SEM is not usually as high resolution as TEM, it is very useful in imaging larger aggregates, roughness of surfaces, and tracking nanoparticle interaction with the surface (Chadha et al., 2024). Atomic Force Microscopy (AFM) goes beyond the electron microscope by investigating the morphology of surfaces using a nanoscale tip that is scanning over the sample. It provides topographic maps in aqueous or air atmosphere, which is advantageous in the study of nanoparticle interaction with biomolecules, membranes, or soft matter without the need for extensive sample preparation (Moraille et al., 2022).

7.2. Spectroscopy

UV-Visible Spectroscopy has been widely used to observe the progress of nanoparticle synthesis and characterize optical properties, e.g., surface plasmon resonance (SPR) in metallic nanoparticles. The shift in the location and amount of the absorbance peaks is an indirect indication of the particle size, aggregation, and the modifications of the surface (Dixit et al., 2025). Fourier Transform Infrared Spectroscopy (FTIR) can be used to identify functional groups and to analyze the chemistry of a surface. FTIR can be used to assess vibrational transitions to confirm functionalization of nanoparticles with ligands or polymers, or biomolecules, which is key to biomedical applications (Pasieczna-Patkowska et al., 2025b).

Raman Spectroscopy & Raman Spectroscopy, including surface-enhanced Raman scattering (SERS), is are chemically fingerprinting molecules of nanoparticles as well as the adsorbed species. Raman is commonly used to characterize carbon-based nanostructures, including graphene and carbon nanotubes, to provide structural and/ ties bonding detail (Kharlamova, 2023). X-ray Photoelectron Spectroscopy (XPS) allows accurate surface chemical composition and oxidation states of nanoparticle elements, having a detection depth that is a few nanometers. This method is vital in the verification of the surface layer, charge levels, and determinants of catalytic activities (Foster et al., 2025).

7.3. Structural Characterization

X-ray Diffraction (XRD) is still the most widespread method in order to identifying the crystalline structure, phase purity, and average crystallite size of nanoparticles. By relying on the Scherrer equation, it is possible to obtain crystallite sizes, or lattice strain can be measured, as well as polymorphism (Shishir et al., 2025). Small-Angle X-ray Scattering (SAXS) is a complement of XRD since it aims at characterizing nanostructures in solutions. AXS can provide the size distributions of the particles, their shapes, and show internal porosity to allow in situ investigation of self-assembly and hybrid nanostructures (Yaghmur & Hamad, 2022).

7.4. Dynamic and Electrokinetic Analysis

Dynamic Light Scattering (DLS) is most typically applied in the measurement of nanoparticle size distribution in suspension. DLS can give an accurate measurement of the hydrodynamic diameter and polydispersity index by precise measurement of light fluctuations caused by Brownian motion but may be biased to large aggregates (Singh et al., 2024). Nanoparticle Tracking Analysis (NTA) has been developed to step forward, where individual motion is tracked, permitting size distribution and particle concentration to be measured with enhanced resolution over DLS (Foreman-Ortiz et al., 2022). The electrokinetic potential of nanoparticles in a solution is measured by the Zeta Potential Analysis and can be used as an indicator of colloidal stability. Very high or negative zeta potentials indicate repelling forces that inhibit aggregation, and near-neutral values tend to indicate instability. Such a property also indicates the interaction of nanoparticles with biological lipid films and environmental media (Ferreyra Maillard et al., 2021).

7.5. Thermal and Surface Properties

Thermogravimetric analysis (TGA) is used to quantify surface coatings, water content, or the thermal stability of nanoparticles by measuring a mass change with respect to temperature. It is especially significant for hybrid and polymer-covered nanostructures (Rami et al., 2020). Differential Scanning Calorimetry (DSC) can detect thermal transitions like melting, crystallization, and glass transition, as is the case with polymeric and lipid nanoparticles (Hussain et al., 2024). The Brunauer -Emmett- Teller (BET) Analysis measures the surface area and porosity in relation to the nitrogen adsorption- desorption isotherms. The effect of a surface area on genetic efficiency and adsorption capacity cannot be underestimated, so the BET measurements of the nanoparticle are essential to tailor it for energy storage and remediation (Vyawahare et al., 2022).

7.6. Correlative and Multi-Technique Approaches

No individual method can fully embrace nanoparticles' dynamic properties; thus, correlative approaches that combine a versatile toolset are commonly embraced. As an example, TEM and DLS could be used to cross-correlate particle size between dried and hydrated conditions, whereas XRD and Raman can collectively offer information on the structure of crystalline and amorphous phases (Basha et al., 2025a). In biomedical research, correlative pipelines can even combine microscopy, e.g., with proteomics and spectroscopy to characterize nanoparticle-protein interactions in detail (Jagadeesh et al., 2024). Nanoscience research is being motivated by the advances in high-throughput characterization platforms and in situ monitoring strategies. Artificial intelligence and machine learning are also being investigated as a way of automatically analyzing large volumes of data produced by such instruments and thus advancing the discovery-to-application pipeline (Basha et al., 2025b).

Table 2. Toxicological assessment framework for nanoparticles

Assessment Level	Experimental Models / Systems	Endpoints Evaluated	Limitations & Challenges	References
In vitro (Cell-based)	Human/animal cell lines, organoids	Cytotoxicity (MTT, LDH release), oxidative stress (ROS, GSH), genotoxicity (Comet assay, micronucleus test), uptake/ localization (confocal microscopy, TEM)	Cell line-specific responses; lack of systemic interactions; protein corona effects not fully captured	(Takáč et al., 2025; Teixeira et al., 2024; Xavier et al., 2021)
In vivo (Animal models)	Rodents, zebrafish, Drosophila	Biodistribution, bioaccumulation, organ toxicity, immunological response, reproductive/developmental toxicity	Ethical concerns, interspecies variability, and high cost	(D'amora et al., 2021; Elizalde-Velázquez et al., 2023; Mukherjee et al., 2022)
Environmental Models	Soil microcosms, aquatic systems, and plants	Bioavailability, ecotoxicity (growth inhibition, photosynthetic efficiency, soil respiration), and trophic transfer	Environmental complexity, lack of standardized assays	(Nthunya et al., 2025; Pathak et al., 2025; Xu et al., 2022)
Omics Approaches	Transcriptomics, proteomics, metabolomics	Mechanistic insights, pathways altered by NP exposure	High data complexity, reproducibility issues	(Akhter Bhat et al., 2025; Y. Li et al., 2024; Sanches et al., 2024)
Computational / Predictive Models	QSAR (Quantitative Structure–Activity Relationship), molecular docking, nanoinformatics	Toxicity prediction, safe-by-design strategies, hazard ranking	Limited training datasets need validation with experimental data	(Belfield et al., 2023; J. Li et al., 2022; Thomas et al., 2025; Varsou et al., 2024)

8. SMART AGRICULTURE APPLICATIONS OF NANOTECHNOLOGY

The integration of nanotechnology into agriculture has opened new avenues for improving crop productivity, resource efficiency, and environmental sustainability. By leveraging the unique physicochemical properties of nanoparticles, such as their high surface area, reactivity, and controlled-release potential, scientists have developed innovative tools that align with the goals of smart and climate-resilient agriculture. Smart agriculture focuses on optimizing input use, monitoring environmental conditions, and automating decision-making processes through data-driven technologies. In this context, nanotechnology contributes by enabling precise and efficient agricultural interventions, ranging from targeted nutrient delivery to real-time sensing of soil and plant health (Yadav et al., 2023).

6.1 Nano-Enabled Inputs and Delivery Systems

Nanoparticles serve as carriers for fertilizers, pesticides, and growth regulators, enabling controlled and site-specific release of active ingredients. Such nano-fertilizers and nano-pesticides minimize nutrient losses through leaching or volatilization, enhance uptake efficiency, and reduce environmental contamination. Nanocarriers can be engineered to respond to external stimuli such as pH, moisture, or temperature, ensuring that nutrient release coincides with plant demand, an essential principle of precision agriculture (Ullah et al., 2024).

6.2 Nanosensors for Precision Monitoring

Nanosensors are among the most promising nanotechnology-based tools for smart farming. These sensors can detect and quantify changes in soil moisture, nutrient content, temperature, and pest activity at the micro-level. They enable real-time monitoring and data-driven decision-making, allowing farmers to manage resources efficiently. Integration of nanosensors with the Internet of Things (IoT) and artificial intelligence (AI) facilitates remote diagnostics and predictive analytics, promoting sustainable and high-yield farming practices (Miller et al., 2025).

6.3 Environmental and Resource Management

Nanomaterials also play an important role in soil and water remediation. Engineered nanoparticles, such as iron oxide and titanium dioxide, can remove heavy metals, pesticides, and other contaminants from soil and irrigation systems. Furthermore, nano-enabled films and coatings improve water retention, seed germination, and protection against abiotic stresses like drought and salinity. Collectively, these technologies enhance ecosystem resilience and contribute to the long-term sustainability of agricultural production systems (Mukhopadhyay et al., 2022).

9. CHALLENGES AND FUTURE PERSPECTIVE IN NANOTECHNOLOGY

Nanotechnology is growing its horizons, but the process of its transfer to general use, with full responsibility, is predetermined by various problems of current concern. Sustainability and safety are of the foremost concerns since the same features that make nanoparticles unique, namely their high surface reactivity, tunable physicochemical properties, and their high bio java are what can pose toxicity hazards to human beings, ecosystems, and non-target organisms (Xuan et al., 2023). It is this duality that highlights the need to use safe-by-design approaches, where a hazard analysis, life cycle thinking and environmentally friendly synthesis methods are incorporated into the development process early in the design process. More so, the fast-growing application of nanomaterials in varying fields, such as agriculture and medicine, raises the need to implement the concept of circular economy that will minimize wastes, increase recycling, and improve the presence of environmental impacts (Saxena et al., 2025). A parallel issue is that of weak and unstandardized methodologies of the characterization and reporting of the characteristics of nanoparticles. Such initiatives as MIAN (the Minimum Information About Nanomaterials) and the FAIR (Findable, Accessible, Interoperable, Reusable) data concepts are paramount in providing reproducibility, transparency, and cross-comparability of results. Nonetheless, their acceptance is not uniform in the various laboratories and industries, resulting in a bottleneck in regulatory approval, safety assessments and within the acceptance by people (Serrano et al., 2024).

Nanotechnology combined with the new frontiers are the future and it will be incredible. This has been enabled by means of nano informatics and AI guided design, and will increasingly provide predictive designs for NP activity behavior, the pathways tapped for synthesis changes in these (and emergence of new) as well as population of what can be created (easier than ever before) within the discovery process at an increasing clip (Shirzad et al., 2025).). Likewise, alignment with an emerging field of synthetic bioengineering propounds paths to more precise and biodegradable biomimetic and bioengineered NPs. But this progress also needs to be looked at from the perspective of ethics and governance around the world. However, there is no harmony in international regulations, and the limits of nanomaterial toxicity or occupational risk, or premarket information requirements are different for products. Lack of standardized methodologies adversely affects innovation and has wider implications in terms of equity and safety, particularly within less well-regulated settings such as many developing countries (Ali et al., 2023). The answer to these deficiencies has no other name than global governance and should combine scientific knowledge, an ethical approach, participation of citizens, and policy coherence. Finally, the future for nano squared is innovation and responsibility, discovery of sustainable (economical, environmentally safe, and socially acceptable) well-characterization natural materials which will act as a contributing element for scientific and industrial development, protecting human health and faunal habitat.

10. CONCLUSION

Nanotechnology is rapidly advancing as a key field in the 21st century, offering transformative applications in medicine, agriculture, energy, and environmental protection due to the unique properties of nanomaterials. These materials, with their size effects and high surface-to-volume ratio, exhibit remarkable structural, optical, and catalytic properties. However, the same properties that make them useful also introduce risks, including toxicity, persistence, and bioaccumulation, which can have unforeseen effects on living systems and the environment. This dual nature of nanotechnology underscores the need for careful consideration of safety and sustainability in its development. Future progress must be aligned with eco-friendly, safety-by-design approaches and standardized nanoparticle characterization to ensure responsible use. Green synthesis, lifecycle assessments, and biodegradability should be integral to nanomaterial design. Additionally, standardized reporting and reproducible research practices are essential for regulatory approval and broad acceptance. Advanced tools such as microscopy and spectroscopy will be critical in addressing concerns related to the safety, quality, and functionality of nanomaterials.

REFERENCES

Abbas, R., Luo, J., Qi, X., Naz, A., Khan, I. A., Liu, H., Yu, S., & Wei, J. (2024). Silver Nanoparticles: Synthesis, Structure, Properties and Applications. Nanomaterials, 14(17), 1425. DOI: 10.3390/NANO14171425

Abbasi, A. J., Anas, M., Elahi, M., Khan, A., Khattak, W. A., Saleem, M. H., Fahad, S., Elansary, H. O., Mahmoud, E. A., & Ahmed, T. (2025). Restoring wheat productivity and nutrient balance under cadmium stress through reducing toxicity, metal uptake, and oxidative damage using selenium nanoparticles. *Journal of Trace Elements in Medicine and Biology*, *89*, 127644. PMID: 40233590

Aftab, S., Shah, A., Erkmen, C., Kurbanoglu, S., & Uslu, B. (2021). Quantum dots: Synthesis and characterizations. Electroanalytical Applications of Quantum Dot-Based Biosensors, 1–35. DOI: 10.1016/B978-0-12-821670-5.00005-1

Agosti, E., Zeppieri, M., Antonietti, S., Battaglia, L., Ius, T., Gagliano, C., Fontanella, M. M., & Panciani, P. P. (2024). Navigating the Nose-to-Brain Route: A Systematic Review on Lipid-Based Nanocarriers for Central Nervous System Disorders. *Pharmaceutics*, *16*(3). Advance online publication. DOI: 10.3390/PHARMACEUTICS16030329 PMID: 38543223

Ahmad, F., Salem-Bekhit, M. M., Khan, F., Alshehri, S., Khan, A., Ghoneim, M. M., Wu, H. F., Taha, E. I., & Elbagory, I. (2022). Unique Properties of Surface-Functionalized Nanoparticles for Bio-Application: Functionalization Mechanisms and Importance in Application. *Nanomaterials (Basel, Switzerland)*, *12*(8), 1333. DOI: 10.3390/NANO12081333 PMID: 35458041

Akhter Bhat, T., Rather, R. A., Bashir, S., Mishra, N.-S., & John, R. (2025). Nanoparticles in plant systems: Omics-based perspectives on stress adaptation and toxicological implications. *Plant Nano Biology*, *13*, 100181. DOI: 10.1016/J.PLANA.2025.100181

Al-Amrani, S., Al-Jabri, Z., Al-Zaabi, A., Alshekaili, J., & Al-Khabori, M. (2021). Proteomics: Concepts and applications in human medicine. *World Journal of Biological Chemistry*, *12*(5), 57. DOI: 10.4331/WJBC.V12.I5.57 PMID: 34630910

Al-Anil, A. W., Jeberl, J. N., & Elewil, A. S. (2021). Development of a nanostructured double-layer coated tablet based on polyethylene glycol/gelatin as a platform for hydrophobic molecules delivery. *Egyptian Journal of Chemistry*, *0*(0), 1759–1767. DOI: 10.21608/EJCHEM.2021.52019.3066

Al-Fakeh, M. S., & Alsaedi, R. O. (2021). Synthesis, Characterization, and Antimicrobial Activity of CoO Nanoparticles from a Co (II) Complex Derived from Polyvinyl Alcohol and Aminobenzoic Acid Derivative. *TheScientificWorldJournal*, *2021*. Advance online publication. DOI: 10.1155/2021/6625216 PMID: 33994882

Ali, F., Neha, K., & Parveen, S. (2023). Current regulatory landscape of nanomaterials and nanomedicines: A global perspective. *Journal of Drug Delivery Science and Technology*, *80*. Advance online publication. DOI: 10.1016/J.JDDST.2022.104118

Altammar, K. A. (2023). A review on nanoparticles: Characteristics, synthesis, applications, and challenges. *Frontiers in Microbiology*, *14*, 1155622. DOI: 10.3389/FMICB.2023.1155622 PMID: 37180257

Ambade, R. B., Lee, K. H., Kang, D. J., & Han, T. H. (2023). Advances in Porous Graphene and Scalable Wet-Spinning Fiber Assembly. *Accounts of Materials Research*, *4*(5), 389–402. DOI: 10.1021/ACCOUNTSMR.2C00186

An, C., Sun, C., Li, N., Huang, B., Jiang, J., Shen, Y., & Wang, Y. (2022). Nanomaterials and nanotechnology for the delivery of agrochemicals: Strategies towards sustainable agriculture. *Journal of Nanobiotechnology*, *20*(1), 11. PMID: 34983545

Anas, M., Bibi, S., Farwa, U., Khan, A., & Quraishi, U. M. (2024). Manganese oxide nanoparticles elicit cadmium tolerance in wheat seedlings by ionomic and phenomic adjustment, regulation of AsA–GSH pathway, cellular thickness recovery, and antioxidant modulation. *Environmental Science. Nano*, *11*(11), 4588–4605.

Anas, M., Khan, A., Falak, A., Shinwari, Z. K., Mohamed, H. E. A., Hkiri, K., Maaza, M., Fahad, S., & Quraishi, U. M. (2025). Harnessing Biomedical Applications of Ascorbic Acid, Berberine, and Glutathione-Capped Quantum Dots Through Multifunctional Physico-chemical, Biochemical, and Antimicrobial Assays. *BioNanoScience*, *15*(2), 259.

Badir, A., Refki, S., & Sekkat, Z. (2025). Utilizing gold nanoparticles in plasmonic photothermal therapy for cancer treatment. *Heliyon*, *11*(4), e42738. DOI: 10.1016/J.HELIYON.2025.E42738 PMID: 40084020

Basha, S., Mukunda, D. C., Pai, A. R., & Mahato, K. K. (2025a). Assessing amyloid fibrils and amorphous aggregates: A review. *International Journal of Biological Macromolecules*, *311*, 143725. DOI: 10.1016/J.IJBIOMAC.2025.143725 PMID: 40324497

Basha, S., Mukunda, D. C., Pai, A. R., & Mahato, K. K. (2025b). Assessing amyloid fibrils and amorphous aggregates: A review. *International Journal of Biological Macromolecules*, *311*, 143725. DOI: 10.1016/J.IJBIOMAC.2025.143725 PMID: 40324497

Batisai, E., & Mbonzhe, L. (2023). CO2 sorption studies of bis-pyridyldiimide mixed ligand metal–organic frameworks. *Acta Crystallographica Section a Foundations and Advances*, *79*(a2), C1325–C1325. DOI: 10.1107/S2053273323082967

Behrens, M. A., Franzén, A., Carlert, S., Skantze, U., Lindfors, L., & Olsson, U. (2025). On the Ostwald ripening of crystalline and amorphous nanoparticles. *Soft Matter*, *21*(12), 2349–2354. DOI: 10.1039/D4SM01544D PMID: 40019147

Belew, A. A., & Assege, M. A. (2025). Solvothermal synthesis of metal oxide nanoparticles: A review of applications, challenges, and future perspectives. *Results in Chemistry*, *16*, 102438. DOI: 10.1016/J.RECHEM.2025.102438

Belfield, S. J., Firman, J. W., Enoch, S. J., Madden, J. C., Erik Tollefsen, K., & Cronin, M. T. D. (2023). A review of quantitative structure-activity relationship modelling approaches to predict the toxicity of mixtures. *Computational Toxicology (Amsterdam, Netherlands)*, *25*, 100251. DOI: 10.1016/J.COMTOX.2022.100251

Bi, Y., Xie, S., Li, Z., Dong, S., & Teng, L. (2025). Precise nanoscale fabrication technologies, the “last mile” of medicinal development. *Acta Pharmaceutica Sinica. B*, *15*(5), 2372–2401. DOI: 10.1016/J.APSB.2025.03.040 PMID: 40487646

Bibi, F., Khan, A., Iqrar, I., Iqbal, M., & Shinwari, Z. K. (2024). Therapeutic applications and mechanism of action of plant-mediated silver nanoparticles. *Pakistan Journal of Botany*, *56*(2), 531–540.

Blagojevic, L., & Kamaly, N. (2023). Long-lasting Cross-linked PLGA-Inspired Nanoparticles from One-pot Nanopolymerization of Precisely Sequenced Short Oligolactoglycolic Acid Dimethacrylates. DOI: 10.26434/CHEMRXIV-2023-7V4V4

Borrajo, M. L., Lou, G., Anthiya, S., Lapuhs, P., Moreira, D. J. F., Tobío, A., Loza, M. I., Vidal, A., & Alonso, M. J. (2024). Nanoemulsions and nanocapsules as carriers for the development of intranasal mRNA vaccines. DOI: 10.21203/RS.3.RS-4204651/V1

Bresinsky, M., & Goepferich, A. (2025). Control of biomedical nanoparticle distribution and drug release in vivo by complex particle design strategies. *European Journal of Pharmaceutics and Biopharmaceutics*, *208*, 114634. DOI: 10.1016/J.EJPB.2025.114634 PMID: 39826847

Bulavchenko, O. A., & Vinokurov, Z. S. (2023). In Situ X-ray Diffraction as a Basic Tool to Study Oxide and Metal Oxide Catalysts. Catalysts, 13(11), 1421. DOI: 10.3390/CATAL13111421

C, F. C., & T, K. (2024). Advances in stabilization of metallic nanoparticle with biosurfactants- a review on current trends. Heliyon, 10(9), e29773. DOI: 10.1016/J.HELIYON.2024.E29773

Chadha, V., Miller, N. C., Ding, R., Beschorner, K. E., & Jacobs, T. D. B. (2024). Evaluating scanning electron microscopy for the measurement of small-scale topography. *Surface Topography: Metrology and Properties*, *12*(3). Advance online publication. DOI: 10.1088/2051-672X/AD49B9 PMID: 39896093

Chaikittisilp, W., Yamauchi, Y., & Ariga, K. (2022). Material Evolution with Nanotechnology, Nanoarchitectonics, and Materials Informatics: What will be the Next Paradigm Shift in Nanoporous Materials? *Advanced Materials*, *34*(7). Advance online publication. DOI: 10.1002/ADMA.202107212 PMID: 34637159

Chaikittisilp, W., Yamauchi, Y., & Ariga, K. (2022). Material evolution with nanotechnology, nanoarchitectonics, and materials informatics: What will be the next paradigm shift in nanoporous materials? *Advanced Materials*, *34*(7), 2107212. PMID: 34637159

Chaupoo, A. S., Dhakad, A., Beese, S., Kumar, A., Kumar, P. K., Pandey, S. K., & Kurdekar, A. K. (2023). Transforming Horticulture: The Influence of Nanotechnology on Crop Enhancement. *International Journal of Environment and Climate Change*, *13*(11), 3381–3400. DOI: 10.9734/IJECC/2023/V13I113512

Chehelgerdi, M., Chehelgerdi, M., Allela, O. Q. B., Pecho, R. D. C., Jayasankar, N., Rao, D. P., Thamaraikani, T., Vasanthan, M., Viktor, P., Lakshmaiya, N., Saadh, M. J., Amajd, A., Abo-Zaid, M. A., Castillo-Acobo, R. Y., Ismail, A. H., Amin, A. H., & Akhavan-Sigari, R. (2023). Progressing nanotechnology to improve targeted cancer treatment: Overcoming hurdles in its clinical implementation. *Molecular Cancer*, *22*(1), 169. DOI: 10.1186/S12943-023-01865-0 PMID: 37814270

Chou, W. C., & Lin, Z. (2023). Impact of protein coronas on nanoparticle interactions with tissues and targeted delivery. *Current Opinion in Biotechnology*, *85*, 103046. DOI: 10.1016/J.COPBIO.2023.103046 PMID: 38103519

D'amora, M., Raffa, V., De Angelis, F., & Tantussi, F. (2021). Toxicological Profile of Plasmonic Nanoparticles in Zebrafish Model. International Journal of Molecular Sciences, 22(12), 6372. DOI: 10.3390/IJMS22126372

De, T., Trieu, V., Myers, S., Qazi, S., Saund, S., & Lee, C. (2025). Sub-15 nm Nanoparticles for Drug Delivery: Emerging Frontiers and Therapeutic Potential. *International Journal of Molecular Sciences*, *26*(22), 10842. PMID: 41303330

Dehghani, M., Hajipour-Verdom, B., & Abdolmaleki, P. (2024). Effect of precipitating agent, N2 gas, extract volume and pH on the magnetic properties of magnetite nanoparticles by green synthesis from aqueous pomegranate peel extract. *Frontiers in Chemistry*, *12*. Advance online publication. DOI: 10.3389/FCHEM.2024.1413077 PMID: 39114264

Dhahi, T. S., Yousif Dafhalla, A. K., Tayfour, O. E., Mubarakali, A., Alqahtani, A. S., Tayfour Ahmed, A. E., Elobaid, M. E., Adam, T., & Gopinath, S. C. B. (2024). Advances in nano sensors for monitoring and optimal performance enhancement in photovoltaic cells. *iScience*, *27*(4). Advance online publication. DOI: 10.1016/J.ISCI.2024.109347 PMID: 38550986

Díez, A. G., Rincón-Iglesias, M., Lanceros-Méndez, S., Reguera, J., & Lizundia, E. (2022). Multi-component magnetic nanoparticle engineering: The role of structure-property relationship in advanced applications. *Materials Today. Chemistry*, *26*, 101220. DOI: 10.1016/J.MTCHEM.2022.101220

Díez-Pascual, A. M. (2021). Carbon-Based Nanomaterials. International Journal of Molecular Sciences, 22(14), 7726. DOI: 10.3390/IJMS22147726

Dixit, A., Sonwal, S., Upadhyay, A., Bajpai, V. K., Huh, Y. S., & Shukla, S. (2025). Development and characterization of gold nanoparticle-based colorimetric sensing assay functionalized with L-cysteine for the detection of calcium carbide in ripened mango and banana fruits. *Heliyon*, *11*(4), e42614. DOI: 10.1016/J.HELIYON.2025.E42614 PMID: 40066027

Domb, A. J., Sharifzadeh, G., Nahum, V., & Hosseinkhani, H. (2021). Safety Evaluation of Nanotechnology Products. *Pharmaceutics*, *13*(10). Advance online publication. DOI: 10.3390/PHARMACEUTICS13101615 PMID: 34683908

Doskocz, N., Affek, K., Matczuk, M., Drozd, M., & Załęska-Radziwiłł, M. (2025). Nanoparticles in wastewater: A comprehensive approach to understanding their ecotoxicity and genotoxicity. *Desalination and Water Treatment*, *321*, 100988. DOI: 10.1016/J.DWT.2025.100988

Ehsan, M., Hernández, V. E., Hassan, D., Sani, A., Viveros, F. M. L., López, J. L. B., Trejo, E. B., Gulzar, R., Ali, N., & Bashir, S. (2025). A greenhouse pot experiment assessing the zinc-accumulating behavior of Lupinus uncinatus Schldl. *Asian Journal of Water, Environment and Pollution*, 025140101.

Elizalde-Velázquez, G. A., Herrera-Vázquez, S. E., Elizalde-Velázquez, G. A., & Herrera-Vázquez, S. E. (2023). *Zebrafish as Model Organism in Aquatic Ecotoxicology: Current Trends and Future Perspectives*. Zebrafish Research - An Ever-Expanding Experimental Model. DOI: 10.5772/INTECHOPEN.1002731

Ewii, U. E., Attama, A. A., Olorunsola, E. O., Onugwu, A. L., Nwakpa, F. U., Anyiam, C., Chijioke, C., & Ogbulie, T. (2025). Nanoparticles for drug delivery: Insight into in vitro and in vivo drug release from nanomedicines. *Nano TransMed*, *4*, 100083. DOI: 10.1016/J.NTM.2025.100083

Ezenkwa, O. E., Hassan, A., & Samsudin, S. A. (2022). Comparison of mechanical properties and thermal stability of graphene-based materials and halloysite nanotubes reinforced maleated polymer compatibilized polypropylene nanocomposites. *Polymer Composites, 43*(3), 1852–1863. DOI: 10.1002/PC.26503

Falak, A., Anas, M., Khan, A., Hayat, A., Shaheen, Z., Saleem, M. H., Fahad, S., & Quraishi, U. M. (2025). Efficacy of ascorbic acid coated quantum dots in alleviating lead-induced oxidative damage and enhancing growth parameters in rice (Oryza sativa L.) for sustainable cultivation. *Journal of Trace Elements in Medicine and Biology, 88*, 127603. https://doi.org/https://doi.org/10.1016/j.jtemb.2025.127603 PMID: 39847985

Fatima, R., Katiyar, P., & Kushwaha, K. (2025). Recent advances in mesoporous silica nanoparticle: Synthesis, drug loading, release mechanisms, and diverse applications. *Frontiers in Nanotechnology, 7*, 1564188. DOI: 10.3389/FNANO.2025.1564188/XML

Ferreyra Maillard, A. P. V., Espeche, J. C., Maturana, P., Cutro, A. C., & Hollmann, A. (2021). Zeta potential beyond materials science: Applications to bacterial systems and to the development of novel antimicrobials. Biochimica et Biophysica Acta (BBA) -. *Biomembranes, 1863*(6), 183597. DOI: 10.1016/J.BBAMEM.2021.183597 PMID: 33652005

Foreman-Ortiz, I. U., Fung Ma, T., Hoover, B. M., Wu, M., Murphy, C. J., Murphy, R. M., & Pedersen, J. A. (2022). Nanoparticle tracking analysis and statistical mixture distribution analysis to quantify nanoparticle–vesicle binding. *Journal of Colloid and Interface Science, 615*, 50–58. DOI: 10.1016/J.JCIS.2022.01.141 PMID: 35123359

Foster, J., Lyu, X., Serov, A., Mauger, S., Padgett, E., & Pylypenko, S. (2025). X-ray photoelectron spectroscopy investigation of iridium oxide catalyst layers: Insights into the catalyst-ionomer interface. *Electrochimica Acta, 517*, 145705. DOI: 10.1016/J.ELECTACTA.2025.145705

Garjaei, S. S., Koukabi, N., & Parouch, A. N. (2024). Enhancing Photocatalytic Activity, Stability, and Separability of TiO2 through Synthesis of Fe3O4@Fibroin-TiO2 Composite. *ChemistrySelect, 9*(18). Advance online publication. DOI: 10.1002/SLCT.202304658

Geszke-Moritz, M., & Moritz, M. (2024). Biodegradable Polymeric Nanoparticle-Based Drug Delivery Systems: Comprehensive Overview, Perspectives and Challenges. Polymers, 16(17), 2536. DOI: 10.3390/POLYM16172536

Githala, C. K., & Trivedi, R. (2023). Review on synthesis method, biomolecules involved, size affecting factors and potential applications of silver nanoparticles. *Biocatalysis and Agricultural Biotechnology, 54*, 102912.

Gonçalves, M. C. (2018). Sol-gel silica nanoparticles in medicine: A natural choice. design, synthesis and products. *Molecules (Basel, Switzerland), 23*(8). Advance online publication. DOI: 10.3390/MOLECULES23082021 PMID: 30104542

Gorelik, R., Asaduzzaman, A., Manga, V. R., Thakur, A., & Muralidharan, K. (2022). A First-Principles Investigation of Lithium and Sodium Ion Diffusion in C60 Molecular Solids. *The Journal of Physical Chemistry. C, Nanomaterials and Interfaces, 126*(9), 4259–4266. DOI: 10.1021/ACS.JPCC.1C09269

Gu, S., & Wang, D. (2024). Electrostatic Interaction-Driven Fabrication of Large-Area, Freestanding Nanoparticle Surfactant Membranes with Controllable Elastic Properties. ACS Applied Materials &Amp. *Interfaces*, *16*(34), 45778–45787. DOI: 10.1021/ACSAMI.4C11820 PMID: 39140693

Gupta, J., Quadros, M., & Momin, M. (2023). Mesoporous silica nanoparticles: Synthesis and multifaceted functionalization for controlled drug delivery. *Journal of Drug Delivery Science and Technology*, *81*, 104305. DOI: 10.1016/J.JDDST.2023.10430

Ham, K. M., Kim, M., Bock, S., Kim, J., Kim, W., Jung, H. S., An, J., Song, H., Kim, J. W., Kim, H. M., Rho, W. Y., Lee, S. H., Park, S. M., Kim, D. E., & Jun, B. H. (2022). Highly Bright Silica-Coated InP/ZnS Quantum Dot-Embedded Silica Nanoparticles as Biocompatible Nanoprobes. *International Journal of Molecular Sciences*, *23*(18). Advance online publication. DOI: 10.3390/IJMS231810977 PMID: 36142888

Hassan, D., & Sani, A. (2025). Chitosan films developed using all-natural resource for fruit preservation and the impact of lemon peel extract mediated nickel ferrite nanoparticles on films' physical and barrier properties. *Food Chemistry*, *482*, 144068. https://doi.org/https://doi.org/10.1016/j.foodchem.2025.144068 PMID: 40215842

Hassan, D., Sani, A., Chanihoon, G. Q., Antonio Pérez, A., Ehsan, M., & Torres Huerta, A. L. (2024). Environmentally sustainable and green polymeric method for chitosan (CH) film synthesis using natural acids and impact of zinc ferrite nanoparticles (NPs) on Water Solubility (WS) and physical properties. *Polymers*, *16*(24), 3466. PMID: 39771318

Hassan, D., Sani, A., & Medina, D. I. (2023). Limitations of Nanocarriers Such as Cell and Tissue Toxicity, Genotoxicity, Scale-Up of Nanomaterials. In Khan, F. A. (Ed.), *Nano Drug Delivery for Cancer Therapy: Principles and Practices* (pp. 149–171). Springer Nature Singapore. DOI: 10.1007/978-981-99-6940-1_8

Hassan Ahmed, H. E., & Soylak, M. (2024). Exploring the potential of carbon quantum dots (CQDs) as an advanced nanomaterial for effective sensing and extraction of toxic pollutants. *Trends in Analytical Chemistry*, *180*. Advance online publication. DOI: 10.1016/J.TRAC.2024.117939

Hussain, K. I., Ashour, A., Yousef, E. S., & Shaaban, E. R. (2024). Thermal stability and crystallization kinetic of Se-Te-Ag glassy alloys and thick films for electronic devices. *Chalcogenide Letters*, *21*(1), 65–80. DOI: 10.15251/CL.2024.211.65

Jagadeesh, P., Rangappa, S. M., & Siengchin, S. (2024). Advanced characterization techniques for nanostructured materials in biomedical applications. *Advanced Industrial and Engineering Polymer Research*, *7*(1), 122–143. DOI: 10.1016/J.AIEPR.2023.03.002

Jangid, H., Joshi, H. C., Dutta, J., Ahmad, A., Alshammari, M. B., Hossain, K., Pant, G., & Kumar, G. (2025). Advancing food safety with biogenic silver nanoparticles: Addressing antimicrobial resistance, sustainability, and commercial viability. *Food Chemistry: X*, *26*, 102298. DOI: 10.1016/J.FOCHX.2025.102298 PMID: 40109906

Joudeh, N., & Linke, D. (2022). Nanoparticle classification, physicochemical properties, characterization, and applications: a comprehensive review for biologists. Journal of Nanobiotechnology, 20(1), 1–29. DOI: 10.1186/S12951-022-01477-8

Junaid, M., & Wang, J. (2021). Interaction of nanoplastics with extracellular polymeric substances (EPS) in the aquatic environment: A special reference to eco-corona formation and associated impacts. *Water Research, 201*. Advance online publication. DOI: 10.1016/J.WATRES.2021.117319 PMID: 34130084

Kalaoglu-Altan, O. I., Karagüzel Kayaoğlu, B., & Trabzon, L. (2025). Enhancing the material properties of recycled poly (ethylene terephthalate) nanofibers through carbon nanotube and carbon nanotube: Graphene incorporation. *Polymer Engineering and Science*, *65*(3), 1070–1080.

Kaymaz, S. V., Nobar, H. M., Sarıgül, H., Soylukan, C., Akyüz, L., & Yüce, M. (2023). Nanomaterial surface modification toolkit: Principles, components, recipes, and applications. *Advances in Colloid and Interface Science*, *322*, 103035. DOI: 10.1016/J.CIS.2023.103035 PMID: 37931382

Khan, A., Anas, M., Bibi, F., Ali, M., Khalil, A. T., Munawar, K. S., Mohamed, H. E. A., Hkiri, K., Maaza, M., & Shinwari, Z. K. (2025). Phytochemical-Mediated Green Synthesis of Silver, Copper, and Ag–Cu Bimetallic Nanoparticles Using Peganum harmala Demonstrating Advanced Catalytic, Antioxidant, and Biomedical Applications. *Applied Biochemistry and Biotechnology*. Advance online publication. DOI: 10.1007/s12010-025-05186-4 PMID: 39964595

Khan, A., Anas, M., Qadri, T. A., Algopishi, U. B., Fayad, E., Binjawhar, D. N., Khan, R. A., Maaza, M., & Shinwari, Z. K. (2025). Eco-friendly synthesis of copper oxide-silver bimetallic nanoparticles using Withania coagulans: Characterization and biomedical potential through antioxidant, antibacterial, and catalytic activities. *Inorganic Chemistry Communications, 115131*. https://doi.org/https://doi.org/10.1016/j.inoche.2025.115131

Khan, A., Anas, M., Ullah, N., Khattak, W. A., Ali, M., Maaza, M., & Shinwari, Z. K. (2025). Harnessing Green Nanotechnology: Exploring the Potential of Silver, Copper, and their Nanocomposites for Biomedical and Agricultural Applications. *Pakistan Journal of Botany*, *57*(6), 6. https://doi.org/http://dx.doi.org/10.30848/PJB2025-6(5)

Khan, A., Ashiq, B., Qadri, T. A., Khattak, W. A., Anas, M., Khan, R. A., Abbas, S., Shinwari, Z. K., & Maaza, M. (2025). Nanotechnology Solutions For Advancing Sustainable Development Goals (SDGs): Integrating Nanomaterials For Global Sustainability. *BioNanoScience*, *15*(4), 590. DOI: 10.1007/s12668-025-02214-9

Khan, A., Younis, T., Anas, M., Ali, M., Shinwari, Z. K., Khalil, A. T., Munawar, K. S., Mohamed, H. E. A., Hkiri, K., Maaza, M., Seleiman, M. F., & Khan, N. (2025). Withania coagulans-mediated green synthesis of silver nanoparticles: Characterization and assessment of their phytochemical, antioxidant, toxicity, and antimicrobial activities. *BMC Plant Biology*, *25*(1), 574. DOI: 10.1186/s12870-025-06533-7 PMID: 40316892

Kharlamova, M. V. (2023). Advances in Surface-Enhanced and Tip-Enhanced Raman Spectroscopy, Mapping and Methods Combined with Raman Spectroscopy for the Characterization of Perspective Carbon Nanomaterials. *Nanomaterials (Basel, Switzerland)*, *13*(17), 2495. DOI: 10.3390/NANO13172495 PMID: 37687003

Khuddus, M., Gavhane, U. A., & Jayakannan, M. (2024). Structural Engineering of <scp>l</scp>-Aspartic Amphiphilic Polyesters for Enzyme-Responsive Drug Delivery and Bioimaging in Cancer Cells. *ACS Polymers Au*, *4*(5), 392–404. DOI: 10.1021/ACSPOLYMERSAU.4C00013 PMID: 39399886

Kianfar, E. (2025). Dendrimers in Medicine: Properties, Drug Encapsulation Mechanisms and Applications. *Trends in Sciences*, *22*(8), 10085. DOI: 10.48048/tis.2025.10085

Kim, Y. N., & Kang, E. B., Park, T. W., & Park, W. I. (2024). Controlled Formation of Porous Cross-Bar Arrays Using Nano-Transfer Printing. *Materials (Basel)*, *17*(22). Advance online publication. DOI: 10.3390/MA17225609 PMID: 39597432

Krucker-Velasquez, E., Bazant, M. Z., Alexander-Katz, A., & Swan, J. W. (2025). Potential of mean force and underscreening of polarizable colloids in concentrated electrolytes. *Nano Letters*, *25*(26), 10362–10368. PMID: 40534272

Kumar, R., Kumar, A., Bhardwaj, S., Sikarwar, M., Sriwastaw, S., Sharma, G., & Gupta, M. (2025). Nanotoxicity unveiled: Evaluating exposure risks and assessing the impact of nanoparticles on human health. *Journal of Trace Elements and Minerals*, *13*, 100252. DOI: 10.1016/J.JTEMIN.2025.100252

Kumari, A., Yadav, S. K., & Yadav, S. C. (2010). Biodegradable polymeric nanoparticles based drug delivery systems. *Colloids and Surfaces. B, Biointerfaces*, *75*(1), 1–18. DOI: 10.1016/J.COLSURFB.2009.09.001 PMID: 19782542

Kumari, S., Raturi, S., Kulshrestha, S., Chauhan, K., Dhingra, S., András, K., Thu, K., Khargotra, R., & Singh, T. (2023). A comprehensive review on various techniques used for synthesizing nanoparticles. *Journal of Materials Research and Technology*, *27*, 1739–1763. DOI: 10.1016/J.JMRT.2023.09.291

Lee, Y. K., Choi, E. J., Webster, T. J., Kim, S. H., & Khang, D. (2014). Effect of the protein corona on nanoparticles for modulating cytotoxicity and immunotoxicity. *International Journal of Nanomedicine*, *10*, 97–113. DOI: 10.2147/IJN.S72998 PMID: 25565807

Li, J., Wang, C., Yue, L., Chen, F., Cao, X., & Wang, Z. (2022). Nano-QSAR modeling for predicting the cytotoxicity of metallic and metal oxide nanoparticles: A review. *Ecotoxicology and Environmental Safety*, *243*. Advance online publication. DOI: 10.1016/J.ECOENV.2022.113955 PMID: 35961199

Li, Q., Jin, Q., Gao, C., Zhang, X., Zhao, X., He, Y., Chu, D., & Bai, W. (2025). Advances in data-driven integrated design synthesis optimization and prediction of carbon nanotube. *Carbon Letters*, *35*(5), 1893–1931. DOI: 10.1007/s42823-025-00952-0

Li, Y., Vulpe, C., Lammers, T., & Pallares, R. M. (2024). Assessing inorganic nanoparticle toxicity through omics approaches. *Nanoscale*, *16*(34), 15928–15945. DOI: 10.1039/D4NR02328E PMID: 39145718

Li, Z., Li, S., Liu, J., Zhou, L., Liang, R., Liu, F., & Tang, N. (2025). Realization of Near-Room-Temperature Superparamagnetism in Graphene Quantum Dots by High-Concentration Semi-Ionic Fluorine Functionalization. *The Journal of Physical Chemistry Letters*, *16*(20), 4875–4882. DOI: 10.1021/ACS.JPCLETT.5C00614 PMID: 40340512

Liang, X., Mi, L., Yu, Z., Wang, M., Hu, Y., Zheng, X., Shao, Y., Zhu, Z., & Shao, Y. (2021). Multifunctional polyethyleneimine for synthesis of core-shell nanostructures and electrochemiluminescent detection of three AMI biomarkers. *Science China. Chemistry*, *64*(12), 2230–2238. DOI: 10.1007/S11426-021-1080-6

López-Pérez, L., Zarubina, V., & Melián-Cabrera, I. (2021). The Brunauer–Emmett–Teller model on alumino-silicate mesoporous materials. How far is it from the true surface area? *Microporous and Mesoporous Materials*, *319*. Advance online publication. DOI: 10.1016/J.MICROMESO.2021.111065

Lyle, V. (2022). pH-Based Cancer Detection by Graphene Quantum Dots. *The Boller Review*, *6*. Advance online publication. DOI: 10.18776/TCU/BR/6/149

Ma, X., Tian, Y., Yang, R., Wang, H., Allahou, L. W., Chang, J., Williams, G., Knowles, J. C., & Poma, A. (2024). Nanotechnology in healthcare, and its safety and environmental risks. Journal of Nanobiotechnology, 22(1), 1–81. DOI: 10.1186/S12951-024-02901-X

Maaza, M., Henini, M., Ezema, F., Manikandan, E., Kennedy, J., Bouziane, K., Chaker, M., Gibaud, A., Haque, A., & Nuru, Z. (2022). Peculiar size effects in nanoscaled systems. Nano-Horizons: Journal of Nanosciences and Nanotechnologies, 1.

Mahanta, S., Shree, J., Santra, S. C., Moulick, D., & Hossain, A. (2023). Deciphering of mycogenic nanoparticles by spectroscopic methods. Myconanotechnology and Application of Nanoparticles in Biology: Fundamental Concepts, Mechanism and Industrial Applications, 93–117. DOI: 10.1016/B978-0-443-15262-7.00003-6

Mahdi, Z. S., Aljeboree, A. M., Rasen, F. A., Salman, N. A. A., & Alkaim, A. F. (2023). Synthesis, Characterization, and Regeneration of Ag/TiO2 Nanoparticles: Photocatalytic Removal of Mixed Dye Pollutants. *RAiSE*, *59*(1). Advance online publication. DOI: 10.3390/ENGPROC2023059216

Mahmoudi, M., Landry, M. P., Moore, A., & Coreas, R. (2023). The protein corona from nanomedicine to environmental science. *Nature Reviews. Materials*, *8*(7), 1. DOI: 10.1038/S41578-023-00552-2 PMID: 37361608

Mane, P. V., Rego, R. M., Yap, P. L., Losic, D., & Kurkuri, M. D. (2024). Unveiling cutting-edge advances in high surface area porous materials for the efficient removal of toxic metal ions from water. *Progress in Materials Science*, *146*, 101314. DOI: 10.1016/J.PMATSCI.2024.101314

Mani, S., Sachdeva, S., Mani, A., R Vora, H., S. Gholap, S., & Kaur Sodhi, J. (2021). Nano-robotics: The future of health and dental care. IP International Journal of Periodontology and Implantology, 6(1), 6–10. DOI: 10.18231/J.IJPI.2021.002

Manzoor, Y., Hasan, M., Zafar, A., Dilshad, M., Ahmed, M. M., Tariq, T., Hassan, S. G., Hassan, S. G., Shaheen, A., Caprioli, G., & Shu, X. (2022). Incubating Green Synthesized Iron Oxide Nanorods for Proteomics-Derived Motif Exploration: A Fusion to Deep Learning Oncogenesis. *ACS Omega*, *7*(51), 47996–48006. DOI: 10.1021/ACSOMEGA.2C05948 PMID: 36591177

Maumela, M., & Bingwa, N. (2023). *Leveraging Dendrimer Macromolecules for the Encapsulation and Stabilisation of Nano-Sized Ruthenium Catalysts: Evaluation of Catalytic Reaction Kinetics in the Reduction of Pollutants Organic Dyes, Oxidation of Alcohols and Alkenes as Well as Hydrogenation Reactions*. Ruthenium - Materials Properties, Device Characterizations, and Advanced Applications. DOI: 10.5772/INTECHOPEN.111453

Mcoyi, M. P., Mpofu, K. T., Sekhwama, M., & Mthunzi-Kufa, P. (2025). Developments in Localized Surface Plasmon Resonance. *Plasmonics*, *20*(7), 5481–5520. DOI: 10.1007/S11468-024-02620-X/FIGURES/15

Mehta, M., Bui, T. A., Yang, X., Aksoy, Y., Goldys, E. M., & Deng, W. (2023). Lipid-Based Nanoparticles for Drug/Gene Delivery: An Overview of the Production Techniques and Difficulties Encountered in Their Industrial Development. *ACS Materials Au*, *3*(6), 600–619. DOI: 10.1021/ACSMATERIALSAU.3C00032/ASSET/IMAGES/LARGE/MG3C00032_0004.JPEG PMID: 38089666

Mekuye, B., & Abera, B. (2023). Nanomaterials: An overview of synthesis, classification, characterization, and applications. *Nano Select*, *4*(8), 486–501. DOI: 10.1002/NANO.202300038

Miller, T., Mikiciuk, G., Durlik, I., Mikiciuk, M., Łobodzińska, A., & Śnieg, M. (2025). The IoT and AI in Agriculture: The Time Is Now—A Systematic Review of Smart Sensing Technologies. *Sensors (Basel)*, *25*(12), 3583. PMID: 40573470

Mittal, P., Saharan, A., Verma, R., Altalbawy, F. M. A., Alfaidi, M. A., Batiha, G. E. S., Akter, W., Gautam, R. K., Uddin, M. S., & Rahman, M. S. (2021). Dendrimers: A New Race of Pharmaceutical Nanocarriers. *BioMed Research International*, *8844030*. Advance online publication. DOI: 10.1155/2021/8844030 PMID: 33644232

Mohseni, N., & Bahram, M. (2024). A near-infrared plasmonic biosensor for detection of morphine and codeine in biological samples based on the end-to-end assembly of modified gold nanorods. *Analytical Methods*, *16*(26), 4247–4254. DOI: 10.1039/D4AY00442F PMID: 38904334

Moradialvand, Z., Parseghian, L., & Rajabi, H. R. (2025). Green synthesis of quantum dots: Synthetic methods, applications, and toxicity. *Journal of Hazardous Materials Advances*, *18*, 100697. DOI: 10.1016/J.HAZADV.2025.100697

Moraille, P., Abdali, Z., Ramkaran, M., Polcari, D., Patience, G. S., Dorval Courchesne, N. M., & Badia, A. (2022). Experimental methods in chemical engineering: Atomic force microscopy – AFM. *Canadian Journal of Chemical Engineering*, *100*(10), 2778–2806. DOI: 10.1002/CJCE.24407

Mukherjee, P., Roy, S., Ghosh, D., & Nandi, S. K. (2022). Role of animal models in biomedical research: A review. *Laboratory Animal Research*, *38*(1), 18. DOI: 10.1186/S42826-022-00128-1 PMID: 35778730

Mukhopadhyay, R., Sarkar, B., Khan, E., Alessi, D. S., Biswas, J. K., Manjaiah, K. M., & Ok, Y. S. (2022). Nanomaterials for sustainable remediation of chemical contaminants in water and soil. *Critical Reviews in Environmental Science and Technology*, *52*(15), 2611–2660.

Nguyen, B. A., Purnamasari, D. A., Khan, N., & Park, J. S. (2025). A review of state-of-the-art strategies for encapsulating hydrophilic drugs into lipid nanocarriers. *Journal of Pharmaceutical Investigation*, 1–17. DOI: 10.1007/S40005-025-00765-3/TABLES/6

Nogueira, S. S., Samaridou, E., Simon, J., Frank, S., Beck-Broichsitter, M., & Mehta, A. (2024). Analytical techniques for the characterization of nanoparticles for mRNA delivery. *European Journal of Pharmaceutics and Biopharmaceutics, 198*, 114235. DOI: 10.1016/J.EJPB.2024.114235 PMID: 38401742

Nthunya, L. N., Mosai, A. K., López-Maldonado, E. A., Bopape, M., Dhibar, S., Nuapia, Y., Ajiboye, T. O., Buledi, J. A., Solangi, A. R., Sherazi, S. T. H., Ndungu, P. N., Mahlangu, O. T., & Mamba, B. B. (2025). Unseen threats in aquatic and terrestrial ecosystems: Nanoparticle persistence, transport and toxicity in natural environments. *Chemosphere, 382*, 144470. DOI: 10.1016/J.CHEMOSPHERE.2025.144470 PMID: 40378499

Nyabadza, A., McCarthy, É., Makhesana, M., Heidarinassab, S., Plouze, A., Vazquez, M., & Brabazon, D. (2023). A review of physical, chemical and biological synthesis methods of bimetallic nanoparticles and applications in sensing, water treatment, biomedicine, catalysis and hydrogen storage. *Advances in Colloid and Interface Science, 321*, 103010. DOI: 10.1016/J.CIS.2023.103010 PMID: 37804661

Padinhattath, S. P., Shaibuna, M., & Gardas, R. L. (2025). Ionic liquids and deep eutectic solvents: A brief prologue and their applications in sustainable extraction and separation processes. *Journal of the Indian Chemical Society, 102*(4), 101638. DOI: 10.1016/J.JICS.2025.101638

Panico, S., Capolla, S., Bozzer, S., Toffoli, G., Dal Bo, M., & Macor, P. (2022). Biological Features of Nanoparticles: Protein Corona Formation and Interaction with the Immune System. *Pharmaceutics, 14*(12), 2605. DOI: 10.3390/PHARMACEUTICS14122605 PMID: 36559099

Parveen, S., Gupta, P., Kumar, S., & Banerjee, M. (2023). Lipid polymer hybrid nanoparticles as potent vehicles for drug delivery in cancer therapeutics. *Medicine in Drug Discovery, 20*, 100165. DOI: 10.1016/J.MEDIDD.2023.100165

Pasieczna-Patkowska, S., Cichy, M., & Flieger, J. (2025b). Application of Fourier Transform Infrared (FTIR) Spectroscopy in Characterization of Green Synthesized Nanoparticles. Molecules, 30(3), 684. DOI: 10.3390/MOLECULES30030684

Pathak, G., Mangla, S., Guddu, Gupta, K., Veer Bhan, & Kapoor, R. K. (2025). Toxicological assessment and risk management of nanoparticles mediated composite materials-critical review: state of the art. Discover Polymers, 2(1), 1–27. DOI: 10.1007/S44347-025-00023-7

Pathiraja, G., Herr, D. J. C., & Rathnayake, H. (2022). Nanoscopic Insight into Sol–Gel Chemical Kinetics of Oriented Attachment Crystal Growth in Anisotropic Copper Hydroxide Nanowires. Crystal Growth &Amp. *Design, 22*(5), 2889–2902. DOI: 10.1021/ACS.CGD.1C01247

Qin, K., Zheng, Z., Wang, J., Pan, H., & Tang, R. (2024). Biomineralization strategy: From material manufacturing to biological regulation. *Giant (Oxford, England), 19*, 100317. DOI: 10.1016/J.GIANT.2024.100317

Rami, J. M., Patel, C. D., Patel, C. M., & Patel, M. V. (2020). Thermogravimetric analysis (TGA) of some synthesized metal oxide nanoparticles. *Materials Today: Proceedings, 43*, 655–659. DOI: 10.1016/J.MATPR.2020.12.554

Rizzi, F., Castaldo, R., Latronico, T., Lasala, P., Gentile, G., Lavorgna, M., Striccoli, M., Agostiano, A., Comparelli, R., Depalo, N., Curri, M. L., & Fanizza, E. (2021). High surface area mesoporous silica nanoparticles with tunable size in the sub-micrometer regime: Insights on the size and porosity control mechanisms. *Molecules (Basel, Switzerland)*, *26*(14). Advance online publication. DOI: 10.3390/MOLECULES26144247 PMID: 34299522

Rodriguez-Loya, J., Lerma, M., & Gardea-Torresdey, J. L. (2023). Dynamic Light Scattering and Its Application to Control Nanoparticle Aggregation in Colloidal Systems: A Review. Micromachines, 15(1), 24. DOI: 10.3390/MI15010024

Roucoux, A., & Philippot, K. (2021). New trends in the design of metal nanoparticles and derived nanomaterials for catalysis. Nanoparticles in catalysis: Advances in synthesis and applications, 1-11.

Rozhin, A., Batasheva, S., Kruychkova, M., Cherednichenko, Y., Rozhina, E., & Fakhrullin, R. (2021). Biogenic Silver Nanoparticles: Synthesis and Application as Antibacterial and Antifungal Agents. *Micromachines*, *12*(12). Advance online publication. DOI: 10.3390/MI12121480 PMID: 34945330

Sachdeva, S., Mani, A., Mani, S. A., Vora, H. R., Gholap, S. S., & Sodhi, J. K. (2021). Nano-robotics: The future of health and dental care. *IP Int J Periodontol Implantol*, *6*(1), 6–10.

Saleh, H. M., & Hassan, A. I. (2023). Synthesis and Characterization of Nanomaterials for Application in Cost-Effective Electrochemical Devices. Sustainability, 15(14), 10891. DOI: 10.3390/SU151410891

Sanches, P. H. G., de Melo, N. C., Porcari, A. M., & de Carvalho, L. M. (2024). Integrating Molecular Perspectives: Strategies for Comprehensive Multi-Omics Integrative Data Analysis and Machine Learning Applications in Transcriptomics, Proteomics, and Metabolomics. Biology, 13(11), 848. DOI: 10.3390/BIOLOGY13110848

Sanchis-Gual, R., Coronado-Puchau, M., Mallah, T., & Coronado, E. (2023). Hybrid nanostructures based on gold nanoparticles and functional coordination polymers: Chemistry, physics and applications in biomedicine, catalysis and magnetism. *Coordination Chemistry Reviews*, *480*, 215025. DOI: 10.1016/J.CCR.2023.215025

Sani, A., Hassan, D., Chanihoon, G. Q., Melo Máximo, D. V., & Sanchez-Rodriguez, E. P. (2024). Green Chemically Synthesized Iron Oxide Nanoparticles–Chitosan Coatings for Enhancing Strawberry Shelf-Life. *Polymers*, *16*(23), 3239. PMID: 39683984

Sani, A., Hassan, D., Ehsan, M., Sánchez-Rodríguez, E. P., & Melo-Máximo, D. V. (2025). Improving strawberry shelf life using chitosan and zinc oxide nanoparticles from ginger-garlic extracts. *Applied Food Research*, *5*(1), 100765. https://doi.org/https://doi.org/10.1016/j.afres.2025.100765

Saxena, S., Moharil, M. P., Jadhav, P. V., Ghodake, B., Deshmukh, R., & Ingle, A. P. (2025). Transforming waste into wealth: Leveraging nanotechnology for recycling agricultural byproducts into value-added products. *Plant Nano Biology*, *11*, 100127. DOI: 10.1016/J.PLANA.2024.100127

Serna-Gallén, P., & Mužina, K. (2024). *Metallic nanoparticles at the forefront of research: Novel trends in catalysis and plasmonics*. Nano Materials Science. DOI: 10.1016/J.NANOMS.2024.10.009

Serrano, B. A., Gheorghe, L. C., Exner, T. E., Resch, S., Wolf, C., Himly, M., Falk, A., Bossa, N., Vazquez, S., Papadiamantis, A. G., Afantitis, A., Melagraki, G., Maier, D., Saramveis, H., Willighagen, E., Lobaskin, V., Oldfield, J. D., & Lynch, I. (2024). The role of FAIR nanosafety data and nanoinformatics in achieving the UN sustainable development goals: The NanoCommons experience. *RSC Sustainability*, *2*(5), 1378–1399. DOI: 10.1039/D3SU00148B

Shah, S. I. A., Ahmad, W., Anwar, M., Shah, R., Khan, J. A., Shah, N. S., Al-Anazi, A., & Han, C. (2025). Synthesis, properties, and applications of Fe3O4 and Fe3O4-based nanocomposites: A review. *Applied Catalysis, O, Open*, *203*, 207049. DOI: 10.1016/J.APCATO.2025.207049

Shahazi, R., Majumdar, S., Saddam, A. I., Mondal, J., Rahman, M. M., & Alam, Md. M. (2023). Carbon nanomaterials for biomedical applications: A comprehensive review. *Nano Carbons*, *1*(1), 448. DOI: 10.59400/N-C.V1I1.448

Shakiba, S., Astete, C. E., Paudel, S., Sabliov, C. M., Rodrigues, D. F., & Louie, S. M. (2020). Emerging investigator series: polymeric nanocarriers for agricultural applications: synthesis, characterization, and environmental and biological interactions. *Environmental Science. Nano*, *7*(1), 37–67.

Sheikhzadeh, E., Beni, V., & Zourob, M. (2021). Nanomaterial application in bio/sensors for the detection of infectious diseases. *Talanta*, *230*. Advance online publication. DOI: 10.1016/j.talanta.2020.122026 PMID: 33934756

Shen, Y., Su, S., Zhao, W., Cheng, S., Xu, T., Yin, K., Chen, L., He, L., Zhou, Y., Bi, H., Wan, S., Zhang, Q., Wang, L., Ni, Z., Banhart, F., Botton, G. A., Ding, F., Ruoff, R. S., & Sun, L. (2021). Sub-4 nm Nanodiamonds from Graphene-Oxide and Nitrated Polycyclic Aromatic Hydrocarbons at 423 K. *ACS Nano*, *15*(11), 17392–17400. DOI: 10.1021/ACSNANO.1C00209 PMID: 34128643

Shinde, S. (2023). Nanoengineered Drug Delivery Systems for Targeted Treatment of Respiratory Diseases: A Review Study. *International Research Journal of Modernization in Engineering Technology and Science*. Advance online publication. DOI: 10.56726/IRJMETS44660

Shipunova, V. O., Belova, M. M., Kotelnikova, P. A., Shilova, O. N., Mirkasymov, A. B., Danilova, N. V., Komedchikova, E. N., Popovtzer, R., Deyev, S. M., & Nikitin, M. P. (2022). Photothermal therapy with HER2-targeted silver nanoparticles leading to cancer remission. *Pharmaceutics*, *14*(5), 1013. PMID: 35631598

Shirzad, M., Shaban, M., Mohammadzadeh, V., Rahdar, A., Fathi-karkan, S., Hoseini, Z. S., Najafi, M., Kharaba, Z., & Aboudzadeh, M. A. (2025). Artificial Intelligence-Assisted Design of Nanomedicines for Breast Cancer Diagnosis and Therapy: Advances, Challenges, and Future Directions. *BioNanoScience*, *15*(3). Advance online publication. DOI: 10.1007/S12668-025-01980-W

Shishir, M. K. H., Islam, M. M., Islam, M. T., Rahaman, M. A., Mustak, M. H., Mostafa, M. G., & Alam, M. A. (2025). Crystallographic benchmarking of powder X-ray line diffraction pattern profiling of monoclinic sucrose nanocrystal. *Results in Surfaces and Interfaces*, *20*, 100595. DOI: 10.1016/J.RSURFI.2025.100595

Shourni, S., Javadi, A., Hosseinpour, N., Bahramian, A., & Raoufi, M. (2022). Characterization of protein corona formation on nanoparticles via the analysis of dynamic interfacial properties: Bovine serum albumin - silica particle interaction. *Colloids and Surfaces. A, Physicochemical and Engineering Aspects*, *638*, 128273. DOI: 10.1016/J.COLSURFA.2022.128273

Shrestha, A., Gao, X., Hicks, J. C., & Paolucci, C. (2021). Nanoparticle Size Effects on Phase Stability for Molybdenum and Tungsten Carbides. *Chemistry of Materials*, *33*(12), 4606–4620. DOI: 10.1021/ACS.CHEMMATER.1C01120

Singh, B. V., Girase, I. P., Sharma, M., Tiwari, A. K., Baral, K., & Pandey, S. K. (2024). Nanoparticle-enhanced approaches for sustainable agriculture and innovations in food science. *International Journal of Environment and Climate Change*, *14*(1), 293–313.

Siyalo, S., Etefa, H. F., & Dejene, F. B. (2025). Enhancing structural and optical properties of CuO thin films through gallium doping: A pathway to enhanced photoluminescence and predict for solar cells applications. *Chemical Physics Impact*, *10*, 100832. DOI: 10.1016/J.CHPHI.2025.100832

S.T, H. (2024). Thresholds and Delimitations of Quantum Confinement in Spherical Gallium Nitride and Gallium Arsenide Quantum Dots. *International Journal of Research Publication and Reviews*, *5*(5), 6770–6774. DOI: 10.55248/GENGPI.5.0524.1288

Stiufiuc, G. F., & Stiufiuc, R. I. (2024). Magnetic Nanoparticles: Synthesis, Characterization, and Their Use in Biomedical Field. Applied Sciences, 14(4), 1623. DOI: 10.3390/APP14041623

Sukarman, Kristiawan, B., Budiana, E. P., Khoirudin, & Abdulah, A. (2025). Multi-technique characterization of TiO2 nanoparticles: Crystallite size, microstrain, and phase analysis for nanomaterial applications – a review. *Hybrid Advances*, *11*, 100523. DOI: 10.1016/J.HYBADV.2025.100523

Sukmanee, T., Szuster, M., Gorski, A., Hołdyński, M., & Gawinkowski, S. (2023). Tunable-wavelength nanosecond laser tailoring of plasmon resonance spectra of gold nanoparticle colloids. *Nanoscale Advances*, *5*(14), 3697–3704. DOI: 10.1039/D3NA00225J PMID: 37441263

Sun, Y., Zhou, Y., Rehman, M., Wang, Y. F., & Guo, S. (2024). Protein Corona of Nanoparticles: Isolation and Analysis. *Chem & Bio Engineering*, *1*(9), 757–772. DOI: 10.1021/CBE.4C00105/ASSET/IMAGES/LARGE/BE4C00105_0005.JPEG PMID: 39974182

Takáč, P., Michalková, R., Čižmáriková, M., Bedlovičová, Z., Balážová, Ľ., Laca Megyesi, Š., Mačeková, Z., Takáčová, G., Moreno-Borrallo, A., Ruiz-Hernandez, E., Isakov, L., & Takáč, P. (2025). Do We Know Enough About the Safety Profile of Silver Nanoparticles in Oncology? A Focus on Novel Methods and Approaches. International Journal of Molecular Sciences, 26(11), 5344. DOI: 10.3390/IJMS26115344

Taniguchi, A., Miyata, R., Ohtani, M., & Kobiro, K. (2022). Insights into the solvothermal reaction for synthesizing tin(<scp>iv</scp>) oxide porous spheres. *RSC Advances*, *12*(35), 22902–22910. DOI: 10.1039/D2RA03403D PMID: 36105993

Teixeira, M. I., Lopes, C. M., Amaral, M. H., & Costa, P. C. (2024). Navigating Neurotoxicity and Safety Assessment of Nanocarriers for Brain Delivery: Strategies and Insights. *Acta Biomaterialia*, *189*, 25–56. DOI: 10.1016/J.ACTBIO.2024.09.027 PMID: 39307261

Thomas, J., Kumar, V., Sharma, N., John, N., Umesh, M., Kumar Dasarahally Huligowda, L., Kaur, K., & Utreja, D. (2025). Recent approaches in nanotoxicity assessment for drug delivery applications: Challenges and prospects. *Medicine in Drug Discovery*, *25*, 100204. DOI: 10.1016/J.MEDIDD.2025.100204

Tsukamoto, H. (2023). Tensile strength and anisotropy of repeated hot-rolled carbon nanotube/aluminum matrix composites. *Journal of Composite Materials*, *57*(7), 1223–1234. DOI: 10.1177/00219983231153247

Udoisoh, M. G., Arewa, A. J., Babalola, A. V., & Amaechi, C. J. (2025). Energy Regimes, Thresholds and Delimitation in Ellipsoidal Quantum Dots with a Modified Hellmann Potential. European Journal of Applied Science. *Engineering and Technology*, *3*(2), 304–320.

Ullah, N., Khan, A., Younis, T., Maqbool, H., Iqbal, M., Ali, M., & Shinwari, Z. K. (2025). Advances in Nanocomposite-Based Fertilizers for Sustainable Agricultural Practices. Proceedings of the Pakistan Academy of Sciences: B. Life and Environmental Sciences, 62(1), 1-18.

Ullah, Z., Iqbal, J., Abbasi, B. A., Ijaz, S., Ahmad, S., Khan, S., & Mahmood, T. (2024). Nano-formulation for Agriculture Applicability. In *Revolutionizing Agriculture: A Comprehensive Exploration of Agri-Nanotechnology* (pp. 325–367). Springer Nature Switzerland.

Varsou, D. D., Kolokathis, P. D., Antoniou, M., Sidiropoulos, N. K., Tsoumanis, A., Papadiamantis, A. G., Melagraki, G., Lynch, I., & Afantitis, A. (2024). In silico assessment of nanoparticle toxicity powered by the Enalos Cloud Platform: Integrating automated machine learning and synthetic data for enhanced nanosafety evaluation. *Computational and Structural Biotechnology Journal*, *25*, 47–60. DOI: 10.1016/J.CSBJ.2024.03.020 PMID: 38646468

Vilanova, O., Martinez-Serra, A., Monopoli, M. P., & Franzese, G. (2024). Characterizing the hard and soft nanoparticle-protein corona with multilayer adsorption. *Frontiers in Nanotechnology*, *6*, 1531039. DOI: 10.3389/FNANO.2024.1531039/BIBTEX

Vinukonda, A., Bolledla, N., Jadi, R. K., Chinthala, R., & Devadasu, V. R. (2025). Synthesis of nanoparticles using advanced techniques. *Next Nanotechnology*, *8*, 100169. DOI: 10.1016/J.NXNANO.2025.100169

Vyas, R., Manisha, M., Singh, A., & Mathur, N. (2024). Mycofabrication of Iron Nanoparticles: Applications and Future Prospectus. *Nanofabrication*, *9*. Advance online publication. DOI: 10.37819/NANOFAB.9.1808

Vyawahare, P., Tun, H., Vaughn, M. W., & Chen, C. C. (2022). From Langmuir isotherm to Brunauer–Emmett–Teller isotherm. *AIChE Journal. American Institute of Chemical Engineers*, *68*(3). Advance online publication. DOI: 10.1002/AIC.17523

Wilson, B. K., & Prud'homme, R. K. (2021). Nanoparticle size distribution quantification from transmission electron microscopy (TEM) of ruthenium tetroxide stained polymeric nanoparticles. *Journal of Colloid and Interface Science*, *604*, 208–220. DOI: 10.1016/J.JCIS.2021.04.081 PMID: 34265681

Xavier, M., Parente, I. A., Rodrigues, P. M., Cerqueira, M. A., Pastrana, L., & Gonçalves, C. (2021). Safety and fate of nanomaterials in food: The role of in vitro tests. *Trends in Food Science & Technology*, *109*, 593–607. DOI: 10.1016/J.TIFS.2021.01.050

Xu, Z., Long, X., Jia, Y., Zhao, D., & Pan, X. (2022). Occurrence, transport, and toxicity of nanomaterials in soil ecosystems: A review. *Environmental Chemistry Letters*, *20*(6), 3943–3969. DOI: 10.1007/S10311-022-01507-9 PMID: 36467872

Xuan, L., Ju, Z., Skonieczna, M., Zhou, P. K., & Huang, R. (2023). Nanoparticles-induced potential toxicity on human health: Applications, toxicity mechanisms, and evaluation models. *MedComm*, *4*(4), e327. DOI: 10.1002/MCO2.327 PMID: 37457660

Yadav, A., Yadav, K., Ahmad, R., & Abd-Elsalam, K. A. (2023). Emerging frontiers in nanotechnology for precision agriculture: Advancements, hurdles and prospects. *Agrochemicals*, *2*(2), 220–256.

Yadav, H. O. S. (2024). Three-body interaction of gold nanoparticles: The role of solvent density and ligand shell orientation. *Physical Chemistry Chemical Physics*, *26*(15), 11558–11569. DOI: 10.1039/D3CP06334H PMID: 38533797

Yaghmur, A., & Hamad, I. (2022). Microfluidic Nanomaterial Synthesis and In Situ SAXS, WAXS, or SANS Characterization: Manipulation of Size Characteristics and Online Elucidation of Dynamic Structural Transitions. *Molecules (Basel, Switzerland)*, *27*(14). Advance online publication. DOI: 10.3390/MOLECULES27144602 PMID: 35889473

Yan, Y., Zhang, G., Wu, C., Ren, Q., Liu, X., Huang, F., Cao, Y., & Ye, W. (2022). Structural Exploration of Polycationic Nanoparticles for siRNA Delivery. ACS Biomaterials Science &Amp. *Engineering*, *8*(5), 1964–1974. DOI: 10.1021/ACSBIOMATERIALS.2C00196 PMID: 35380797

Zeng, S., Chen, Y., Zhou, F., Zhang, T., Fan, X., Chrzanowski, W., Gillies, M. C., & Zhu, L. (2023). Recent advances and prospects for lipid-based nanoparticles as drug carriers in the treatment of human retinal diseases. *Advanced Drug Delivery Reviews*, *199*, 114965. DOI: 10.1016/J.ADDR.2023.114965 PMID: 37315899

Zhang, P., Chen, D., Li, L., & Sun, K. (2022). Charge reversal nano-systems for tumor therapy. *Journal of Nanobiotechnology*, *20*(1), 1–27. DOI: 10.1186/S12951-021-01221-8/TABLES/3 PMID: 35012546

Zhao, W., Jia, W., Zhou, J., Zhai, T., Wu, Y., Rana, Z., Sun, P., Liu, Y., Zhou, S., Xiang, G., & Wang, X. (2025). Nanoscale Grain Boundary-Weakened Ce–O Covalency and Surface Confinement Intrinsically Boosting Ceria Surface Oxygen Reactivity. *Journal of the American Chemical Society*, *147*(15), 13050–13058. DOI: 10.1021/JACS.5C03536 PMID: 40165509

Zheng, J., Sun, Y., Shen, Y., & Zhou, Z. (2025). Surface engineering of nanoparticles for precision medicine. *Precision Medicine and Engineering*, *2*(3), 100037. DOI: 10.1016/J.PREME.2025.100037

Zhou, Y., & Dong, W. (2025). Epoxy functionalized small particle size superparamagnetic iron oxide nanoparticles for CD4+T cell sorting. DOI: 10.21203/RS.3.RS-6039824/V1

Zuhrotun, A., Oktaviani, D. J., & Hasanah, A. N. (2023). Biosynthesis of Gold and Silver Nanoparticles Using Phytochemical Compounds. *Molecules (Basel, Switzerland)*, *28*(7), 3240. DOI: 10.3390/MOLECULES28073240 PMID: 37050004

Chapter 3
Nanoparticle-Plant Interactions: Uptake, Mobility, and Biocompatibility

Sameel Hassan
https://orcid.org/0009-0001-7559-5550
University of Tuscia, Italy

Sampath Chinnam
M.S. Ramaiah Institute of Technology, India

Syeda Anber Zahra
https://orcid.org/0009-0009-2933-9001
Shanghai Zhongqiao Vocational and Technical University, China

Damla Nur
TURKSOY, Selcuk University, Turkey

Nizakat Bibi
School of Life Science, Kaifeng, China

Alvina Hayat
University of Otago, New Zealand

Banzeer Ahsan Abbasi
https://orcid.org/0000-0003-3036-4481
Rawalpindi Women University, Pakistan

ABSTRACT

In the era of climate challenges and emerging plant diseases, nanotechnology offers new solutions for crop protection. The chapter deals with complex behavior of NPs in contact with plants, especially when acting against viral diseases. NPs can enter plant systems by several routes: leaves, root, seeds. The chapter explores how NPs interact with plants, focusing on antiviral roles. NPs enter via leaves, roots, seeds with uptake influenced by size, shape, surface charge, and coatings. Once inside, they move through apoplast, symplast, and vascular tissues for systemic distribution. Environmental factors like soil pH, moisture, and microbes affect NP behavior. Biocompatibility is highlighted as low doses can boost immunity, suppress viral replication, while excess causes oxidative stress and phytotoxicity. The role of ROS, antioxidant defenses, and dose-dependent effects is examined. Case studies of silver, zinc oxide, silica, and chitosan NPs illustrate applications in cereals, legumes. The chapter concludes with strategies for safe, efficient NPs use, and prospects for sustainable agriculture.

INTRODUCTION

Nanotechnology has unlocked new frontiers in plant science by enabling precise control over materials at the nanoscale, allowing for innovative solutions in agriculture and crop protection. (Abbasi et al., 2025; Khan et al., 2025; Ullah et al., 2025). Among these developments, nanoparticles (NPs) engineered

DOI: 10.4018/979-8-3373-4913-8.ch003

materials, usually under 100 nanometers in diameter, are becoming promising as they can engage with plant systems in highly specific and dynamic ways. (Anas et al., 2024; Ullah et al., 2025). Such exchanges are especially useful when spraying agrochemicals becomes insufficient against climate change and the emergence of new viral pathogens, and the overuse of conventional agrochemicals. (Ali et al., 2023; Fu et al., 2020). In contrast to traditional inputs, the nanoparticle can be designed with the desired size, structure, ligand, and functionality, which provide a specification of delivery, prolonged release, and directed movement through plant organs. (Falak et al., 2025; Ullah et al., 2023). Nanoparticles play multi-functional roles in plants. IIaving different compositions and functionalization, nanoparticles can be used as the carriers of micronutrients, growth stimulators, and antimicrobial agents (Elahi et al., 2025; Semenova et al., 2024). In plant virology, common examples of NPs that have good antiviral inhibitory action include metallic NPs like silver and zinc oxide, which may impair viral replication, promote systemic acquired resistance, or even inactivate virions. (Vargas-Hernandez et al., 2020). Such gains cannot only be quantified in terms of defense but also in plant growth pathways, hormonal signaling, and nutrient uptake, potentially increasing resilience and productivity in stressful environments. (Abasi et al., 2024; Vargas-Hernandez et al., 2020). Nevertheless, the interface of the plant and the NP is strongly multi-factorial. The major ways by which nanoparticles can be infiltrated, either by the foliar or root, or the seed transmission, also depend on physical and chemical properties of the materials and even on conditions of the plant (Avellan et al., 2021). Foliar products have to cross cuticular barriers and stomatal pores, and root-take models depending on plant interactions in the rhizosphere and root membrane permeability. After internalization, NPs can be transported to their target tissues either apoplastically or symplastically to the other tissues, including infected leaves and other reproductive organs (Avellan et al., 2022; Avellan et al., 2021).

Movement of nanoparticles to the remote parts of the plant is facilitated by them vascular delivery of the nanoparticles via the xylem and phloem. (Su et al., 2019). It is important transportation of antiviral agents or micronutrients. (Cojocaru et al., 2020). Flows like endocytosis and transport across plasmodesmata are significant in cell movements of NPs on a cellular-scale in plants to highlight the complexity of this plant cell interaction. (Goh, 2022). But the same mobility leads to the prospects of bioaccumulation and intrusion into the native cell processes which necessitates the balance of activity and toxicity (Zhang et al., 2018). In the use of NPs in agriculture, biocompatibility has been a significant issue. (Kumar et al., 2024). Whilst certain nanomaterials trigger potentially beneficial physiological processes, such as photosynthesis, enzyme activity, and/or stress tolerance, toxic impacts of phytotoxicity, including membrane permeability and damage, oxidative stress, and growth impairment, may be present also at higher concentrations, or with longer exposure. (Ali et al., 2021; Dev et al., 2018). The boundary to cross between the biocompatible and toxic doses is extremely thin and largely relies on the type of NP, species of plants, the developmental stage, and the environmental conditions. Therefore, the actual dose-response relations should be known so that their usage in agriculture is safe. They may include adjusting the curriculum and modifying the structure of lessons (Oberdorster et al., 2007; Servin and White, 2016). The chapter addresses the uptake mode, transport pathways, and biocompatibility of nanoparticles in the plant-based systems with regards to interactions of the viral infection. It draws on recent research and empirical findings to describe the behavior of nanoparticles in plants and how the structural and surface characteristics of nanoparticles may affect their mobility and toxicity, with flow-on effects on the health of plants(Kousar et al., 2025). It also gives case studies of both metallic and non-metallic NPs in crops and concludes by giving practice directions-oriented practices on how to ensure the sustainability of nanoparticle in plant virology incorporation.

NANOPARTICLE UPTAKE IN PLANTS

Plant uptake of nanoparticles is a significant process in the determination of the effectiveness of nanotechnology as well as the safety of nanoparticles in farming. (Bouhadi et al., 2025). The modes of entry into the plant systems of nanoparticles (NPs) are the foliar, root, and seed-mediated entry and all these pathways are based on the anatomical structure of a given plant as well as the physicochemical properties of various nanoparticles on the surface. (Fatima et al., 2020). Once nanoparticles are exposed to the surfaces of the plant, they can either diffuse or enter through the natural orifices passively or actively through the natural orifices or into the cells. Understanding these routes of intake is crucial in the development of nanoparticle formulations which are ideal to transport its content, with the minimal risk to the environment with minimum phytotoxicity. (Fornaguera et al., 2020; Vurro et al., 2019).

Foliar Uptake

Foliar spraying is one of the most widespread methods of applying agro-nanoparticles, particularly antiviral compounds and micronutrient supplements. The surface area of the contact relative to leaves is large, but the uptake via the foliar path is inhibited by the structural barriers, i.e., cuticle, trichomes, and walls of the epidermal cells. (Ghorbanpour & Shahid, 2022). Nanoparticles can further enter through the leaf surface through the cuticular cracks, stomatal openings, or by entering the leaves through the pores rendered by water in high moisture atmospheres. (Bhati et al., 2025). Foliar uptake efficiency is also greatly improved in the presence of nanoparticle coating with surfactants or other hydrophilic ligands that aid in adhesion and increased permeability. (Avellan et al., 2019; Hong et al., 2021). The composition and thickness of the cuticle are also species- and developmental stage-specific and may therefore alter the efficiency of penetration of nanoparticles. Upon their internalization, nanoparticles can penetrate mesophyll cells and then translocate (symplastic or apoplastic) to different plant locations (Avellan et al., 2021). The applications of silver and zinc oxide nanoparticles on plant leaves have demonstrated the ability to translocate systemically on the leaves, and the nanoparticles have the potential to cause systemic resistance pathways that lead to their antiviral properties(Kamal et al., 2025). Nevertheless, the build-up of NPs on the leaf surfaces over a long period of time can also interfere with gas circulation and light absorption in the case that they are not properly metabolized or redistributed (Cai et al., 2019; Faisal et al., 2024). In addition, nanoparticles applied to leaves are likely to react with sunlight and gases in the atmosphere and can possibly change their reactivity and toxicity in this process. This is crucial to address the environmental factors like UV and humidity when going through with an application design. The coupling of nanocarriers with controlled-release is also capable of enhancing the stability of NPs, along with minimizing unwelcome interactions on the phylloplane. (Huang et al., 2018; Li et al., 2021).

Root and Seed Uptake

Considering the nanoparticle uptake process, root uptake is a widely studied pathway, especially in soil or hydroponics systems. Root tissues can also be reached by nanoparticles via the rhizodermis, through root hairs, or by apoplastic/symplastic pathways of the root apex, where the cell walls are thinner and more permeable. It has been suggested that the rhizosphere, which is rich in root exudates and microbial activities, has a large influence on the bioavailability, reactivity, and efficacy of NPs. Nitrogen improvable and NPs move across root barriers through the influence of organic acids, enzymes, and microbial metabolites. (Q. Wang et al., 2023; X. Wang et al., 2023). When adsorbed on the root surface, the nanoparticles can infiltrate the root epidermal and cortical tissues before making their way to the stele, and along with the delivery of the nanoparticles in the xylem to be transported into the rest of the plant. Some of the research findings have shown that nanoparticles, cerium oxide, iron oxide, and chitosan-based formulation improve nutrient uptake and resistance to stress when taken up by roots. The root membrane integrity, water, and ion transport may be impaired, and excessive concentrations or chronic exposure may disrupt the functional responses of the aquaporins. (Rasheed et al., 2022; X. Wang et al., 2023). The other key path for nanoparticle entry is the seed pathway. Nanoparticle-primed seeds are found to germinate better and lead to more seedling growth and stress tolerance to abiotic stress. During imbibition, nanoparticles have the potential to be accessible through the seed coat and the endosperm and translocate to the embryo to trigger premature physiological responses in the seed. Nevertheless, effects on the long-run gene expression and metabolite flux, as well as seedling growth, are little studied and remain to be investigated. (Djanaguiraman et al., 2024).

Figure 1. Nanoparticle uptake in plants occurs via three routes: seed uptake through seed coat and endosperm, foliar uptake through cuticle cracks or stomata with systemic translocation, and root uptake via rhizodermis and root hairs into the xylem

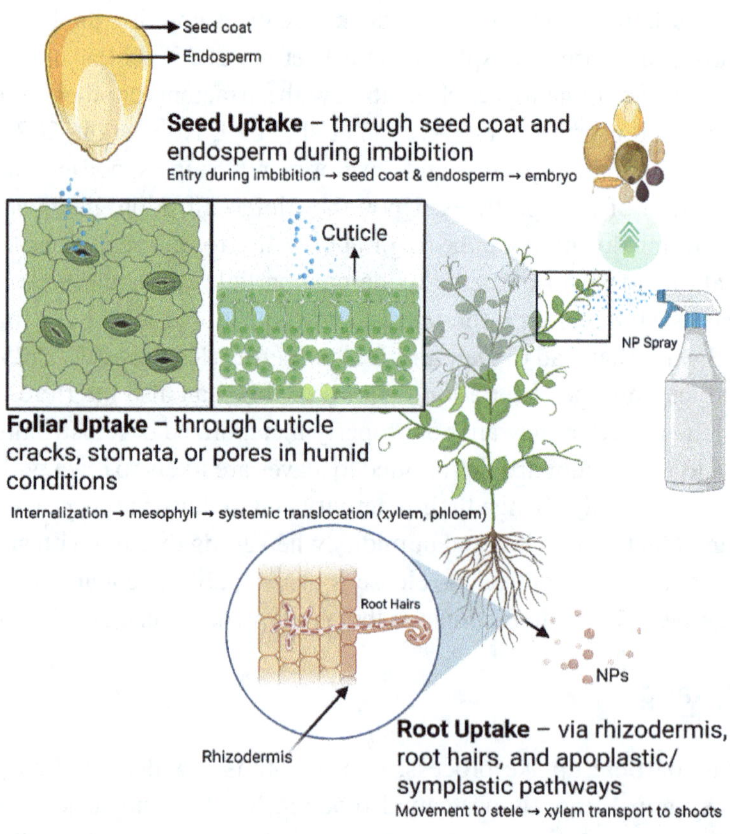

Factors Affecting Uptake Efficiency

Nanoparticle uptake is not only an anatomical response but is also highly dependent upon the characteristics of nanoparticles and environmental conditions. Nanoparticle size is an instrumental factor, as smaller nanoparticles (<50 nm) have a higher chance of entering the cells through wall penetration and pores in comparison to large structures. Surface charge also influences the plant cell wall interactions, which are generally negative; therefore, positively charged NPs participate more effectively in cell wall adherence and penetration. However, uptake can be improved by functionalization, i.e., coating the nano-pharmaceuticals with biocompatible materials, such as chitosan or polyethylene glycol (PEG), to reduce aggregation and improve solubility. (Mosquera et al., 2018; Sanità et al., 2020). Another source of variation is plant species and stages of development. Monocots and dicots differ in terms of stomatal density, root morphology, and cuticle thickness, hence the route of NPs absorption. Young seedlings also tend to be more permeable than established plants, which will make them more receptive to nano-priming approaches. Also, roots can experience changes in architecture and an increase in membrane permeability with stress conditions like drought or salinity, which indirectly influence NP uptake (Ali et al., 2021). Other environmental parameters, including pH, ionic strength, and the presence of organic matter, control the behavior of NPs. In alkaline soils, metal-based NPs can precipitate, lowering their bioavailability, whereas in acidic soils, they can increase their dissolution rates and release, resulting in potential toxicity. When it interacts with soil colloids and microbial exudates, NPs may also experience a change of mobility in the soil and aggregation state, which emphasizes the need for context-specific assessments before use in the field (Liu et al., 2022).

Table 1. Key factors influencing nanoparticle uptake efficiency in plants, highlighting the roles of physicochemical properties, plant traits, environmental conditions, and application methods

Factor	Description	Effect on Uptake	Citations
Particle Size	Smaller NPs (<50 nm) cross cell walls more easily	Higher uptake efficiency	(Mosquera et al., 2018)
Surface Charge	Positive NPs bind to negatively charged cell walls	Enhances membrane penetration	(Balog et al., 2024)
Surface Coating	Chitosan, PEG, or ligand coatings improve dispersion	Prevents aggregation, boosts mobility	(Wang & Mattoussi, 2020)
Shape	Rods vs. spheres influence interaction with tissues	Rods often penetrate faster	(X. Wang et al., 2023)
Solubility	High solubility promotes dissolution and absorption	Enhances ion-mediated uptake	(Jampílek & Kráľová, 2017)
Aggregation	Large NP clusters reduce bioavailability	Lower internalization	(Duan & Li, 2013)
Plant Species	Monocots vs. dicots vary in cuticle thickness & stomata	Species-dependent uptake	(X. Wang et al., 2023)
Development Stage	Young tissues are more permeable than mature tissues	Better absorption in seedlings	(X. Wang et al., 2023)
Root Architecture	Root hairs & surface area affect contact with NPs	Dense hairs improve entry	(Singh & Kumar, 2019)
Leaf Morphology	Stomatal density and cuticle thickness affect foliar uptake	More stomata, better entry	(Avellan et al., 2021)

continued on following page

Table 1. Continued

Factor	Description	Effect on Uptake	Citations
Environmental pH	Acidic vs. alkaline soils alter NP dissolution	pH affects bioavailability	(Dimkpa, 2018)
Organic Matter	Soil humics can bind NPs, changing reactivity	High humics reduce uptake	(Kulikova et al., 2020)
Ionic Strength	Competing ions modify NP stability and solubility	High salt levels limit absorption	(Wahid et al., 2022)
Microbial Activity	Rhizosphere microbes modify NP chemistry	Microbes can enhance uptake	(Zhang et al., 2024)
Stress Conditions	Drought, salinity, or injury changes membrane permeability	Stress can accelerate uptake	(Rasheed et al., 2022)
Application Method	Foliar spray, seed priming, or soil drench differ in efficiency	Delivery mode determines the route	(I. Khan et al., 2022)

INTERNAL TRANSPORT AND MOBILITY

After entry into the plants through leaves, roots, or seeds, the journey of NPs does not end. To play a functional role, such as antiviral activity, release of micronutrients, or stress relief, such nanomaterials need to be delivered effectively to remote tissue sites and cells. (Su et al., 2019). Inner movement is therefore a vital aspect of nanoparticle-plant interaction and fundamentally drives the success, as well as the safety outcomes of nano-enabled agricultural activities. (Wu & Li, 2022; Xin et al., 2020). Transportation within the plant occurs under the conduction of three major routes: the apoplastic and symplastic pathways, the vascular tissues, and the xylem and the phloem, and intercellular trafficking mechanisms (including the use of plasmodesmata and endocytosis). These pathways mimic the natural transport mechanisms of the plant and are adapted to allow transit of synthetic particles depending on their size, surface chemistry, and compatibility with the biosurface (Avellan et al., 2021; X. Wang et al., 2023).

Apoplastic and Symplastic Movement

For short distances, the apoplast and the symplast are the two basic tissue routes for nanoparticle transportation inside the plants. Apoplast is the material outside the cell walls, intercellular space, and xylem vessels. This is a route facilitating the passive transfer of water and solutes without entering a cell. (M. A. U. Khan et al., 2022). Apoplastic NPs are then frequently transported by diffusion or bulk water flow, especially under transpiration-driven pressure, in roots or leaves. Experiments have indicated that small hydrophilic nanoparticles, carbon dots, and some metal oxide NPs can easily diffuse through the pores of a cell wall. Yet, larger or aggregated particles with hydrophobic or neutral coating may be hindered due to the size exclusion limit of cell walls, which is generally between 5 and 20 nm. (Su et al., 2019). The symplastic pathway instead passes through the cytoplasm of living cells and is linked to one another via plasmodesmata. To get into the symplast, the NPs would have to breach the plasma membrane. This can be normally accomplished through active uptake, e.g., by the clathrin-mediated endocytosis (Varma et al., 2022). Inside the cell, nanoparticles can travel laterally between cells by symplastic transport without visiting the extracellular space. It is especially essential that the enzymatic degradation in the apoplast cannot occur owing to the symplastic route of systemic distribution. This pathway is size-selective, though, and particles having a size over 10nm might be excluded unless their coating mediated the integration into the membrane or some sort of vesicular trafficking (Avellan et al., 2021; Sembada & Lenggoro, 2024). It has been previously demonstrated that it is possible to overcome this drawback through altering the surface chemistry of nanoparticles by coating them with amino acids or peptides, or even plant hormones that can increase the symplastic mobility of nanoparticles. These surface-functionalized NPs are more compatible with the plant cellular machinery, less-aggregative, and directionally moving to given cellular organelles or tissue type. This optimization of the nano-bi interface is key to applications that involve delivery to metabolically active areas such as meristems or vascular cambium. (Nair et al., 2010).

Vascular Transport: Xylem and Phloem

After entering plants, nanoparticles often move into the vascular system, enabling their transport over long distances. Xylem and phloem are the two main types of vascular borders, which have different functional decoding, but both play an extremely important role in distributing NPs systemically. Xylem is one-way transportation of water and mineral nutrition sources of roots to the biomass, and phloem distributes organic nutrients like sugars and hormones in the opposite direction between source and sink tissues. (Ma et al., 2017). Roots pick up the nanoparticles, which can enter the xylem across the root cortex and endodermis. In the xylem vessels, the sap flows when transpiration flow carries it upwards. Exposure of silver, zinc oxide, and iron oxide nanoparticles in root systems is transported through the xylem up into the leaves and/or stem tissue. Xylem vessels have different diameters and flow dynamics; therefore, transport can be more or less efficient across species. Greater or conglomerated nanoparticles can be trapped or stick on vessel walls, thus losing mobility or becoming embolic in some cases. (Su et al., 2019). Transport through the phloem is more discriminating and complicated. It includes the loading of NPs into companion cells and into sieve elements, which is frequently necessitated by the surface modification or binding to certain plant metabolites (Su et al., 2019; Wang et al., 2012). Nanoparticles applied through foliar application are able to enter the phloem after traversing a few cell layers, at which point they can attach to sugars or proteins that aid active loading. This pathway is especially useful to deliver nanoparticles to routes or reproductive organs, which do not have access to the xylem. Since sieve plates on phloem vessels are very small (their pores are usually less than 1,100 nm), only particles with low hydrodynamic diameters (20nm or less) or those in biodegradable carriers are successfully translocated (X. Wang et al., 2023). The combination of xylem and phloem transport further increases the spatial distribution capacity of nanoparticles and enables them to deliver them to either targeted tissues or systemically. As an example, during foliar application, anti-viral NPs would need to enter the phloem to alleviate phloem-restricted viral infections; in contrast, nutrient micronutrients delivered to the root can be transported through the xylem to supply shoots. Dual-affinity-coated nanoparticle formulations that can be distributed in xylem and phloem streams are under development as smart carriers to deliver integrated pathogen management. (Ma et al., 2017).

Role of Plasmodesmata and Endocytosis

Nanoparticles that penetrate cells require intracellular transport to their biological outcome. This is supported by plasmodesmata, nanometer-scale cytoplasmic links between neighboring cells, and endocytic routes that organize vesicle-facilitated transport between and within cells. (Gruner & McManus,

2021). Plasmodesmata are an essential avenue of nanoparticle invasion within the cytoplasm, especially symplastic transportation. The typical size exclusion limit of these structures is 1-3nm and is regulated dynamically in response to developmental cues, pathogen attacks, or other external triggers. Select surface-modified nanoparticles, notably those bearing biological motifs or plant-derived ligands, can cause temporary opening of plasmodesmata through which they can enter adjacent cells. The latter type of intercellular transport plays an important role in the distribution of therapeutic NPs within tissues (e.g., epidermis to vascular tissue or parenchyma to phloem) (Rajesh, 2025). Endocytosis is another important process that can be used by plant cells to take up nanoparticles into their structure. The clearer mechanism involves that of endocytosis, which is mediated by clathrin, which encompasses the invagination of the plasma membrane to form endocytic vesicles. Upon endocytic internalization, these vesicles carry the NPs to the early endosomes through which they can be directed to the trans-Golgi network, vacuoles, or even be recycled back to the plasma membrane. Endocytosis is specific, and in many cases requires the interaction of NPs with a receptor on the membrane. Nanoparticles that resemble endogenous ligands, e.g., auxins or pathogen effectors, have been shown to be taken up in an endocytic manner in several plant models (He et al., 2020). Controlled intracellular trafficking is one more benefit of endocytic uptake. Nanoparticles can also be targeted to a particular organelle, e.g., to the chloroplast, the nucleus, or the mitochondrion, depending upon their surface chemistry and cargo. It allows very precise administration of genes, micronutrients, or antiviral substances to the point of action. Nonetheless, another surety of endocytic trafficking is its connection to off-target consequences, a perturbation of the native vesicular transport systems, and unwanted oxidative stress-signaling. Therefore, one of the main challenges of nano-enabled agriculture is the enhancement of the specificity of nanoparticle formulations and alerting to the toxicity. (Soukar et al., 2025).

Figure 2. Internal transport and mobility of nanoparticles in plants. NPs move short distances via the apoplastic pathway (cell walls, intercellular spaces, xylem) or the symplastic pathway (cytoplasm via plasmodesmata, endocytosis). Long-distance movement occurs through vascular transport: xylem (upward flow) and phloem (bidirectional flow). Plasmodesmata and endocytosis further enable intercellular and organelle-specific delivery, though risks of off-target effects and oxidative stress remain

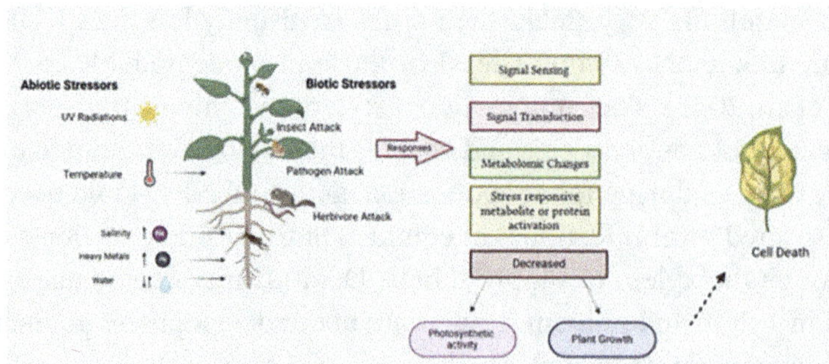

BIOCOMPATIBILITY AND PHYTOTOXICITY

The currently increasing use of nanoparticles (NPs) in agriculture requires a clear concept of how they apply to the plants and are biocompatible. Although a broad range of beneficial agronomic effects can be attributed to NP effects-i.e., increasing nutrient use efficiency, pathogen control, and stress amelioration effects- their interaction with plant physiology is extremely complex. (Wu & Li, 2022). In non-toxic, where possible, optimized concentrations, a range of nanoparticles can promote growth, induce defense responses, and aid nutrient acquisition. Nevertheless, in excess, the same particles can disrupt important physiological mechanisms and redox homeostasis and lead to phytotoxicity. This duality is the reason why it is essential to determine safe range dosages and get to know how specific species of plants react to specific NPs (Verma et al., 2024). Several parameters are normally used to determine the biocompatibility of nanoparticles, such as frequency and concentration of nanoparticles, size, surface chemistry, solubility, and the ability to produce reactive oxygen (ROS). Also, plant-specific variables, e.g., type of tissue, development, and environmental conditions, are critical issues in mediating the toxicity responses. An example includes the fact that young seedlings might respond with increased growth when exposed to some NPs, such as zinc oxide or silica, whereas mature plants exposed to the same under higher doses would respond either by growth retardation, chlorosis, or even necrosis. As such, it is also important to consider a more holistic toxicity assessment at cellular and tissue levels, as well as at whole-plant and ecological levels. (González-Grandío et al., 2021).

Effects on Plant Growth and Physiology

A change in growth patterns is the first indicator of nanoparticles' interaction with plants. Some NPs have been shown to promote seed germination, root growth, and growth in biomass at sub-toxic concentrations. As an example, chitosan nanoparticles resulted in good growth promotion in tomato and wheat because of their biostimulatory properties and ability to stimulate hormonal signaling, such as auxin and gibberellin production. In a similar manner, nano-silicon has also been linked with enhanced water-use efficiency and cell-wall strengthening effects that offer greater resistance to abiotic stress. (Jalil & Ansari, 2019). Constant over-exposure or chronic exposure to NPs tends to suppress growth, which results into the disturbance in cellular metabolism. The heavy doses of metal-based NPs, e.g., silver, copper oxide, or aluminum oxide, have the potential to adversely affect root growth due to alteration of the cell division in the meristematic region and callose deposition to block nutrient flow. Another major area through which NP-induced stress acts is photosynthesis. Disruption of chlorophyll synthesis and rupture of the thylakoid membranes and also inhibition of key photosynthetic enzymes like Rubisco has been reported in the presence of various nanoparticles. This is because it leads to reduced photosynthetic efficiency, reduced number of energy assimilations and finally reduced productivity. (Xu et al., 2025). NPs toxicity is associated with water relationships and nutrient uptake as well. High concentrations of titanium dioxide nanoparticles can modify the expression of aquaporin leading to a reduction in the processes of transpiration and wilting symptoms. Moreover, NPs can increase or decrease ionic transport, in which physicochemical properties of NPs are determinants. These physiological imbalances are the reasons why it is essential to fine-tune the development of NPs in order to make them compatible with target crops, as well as conditions under which they are cultivated (Taylor et al., 2014).

Oxidative Stress and Antioxidant Responses

One of the main mechanisms by which nanoparticles affect plant health positively or negatively is their ability to alter the redox homeostasis. The high surface reactivity and redox activity of many metal-based NPs can potentially result in or form reactive oxygen species (ROS), like hydrogen peroxide (H_2O_2), superoxide anions (O^{2-}), and hydroxyl radicals (OH), either directly or by interacting with other cellular organelles, such as chloroplasts and mitochondria. (Li et al., 2016). At medium levels, ROS functions as signal molecules that lead to protective responses, i.e., an increase in antioxidant enzymes and secondary metabolites. Blood cells challenged with ROS fail to reduce the levels of ROS because of their overproduction, which overwhelms the detoxification system; this results in oxidative stress and the eventual destruction of proteins, lipids, DNA, and membranes. (Schieber & Chandel, 2014). The plants have developed an elaborate antioxidant defense mechanism consisting of both the non-enzymatic and enzymatic (e.g., superoxide dismutase SOD, catalase CAT, peroxidase, ascorbate peroxidase APX) components. All these systems are regulated in the bid to reduce the excessive ROS and achieve cellular homeostasis. When used in low concentrations, NPs such as ZnO and SiO2 have been reported to stimulate numerous actions of antioxidant enzymes and the property can overcome the impact of diverse environmental changes including drought and salinity. This implication means that nanoparticles can be introduced into plants as priming molecules that protect plants against even more severe stresses. (Ishtiaq et al., 2023). In contrast, strong concentrations or repeated doses of particular nanoparticles may overcome the antioxidant defense. This leads to lipid peroxidation, electrolyte leakage, and dysfunction of important metabolic enzymes. There is increased MDA release, a marker of damaged membranes, that is regularly high during nanoparticle oxidative stress. Besides that, impaired effect on electron transport chains due to NP interference may result in ATP depletion and cell death. The levels of oxidative damage are frequently nanoparticle-dependent and tightly associated with the solubility, rate of dissolution, and tendency to liberate metal ions (Sharma et al., 2019). Additional omics technologies, such as transcriptomics and proteomics, have also verified the up- or down-regulation of redox-associated genes in NP-obligated plants. Molecular Analyses of the genes responsible for glutathione metabolism, ROS detoxification, and redox signaling demonstrate an intricate, regulated response to exposure to nanoparticles. The results illustrate a possible combination of nanotechnology and molecular plant biology to tune the plant redox network to enhance stress tolerance. (Wang et al., 2025).

Toxicity Thresholds and Risk Factors

Another factor that is of utmost importance, but the hardest to integrate, is the definition of toxicity levels in the context of safe and effective nanoparticle-based products in agriculture. Various conditions, such as dosage, days of exposure, plant species, and environmental parameters, affect the level at which a nanoparticle changes its character to become harmful. The thresholds are normally narrow and can hugely differ depending on the type of NPs. To exemplify, although 10 and 50 mg/L of ZnO nanoparticles have been reported to increase root growth, it was found that root growth, biomass, and chlorophyll are reduced at concentrations above 100 mg/L in maize and wheat. (Korniyenko et al., 2024). The significance of aggregation behavior of the nanoparticles in toxicity is also very significant. NPs in aqueous solutions or a soil matrix will be agglomerated and this can alter their surface area reactivity and plant interactions. Polysaccharide-coated nanoparticles, polysaccharide-surfactant-coated nanoparticles, and polysaccharide-biomolecule-coated nanoparticles, will be biocompatible in suspension and less toxic since they have reduced ion dissolution and mitigation. On the contrary, free or poorly dispersed NPs have a tendency to cause or release metal ions at a very high rate especially in acidic or salty conditions. (Chen, 2018). The other decisive aspect is the exposure route. The behavior of nanoparticles on foliage might be different from what happens when nanoparticles are absorbed in the root, where they have to traverse the network of the rhizosphere and root structure. In seed-priming uses the exposure time is normally short, and beyond some point can repress or kill the embryo and end up in poor seedlings. In addition, the same compounds used at similar concentrations may have varying effects upon exposure based upon differences in tissue permeability, the density of stomata, and the thickness of any cuticle (Djanaguiraman et al., 2024). There are issues of cumulative exposure and ecological persistence. Repeated use of nanoparticles in fields possibly causes the storage of nanoparticles in plant tissues, soil, and groundwater water and it is not yet known what the long-term effects of this process may be. More specifically, metallic NPs that are not easily degraded could remain in the indicated ecosystems and cause harm to soil microbial flora, earthworms, and even other plant species via rhizospheric interactions. Any agricultural application of nanoparticles, therefore, has to be preliminarily assessed through ecotoxicological and field trials coupled with any modeling work needed, lest cumulative exposure introduces an ecologically unsafe amount to any such use (Wiesner et al., 2009).

Figure 3. Biocompatibility and phytotoxicity of nanoparticles in plants. At low doses, NPs enhance growth, nutrient uptake, and stress tolerance, while excessive exposure disrupts physiology, induces oxidative stress, and damages cellular structures. Antioxidant defenses can mitigate effects, but toxicity thresholds vary with nanoparticle type, dose, plant species, and environment, underscoring the need for safe application

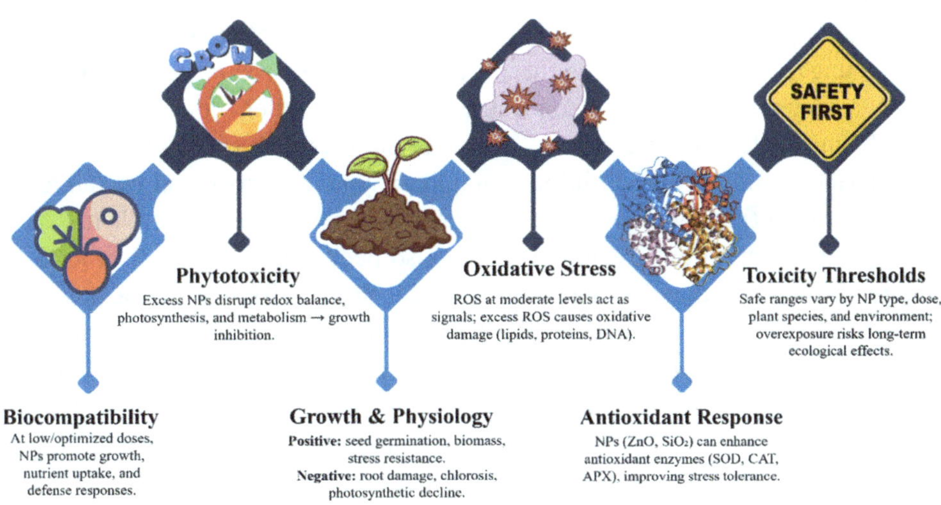

CASE STUDIES AND FUTURE PERSPECTIVES

Practical use of nanoparticles (NPs) in plant systems has been on an increasing trend over the last twenty years, especially as researchers have shifted away from proof-of-concept studies to actual trials. Practical applications have been investigated in case studies of several different crop species, cereals, legumes, fruits, and vegetables, where the use of both metallic and non-metallic nanoparticles has been shown to benefit the health of plants by different methods: increasing plant growth, increasing stress tolerance, activating plant defense mechanisms, or even inactivating pathogens, such as viruses. The evidence of the level of efficacy and safety is, however, relative to the nanoparticle formulation, the species, and the stage of development of the plants and environmental conditions. Some example case studies that detail the use of nanoparticles in crop systems will be discussed, in addition to future directions on how this can be safely applied in sustainable farming.

Examples From Metallic and Non-Metallic NPs

Silver nanoparticles (AgNPs) have been the most comprehensively studied nanoparticles, both because of their antimicrobial and antiviral activity (Luceri et al., 2023; Salleh et al., 2020). Silver nanoparticles, especially, have attracted much attention due to their worthwhile biocidal actions. AgNPs show antiviral effects since they act to inhibit the Tomato Mosaic Virus (ToMV) in tomato (*Solanum lycopersicum*) by suppressing viral replication and activating systemic acquired resistance (SAR) pathways. Plants treated with compounds had a lower symptom severity and physiological parameters than untreated controls. Similarly, in cowpea (*Vigna unguiculata*), foliar-applied AgNPs showed a big decrease in Cowpea Mosaic Virus (CPMV) burden, probably by interfering with viral coat protein and RNA synthesis (Naumenko et al., 2023). Zinc oxide nanoparticles, however, have been found to play a bifaceted role of acting as sources of micronutrients as well as disease suppressants. In wheat (*Triticum aestivum*), ZnO NPs increased chlorophyll concentration, antioxidant enzymes, and yield under drought stress, besides reducing the leaf rust incidence. The role of these particles is believed to be by stimulating the availability of zinc (which is critical in protein and enzyme activity), and by inducing secondary defense mechanisms, such that pathogen establishment may be curtailed. Nonetheless, it is important to be cautious, because high doses of ZnO NPs have also been shown to result in ROS build-up and oxidative damage in spinach and maize (Kang et al., 2024; Singh, 2019). Nanoparticles of iron oxides (Fe_3O_4) are attractive due to their magnetic properties, which can be used in targeting and imaging. In rice (*Oryza sativa*), citric acid-coated magnetite nanoparticles promoted germination and iron nutrition as well as tolerance to salinity and iron deficiency. It is of particular note that these nanoparticles were able to translocate quite effectively both through xylem and phloem with minimal tissue damage, and this is a considerable testament to biocompatible surface modification (Tang et al., 2025). Nanomaterials of non-metallic origin also find application, particularly, chitosan, silica, and carbon nanomaterials. Natural polysaccharides of chitin origin provide chitosan nanoparticles, which have received a lot of attention due to their low toxicity, biodegradation capabilities, and immune-modulating capabilities. In tobacco, the NPs chitosan increased tolerance to the Tobacco Mosaic Virus (TMV) by expressing PR (pathogenesis-related) genes and the formation of callose deposits at the points of infection. It is thought that the effects of these are due to both direct antiviral action and training of the plant's innate immune system (Ahmad et al., 2022). Silicon dioxide (SiO_2) nanoparticles have been shown to enhance structural and drought-resistant properties as well as protection against viruses. In cucumber (*Cucumis sativus*), foliar application of SiO_2 NPs resulted in better stomatal conductance, chlorophyll retention, and less CMV getting afflicted. Carbon nanomaterials, including fullerenes, graphene oxide, and carbon dots, are becoming vehicles in the delivery of genes and biosensing. Carbon dots made using citric acid have demonstrated the capability of penetrating through plant cells and acting as nanocarriers to carry dsRNA, with possible use in gene silencing and virus inhibition (Devi et al., 2025; Mostafavi & Zare, 2022). As demonstrated in above case studies, nanoparticle efficacy is very sensitive to its design attributes, such as particle size, coating materials, charge, and release kinetics. Further, the plant species and the nature of the pathogen or stressor modulate the response, and it becomes necessary to customize the nanoparticle systems to a particular application.

Recommendations for Safe Application

Since there is a rising interest in the application of nanoparticles in agriculture, there is a need to develop standardized guidelines under which the nanoparticles can be used safely and effectively. This encompasses suggestions on all the phases of the nanoparticle life cycle; that is, manufacture and characterization through field use and monitoring.

1. **Rational Design and Characterization:** The development of nanoparticles must be based on the understanding of the desirable performance, nutrient delivery, and stress alleviation as well as pathogen inhibition. Particle size, zeta potential, shape, dissolution rate and solubility are the most important features, which should be well characterized. It is desirable to use green synthesis pathways flighted with the use of biological materials (e.g., plant extracts, microbes) that will help minimize the release of toxic byproducts into the environment and improve environmental friendliness (Buchman et al., 2019).
2. **Optimization of Concentration and Dose Response Testing:** It is necessary to conduct a comprehensive dose-response study to determine the lowest concentration that is effective, and the dose that is phytotoxic before applying the chemical to the field. The number of crop species, soil conditions, and weather conditions that should be considered in such studies should be varied to introduce some of the real-life characterization. They are supposed to be determined by short-duration physiological indicators (e.g., chlorophyll, photosynthesis rate) and long-term consequences (e.g., yield, reproductive success) (Iavicoli et al., 2018; Summers et al., 2011).
3. **Safe Delivery Systems:** Due to the plants used and distribution requirements, mode of delivery, namely priming the seed, foliar, root drench, or nano-encapsulation, should be chosen. The incorporation into liposomes, appropriate choice of polymeric carriers, or encapsulation within hydrogel can be used to increase stability and release and decrease undesirable interactions with untargeted tissues (Gill et al., 2007).
4. **Penetration with the Crop management Practices:** Nanoparticles ought to be integrated with integrated crop management (ICM) and entail appropriate irrigation, nutrient control and pest control. Such synergistic plans can provide the greatest benefits and reduce the risks of damaging the environment and the development of health risks. An increase in efficacy can also be achieved with the help of a combination with biofertilizers or biocontrol agents and less dependence on synthetic agrochemicals (Khandelwal et al., 2016).
5. **Assessment and regulation of environmental risk:** Monitoring of nanoparticles after use is among the primary issues in an attempt to determine the presence of accumulated nanoparticles in soil and

condensed waterflow and plant runs. Integrated principles in the examination of soil microbiome, leachate and particulate NP decomposition sequencing must be embraced. There is the need to come up with regulatory provisions with special safety assessment provisions as far as nano is concerned and especially the foods crops and edible tissue. The need of transparency in labeling and communicating the use of the nanoparticles must become a requirement in providing safety upon the consumer and being a good steward of the environment. (Singh & Kumar, 2023).

RESEARCH GAPS AND FUTURE DIRECTIONS

Even though the research on nanoparticle-plant interactions has advanced rapidly, some substantial gaps in the knowledge still exist. Firstly, it is not clear what nanoparticles long-term destiny could be in plant tissue and soil ecospheres. Short-term physiological responses have been studied extensively, but transgenerational effects of nanoparticle exposure on seed viability and germination, as well as epigenetic changes, are not well understood. Secondly, there is a need for more mechanistic knowledge at the molecular level. Although transcriptomic and proteomic studies provided data on thousands of differently expressed genes in response to NP exposure, signaling pathways and gene networks maintaining nanoparticle sensing, trafficking, and regulation of responses remain devoid of an in-depth analysis. Functional genomic studies based on CRISPR-Cas9 knockouts and RNAi silencing can be used to determine critical regulators of NP uptake and processing. Thirdly, one should investigate in greater detail the impact of nanoparticles on plant microbiomes, in the rhizosphere, and on the phyllosphere. It has been shown that nanoparticles increase the activity of good microorganisms or reduce that of bad ones but other studies suggest that the restriction of the microbial diversity and nitrogen cycling is interfered and derailed. A joint use of metagenomics, metabolomics, and spatial mapping can give more information about the complexity of these interactions in the new field in systems biology. Fourthly, it is urgently required to possess field-level validations. Majority of the study on nanoparticles is usually done under controlled conditions (greenhouse or laboratory). When scaling these technologies, they should be tested in various environmental conditions- different temperature, rain, soil pH levels, and pressure of pathogens. Participatory research and multisite work that will involve the farmer will be critical in ensuring that it becomes relevant and finally adopted in the field. Finally, the need to develop so-called smart nanoparticles engineered systems capable of sensing environmental signals and capable of changing properties in response to stimuli, as well as, the release of cargo in a controlled fashion, is on the rise. Cases in point are pH-sensitive nanoparticles in order to release nutrients in acidic soil and light-mediated antiviral nanoparticles. The future of these systems would be the new generation which would be granted the interdisciplinary collaboration of the plant scientists, nanotechnologists, material chemists and environmental toxicologists.

CONCLUSION

Nanotechnology has transformative potential in agriculture, especially when nanoparticles (NP) can interact with plants. It has described how NPs are acquired through foliar, root, and seed routes, how they traverse through apoplastic and symplastic, and how they are transported systemically through the xylem and phloem. These nanoscale materials can improve nutrient transport, trigger plant defense mechanisms, and suppress virus replication. When used in the right quantities and at the right density, NPs are known to cause a significant improvement in plant physiological activities, stress mitigation, and production of various kinds of crops. Pros and cons of nanoparticles should be addressed accordingly. High doses are able to tip the redox balance, inhibit photosynthesis, and cause oxidative damage. Response to NPs also varies greatly based on particle size and coating, as well as concentrations and the specific plant species. Much is, therefore, necessary to know about toxicity thresholds, to optimize formulations, and adjust applications to environmental situations to ensure safe application. In the future, there is a need to validate it in the field on a long-term basis, to determine environmental fate, and examination of plant microbiome interactions. New areas in precision crop management can be unlocked by the development of intelligent, stimuli-responsive nanomaterials and the convergence of these materials with genomic tools. Technologies built on nanoparticles can have very good potential, with proper safeguards, to enhance sustainable and resilient agriculture in the context of mounting environmental and biological pressures.

REFERENCES

Abasi, F. (2024). Nanomaterials for plant resilience: A green approach to combating biotic stress. *BioNanoScience, 14*(4), 3934–3949.

Abbasi, A. (2025). Restoring Wheat Productivity and Nutrient Balance Under Cadmium Stress Through Reducing Toxicity, Metal Uptake, and Oxidative Damage Using Selenium Nanoparticles. *Journal of Trace Elements in Medicine and Biology, 127644*. https://doi.org/https://doi.org/10.1016/j.jtemb.2025.127644

Ahmad, A. (2022). Metallic and Non-metallic Nanoparticles against Viruses. In *Nanomaterials in the Battle Against Pathogens and Disease Vectors* (pp. 205–228). CRC Press.

Ali, S. (2021). Uptake, translocation, and consequences of nanomaterials on plant growth and stress adaptation. *Journal of Nanomaterials, 2021*(1), 6677616.

Ali, S. (2023). Nano-agrochemicals as substitutes for pesticides: Prospects and risks. *Plants, 13*(1), 109. PMID: 38202417

Anas, M. (2024). Manganese oxide nanoparticles elicit cadmium tolerance in wheat seedlings by ionomic and phenomic adjustment, regulation of AsA–GSH pathway, cellular thickness recovery, and antioxidant modulation []. *Environmental Science: Nano, 11*(11), 4588-4605. DOI: 10.1039/D4EN00490F

Avellan, A. (2019). Nanoparticle size and coating chemistry control foliar uptake pathways, translocation, and leaf-to-rhizosphere transport in wheat. *ACS Nano, 13*(5), 5291–5305. PMID: 31074967

Avellan, A. (2021). Critical review: Role of inorganic nanoparticle properties on their foliar uptake and in planta translocation. *Environmental Science & Technology, 55*(20), 13417–13431. PMID: 33988374

Avellan, A. (2022). Biological barriers, processes, and transformations at the soil–plant–atmosphere interfaces driving the uptake, translocation, and bioavailability of inorganic nanoparticles to plants. In *Inorganic nanopesticides and nanofertilizers: A view from the mechanisms of action to field applications* (pp. 123–152). Springer.

Balog, S. (2024). Does the surface charge of the nanoparticles drive nanoparticle–cell membrane interactions? *Current Opinion in Biotechnology, 87*, 103128. PMID: 38581743

Bhati, V. S. (2025). Foliar uptake and translocation of nanoparticles in the plants: A review. *Journal of Plant Nutrition*, 1–38.

Bouhadi, M. (2025). Nanoparticles for sustainable agriculture: Assessment of benefits and risks. *Agronomy (Basel), 15*(5), 1131.

Buchman, J. T. (2019). Understanding nanoparticle toxicity mechanisms to inform redesign strategies to reduce environmental impact. *Accounts of Chemical Research, 52*(6), 1632–1642. PMID: 31181913

Cai, L. (2019). Preventing viral disease by ZnONPs through directly deactivating TMV and activating plant immunity in Nicotiana benthamiana. *Environmental Science. Nano, 6*(12), 3653–3669.

Chen, H. (2018). Metal based nanoparticles in agricultural system: Behavior, transport, and interaction with plants. *Chemical Speciation and Bioavailability, 30*(1), 123–134.

Cojocaru, F.-D. (2020). Nanomaterials designed for antiviral drug delivery transport across biological barriers. *Pharmaceutics*, *12*(2), 171. PMID: 32085535

Dev, A. (2018). Nanomaterial toxicity for plants. *Environmental Chemistry Letters*, *16*(1), 85–100.

Devi, H. (2025). Genetic Modifications and Gene Delivery Using Nanobiotechnology for Improving Yield and Stress Tolerance. In *Advances in Seed Quality Evaluation and Improvement* (pp. 309–330). Springer.

Dimkpa, C. O. (2018). Soil properties influence the response of terrestrial plants to metallic nanoparticles exposure. *Current Opinion in Environmental Science & Health*, *6*, 1–8.

Djanaguiraman, M. (2024). Uptake, translocation, toxicity, and impact of nanoparticles on plant physiological processes. *Plants*, *13*(22), 3137. PMID: 39599346

Duan, X., & Li, Y. (2013). Physicochemical characteristics of nanoparticles affect circulation, biodistribution, cellular internalization, and trafficking. *Small*, *9*(9-10), 1521–1532. PMID: 23019091

Elahi, M. (2025). Nitrogen Nano-fertilizers Enhance Wheat Productivity, Improve Nitrogen Use Efficiency, and Mitigate Soil Nitrogen Losses While Maintaining Grain Quality Compared to Conventional Fertilizers. *BioNanoScience*, *15*(3), 477. DOI: 10.1007/s12668-025-02065-4

Faisal, M. (2024). Metallic allies in drought resilience: Unveiling the influence of silver and zinc oxide nanoparticles on enhancing tomato (Solanum lycopersicum) resistance through oxidative stress regulation. *Plant Physiology and Biochemistry*, *212*, 108722. PMID: 38761543

Falak, A. (2025). Efficacy of ascorbic acid coated quantum dots in alleviating lead-induced oxidative damage and enhancing growth parameters in rice (Oryza sativa L.) for sustainable cultivation. *Journal of Trace Elements in Medicine and Biology*, *88*, 127603. DOI: 10.1016/j.jtemb.2025.127603 PMID: 39847985

Fatima, A. (2020). Interaction between copperoxide nanoparticles and plants: Uptake, accumulation and phytotoxicity. In *Sustainable Agriculture Reviews 41: Nanotechnology for Plant Growth and Development* (pp. 143-161). Springer.

Fornaguera, C. (2020). *Unraveling polymeric nanoparticles cell uptake pathways: two decades working to understand nanoparticles journey to improve gene therapy. Cell Biology and Translational Medicine* (Vol. 9). Stem Cell-Based Therapeutic Approaches in Disease.

Fu, L. (2020). Nanotechnology as a new sustainable approach for controlling crop diseases and increasing agricultural production. *Journal of Experimental Botany*, *71*(2), 507–519. PMID: 31270541

Ghorbanpour, M., & Shahid, M. A. (2022). *Nano-enabled Agrochemicals in Agriculture*. Academic Press.

Gill, S. (2007). Nanoparticles: Characteristics, mechanisms of action, and toxicity in pulmonary drug delivery—a review. *Journal of Biomedical Nanotechnology*, *3*(2), 107–119.

Goh, N. S. (2022). *Advancing Nanomaterial Usage in Agriculture: Engineering Nanomaterials towards Nanomaterial-Mediated Plant Genetic Engineering*. University of California.

González-Grandío, E. (2021). Carbon nanotube biocompatibility in plants is determined by their surface chemistry. *Journal of Nanobiotechnology*, *19*(1), 431. PMID: 34930290

Gruner, H. N., & McManus, M. T. (2021). Examining the evidence for extracellular RNA function in mammals. *Nature Reviews. Genetics*, *22*(7), 448–458. PMID: 33824487

He, Z. (2020). Cold atmospheric plasma stimulates clathrin-dependent endocytosis to repair oxidised membrane and enhance uptake of nanomaterial in glioblastoma multiforme cells. *Scientific Reports*, *10*(1), 6985. PMID: 32332819

Hong, J. (2021). Foliar application of nanoparticles: Mechanisms of absorption, transfer, and multiple impacts. *Environmental Science. Nano*, *8*(5), 1196–1210.

Huang, B. (2018). Advances in targeted pesticides with environmentally responsive controlled release by nanotechnology. *Nanomaterials (Basel, Switzerland)*, *8*(2), 102. PMID: 29439498

Iavicoli, I. (2018). Nanoparticle exposure and hormetic dose–responses: An update. *International Journal of Molecular Sciences*, *19*(3), 805. PMID: 29534471

Ishtiaq, M. (2023). Seed priming with the selenium nanoparticles maintains the redox status in the water stressed tomato plants by modulating the antioxidant defense enzymes. *Plants*, *12*(7), 1556. PMID: 37050182

Jalil, S. U., & Ansari, M. I. (2019). Nanoparticles and abiotic stress tolerance in plants: synthesis, action, and signaling mechanisms. *Plant signaling molecules*, 549-561.

Jampílek, J., & Kráľová, K. (2017). Nanomaterials for delivery of nutrients and growth-promoting compounds to plants. *Nanotechnology: An agricultural paradigm*, 177-226.

Kamal, A., Yang, J., Batool, M., Ara, U., Khattak, W. A., Touhami, D., Aljowaie, R. M., & Elshikh, M. S. (2025). Green synthesis of zinc oxide nanoparticles, characterization and its applications in inducing disease resistance against spot blotch disease in wheat by enhancing its physiological and biochemical parameters. *Journal of Crop Health*, *77*(1), 5.

Kang, M. (2024). A critical review on the toxicity regulation and ecological risks of zinc oxide nanoparticles to plants. *Environmental Science. Nano*, *11*(1), 14–35.

Khan, A. (2025). Eco-friendly synthesis of copper oxide-silver bimetallic nanoparticles using Withania coagulans: Characterization and biomedical potential through antioxidant, antibacterial, and catalytic activities. *Inorganic Chemistry Communications*, *115131*. https://doi.org/https://doi.org/10.1016/j.inoche.2025.115131

Khan, I. (2022). Nanoparticle's uptake and translocation mechanisms in plants via seed priming, foliar treatment, and root exposure: A review. *Environmental Science and Pollution Research International*, *29*(60), 89823–89833. PMID: 36344893

Khan, M. A. U. (2022). Transport mechanism from Quantum dots to plant systems. In *Quantum Dots for Plant Systems* (pp. 103–136). Springer.

Khandelwal, N. (2016). Budding trends in integrated pest management using advanced micro-and nanomaterials: Challenges and perspectives. *Journal of Environmental Management*, *184*, 157–169. PMID: 27697374

Korniyenko, V. I. (2024). Nanoparticles: Definition, toxicity, approaches to regulation, migration routes in the environment. *Publishing House "Baltija Publishing"*.

Kousar, S., Ahmad, D., Khattak, W. A., Anwar, M. M., Majeed, M., Shah, G. M., & Muhammad, M. (2025). Environmental resilience: Navigation of lead and beneficial elements in soil. In *Beneficial Elements for Remediation of Heavy Metals in Polluted Soil* (pp. 221–240).

Kulikova, N. A. (2020). Silver nanoparticles stabilized by humic substances adversely affect wheat plants and soil. *Journal of Nanoparticle Research*, *22*(5), 100.

Kumar, P. (2024). Biocompatibility, toxicity concerns, environmental and safety considerations, and legal aspects of functionalized magnetic nanoparticles. *Functionalized magnetic nanoparticles for theranostic applications*, 533-558.

Li, M. (2016). Brassinosteroid ameliorates zinc oxide nanoparticles-induced oxidative stress by improving antioxidant potential and redox homeostasis in tomato seedling. *Frontiers in Plant Science*, *7*, 615. PMID: 27242821

Li, N. (2021). Advances in controlled-release pesticide formulations with improved efficacy and targetability. *Journal of Agricultural and Food Chemistry*, *69*(43), 12579–12597. PMID: 34672558

Liu, Y. (2022). Stability and interaction of biochar and iron mineral nanoparticles: Effect of pH, ionic strength, and dissolved organic matter. *Biochar*, *4*(1), 47.

Luceri, A. (2023). Silver nanoparticles: Review of antiviral properties, mechanism of action and applications. *Microorganisms*, *11*(3), 629. PMID: 36985203

Ma, Y. (2017). Xylem and phloem based transport of CeO2 nanoparticles in hydroponic cucumber plants. *Environmental Science & Technology*, *51*(9), 5215–5221. PMID: 28383248

Mosquera, J. (2018). Cellular uptake of nanoparticles versus small molecules: A matter of size. *Accounts of Chemical Research*, *51*(9), 2305–2313. PMID: 30156826

Mostafavi, E., & Zare, H. (2022). Carbon-based nanomaterials in gene therapy. *OpenNano*, *7*, 100062.

Nair, R. (2010). Nanoparticulate material delivery to plants. *Plant Science*, *179*(3), 154–163.

Naumenko, K. (2023). Antiviral activity of silver nanoparticles against the influenza A virus. *Journal of Virus Eradication*, *9*(2), 100330. PMID: 37416089

Oberdörster, G. (2007). Concepts of nanoparticle dose metric and response metric. *Environmental Health Perspectives*, *115*(6), A290–A290. PMID: 17589571

Rajesh, P. (2025). Nanocarriers for Targeted Delivering and Translocation of Chemical Cargoes in Plant Cells. In *Nanocarriers in Plant Science and Agriculture* (pp. 125–164). IGI Global Scientific Publishing.

Rasheed, A. (2022). The role of nanoparticles in plant biochemical, physiological, and molecular responses under drought stress: A review. *Frontiers in Plant Science*, *13*, 976179. PMID: 36507430

Salleh, A. (2020). The potential of silver nanoparticles for antiviral and antibacterial applications: A mechanism of action. *Nanomaterials (Basel, Switzerland)*, *10*(8), 1566. PMID: 32784939

Sanità, G. (2020). Nanoparticle surface functionalization: How to improve biocompatibility and cellular internalization. *Frontiers in Molecular Biosciences*, *7*, 587012. PMID: 33324678

Schieber, M., & Chandel, N. S. (2014). ROS function in redox signaling and oxidative stress. *Current Biology*, *24*(10), R453–R462. PMID: 24845678

Sembada, A. A., & Lenggoro, I. W. (2024). Transport of nanoparticles into plants and their detection methods. *Nanomaterials (Basel, Switzerland)*, *14*(2), 131. PMID: 38251096

Semenova, N. A. (2024). Fertilizers based on nanoparticles as sources of macro-and microelements for plant crop growth: A review. *Agronomy (Basel)*, *14*(8), 1646.

Servin, A. D., & White, J. C. (2016). Nanotechnology in agriculture: Next steps for understanding engineered nanoparticle exposure and risk. *NanoImpact*, *1*, 9–12.

Sharma, S. (2019). Effect of nanoparticles on oxidative damage and antioxidant defense system in plants. *Molecular plant abiotic stress: Biology and biotechnology*, 315-333.

Singh, D., & Kumar, A. (2019). Understanding the effect of the interaction of nanoparticles with roots on the uptake in plants. In *Environmental Nanotechnology* (Vol. 3, pp. 277–304). Springer.

Singh, R., & Kumar, S. (2023). Regulatory and safety concerns regarding the use of active nanomaterials in food industry. In *Nanotechnology advancement in agro-food industry* (pp. 269–306). Springer.

Singh, S. (2019). Zinc oxide nanoparticles impacts: Cytotoxicity, genotoxicity, developmental toxicity, and neurotoxicity. *Toxicology Mechanisms and Methods*, *29*(4), 300–311. PMID: 30489211

Soukar, J. (2025). Organelle-targeting nanoparticles. *Advancement of Science*, *12*(7), 2411720. PMID: 39806939

Su, Y. (2019). Delivery, uptake, fate, and transport of engineered nanoparticles in plants: A critical review and data analysis. *Environmental Science. Nano*, *6*(8), 2311–2331.

Summers, H. D. (2011). Statistical analysis of nanoparticle dosing in a dynamic cellular system. *Nature Nanotechnology*, *6*(3), 170–174. PMID: 21258333

Tang, Y. (2025). Enhancing maize stress tolerance with nickel ferrite nanoparticles: A sustainable approach to combat abiotic stresses. *Environmental Science. Nano*, *12*(1), 302–314.

Taylor, A. F. (2014). Investigating the toxicity, uptake, nanoparticle formation and genetic response of plants to gold. *PLoS One*, *9*(4), e93793. PMID: 24736522

Ullah, Z., Gul, F., Iqbal, J., Abbasi, B. A., Kanwal, S., Chalgham, W., & Mahmood, T. (2023). Biogenic synthesis of multifunctional silver oxide nanoparticles (Ag2ONPs) using Parieteria alsinaefolia delile aqueous extract and assessment of their diverse biological applications. *Microorganisms*, *11*(4), 1069. PMID: 37110492

Ullah, Z., Iqbal, J., Abbasi, B. A., Gul, F., Ali, S., Kanwal, S., & Mahmood, T. (2025). Eco-friendly synthesis of iron oxide nanoparticles using parietaria alsinifolia extracts and evaluation of biological applications. *Applied Biochemistry and Biotechnology*, *197*(4), 2328–2355. PMID: 39752127

Vargas-Hernandez, M. (2020). Nanoparticles as potential antivirals in agriculture. *Agriculture, 10*(10), 444.

Varma, S. (2022). Cellular uptake pathways of nanoparticles: Process of endocytosis and factors affecting their fate. *Current Pharmaceutical Biotechnology, 23*(5), 679–706. PMID: 34264182

Verma, K. K. (2024). Synergistic interactions of nanoparticles and plant growth promoting rhizobacteria enhancing soil-plant systems: A multigenerational perspective. *Frontiers in Plant Science, 15*, 1376214. PMID: 38742215

Vurro, M. (2019). Safe nanotechnologies for increasing the effectiveness of environmentally friendly natural agrochemicals. *Pest Management Science, 75*(9), 2403–2412. PMID: 30672106

Wahid, I. (2022). Biosynthesized gold nanoparticles maintained nitrogen metabolism, nitric oxide synthesis, ions balance, and stabilizes the defense systems to improve salt stress tolerance in wheat. *Chemosphere, 287*, 132142. PMID: 34826894

Wang, Q. (2023). Interplay of metal-based nanoparticles with plant rhizosphere microenvironment: Implications for nanosafety and nano-enabled sustainable agriculture. *Environmental Science. Nano, 10*(2), 372–392.

Wang, W., & Mattoussi, H. (2020). Engineering the Bio–Nano Interface Using a Multifunctional Coordinating Polymer Coating. *Accounts of Chemical Research, 53*(6), 1124–1138. PMID: 32427464

Wang, X. (2023). Nanoparticles in plants: Uptake, transport and physiological activity in leaf and root. *Materials (Basel), 16*(8), 3097. PMID: 37109933

Wang, Y. (2025). Multi-omics approaches to decipher the interactions of nanoparticles and biological systems. *Nature Reviews Bioengineering, 3*(4), 333–348.

Wang, Z. (2012). Xylem-and phloem-based transport of CuO nanoparticles in maize (Zea mays L.). *Environmental Science & Technology, 46*(8), 4434–4441. PMID: 22435775

Wiesner, M. R. (2009). Decreasing uncertainties in assessing environmental exposure, risk, and ecological implications of nanomaterials. ACS Publications.

Wu, H., & Li, Z. (2022). Nano-enabled agriculture: How do nanoparticles cross barriers in plants? *Plant Communications, 3*(6). PMID: 35689377

Xin, X. (2020). Nano-enabled agriculture: From nanoparticles to smart nanodelivery systems. *Environmental Chemistry, 17*(6), 413–425.

Xu, L. (2025). Chronic toxic effects of silver nanoparticles on Chlamydomonas reinhardtii: Photosynthesis, antioxidation and internalization. *Journal of Environmental Management, 377*, 124428.

Zhang, H. (2024). Multifaceted impacts of nanoparticles on plant nutrient absorption and soil microbial communities. *Frontiers in Plant Science, 15*, 1497006. PMID: 39606675

Zhang, J. (2018). The effects and the potential mechanism of environmental transformation of metal nanoparticles on their toxicity in organisms. *Environmental Science. Nano, 5*(11), 2482–2499.

Chapter 4
Mechanisms of Nanoparticle-Induced Antiviral Activity in Plants

Waseem Ahmed Khattak
https://orcid.org/0009-0004-1356-2579
Quaid-i-Azam University, Islamabad, Pakistan

Tauqeer Ahmed Qadri
COMSATS University Islamabad, Pakistan

Jamal Abdul Nasir
Kathleen Lonsdale Materials Chemistry, University College London, UK

Banzeer Ahsan Abbasi
https://orcid.org/0000-0003-3036-4481
Rawalpindi Women University, Pakistan

ABSTRACT

Plant viral diseases cause severe yield and quality losses worldwide, and conventional control methods often fail due to viral evolution and environmental limits. Nanoparticles (NPs) offer novel antiviral solutions through unique properties enabling direct actions—such as blocking viral entry, disrupting capsids, and inducing ROS—and indirect actions by activating plant immunity and antioxidant defenses. Though promising, challenges include phytotoxicity, environmental risks, and scalability. Green-synthesized, biodegradable NPs integrated with precision agriculture hold potential for sustainable viral disease management.

INTRODUCTION

Plant viral diseases are a significant and increasing burden in world food production, which has highly diminished the quantity and quality of crops and compromised food security (Strange & Scott, 2005). Among the most troublesome pathogens is the virus as they cannot be treated, particularly not

DOI: 10.4018/979-8-3373-4913-8.ch004

with accompanying curative measures, given that, unlike bacteria or even fungi, viruses are obligate intracellular destages and are known to seize host cell apparatus to eventually reproduce; once a host is infected, it is already too late. In most cases, resistant cultivars, cultural production, and control of vectors are key elements in the management of many plant viruses, which can be restricted by a rapid viral evolution, resistance lifestyle decay, strain-level-based vectors pressure, or environmental bottlenecks (Abdelkader et al., 2025; Liu et al., 2024). Nanotechnology has been promising in recent years as an addition or alternative measure in the context of managing the plant viruses. Nanoparticles (NPs) are materials that are usually 1 to 100 nm in size and have distinct physicochemical properties, such as a large surface-area-to-volume ratio, tunable morphology (size, shape), controllable surface charge, and functionalization or coating (Anas et al., 2024a), rendering them formidable interactions with biological systems when compared with bulk materials. Researchers have indicated that nanoparticles can dampen viral infectivity on various pathways by utilizing various types of nanoparticle metallic nanoparticles silver, copper), metal oxides, biogenic nanoparticles, etc. (Falak et al., 2025; Khan et al., 2025). One of such examples is that AgNPs made with the help of pomegranate peel extract prevented tobacco mosaic virus (TMV) entry in the tomatoes, and postponed viral growth, abated viral concentration in tomatoes and up-regulated pathogenism-related (PR) genes and antioxidant enzymes (Al-Askar et al., 2023). Similarly, copper oxide biosynthesized using orange peel extract nanoparticles (CuO NPs) directly attracted the moment of TMV virions in vitro and induced increased expression of the genes associated with resistance (PR1, PR2, ERF1 etc.) in *Nicotiana benthamiana*, which reduced the accumulation of viruses in vivo (Liu et al., 2024). Nanoparticles have an antiviral effect whose exact mechanisms are not fully comprehended even though the evidence of this action is increasing in number. Other studies identify direct antiviral actions like any one of, yet not limited to, virion disruption or binding to the viral surface proteins; others describe indirect, plant-mediated reactions, such as stimulation of systemic acquired resistance (SAR), regulation of phytohormone signaling (salicylic acid, jasmonic acid, ethylene), induction of antioxidant enzymes, or increase of reactive oxygen species (ROS) to levels lethal to viral components whilst still permissible to the host NPs workability depends on numerous contextual factors: the metal (oxide, biogenic etc.) that constitutes the nanoparticle, the size, form, coating, concentration, timing of use (pre-infection or post infection), the type of plant species, the type of virus (RNA or DNA virus; enveloped and non-enveloped), and environmental conditions, such as light, temperature, humidity (Abdelkader et al., 2025; Luceri et al., 2023). They also include safety and sustainability. This phytotoxic effect of nanoparticles may be very strong at high concentrations, they may be accumulated in vegetable tissues or the soil, exposing non-target organisms to their toxicity, or they have some unknown ecological hazards (Ali and Rehman, 2023; Liu et al., 2024). As such, although nanoparticles promise a lot in regard to antiviral coasting in plants, a mechanistic comprehension of the mechanism would guarantee efficient application: the application would be more successful in boosting benefits than reducing harm, and in field applications and not only in control laboratory or greenhouse settings. Due to the numerous antiviral pathways, the differences in efficiency with nanoparticle and environmental conditions, and safety results, it is opportune and a warrant to synthesize the existing knowledge into a stock frame (Chakravarty & Vora, 2021). This chapter, the aim is to compile the knowledge regarding the actions of nanoparticles against plant viruses and what control this activity and what are the gaps in knowledge. This mechanistic understanding will assist in rational biological plan of attack of nanoparticle based strategies that are biologically, security, and proportional in food arena (Francis et al., 2024).

SCOPE AND OBJECTIVES

The extent of this chapter is to bring out a broad but critical review of the mechanisms through which nanoparticles can cause antiviral activity in plants not only directly acting on viruses but also indirectly mediated through the host. It will look at various emotional classes of the nanoparticles, their physicochemical properties, environmental and use parameters and what is most significant; safety issues. To determine the categories of nanoparticles developed in antiviral plant research (e.g. metallic nanoparticles, such as silver, copper), metal oxides, biogenic nanoparticles, and non-metallic NPs) and to analyze how the physicochemical properties of these types (size, shape, surface charge, functionalization/coating) can affect the antiviral outcome (Luceri et al., 2023; Ali and Rehman, 2023). The evidence on direct antiviral action, including the disruption of viruses, the inhibition of attaching or entering host cells, the introduction of viral replication in the genome, and the inhibition of viral propagation in plant tissues, needs to be formed and synthesized (Liu et al., 2024; Al-Askar et al., 2023). In order to measure indirect and plant-mediated processes, such as activation of plant immune pathways (e.g. systemic acquired resistance, salicylic acid, jasmonic acid, ethylene signalling), augmentation of antioxidant defenses (e.g. superoxide dismutase, catalase, peroxidases), activation of defense-related transcription (PR genes) and resolution of reactive oxygen species (Ali et al., 2021; Liu et al., 2024). So as to evaluate the effect of application parameters and contextual variables (dose, time in relation to infection, soil application, foliar spray, or seed treatment, etc.) and host plant society and a certain virus type, and the environment (light, heat, humidity, and so on) on both mechanism and effect (Abdelkader et al., 2025; Luceri et al., 2023). To investigate safety, toxicity, and environmental impacts of the use of nanoparticles as antivirals in plants: possible phytotoxicity, plant tissue and soil concentration, Found on non-target organisms off-target, persistence, and trade-offs between antiviral activity and risk (Ali and Rehman, 2023; Liu et al., 2024). To gain an understanding of areas of knowledge deficiency and allusiveness to future research directions, including molecular-level mechanistic research (transcriptomics, proteomics, imaging), normalization of experimental protocols to allow inter study comparability, long-term field experiments, and recommendations on safe and scalable application to agricultural systems to unforeseen circumstances (Luceri et al., 2023; Abdelkader et al., 2025).

OVERVIEW OF PLANT–VIRUS INTERACTIONS

Plants and viruses are dynamic and the interactions are two stage processes representing unbroken molecular conversation between an invading virus and its host (Wang, 2015). It is first critical to know the dialogue and then look at the way nanoparticles distort it. Reviews of existing tendencies underline that to be successful a viral infection is tailored by (1) entry or delivery of viral genetic material into susceptible cells (most frequently through insect transmission), (2) disassembly or uncoating of viral genome organization and viral susceptibility of cells to infection, (3) translation of viral proteins and formation of replication complex, (4) movement of infection through cell-to-cell contacts by using plasmodesmata onto migration proteins on the virus and (5) long-range, systemic mobility through the vasculature to these phases are in-extricable linked to host metabolism: viruses rearrange host transcriptional, translational as well as metabolic systems to need nucleotides, ATP, membrane surfaces as well as protein production (Jiang et al., 2024). Synthetic reviews covering these stages which include useful schematic diagrams of the infection cycle and methods of transmission can be found in Jiang et al. (2024) and the recent frontiers review on plant-virus contact (Mandabi et al., 2024); a figure that might be adapted to your chapter could be found in those reviews providing figure panels of entry, transmission, and systemic spread very explicitly (Jiang et al., 2024; Mandabi et al., 2024).

Viral Infection Cycle in Plants

Plant viral infection cycle can be divided into a series of biologically and time separable events that jointly define pathogenic outcome (Jones, 2014). The initial step is transmission: viruses in the vegetative state are most frequently introduced by biotic vectors (aphids, whiteflies, thrips, leafhoppers) in any of multiple types of transmission (non-persistent, semi-persistent, persistent non-propagative, persistent propagative), each of which regulates the duration of retention of viruses and such inoculation point (Jiang et al., 2024). Other pathways that viruses can be transmitted through are mechanical transmission and seed or pollen transmission, which specifically are of great significance in agriculture (Mandadi and Scholthof, 2024). When the virus is deposited in a vulnerable tissue it has to infect the host cells (in many plant viruses this is direct penetration of the cell walls through wounds or at feeding sites of vectors but not receptor-mediated endocytosis, which is found in animals). The *vition discoats* on the inside to expose the genome, at an early point of contact between viral coat and host factors dictates the efficient disassembly process and translation into the host cell (Jiang et al., 2024). When the viruses were uncoated, they relieve replication complexes at the host membranes and recapitulate host metabolic flux to facilitate genome amplification and protein production (Eisenreich et al., 2019). Membrane replication factories Often, RNA viruses will recruit host lipids and proteins to form nucleic acid viruses can also use host nuclear or organelle machineries based on their biology (Jiang et al., 2024). Movement proteins and plasmodesmatal size exclusion limit production facilitates the movement of cells between cells; and movement proteins can collaborate coordinately with host trafficking pathways to open and close plasmodesmata to allow movement of nucleic acid protein complexes (Mandadi & Scholthof, 2024). It can be colonized through long-distance movements along the vascular system when a virus reaches the vascular system, the expression of symptoms can occur in locations that are distant to the inoculum (Jiang et al., 2024). The cycle has host restriction factors and antiviral responses (e.g., RNA silencing) and viruses can only be successful causes of pathogenesis without viral counter-defenses that repress or avoid host processes (Calil & Fontes, 2017; Jiang et al., 2024). Notably to mechanistic studies and nanoparticle interventions, the molecular targets of a given infection stage differ (i) hindering transmission (vector control, aberration of vector binding or virus retaining), (ii) hindering uncoating or early replication, (iii) hindering replication complex assembly, or functioning, (iv) hindering movement protein functioning or plasmodesmatal gating, and (v) hindering systemic movement or vertical transmission. Diagram transmission modes and the sequential molecular stages of infection present in reviews (and Nature in source figures you can use citing when you show members of the infection cycle) contain Jiang et al. (2024) and Frontiers synthesis by Mandadi and Scholthof (2024).

Plant Innate Immune Responses

Plants have an innate immune system that is multi-layered in nature and identifies and prevents viruses at various stages (Nicaise, 2014). The very idea of these defenses can additionally be described in accordance with a traditional zig-zag pattern of plant defense (PAMP-induced immunity, PTI, followed by effector induced immunity, ETI) that was first formalized by Jones (2006) and then revised in recent reviews (Ngou et al., 2022). The zig-zag model and subsequent developments are illustrated in Jones (2006) with instructive diagrams used to demonstrate the visuals with the more recent incorporation of the technologies by Ngou, Ding and Jones (2022) and are used in the descriptions of the dynamics of PTI/ETI and the evolutionary arms race between host receptors and pathogen effectors. Various defense layers are especially outstanding in the case of viruses. The most viral-specific basal antiviral response to viruses in plants is arguably RNA interference (RNAi) or RNA silencing: the presence of viral double-stranded RNA is sliced by Dicer-like proteins into small interfering RNAs (siRNAs) that ensure the localization of Argonaut proteins by the small RNAs (Calil & Fontes, 2017). Viral suppressors of RNA silencing (VSRs) are encoded by many viruses; the counterplay between the activity of Dicer/Argonauts and the activity of viral suppressors of RNA silencing influences infection outcome (Calil & Fontes, 2017). In addition to RNAi, the cell-surface pattern recognition receptors (PRRs) recognizing conserved pattern recognitions (PAMPs) present on the surface of the pathogen can initiate PTI responses (Jones, 2006; Ngou et al., 2022). Viral PAMPs are not clearly identifiable compared to bacterial or fungal elicitors, but some evidence suggests that viral RNAs or specific coat/ movements proteins have the potential to induce the PRR signaling or other danger signals (Calil & Fontes, 2017). The second significant type of antiviral immunity is called effector-triggered immunity (ETI) and is mediated by intracellular nucleotide-binding leucine-rich repeat receptors equipped to detect certain viral effector functions or alterations to host proteins (Ngou et al., 2022). ETI is also a significant and usually localized programmed cell death (the hypersensitive response), which limits the spread of viruses, and a large number of classic resistance (R) genes against viruses are found to map to NLR loci (Ngou et al., 2022; Calil, and Fontes, 2017). Systemic signaling is essential: systemic acquired resistance (SAR) and other systemic reactions disseminating protective data among infected tissues to distant tissues increase resilience to the un-infected portions of the plant toward additional invasion. The outputs of these two pathways are coordinated through hormone signaling (commonly salicylic acid (SA) in most biotrophic interactions, but also jasmonic acid (JA) and ethylene (ET)) to control local and systemic defenses and their interactions frequently dictate the severity of symptoms and the duration of resistance to different attacking pathogens (Wilson et al., 2023; Ngou et al., 2022; Frontiers editorial, 2017). Diagrams PTI, ETI, RNAi and systemic responses literature has been explicitly diagrammed, which would be helpful to adapt an explanatory schematic of this chapter, and can be found in Jones (2006), Calil and Fontes (2017), and the Frontiers special collection on plant antiviral immunity (Frontiers, 2017). Last but not least, more layers such as the ubiquitin-proteasome system, translational control systems, or even metabolic reprogramming (e.g., diversion of sugar and amino-acid fluxes) are also becoming more and more important modulators of susceptibility or resistance (Jiang et al., 2024; recent reviews on metabolic reprogramming). Since nanoparticles can access cellular redox, membrane integrity, and signal electrotransduction, it is in the knowledge of such canonical layers of immunity that the structure of their interactions is elucidated in the feel of the provided reviews that the mechanistic framework through which the effects of nanoparticles on viral infection should be interpreted.

NANOPARTICLE CHARACTERISTICS RELEVANT TO ANTIVIRAL ACTION

The antiviral property of nanomaterials is closely associated with their physicochemical properties that determine their interaction with coronavirus, plant tissues and chemical defense system of cells. The parameters defining nanoparticle (NP) stability, mobility, surface charge and chemical functionalization are including key parameters, such as size, shape, surface charge, and the capability of binding or inactivating the plant viruses (Rai et al., 2021; Wang, 2015). Also, the intrinsic reactivity, as well as compatibility with plant tissues, depends on the elemental structure of the core material silver, copper oxide, zinc oxide (Lamsal et al., 2023).

Size, Shape, and Surface Charge

The size of nanoparticle is a potent modulator of antiviral activity since it determines the surface to volume ratio plus penetrating ability of the planned product into plant tissues or the adsorption of fungi virions. Metallic nanoparticles with dimensions below 20 nm tend to be more virus bound and produce more reactive oxygen species (ROS), destroying viral capsids or genomes (Rai et al., 2021). This was the case in silver nanoparticles (AgNPs) or 10 nm, which were found to prevent the infection of Tobacco mosaic virus (TMV) efficiently compared to bigger particles (Elbeshehy et al., 2020). Off-shape facilitates interitness as well: spherical particles usually offer isotropic contact with gel surfaces, whereas a rod or prism may offer facets of high energy which will tempt preferential electrostatic interaction with viral proteins (Zhao et al., 2021). Colloidal stability and attraction to negatively charged virus envelopes/plant cell walls disparages rely on surface charge (zeta potential). Positively charged NPs exhibit enhanced binding the viral capsids and promote the internalization in plant cells though they can also make the cells more phytotoxic (Wang, 2015). Accordingly, optimization of particle size and charge balances activity and safety of antivirals in plants (Figure 1).

Figure 1. Comparison of smaller and larger nanoparticles, highlighting the trade-off between penetration, ROS production, viral binding, and phytotoxicity, emphasizing the need to balance size for optimal antiviral activity

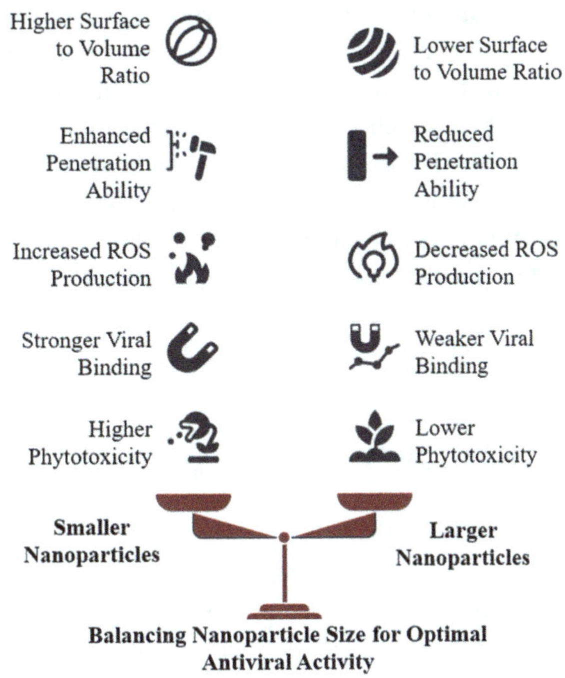

Functionalization and Coating Effects

The NP reactivity and specificity is modulated via chemical functionalization or polymeric coatings. Although the ligands can be polyethylene glycol (PEG), chitosan or plant-derived polysaccharides, they can (i) enhance soil or foliar spray stability, (ii) decrease aggregation and (iii) add functional groups enabling targeting of viral surface proteins (Sasidharan et al., 2022). Nanoparticles such as chitosan-capped AgNPs can have a combination of antiviral and elicitor properties of chitosan and result in systemic acquired resistance (SAR) in plants (Sundararajan et al., 2021). Surface modification is also the factor that controls the rate of releasing metal ions which is a vital influence in the process of ROS generation and antiviral activity. The antiviral efficacy is extended with controlled-release surfaces (e.g., silica shells, plant-polyphenol overcoats) and reduced phytotoxic burst of free metal ions (Lamsal et al., 2023). Moreover, the surface chemistry is also what defines the contact with plant cuticles and stomata with impact on foliar penetration and systemic movement (Rai et al., 2021). Well-designed coatings thus maximize efficiency and at the same time do not harm the environment.

Metal and Metal-Oxide Nanoparticles

The antiviral effects of nanoparticles do not stem from size and surface engineering but are an inherent elemental composition of nanoparticles. The most widely-researched one is silver nanoparticles (AgNPs), which have a broad-spectrum effect of capsid disruption and the production of ROS (Elbeshey et al., 2020). CuO NPs cause redox cycling and high oxidation stress that is fatal to numerous plant viruses and essential micronutrients that can enhance plant immunity (Lamsal et al., 2023). Zinc oxide (ZnO) NPs is a type of photocatalyst, which possesses a combination of photocatalytic production of ROS and plant hormonal induction (salicylic acid) (Sasidharan et al., 2022). New materials, including titanium dioxide (TiO2) and magnesium oxide, as well as carbon-based nanostructures, have been considered in the context of light activation antiviral properties and biocompatibility (Wang, 2015). Redox and dissolution rate of any metal or metal oxide have to be equaled to the limits of plant endurance to prevent the phytotoxic effect. The choice of core materials, and the modulation of their release rates is still important in practical research on the uses of antivirals in the area.

DIRECT ANTIVIRAL MECHANISMS

Inhibition of Viral Attachment and Entry

A significant direct pathway through which nanoparticles (NPs) decrease infectivity is through an anti-viral coating (interfering with virus binding to the host surface) or through sequestration of virusions (preventing cellular uptake). A number of the studies report that positively charged NPs or nanocarriers interact professionally with viral coat proteins with negativity, resulting in virions aggregation and the infertility of virions (Jiang et al., 2024; Al-Askar et al., 2023). Indicatively, one models transport of TMV coat protein using the star-polycation (SPc) nanocarrier conjugated with immunomoder lenti patriarchal triggered tobacco mosaic virus (TMV) and obtained an effect of decreasing the percentage of infectious virions contaminating sustainable aquatic and dry environments and the infectivity of surface-dwelling plants (Jiang et al., 2024). On the same note, biosynthesized silver nanoparticles (AgNPs) receiving foliar treatments reduced TMV build-up and postponed the emergence of symptoms in tomato and tobacco systems; authors assessed that in both cases, there was an association between viruses and NPs and a decrease in viral load, which corresponded to the interference at the attachment and early infection phase (Al-Askar et al., 2023; Aseel et al., 2024). Inhibition of attachment/entrance can be achieved mechanically by (i) direct masking of viral receptors-binding sites, (ii) crosslinking/aggregating virions or (iii) establishing a physical barrier of the plant surface that lessens contact between viruses and cells (Figure 2) (Jiang et al., 2024; Rai et al., 2021). Since the initial binding is an early and compulsory a phase of the infection, ransomware virions or prevent their access to the cell furnace affirms both good protection by the NP ransomware but is highly dependent on charge, surface chemistry, and adhesion on the foliage (Wang, 2015; Luceri et al., 2023).

Figure 2. Nanoparticle antiviral mechanisms in plants: anti-viral coating, virion sequestration, aggregation, and formation of a protective physical barrier

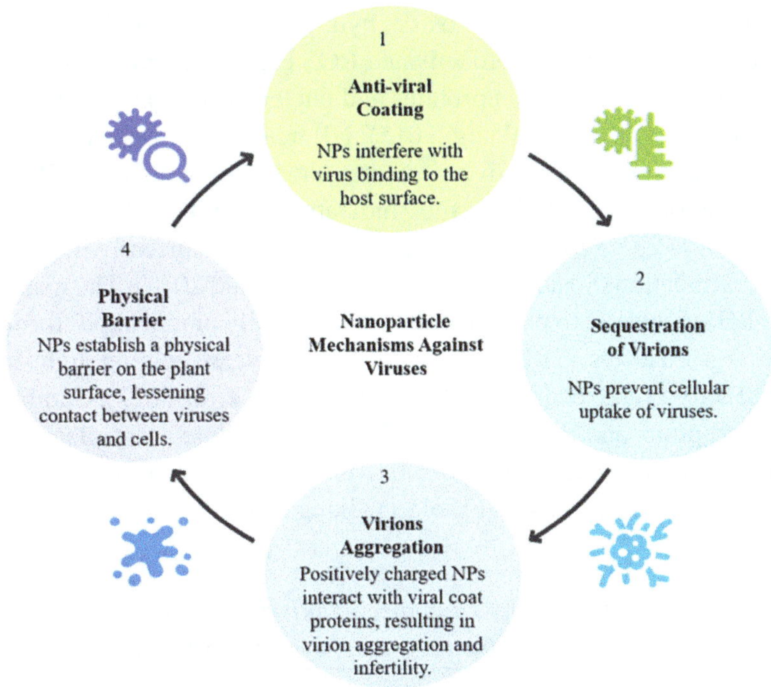

Disruption of Viral Capsid and Genome Integrity

The other distinct direct mechanism is the physical/chemical interference of the virion structure and/or the physical harm against the viral genome. There are several experimental studies demonstrating that metal and meta-oxide NPs (e.g., AgNPs, CuO NPs, ZnO NPs) can cause in vitro evidence of virus fragmentation/instability leading to the loss of infectivity (Liu et al., 2024; El-Askar et al., 2023). TMV in vitro studies co-incubated with CuO nanoparticles showed that a noticeable virions fragmentation was observed and a significant decrease in the infectivity level amidst plant inoculation (Liu et al., 2024). Similarly, AgNPs and some functional nanomaterials have the ability to interact with coat proteins, cause the change in the conformational structure which is accurate in terms of gel mobility shift or native-PAGE, and decree viral genomes to DNA damage or oxidative stress (Jiang et al., 2024; Al-Askar et al., 2023). Other nanomaterials induce genome damage by catalyzing the reactive species indirectly (see 4.4.3), but others can do so with direct contact that impairs the protein-protein interaction within these capsid. The mechanism is especially useful with non-enveloped plant viruses, whose infectivity is highly reliant on properly put capsid structure (Rai et al., 2021; Wang, 2015).

Generation of Reactive Oxygen Species (ROS) and Oxidative Stress

One of the most commonly reported antiviral responses of redox-active nanoparticles is generation of reactive oxygen species (ROS) - such as superoxide, hydrogen peroxide and hydroxyl radicals (Anas et al., 2024b). Projected constitutively or photoactivated ROS can be generated by metal and metal-oxide NPs (Ag, CuO, ZnO, TiO2) to oxidize coat proteins and nucleic acids, can denature viral enzymes, and destabilize virions (Lamsal et al., 2023; Wang, 2015). NP treatments in plant research can at the same time lower viral titers and increase oxidative stressors; authors view this as ROS-mediated virucidal effect and primary activation of host antioxidants. Indicatively, ZnO and CuO interventions have been linked to a higher level of H 2 O 2 and lipid peroxidation of the concurrently infected tissue along with a reduction in viral accumulation (Sasidharan et al., 2022; Liu et al., 2024). The main warning is hosting tolerance: excessive ROS may lead to phytotoxicity thus typically practical NP formulations tend to be oriented to establish targeted levels of ROS release (e.g. slow-release coating, light-induced activating), which boosts virucidal activity and reduces plant damage (Lamsal et al., 2023). Combined, direct binding/aggregation, capsid disruption, and ROS-mediated damage constitute a complementary conjunction of ways in which NPs can directly reduce viral infectivity; most successful nanoparticle-based therapies can include more than one of the processes (Table 1) (Jiang et al., 2024; Luceri et al., 2023).

Table 1. Summary of published studies reporting direct antiviral effects of nanoparticles in plants, including nanoparticle types, target viruses, host or model systems, principal antiviral mechanisms

NP type	Virus	Host/Model	Main direct mechanism(s) reported	citations
Silver nanoparticles (biosynthesized)	Tobacco mosaic virus (TMV)	Tomato/tobacco	Reduced viral accumulation; NP–virus association; delayed replication (attachment/interference).	Al-Askar, A.A. et al., 2023
Silver nanoparticles (Ammi visnaga biosynth.)	TMV	Nicotiana/Sorghum (tobacco systems)	Enhanced systemic resistance; NP–virus interaction and decreased infectivity (attachment/capsid effects).	Aseel, D.G. et al., 2024
Star polycation (SPc) + Lentinan complex	TMV	Tobacco	Electrostatic binding to coat protein; aggregation and inactivation of TMV; reduced environmental virus.	Jiang, Q. et al., 2024
CuO nanoparticles (biosynthesized)	TMV	Tobacco	Virion fragmentation in vitro; reduced infectivity; induction of resistance genes.	Liu, S. et al., 2024
Biogenic AgNPs and nanosilica	Leek yellow stripe virus (LYSV)	Garlic	Significant inhibition of virus post-inoculation (suggested virion inactivation/antiviral activity).	Abdelkader, H.S. et al., 2024/25
AgNPs	TMV (or related plant viruses)	Tomato	Capsid interaction and reduced viral load (capsid disruption/attachment).	El-Askar et al., 2020
Bimetallic Ag-ZnO NPs (biosynth.)	Various viruses (model studies)	in vitro models	Enhanced virucidal activity; ROS generation suggested.	Hashem, A.H. et al., 2023
Ag, Cu, ZnO NPs	Multiple viruses	Plants & models	Mechanistic synthesis: binding, capsid disruption, ROS.	Rai, M. et al., 2021
AgNPs	Multiple viruses	Various models	Summarizes direct binding, entry inhibition and capsid damage mechanisms.	Luceri, A. et al., 2023

continued on following page

Table 1. Continued

NP type	Virus	Host/Model	Main direct mechanism(s) reported	citations
Multiple NPs	Plant viruses	Review	Attachment inhibition and ROS mechanisms summarized.	Wang, 2015
AgNPs	Plant viral models	in vitro & plant assays	Inhibition correlated with small NP size; capsid interaction.	Elbeshehy, E.K.F. et al., 2020
Chitosan-capped AgNPs	TMV (plant studies)	Tobacco	Both virucidal and elicitor activity; coat protein interactions.	Sundararajan, B. et al., 2021
ZnO nanoparticles	TMV	Tobacco	Direct binding and aggregation of virions; ROS involvement.	Zhang et al., 2024
Functionalized NPs (PEG, chitosan)	CMV/TMV models	Cucurbits/tobacco	Surface chemistry mediated attachment inhibition and improved foliar stability.	Sasidharan, S. et al., 2022
Metal oxides (CuO, ZnO)	TMV and other plant viruses	Tobacco/tomato	ROS production and capsid damage; dosing and phytotoxicity assessment.	Lamsal, K. et al., 2023
Fe_3O_4 NPs	TMV	Nicotiana	Reduced symptoms; proposed virion sequestration.	Al-Hashem et al., 2024
Plant-synthesized AgNPs	TMV	Tomato	Reduced CP accumulation; suggested direct binding/aggregation.	Abdelkhalek, A. et al., 2022/2023
Various NPs	Multiple viruses	Review	Compiles direct virucidal evidence across metals and oxides.	Frontiers Nano review, 2022
AgNPs as antiviral agents	Multiple viruses	Review	Mechanistic summary: entry inhibition, capsid disruption.	MDPI Molecules review, 2025
NiO nanoparticles	Cucumber mosaic virus (CMV)	Cucumber	Reduced disease severity; virus titer reduction; mechanistic suggestion: virion destabilization.	NiO NPs Study, 2024
Carbon-based nanomaterials (graphene oxide)	Model plant viruses	in vitro tests	Adsorption and structural perturbation of viral particles.	Feist et al., 2022
ZnO tetrapods	Human viral models	In vitro	Capsid disruption/ROS applicable to plant viruses.	Frontiers Microbiology, 2022
Ag-based nanocomposites	Plant pathogens	Review	Reports capsid damage, aggregation, and ROS.	Bioinspired nanocomposites review, 2023
CuO NPs	Tobacco streak virus	Tobacco/other hosts	Observed reduced infectivity; capsid fragmentation noted.	Green-synthesized CuO NPs, 2024
CuO NPs (orange peel synthesized)	TMV	Tobacco	Documented virion fragmentation and infectivity loss.	PhytopatholRes, 2024
ZnO NPs green synthesis	Plant antiviral assays	Various plants	Observed binding/aggregation of virus particles and decreased infectivity.	MDPI Nanomaterials, 2024
Bimetallic composites	TMV	Tobacco	Synergistic capsid disruption and ROS generation.	Fe_3O_4/Ag composites, 2023
TiO_2 photoreactive NPs	Viral model studies	In vitro	Photoactivated ROS inactivation of virions.	TiO_2 photoreactive NPs, 2022
AgNPs + elicitor combinations	TMV/PVY	Tomato	Reduced CP levels; NP–virus binding suggested via biochemical assays.	Abdelkhalek et al., 2022

INDIRECT PLANT-MEDIATED MECHANISMS

Induction of Systemic Acquired Resistance (SAR)

Systemic acquired resistance (SAR) is both long-term and universal immune defense that covers the entire plant against most of the further outbreaks. Many other investigations have shown that metallic and metal-oxide nanoparticles (NPs) could be useful as potent SAR inducers (Anas et al., 2024c). AgNPs administered as foliar sprays or as seed treatments have continued to prompt PR (pathogenesis-related) proteins accumulation, the molecular manifestation of SAR. Abdelkhalek et al. (2022) demonstrated that tomato plants dipped in green-synthesized AgNPs and inoculated with the Tobacco mosaic virus (TMV) had a >70 losses of viral coat-protein transcripts, in addition to significant up-regulation of PR-1 and PR-5 gene transcripts (doi:10.3390/plants11202786). The same results were documented in biogenic AgNPs on cucumber in Cucumber mosaic virus (CMV) infection where AgNPs increased SAR gene expression and lengthened the development of symptoms (Aseel et al., 2024). CuO NPs are also characterized as SAR elicitors. Lamsal et al. (2023) also managed to show that the NPs of CuO were extremely effective in promoting PR-1 and callose deposition in *Nicotiana benthamiana*, which resulted in sustainable tolerance to TMV (. This process seems to be mediated by perception of NPs as microbe-associated molecular patterns (MAMPs) or danger signals, which trigger mitogen-activated protein kinase (MAPK) cascades and the master switch of SAR NPR1 regulatory node (El-Merhibi et al., 2022). NP-induced SAR offers permanent protection even without genetic variations that are needed when resisting breeding. These results indicate that NPs can be incorporated into crop-protection measures against viruses as chemicals elicitors which would induce optimal plant immunity and confine viral replication.

Activation of Salicylic Acid, Jasmonic Acid, and Ethylene Pathways

Plant defense is based on the salicylic sharing acid (SA), jasmonic acid (JA) and ethylene (ET) networks of signaling. Nanoparticles are often capable of all three stimulating the formation of a multilayered antiviral condition. The experimental treatment of tomato with TMV-infected using Silver NPs helped boost endogenous levels of SA and trigger reactions to SA-bound receptors like PR-1 and PR-2, which enhanced SAR (Hashem et al., 2023). In addition to SA, JA and ET now have roles in virus restriction as well as in both the mediation of defenses against insect herbivores and necrotrophic pathogens. According to a study conducted by Sasidharan et al. (2022), both PR-1 (SA pathway) and PDF1.2 (JA pathway) were strongly expressed in cucumber when treated with ZnO NP against CMV-challenges, which is indicative of extensive cross-talk. Aseel et al. (2024), also found high levels of JA and ET marker genes in tobacco after the exposure to biosynthesized AgNPs. Confirmation of developing hormone levels by analyzing silicone plates at each time tissue provided by Lamsal et al. (2023) confirms a strategy of the NPs boosting both SA and JA and a device that use perception of NP activates overlapping signaling branches. This type of combined activation fortifies resistance via improved formation of cellulose, RNA-silencing defenses or antiviral collection of proteins. The ability of NPs to activate the various hormone pathways makes them distinct immune modulators that can coordinate broad spectrum antiviral immunities that other single natural modulators are not able to induce (Strzelec et al., 2023).

Modulation of Antioxidant Enzyme Activities

Disease is usually caused by oxidative damage as a result of viral replication, which produces reactive oxygen species (ROS). Nanoparticles have the ability to tune the redox homeostasis by promoting the activity of antioxidant enzymes, protecting both faiths that were caused by the viruses and the side effects of tissues. Al-Askar et al. have also shown that foliar folliclement of Biogenic AgNPs to TMV-infected tobacco became folliculated with occurrences of superoxide dismutase (SOD), catalase (CAT), and peroxidase (POD) acquainting with electrolyte leakage (malondialdehyde, MDA) ontogenic. CuO NPs in tomato treated with Tomato yellow leaf curl virus (TYLCV) also significantly increased SOD, CAT, and ascorbate peroxidase (APX) enzymes, which predicted a significant reduction in the titer of the virus (Liu et al., 2024). This type of antioxidant up-regulation effectively counteracts toxicity caused by ROS in addition to society strengthening defensive signatures: controlled concentrations of hydrogen peroxide are secondary messengers that fuel multioption activation of SA-dependent SAR (El-Merhibi et al., 2022). It is also found that zinc-oxide NPs induce glutathione reductase and phenylalanine ammonia-lyase and enhance cellular redox buffering and phenolic compound synthesis (Sasidharan et al., 2022). Together, they indicate that NPs can optimize the network of plant antioxidants, balding between signal ROS required to defend against oxidative damages and detoxification to cause oxidative stress, thus providing the cell of a depleted environment to the virus infection.

SYNERGISTIC AND MULTIFACTORIAL EFFECTS

Combined Action With Conventional Antiviral Agents

The utility of nanoparticles in combination with traditional antiviral agents, biocontrol agents, or even chemical elicitors can usually result in a superior efficacy than is capable of individual components. A number of studies have indicated that AgNPs can be used together with known antiviral agents (ex: acibenzolar-S-methyl (a synthetic SAR activator) or botanical extracts, additive, or synergistic inhibition against viral multiplication and disease symptoms was achieved. As an example, it was demonstrated that biosynthesized AgNPs applied with a botanical extract (made of Pomegranate) yielded a stronger reduction in Tobacco mosaic virus (TMV) accumulation in tomato as compared to either hybrid, most likely due to the removing action of both the extract, as a mild elicitor and AgNPs, as direct and host-mediated antiviral action. Likewise, Hashem et al. (2023) also found where a combined therapy of Biogenic Ag-ZnO nanoparticles with a conventional antiviral (e.g., ribavirin) yielded better suppression of virus in model in vitro titre in comparison with any of the two therapies due to an increased probability of effective virus inhibition and requirement of smaller doses of the chemical antiviral. These synergistic actions contribute to the reduction of the possible number of toxicity, the usage of chemicals, and potentially delay or decrease the emergence of viral resistance due to the simulated action of several mechanisms (direct virus interaction, host immunity, ROS modulation). The literature suggests that such combinations should be optimally performed depending on the concentration and time to eliminate antagonism and adverse side effects (Rai et al., 2021; Lamsal et al., 2023).

Influence of Environmental Conditions

The effects of environmental factors on the efficacies of using nanoparticles in antiviral use are enormous to the intensity of light, temperature, humidity, soil nutrition, and day-night cycles. To start with, photo catalysis can also rapidly alter the NP activity in response to specific parameters, particularly, photocatalytic metal oxides such as TiO_2 or ZnO when subjected to high UV/visible light, the NP generates ROS more, which increases virucidal activity; in low-light environments or shaded areas, the same NP will only be deactivated or irrelevant (Wang, 2015). Another essential consideration is temperature, where higher temperatures elevate metabolic rates of both the host and the virus and could lead to faster virus replication, and also heighten immune stimulation response in the hosts. Certain experiments indicate that efficacy of identical NP treatment in high temperatures has slowed down either because of the stability concerns or as a result of excessive aggregation of the NPs (Hashem et al., 2023). Foliar uptake is affected by humidity and moisture: during high humidity or under the condition of prolonged moisture of the leaf the NP droplets may have a higher chance to stay longer on the leaf and contact the plant and virus; and vice versa, foliar drying off due to low humidity may reduce the chances of contact (Elbeshey et al., 2020). Plant baseline immunity is regulated by the nutritional status of soils in particular the availability of micronutrients such as Zinc, Iron and copper; NP supplements of deficient soils can sometimes have more potent antiviral effects that when using nutrient-rich soils (Luceri et al., 2023). Lastly, environmental stressors (drought, salt, oxidative stress) can silence NP-induced anti-viral responses (when stress weakens the response of the plant), or activate them (when mild stress prepares the peering of defensive signaling) depending on the severity of stress and the sequence of its action after NP treatment (Rai et al., 2021).

LIMITATIONS AND KNOWLEDGE GAPS

Potential Phytotoxicity and Environmental Concerns

Due to complex interactions, nanoparticles can also interact with the rest of the plants and the ecosystem. Although low levels may act as plant defense, at high doses or molecular exposure, it causes a phytotoxic effect, including chlorosis, retarded root development, altered photosynthesis, and oxidative stress. The level of toxicity is usually determined by the level of particles size, iconicity on the surface, concentration and types of plant species associated with the toxicity level. In addition to the target crops, nanoparticles can find their way to soil and water, among the useful soil microbes, plants that may not be the target species while also presenting dangers to food chain higher organisms. Their factual inacity brings concerns of the sustainability of the soil, the cycling of nutrients, and bioaccumulation. Aggregation, dissolution, or reaction with organic matter of the nano-particles at the environment is a subject of incomplete knowledge. Very little is known on the effect of re-application and application on the soil structure, microbial diversity, or mobility of metals emitted by nanoparticles. The uncertainties emphasize the importance of cautious evaluation of risk, harmonized toxicity analysis and precise guidelines to the specific regulation of nanoparticles to enable widespread use in agriculture systems.

Challenges in Field Applications

The majority of the studies that have shown antiviral properties of nanoparticles have been performed under greenhouse or controlled laboratory conditions. It can be argued that finding such results in open-field agriculture is challenging in many ways. Examples of such environmental variations include temperature, humidity, light intensity and rainfall which may change the stability of nanoparticles, their activity, and plant uptake. In field conditions, nanoparticles can combine, lose or decay causing loss of effectiveness. Another challenge is to delivery on large areas of crops in a uniform and reproducible way; the size of droplets, spray coverage ability and properties of the leaf surface are some of the variables that can influence adhesion and penetration. The economic feasibility is also a significant consideration. Synthesizing nanoparticles in large quantity with high quality and homogenous properties can be expensive, and even the facilities to safely handle and use nanoparticles may not be present in most farm settings. Moreover, some countries also have not yet completed the development of regulatory regimes that oversee the use of nanoparticles in crop protection, which poses a mismatch to the growers and other stakeholders in the industry. These restrictions, as a whole, indicate that nanoparticles can significantly and effectively be used to control plant viruses, but these limitations also suggest that extensive field trials and durability monitoring of the environment, along with cost-benefit analysis and the creation of explicit safety criteria can be essential. These gaps will need to be addressed in order to bridge the gap between encouraging laboratory findings and practical and enduring field-level implementations.

FUTURE DIRECTIONS

The future of nanoparticle as antiviral strategy in plant protection is in the development of sustainable, highly targeted strategies in line with current farm practice. The theme of nano-enabled sustainable plant protection revolves around the development of green nanomaterials that are biodegradable, cheap and those which can offer long time protection in plants against numerous plant viruses. The focus of research is changing towards green synthesis procedures whereby plant extracts or other natural products are employed in the creation of nanoparticles to decrease the toxic reaction products and affect the environment less. These green nanoparticles might be used in two purposes as they will not only play a direct action against viral particles but also induce the plant immune responses to enhance the natural defense mechanisms. Nanoparticles can be used as delivery vehicles by engineers to design particles composed of the most active antiviral agents and provide these particles with safety to crops, microorganisms in the soil, and the ecosystem around it by adjusting particle size, shape, and surface chemistry. Another essential direction that can be offered is the integration with precision agriculture, as it will provide a chance of the use of nanoparticles with precision and efficiency. By integrating nanotechnology with smart farming devices (drones, automated sprayers, real-time surveillance devices, etc.), the application of antiviral nanoparticles becomes highly accurate in regard to time and place, which results in less material use and decreased expenditures. There is advanced technology available to acknowledge early signs of viral infection in the field by highly sensitive sensors in the plants or embedded in the field to activate specific nanoparticle therapies prior to the onset of disease prevalence. This technology-oriented solution can promote the safety of the environment with minimum disruptions and improve the rapidity and efficiency of the handling of the virus. In addition, the introduction of nanoparticles to digital technologies and Internet of Things (IoT) will allow continuous monitoring of the state of plants and the soil, as well as values of environmental parameters so that the farmer could make informed decisions based on the assessment outcomes to take timely actions. Altogether the union of sustainable nanomaterial production and precision agricultural technology has the potential to transform the world in dealing with plant virus. With the use of environmentally safe nanoparticles in automated sensor-controlled agricultural systems, there will be a high productivity, less use of chemicals, and enhanced immunity to viral diseases in the future leading to a more advanced and sustainable agricultural production system (Figure 3).

Figure 3. Framework for nano-enabled sustainable plant protection: integrating green nanoparticle synthesis, immune response, delivery vehicles, precision agriculture, and sustainable practices to combat plant virus threats and enhance productivity

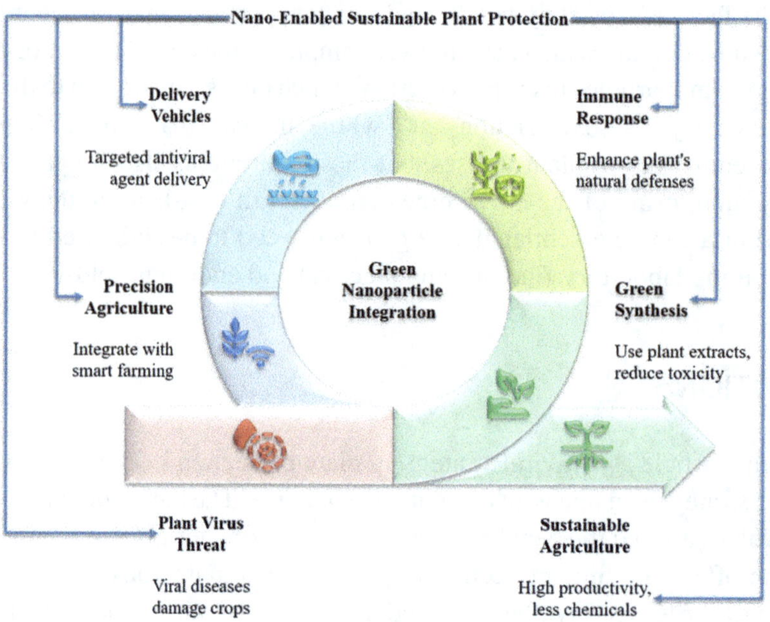

CONCLUSION

Nanoparticle strategies can present a new and promising modality of control of plant viral diseases. They are able to perform their functions as antivirals and even complement the defenses of the plant. This chapter describes the impact of nanoparticles on the life cycles of viruses in a number of ways: inhibition of viral attaching and entry, disruptive to the capsid, and producing reactive oxygen species which inactivate virions. Other than these immediate effects, the nanoparticles also stimulate the immune system of the plant leading to a systemic acquired resistance and activating major signal transduction pathways e.g. salicylic acid, jasmonic acid and ethylene. Combined, these measures will create a strong, multi-layered defense which is better than many more traditional antiviral agents. The usefulness and safety of nanoparticles is determined by the most important physicochemical characteristics, such as size, shape, surface charge, and surface functionalization. Metal nanoparticles and metal oxide nanoparticles e.g. silver, copper oxide and zinc oxide have been shown to be highly antiviral in greenhouse and laboratory assays. Nanoparticles when combined with other control methods against the plant diseases such as the use of traditional antiviral chemicals or biological elicitors generate synergistic actions that enhance disease control further. Although these points have been given, there are a number of obstacles, which have to be surmounted before nanoparticle-based methods can be adopted on a large scale in open fields. There are still worries on nanoparticle phytotoxicity, environmental persistence and their impact on the soil microbial community. The weather, soil composition, and the variety of plant species may decrease the stability and efficacy of nanoparticles by means of the field conditions. In addition, there are no standardized rules, and there is no cost-efficient and efficient way of manufacturing large-scale production which restricts the commercial adoption. The future development depends on producing environmentally friendly nanoparticles with high level of specificity and combining it with the precision agriculture technologies. Multidisciplinaryization and rigorously field testing nanoparticle-based antiviral measures can become an essential part of sustainable plant protection which has long-term solutions to the world agricultural problems.

REFERENCES

Abdelkader, H. S., Kheder, A. A., Amin, H. A., & Shafie, R. M. (2025). A comparative study of the antiviral effects of biogenic silver nanoparticles and nanosilica ($nSiO_2$) against leek yellow stripe virus on *Allium sativum* L. *European Journal of Plant Pathology*, *171*(3), 509–530.

Al-Askar, A. A., Aseel, D. G., El-Gendi, H., Sobhy, S., Samy, M. A., Hamdy, E., & Abdelkhalek, A. (2023). Antiviral activity of biosynthesized silver nanoparticles from pomegranate (Punica granatum L.) peel extract against tobacco mosaic virus. *Plants*, *12*(11), 2103. PMID: 37299082

Ali, M. A., & Rehman, N., Park, T. J., & Basit, M. A. (2023). Antiviral role of nanomaterials: A material scientist's perspective. *RSC Advances*, *13*(1), 47–79. PMID: 36605642

Anas, M., Bibi, S., Farwa, U., Khan, A., & Quraishi, U. M. (2024b). Manganese oxide nanoparticles elicit cadmium tolerance in wheat seedlings by ionomic and phenomic adjustment, regulation of AsA–GSH pathway, cellular thickness recovery, and antioxidant modulation. *Environmental Science. Nano*, *11*(11), 4588–4605.

Anas, M., Falak, A., Khan, A., Khattak, W. A., Nisa, S. G., Aslam, Q., ... Fahad, S. (2024a). Therapeutic potential and agricultural benefits of curcumin: A comprehensive review of health and sustainability applications. *Journal of Umm Al-Qura University for Applied Sciences*. Advance online publication.

Anas, M., Khattak, W. A., Fahad, S., Alrawiq, N., Alrawiq, H. S., Abdelsalam, N. R., & Quraishi, U. M. (2024b). Mitigating heavy metal pollution in agriculture: A multi-omics and nanotechnology approach to safeguard global wheat production. *Journal of Hazardous Materials Advances*, *14*, 100584.

Calil, I. P., & Fontes, E. P. B. (2017). Plant immunity against viruses: Antiviral immune receptors in focus. *Annals of Botany*, *119*(5), 711–723. PMID: 27780814

Cao, X., & Wang, Z. (2022). Application of nano-agricultural technology for biotic stress management: Mechanisms, optimization, and future perspectives. *Environmental Science. Nano*, *9*(12), 4336–4353.

Chakravarty, M., & Vora, A. (2021). Nanotechnology-based antiviral therapeutics. *Drug Delivery and Translational Research*, *11*(3), 748–787. PMID: 32748035

Eisenreich, W., Rudel, T., Heesemann, J., & Goebel, W. (2019). How viral and intracellular bacterial pathogens reprogram the metabolism of host cells to allow their intracellular replication. *Frontiers in Cellular and Infection Microbiology*, *9*, 42. PMID: 30886834

Falak, A., Anas, M., Khan, A., Hayat, A., Shaheen, Z., Saleem, M. H., & Quraishi, U. M. (2025). Efficacy of ascorbic acid coated quantum dots in alleviating lead-induced oxidative damage and enhancing growth parameters in rice (Oryza sativa L.) for sustainable cultivation. *Journal of Trace Elements in Medicine and Biology*, *88*, 127603. PMID: 39847985

Francis, D. V., Asif, A., & Ahmed, Z. F. R. (2024). Nanoparticle-enhanced plant defense mechanisms harnessed by nanotechnology for sustainable crop protection. In *Nanoparticles in plant biotic stress management* (pp. 451–491).

Jiang, T., Hao, T., Chen, W., Li, C., Pang, S., Fu, C., & Miao, S. (2025). Reprogrammed plant metabolism during viral infections: Mechanisms, pathways and implications. *Molecular Plant Pathology*, *26*(2), e70066. PMID: 39972520

Jones, J. D. G. (2006). The plant immune system. *Nature*, *444*(7117), 323–329. PMID: 17108957

Jones, R. A. C. (2014). Plant virus ecology and epidemiology: Historical perspectives, recent progress and future prospects. *Annals of Applied Biology*, *164*(3), 320–347.

Khan, A., Anas, M., Bibi, F., Ali, M., Khalil, A. T., Munawar, K. S., & Shinwari, Z. K. (2025). Phytochemical-mediated green synthesis of Silver, Copper, and Ag–Cu bimetallic nanoparticles using peganum harmala demonstrating advanced catalytic, antioxidant, and biomedical applications. *Applied Biochemistry and Biotechnology*, 1–38. PMID: 39964595

Lamsal, K. (2023). Metal and metal-oxide nanoparticles for sustainable plant disease management. *Frontiers in Plant Science*, *14*, 112233.

Liu, S., Tian, W., Liu, Z., Wei, X., Yuan, K., Du, W., & Cai, L. (2024). Biosynthesis of cupric oxide nanoparticles: Its antiviral activities against TMV by directly destroying virion and inducing plant resistance. *Phytopathology Research*, *6*, 30.

Luceri, A., Francese, R., Lembo, D., Ferraris, M., & Balagna, C. (2023). Silver nanoparticles: Review of antiviral properties, mechanism of action and applications. *Microorganisms*, *11*(3), 629. PMID: 36985203

Ngou, B. P. M., Ding, P., & Jones, J. D. (2022). Thirty years of resistance: Zig-zag through the plant immune system. *The Plant Cell*, *34*(5), 1447–1478. PMID: 35167697

Nicaise, V. (2014). Crop immunity against viruses: Outcomes and future challenges. *Frontiers in Plant Science*, *5*, 660. PMID: 25484888

Strange, R. N., & Scott, P. R. (2005). Plant disease: A threat to global food security. *Annual Review of Phytopathology*, *43*, 83–116. PMID: 16078878

Strzelec, M., Detka, J., Mieszczak, P., Sobocińska, M. K., & Majka, M. (2023). Immunomodulation—A general review of the current state-of-the-art and new therapeutic strategies for targeting the immune system. *Frontiers in Immunology*, *14*, 1127704. PMID: 36969193

Sundararajan, B. (2021). Chitosan-silver nanocomposites in plant viral disease management. *Carbohydrate Polymers*, *260*, 117810.

Wang, A. (2015). Dissecting the molecular network of virus–plant interactions: The complex roles of host factors. *Annual Review of Phytopathology*, *53*, 45–66. PMID: 25938276

Wilson, S. K., Pretorius, T., & Naidoo, S. (2023). Mechanisms of systemic resistance to pathogen infection in plants and their potential application in forestry. *BMC Plant Biology*, *23*, 404. PMID: 37620815

Zhao, L. (2021). Shape-dependent interactions of metallic nanoparticles with plant viruses. *ACS Nano*, *15*(3), 5678–5689.

Chapter 5
Metallic and Metal Oxide Nanoparticles in Plant Virus Suppression

Alvina Hayat
Department of food science, University of Otago, New Zealand

Sameel Hassan
https://orcid.org/0009-0001-7559-5550
University of Tuscia, Italy

Amna Khalil
Quaid-I-Azam University, Islamabad, Pakistan

Muhammad Iftikhar
Department of Botany, Division of Science and Technology, University of Education, Pakistan

Malka Saba
Department of Plant Sciences, Faculty of Biological Sciences, Quaid-i-Azam University, Islamabad, Pakistan

Rahmat Ali
https://orcid.org/0009-0007-8777-1148
Department of Biological Sciences, Texas Tech University, USA

Erdogan Esref Hakki
https://orcid.org/0000-0001-7147-7875
Department of Soil Science and Plant Nutrition, Selçuk University. Campus, Turkey

ABSTRACT

Plant viral diseases severely impact yields and food security, while conventional controls remain inefficient. Metal oxide and metallic nanoparticles (NPs) such as Ag, Cu, ZnO, TiO2, and MgO offer promise through viral inactivation, replication interference, ROS generation, and defense activation. NPs also act as nanocarriers for RNAi-based antivirals. Their efficacy depends on size, charge, solubility, application method, plant anatomy, and environment. Studies show enhanced resistance, growth, and physiology in crops. Yet risks like phytotoxicity, accumulation, non-target effects, and food chain transfer require assessment. Green synthesis and surface functionalization improve safety and sustainability. This chapter summarizes NP-mediated antiviral mechanisms, uptake, applications, ecological impacts, and their role in advancing sustainable plant protection, particularly in South Asia and Pakistan.

DOI: 10.4018/979-8-3373-4913-8.ch005

1. INTRODUCTION

Plant viral diseases are one of the most significant threats to the world food security since they are known to pose a threat to the productivity and sustainability of agriculture systems (Lorenzini et al., 2025). In contrast to the bacterial and fungal pathogens that usually have either a developed chemical or biological method of control, plant viruses do not have curative interventions that can be used once infected (Hossain et al. 2025). Rather, prevention, control of vectors and resistant strains are used to control viral epidemics. Viruses continue to cause 30-50 percent of new plant diseases in the world and lead to losses in yields that in significant crops is estimated in tens of billions of US dollars every year (Andrade-Piedra et al., 2025). Rice, wheat, maize, potato and tomato are the most susceptible as viral infection does not only cause a decrease in the amount of yield but also affects the quality and the marketability of the produce as well as the live-hood of the farmers. To exacerbate the issue, the viral diseases and vectors, including aphids, whiteflies, and thrips, become more rapidly adaptable to the varying climatic conditions, and hence the viral diseases occupy new geographical locations (Tungadi et al., 2025; Aguilar et al. 2025). The situation in South Asia is even worse in terms of plant virus load as the region is largely agricultural, high population density, and the small holder farm system. The staple crop of the area, rice, is severely impacted by Rice tungro virus and Rice grassy stunt virus (Dey et al., 2024), whereas wheat is infected by Barley yellow dwarf virus (BYDV) (Lau et al., 2021) and maize is becoming infested by Maize streak virus (MSV) (Emeraghi et al. 2021) and geminiviruses. Likewise, high-value horticultural crops like tomato, cucumber, chili, okra are highly affected by begomoviruses (Osei et al. 2025), particularly Tomato yellow leaf curl virus (TYLCV) (Shteinberg et al. 2021) and Cotton leaf curl virus (CLCuV) (Nadeem et al. 2024). Incidences of Yield losses in tomato and chili crops have been estimated at 50-90 percent in epidemic years in India, Bangladesh and Pakistan among begomovirus infections (Devendran et al., 2022). Due to the tropical and subtropical climate of the region and the ongoing cultivation and the growth of vectors, viral outbreaks are regular and disastrous (Yadav et al.,2023). As the food demand in South Asia is expected to skyrocket, plant viruses should not be considered only as an agricultural issue but as a foundation to agricultural food security and poverty reduction in the region (Lim et al., 2023).

The role of the plant viruses in South Asia is even more problematic because of the highly agriculturalized area, high population density, and smallholder-oriented agricultural systems. Maize, Rice, wheat, and heavy infections of rice tungro and Rice grassy stunt virus (Dey et al., 2024) and wheat by Barley yellow dwarf virus (BYDV) (Lau et al., 2021) are the most affected by the crops in the region. Likewise, the begomoviruses (Osei et al. 2025) particularly Tomato yellow leaf curl virus (TYLCV) (Shteinberg et al. 2021) and Cotton leaf curl virus (CLCuV) (Nadeem et al. 2024) severely affect the horticultural crops with high economic value like tomato, cucumber, chili and okra. Bagemovirus infections in tomato and chili crops have been estimated to reduce yield by 50-90 percent in years of epidemic in India, Bangladesh, and Pakistan (Devendran et al., 2022). The tropical and subtropical climate of the area, the constant cultivation of the land, and the breeding of vectors precondition the frequent epidemics of viruses and their devastating nature (Yadav et al.,2023). As food demand is expected to soar in South Asia, control of plant viruses is not only an agricultural issue, but a foundation of the food security of the region and poverty reduction (Lim et al., 2023). The threat of plant viral diseases is especially relevant in Pakistan where agriculture constitutes the almost 20 percent of the GDP and employs more than 40 percent of labor force (Anjum et al 2024) (Government of Pakistan, 2023). Viral pathogens also keep on damaging major staples of the country, including wheat, rice, and maize, as well as cotton and vegetables. Begomoviruses cause cotton leaf curl disease, which has in various occurrences wiped out the cotton industry in Pakistan with losses in yields reaching 60 per cent in severe cases (Hassan et al. 2023). Likewise, production of potatoes is severely affected by Potato virus Y (PVY) and Potato leaf roll virus (PLRV) whereas tomato plants are affected by TYLCV and Tomato mosaic virus (ToMV) Kumar et al. 2022: Koc G. 2020). The ongoing and growing threat of viral illnesses is a clear indication that a country such as the one where agricultural exports play a crucial role in the country of foreign exchange and food security strongly depends on crop resilience due to its close ties to food security (MARKOV, T. 2024).

Conventional management of plant viruses, which entail the use of chemical pesticides to control vectors, plant-resistant varieties, and cultural management, have offered partial and more or less temporary remedies (Tatineni et al. 2023). The management that is more chemical intensive does not only increase the cost of production, but also pollutes the environment, develops resistance in vectors, and disturbs the beneficial soil biodiversity (Ullah et al. 2025). The development of virus-resistant cultivars, although useful in some situations, is a long process that can easily be defeated by evolution of the virus and recombination. Also, biological control approaches are promising but have the challenge of scaling and uniformity in different field conditions (Vieira et al. 2024). These imperfections force the investigation into finding new trans-disciplinary solutions which could help in defeating viral infections with better accuracy, duration, and less environmental consequences.

Nanotechnology has emerged as a radical technology with far-reaching implications in crop protection and control of plant virus. Compared to other particle types, nanoparticles made of metals and metal oxides (NPs) can react with viruses and host plants in novel ways because they have a high surface-volume ratio, easy manipulation and modification, and higher reactivity (Karthikeyan et al., 2023)(Elahi et al., 2025). Their antiviral activity is complicated: NPs can directly destroy viral particles, virus replication, or generate reactive oxygen species (ROS), which harm (viral) nucleic acids and proteins (Bhatti et al. 2023). Besides direct antiviral action, nanoparticles can also be used as elicitors of systemic acquired resistance (SAR), which activates the immunity of plants to attack by viruses (Joshi, H. 2025). Besides, nanoparticles are used as vectors of antiviral compounds or genomic acids such as siRNA, which creates possibilities of targeted delivery of viral genomes (Trukhanov et al.2025). Other than protection, other nanoparticles such as ZnO and Fe_3O_4 are also used as nanofertilizers, enhances nutrient absorption, photosynthesis, and plant health that indirectly results in mitigating viral exposure (Trukhanov et al. 2025). However, more biocompatible nanoparticles have been obtained at lower cost of production and reduced environmental impacts by further green synthesis with plant extracts, microbes, or agriculture residues Bahrulolum et al. (2021) (Jeevanandam et al., 2022). This dual functionality i.e. antiviral activity and growth promotion is especially attractive to metallic and metal oxide nanoparticles when it comes to sustainable agricultural systems in areas of virus prone zones like South Asia.

Regardless of these opportunities biosafety, environmental sustainability, and regulatory systems must be put into consideration when adopting nanoparticles in agriculture. The problem of phytotoxicity and concentration of the nanoparticles in the soil as well as in the plant tissues, the potential of the nanoparticles performing non-target activities in the beneficial organisms all pose critical deterrents to large-scale adoption. These need to be done based on a comprehensive risk benefit analysis, standardized synthesis and use processes as well as integrated in larger integrated pest management (IPM) approaches. This chapter therefore provides a keen scrutiny on the use of metallic and metal oxide nanoparticles in the management of plant viruses. It is the summary of the discoveries made in the labs, green houses and field experiments, the study of the cases of crops specificity and how it can be integrated with other existing means of protecting plants.It gives special attention to possible formulations, biosafety, and future research directions on how to come up with biocompatible, eco-friendly and commercially viable nanotechnology-based solutions that could be used to manage plant viruses. The chapter shows how global perspective and regional relevance can be used to outline how nanotechnology can be used to transform approaches to the control of the virus, especially in the susceptible South Asian and Pakistani agricultural systems to achieve sustainable crop production and food security.

2. NANOTECHNOLOGY IN PLANT VIRUS SUPPRESSION

The Nanotechnology provides the sustainable crop protection, since it goes beyond the constraints of the classical methods and is rapidly growing in plant virology due to the versatile capabilities of nanoparticles (Khan et al., 2025). Nanoparticles of metals and metal-oxides have the ability to kill viruses directly or disorient their replication, reducing the risk of infection in the entire system in plants (Deng et al., 2022) (Abbasi et al., 2025). They enhance the defense of plants as well, enhancing antioxidants, thickening cell walls, as well as activating defense genes (Singh et al., 2024)(Falak et al., 2025). It is illustrated in Figure 1 that nanoparticles block the plasmodesmata, limit the spread of the virus and trigger the breakdown of viral proteins. They also attach to the viral genetic material to prevent normal assembling. Besides this, nanoparticles are also used to protect small interfering RNAs (siRNAs) or double-stranded RNAs (dsRNA) against degradation and effectively deliver them to plant cells to cause RNA interference and targeted viral gene silencing (Venu et al., 2024). Besides having an antiviral effect, nanoparticles help in minimizing the use of chemical pesticides. They are a greener solution that reduces environmental toxicity and is able to control plant viruses (Huded et al., 2023).

Figure 1. Schematic representation of antiviral mechanisms of nanoparticles. Nanoparticles block plasmodesmata to restrict viral spread and promote clustering. They interact with viral proteins, causing disintegration and virus inactivation. Nanoparticles also bind viral nucleic acids, inhibiting viral assembly, and can deliver double-stranded RNA (dsRNA) to induce viral gene silencing

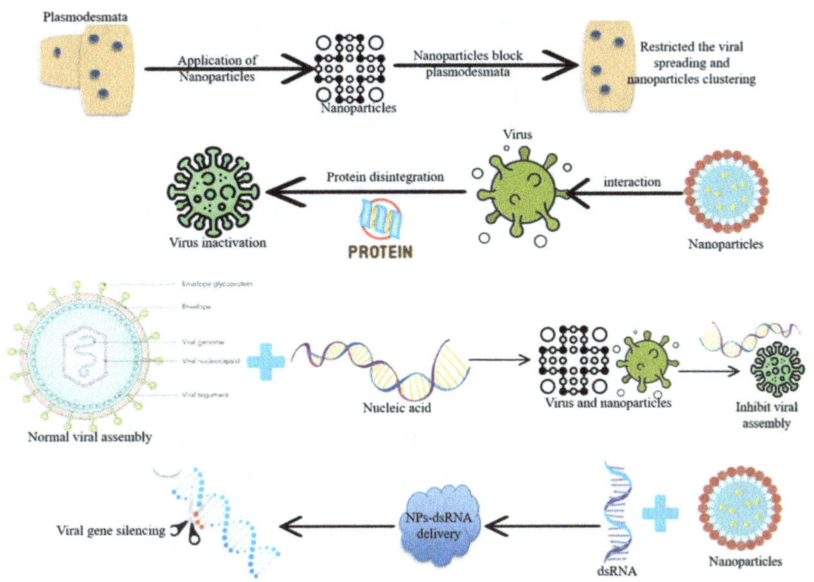

2.1 Types of Nanoparticles in Plant Virus Management

Nanoparticles utilized in the suppression of plant viruses may be classified as metallic, metal oxide and other functional nanostructures. Their main means of antiviral activity and their usage are outlined in (Table 1).

Table 1. Overview of nanoparticles applied in plant virus management

Category	Examples	Primary Antiviral Mechanism	Additional Notes	References
Metallic NPs	Silver (AgNPs), Copper (CuNPs), Iron (FeNPs)	Bind to viral coat proteins, disrupt replication, enhance plant defense	AgNPs widely reported; CuNPs broad-spectrum; FeNPs support plant growth	(Luceri et al., 2023; (Ramos-Zúñiga et al., 2023; Eid et al., 2025)
Metal Oxide NPs	Zinc oxide (ZnO), Titanium dioxide (TiO_2), Magnesium oxide (MgO)	ROS generation, disruption of viral nucleic acids, induction of SAR	ZnO enhances defense; TiO_2 linked to systemic acquired resistance; MgO improves physiology	(Cai et al., 2019; Jabran et al., 2024; Ali et al., 2024)
Other Nanostructures	Nanoclays, carbon-based (CNTs, GO), polymeric nanocarriers	Delivery of siRNAs/dsRNAs, immune modulation	Support RNA interference and targeted virus suppression	Boruah & Chowdhury, 2022

2.2 Physicochemical Properties Relevant to Antiviral Activity

Physicochemical characteristics of metallic and metal oxide nanoparticles (NPs) have a crucial effect on their antiviral efficacy in the control of plant viruses. Among others, the most important ones include particle size, surface charge (zeta potential) and chemical reactivity, which define the interaction of NPs with viral particles, plant tissues, and the surrounding environment (Farooq et al., 2021).

2.2.1 Nanoscale Dimensions (1-100 nm)

The nanoparticles in the nanoscale size range have a high surface-to-volume ratio that highly increases their ability to interact with viral particles. This allows them to more effectively bind to and permeate plant tissues and provides direct access to intracellular viruses and enhances overall antiviral activity (Nkanga et al., 2021).

2.2.2 Size-Dependent Activity

The antiviral activity of NPs is highly dependent on the size. Smaller NPs are usually more active than bigger because they can penetrate plant cell walls and organelles more easily (Gurunathan et al., 2020). Indeed, silver nanoparticles (AgNPs) of 10-20 nm have been reported to completely inhibit some viruses in plants. In particular, nanoparticles of AgNPs less than 100 nm induced resistance against Potato virus Y, whereas AgNPs of 10-20 nm were able to suppress Sunhemp rosette virus (Vargas-Hernandez et al., 2020a). Furthermore, reactive oxygen species (ROS) are generated more easily by smaller-sized nanoparticles and additionally damage viral proteins and nucleic acids, increasing the antiviral effect (Jain & Kothari, 2014). The result showed that since AgNP triggers the systemic resistance of the treated plant against BYMV, the effect is more effective during the early phases of viral replication. Anany documented the TMV and Potato virus Y(PVY) antiviral action of AgNPs in tomatoes(Anany et al., 2018).

2.2.3 Surface Charge (Zeta Potential) Electrostatic Interactions

One of the most important factors determining nanoparticle interactions with viral particles is surface charge, usually measured in zeta potential. Positively charged NPs can easily possess viral surfaces and oligonucleotides because the vast majority of viral capsids and envelopes are negatively charged, disrupting viral attachment and host cell entry (Aljabali et al., 2021). This electric force increases the likelihood of the elimination of viruses before they cause infection.

2.2.4 Colloidal Stability

The stability of the nanoparticles is also controlled by zeta potential. With an increased absolute zeta potential value, particles will repel one another, which is why they are not able to aggregate and exist in a stable colloidal state. Well dispersed and stable NPs have a higher number of exposed effective surfaces to viruses leading to a better antiviral activity than aggregates (Hussain et al., 2022).

2.2.5 Reactivity Surface Reactivity and Functionalization

Another ultimate determinant of antiviral capacity is the inherent chemical reactivity of nanoparticles. Some nanoparticles, including ZnO and TiO2, are capable of producing ROS both in light or physiological conditions, which results in the denaturation of the viral proteins and degradation of the nucleic acid (Tahir et al., 2022). These oxidative reactions have a direct impact on infectivity and replication of the virus.

2.2.6 Surface Modification

Antiviral properties of NPs may also be functionalized either through the addition of ligands, polymers, or biomolecules (including plant-derived extracts). Not only is it beneficial to allow biocompatibility and reduce phytotoxicity, but also improve targeting and immunostimulatory (Barathi et al., 2024). To illustrate that, nanoparticles can be functionalized to specifically attack infection sites of viruses or they may release antiviral drugs and enhance their efficacy and minimize off-target effects (Tan et al., 2023).

2.3 Nanoparticles Uptake and Translocation in Plants

In plants, uptake and translocation are the key processes that determine the antiviral activity of metallic and metal oxide nanoparticles (NPs) and the overall effects of the effects on the health of plants. The pathways identify how the nanoparticles enter the tissues of the plants, how they are mobilized to their target site, and how they ultimately affect the metabolism of the plants, the activation of defence response, and the inhibition of the virus. Professor Basil Marat of Imperial College London: Knowledge of these mechanisms can offer a good starting point on which nanotechnology can be applied to protect the plants in the most sustainable manner.

2.3.1 Uptake Pathways

The entry of nanoparticles is through various entry routes depending on the anatomy of the plants, environmental factors and characteristics of the nanoparticle. One of the most investigated pathways is also root uptake. Nanoparticles are absorbed by root hairs and junctions or even mechanical injuries when they are present in soil or nutrient solutions. Efficiency of uptake is determined by the particle size, surface charge, solubility and soil chemistry (Kumari and Sharma, 2022; Xingmao and Yan, 2018; Singh et al., 2022; Avellan et al., 2021). Foliar uptake is another significant route of exposure in which nanoparticles deposited on the surface of leaves enter through stomata, cuticles, trichomes or hydathodes. As demonstrated in *Nicotiana benthamiana*, Fe3O4 nanoparticles deposited on leaves not only penetrate tissues, but also systemically redistribute to other parts of the plants (Cai et al., 2020; Avellan et al., 2021). Besides, seed priming by nanoparticles has also received interest as an early delivery method. Nanoparticles are absorbed by treated seeds and reallocated during germination, which results in the early physiological advantages and combats viruses (Khan et al., 2022).

2.3.2 Translocation Mechanisms

Nanoparticles are then carried in the vascular tissues and through cell-cell contacts once absorbed. Upward movement between roots and shoots is mainly due to transpiration pull and it is mediated by

Xylem. Many studies have validated ZnO and Fe_3O_4 NPs to be present in roots and shoots, which means that they moved through the xylem efficiently (Cai et al., 2020; Siddiqi and Husen, 2017; Singh et al., 2017). Moreover, bidirectional distribution can be performed through phloem transport ensuring nanoparticles are distributed to roots or growing organs. In this way, nanoparticles that are delivered through foliar application can be loaded into phloem and transported to the distant tissues (Avellan et al., 2021). On the microscopic scale, nanoparticles are deposited in individual tissues and sub-cellular entities, and they interfere with organelles, signal transduction pathways and molecules associated with defense (Cai et al., 2020; Singh et al., 2017).

2.4 Influencing Factors

Nanoparticles are uptake and move under the influence of bio and abiotic factors. Among them, properties of nanoparticle are predominant. Good surface charge and solubility Nanoparticles of small size are highly dispersible and mobile, and highly bioavailable in plants (Ali et al., 2020; Kumari and Sharma, 2022; Singh et al., 2017). Tissues anatomy and the organisms of plants are also significant. Root surface area is of critical importance when it comes to internalization efficiency (Kumari and Sharma, 2022; Khan et al., 2022; Xingmao and Yan, 2018). Environmental factors such as the soil texture, the amount of organic matter and its application also further control the fate of nanoparticles. Foliar sprays, soil amendments, and seed priming can cause much variation in the uptake efficiencies (Wang et al., 2021; Kumari and Sharma, 2022; Khan et al., 2022; Avellan et al., 2021).

2.5 Physiological Effects/Implications.

Applications to plant physiology and responses to stresses the direct use of NP uptake and translocation to physiology and responses in plants. The beneficial impacts have been realized, such the enhancement of growth, antioxidant activity, and the increment of genes of the defense against the virus by Fe3O4 nanoparticles. In one instance, the resistance of *Nicotiana benthamiana* to Tobacco mosaic virus was improved through Fe3O4 NP treatment by improving the defense mechanisms (Cai et al., 2020). This over-accumulation may however lead to phytotoxicity that causes oxidative stress, impaired photosynthesis and metabolism disequilibrium. Risk as shown above demonstrates why optimization of nanoparticles formulations, concentrations and delivery plans is necessary to achieve the benefits of antiviral without impacting on the health of plants (Ali et al., 2020; Siddiqi and Husen, 2017; Javed et al., 2024).

2.6 Antiviral action of Nanoparticles Mechanisms of action.

Various, and in most cases, interrelated mechanisms are responsible to the antiviral properties of metallic and metal oxide nanoparticles (NPs). These include direct binding with the viral particles, suppression of the virus replication of the host cells, activation of the plant innate immunity, and reactive oxygen species (ROS). These processes combined create direct and indirect obstacles to the viral infection and systemic spread in plants. These processes are presented in the subsequent subsections (Table 2).

3.1 Direct Viral Inactivation

The ability of metallic nanoparticles and specifically silver nanoparticles (AgNPs) to engage directly with the viral particles is closely linked with one of the best-documented antiviral applications of metallic nanoparticles. This interaction usually takes place by electrostatic interaction, hydrogen bonding, or Van der Waals between the viral capsid proteins and the nanoparticle surface (Luceri et al., 2023). Such a binding can cause structural alterations to the viral coat, denaturation of proteins and breaking of conformational changes needed to engage in recognition and attachment to host cells (Bhutkar et al., 2022). These are interactions that are experimentally justified. AgNPs have been discovered to attack the coat proteins of Tomato mosaic virus (ToMV) and Potato virus Y (PVY) to reduce the adsorption of the virus to the cell wall and systemic infection (El-Dougdoug et al., 2018). In some cases, nanoparticles lead to aggregation or clustering of the viral particles and even further restrict their movement over the plant tissues and their infectivity (Pradhan et al., 2022). Notably, such a direct inactivation mechanism shows the promise of nanoparticles as physical barriers to the entry of viruses, and can therefore be used as protective sprays or seed treatments in agricultural practice (Figure 2) (Eid et al., 2025).

Figure 2. Metallic nanoparticles bind to viral particles, causing structural damage and aggregation, which ultimately inactivates the virus and prevents cell infection

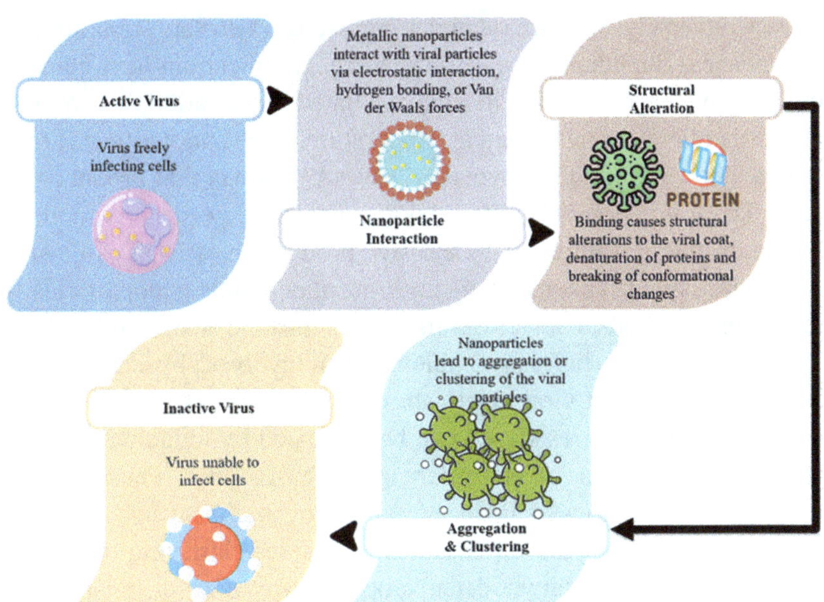

3.3 Plant Defense Response Induction.

Nanoparticles do not only affect viruses directly, but also are the elicitors of immunity in plants. The presence might imitate pathogen-associated molecular patterns (PAMPs), leading to local and systemic defenses (Nabi et al., 2024). This is an immune process, condensing systemic acquired resistance

(SAR), an aggregation of pathogenesis-related (PR) proteins, phenolic compounds, and phytoalexins (El-Dougdoug et al., 2018). The application of foliar AgNPs on tomatoes has been shown to enhance the antioxidant defense by raising the activity of peroxidase (POD), polyphenol oxidase (PPO) and catalase (CAT) activities (Ashraf et al., 2025). These enzymes improve the enhancement of the plant-cell walls, remove the reactive species and reduce the entry and propagation of the viruses (Kaur et al., 2022). Also, the exposure of nanoparticles has been linked to lengthening and deposition of callase which improves structural fortifications to viral invasion. The nanoparticles not only possess their direct action on viruses, but they are also the elicitors of plant immune responses. The presence has the potential to mimic pathogen-associated molecular patterns (PAMPs), which triggers local and systemic responses (Nabi et al., 2024). This immune response implies the stimulation of systemic acquired resistance (SAR) that is the concentration of pathogenesis-related (PR) proteins, phenolic compounds, and phytoalexins (El-Dougdoug et al., 2018).

3.4 Reactive Oxygen Species (ROS) Generation Production.

Another significant antiviral activity is the capacity of nanoparticles to generate ROS either through redox cycling on their surfaces or through the regulation of plant metabolism (Arshad et al., 2023). Extreme amount of ROS can act in a dual mechanism of antiviral defense. On the one hand, ROS lead to oxidative injuries to viral elements, including, their nucleic acids, proteins, and lipid envelopes causing inactivation of viruses. Conversely, ROS are secondary messengers in plant signaling, which induce SAR and localization of responses (Haghpanah et al., 2025). Recent publications have clarified that ZnO, TiO_2 and AgNPs particularly have been effective in ROS-mediated inactivation due to the fact that they are reactive to surfaces at either light or physiological temperature. Such nanoparticles are not only disabling the integrity of viruses, but also initiating signaling waves by tissues that produce systemic resistance (Mohan, P. S. (2021). There is a fine balance in this mechanism however. As moderate production of ROS maximally increases antiviral immunity, excessive production of ROS can cause oxidative stress and phytotoxicity of plant tissues. This mechanism, in turn, demands nanoparticles functionalization, optical dosing, and delivery methods to leverage this and not hurt the health of the plants (Sachdev et al., 2021). The antiviral effect of nanoparticles is complex since it involves physical, molecular, biochemical and immunological processes. Direct viral inactivation can prevent entry and systemic dissemination, replication can be inhibited, the host resistance can be enhanced by inducing immunity, and the ROS can have both first-order antiviral and second-order signaling functions. The mechanisms do not work independently, but they work together to form a very strong multi-layered protection against plant viruses. The further understanding of these mechanisms, besides clarifying the antiviral capability of nanoparticles, sheds light on how safe, targeted, and crop-specific nano-formulations could be designed using safe nanoparticles in the management of the plant viruses sustainably.

Table 2. Mechanisms of antiviral action of metallic and metal oxide nanoparticles in plants

Mechanism	Mode of Action	Representative Findings	References
Direct Viral Inactivation	Binding of NPs to viral capsid proteins via electrostatic or Van der Waals forces; capsid denaturation; viral particle aggregation.	AgNPs bound to coat proteins of *Tomato mosaic virus* and *Potato virus Y*, reducing infectivity.	Arshad et al., 2023;
Inhibition of Replication	Interaction of NPs with viral nucleic acids and enzymes; inhibition of transcription, translation, and genome replication.	AgNPs reduced RNA accumulation and protein synthesis in PVY; disruption of nuclear import pathways.	Ghosh et al., 2022
Induction of Plant Defenses	Elicitation of systemic acquired resistance (SAR); upregulation of antioxidant enzymes and PR proteins:	AgNP foliar sprays in tomato enhanced POD, PPO, CAT activity; increased phenolics and lignification.	Ashraf et al., 2025
ROS Generation	Direct ROS production at NP surfaces or indirect induction through plant metabolic pathways; oxidative damage to viral components.	ZnO, TiO_2, and AgNPs induced ROS that inactivated viral proteins/nucleic acids and triggered SAR.	Sachdev et al., 2021

4. CASE STUDIES AND APPLICATIONS TO THE CROP SYSTEMS.

Metallic nanoparticles and metal oxide nanoparticles (NPs) have been greatly used in the agricultural sector in treating plant viral diseases. In numerous studies, NPs have been reported to have been effective in controlling of the viral infection, enhancement of immunity of the plants, and improved growth parameters of the various crops. These case studies provide an insight into how nanoparticles can be used in crop protection programs without undermining the principles of environmental or biosafety (Table 3).

4.1 Tomato (*Solanum lycopersicum*).

Some viruses that have been discovered to infect tomato plants include Tomato mosaic virus (ToMV) and Tomato spotted wilt virus (TSWV). It has been mentioned that foliar application of silver nanoparticles (AgNPs) could be a significant one as it reduces the viral load by direct contact with the viral particles, as well as by interfering with their reproduction (El-Dougdoug et al., 2018; Luceri et al., 2023). Systemic acquired resistance (SAR) also enhanced tomato in AgNP treatment activating antioxidants enzyme like peroxidase (POD), polyphenol oxidase (PPO) and catalase (CAT) and therefore preparing the plant in advance of future infections. In the same vein, zinc oxide (ZnO) nanoparticles as foliar spray or soil amendment were reported to have a greater capacity to promote viral resistance in tomato plants through the regulation of the expression of genes related to plant defense and the production of reactive oxygen species (ROS) (Singh et al., 2022; Alavi et al., 2022). The studies also point out the potential of NPs as antiviral agents but also as inducers of plant immunity.

4.2 Potato (*Solanum tuberosum*)

Potato virus Y (PVY) is among the main viral pathogens, and it affects the productivity of potatoes in the international environment. The AgNPs of the size range of 10-20 nm have already been reported to suppress the PVY replication since they can combine to bind to the coat proteins of the virus, resulting in the denaturation of the capsid and the prevention of the viral cells entry (Vargas-Hernandez et al., 2020b;

El-Dougdoug et al., 2018). The treatment led to the reduction of systemic infection, more significant expression of pathogenesis-related proteins, and it is suggestive of the direct antiviral action and plant defense mechanism activation, too. Fe3O4 nanoparticles have also been used as foliar sprays to enhance potato to protect against oxidative stress caused by PVY infection because of the enzyme antioxidant activities (Cai et al., 2020; Singh et al., 2022). Such findings demonstrate that the viral infections in tuber crops can be controlled using metallic and metal oxide nanoparticles in a strategic manner.

4.3 *Nicotiana benthamiana*

Nicotiana benthamiana has proved to be an extremely popular model plant to investigate viruses-host interaction. Uptake studies revealed that Fe_3O_4 nanoparticles applied onto the leaves were translocated systemically into the phloem and xylem and reacted with cell organelles and activated defense processes (Avellan et al., 2021; Cai et al., 2020). The nanoparticle treatments enhanced the amount of ROS, antioxidant enzymes and defense related metabolites with enhanced resistance to Tobacco mosaic virus (TMV). Moreover, ZnO and Ag nanoparticles pre-treated seeds resulted in premature NP internalization in the process of germination, and enhanced physiological resilience and virile tolerance at an early stage (Khan et al., 2022; Singh et al., 2017). These studies have demonstrated that nanoparticles offer a two-pronged benefit on growth and antiviral immunity on this agronomic crop.

4.4 Other Crops

Non- tomato, potato and Nicotiana crops have also been noted to have recorded good results of NPs cucumber, pepper and maize. The copper oxide (CuO) and magnesium oxide (MgO) nanoparticles have demonstrated antiviral effects as they have the potential to produce ROS and can induce the denaturation of viral proteins besides raising the production of plant-defensive mechanisms (Vargas-Hernandez et al., 2020b; Singh et al., 2022). Furthermore, small interfering RNAs (siRNAs) have successfully been transported using functionalized nanocarrier into viral genes to provide the correct and targeted inhibition of the viral multiplication (Cruz-Luna et al., 2023; Alavi et al., 2022).

4.5 Incorporation in Crop Protection Strategies.

The opportunities of nanoparticles application in the existing crop protection systems is an opportunity to reduce the reliance on the traditional chemical pesticides and to have sustainable management of the virus. They are applied as foliar sprays, soil amendments, seed priming and vectors of RNAi-based antiviral molecules with the help of NPs. According to pilot-scale experiments, the best doses of nanoparticle can avert diseases (to half), expand the parameters of crop growth, and augment yield without causing severe effects of phytotoxicity (Francis et al., 2024; Raza et al., 2024). Besides, the green-synthesized NPs with the conventional management approach would offer a synergistic effect since both the antiviral and growth-promoting effects of plants may be applied. This type of integrated action is necessary in the sustainability of crop production and environmental sustainability in modern farming. Case studies have been carried out in different crops that have shown that metallic and metal oxide nanoparticles are an effective way of controlling the plants virus. Their diverse capabilities are backed up by the variety of their mechanisms of action including direct viral inactivation and triggering plant defence response as well as their potential application as RNAi delivery vehicles. Nanoparticle-based approaches in combination with the ideal dose and in combination with conventional techniques can assist in protecting the crops, improving crop growth and reducing the usage of hazardous chemical pesticides, which can result in sustainable and resilient agricultural systems.

Table 3. Case studies of metallic and metal oxide nanoparticles for plant virus management in crop systems

Crop	Virus	Nanoparticle Type	Mechanism of Action	Application Method	Key Observations / Benefits	References
Tomato (Solanum lycopersicum)	Tomato mosaic virus (ToMV), Tomato spotted wilt virus (TSWV)	AgNPs	Direct viral inactivation, induction of SAR, increased antioxidant enzymes	Foliar spray	Reduced viral load, enhanced defense enzyme activity (POD, PPO, CAT), primed immunity	El-Dougdoug et al., 2018; Luceri et al., 2023
Tomato	ToMV	ZnO NPs	Modulation of plant defense genes, ROS production	Foliar spray / Soil amendment	Enhanced viral resistance, improved growth	Singh et al., 2022; Alavi et al., 2022
Potato (Solanum tuberosum)	Potato virus Y (PVY)	AgNPs	Capsid binding, inhibition of replication, systemic infection suppression	Foliar spray	Reduced systemic viral infection, increased PR proteins	Vargas-Hernandez et al., 2020a; El-Dougdoug et al., 2018
Potato	PVY	Fe_3O_4 NPs	Activation of antioxidant enzymes	Foliar spray	Mitigation of oxidative stress, enhanced viral tolerance	Cai et al., 2020; Singh et al., 2022
Nicotiana benthamiana	Tobacco mosaic virus (TMV)	Fe_3O_4 NPs	Uptake and systemic translocation, ROS induction	Foliar spray	Increased resistance, systemic defense activation	Avellan et al., 2021; Cai et al., 2020
Nicotiana benthamiana	TMV	AgNPs / ZnO NPs	Seed priming, early-stage uptake, immune priming	Seed priming	Enhanced germination, early viral defense	Khan et al., 2022; Singh et al., 2017
Cucumber / Pepper / Maize	Various plant viruses	CuO, MgO, functionalized nanocarriers	ROS generation, RNAi delivery, viral protein denaturation	Foliar / Soil / Nanocarrier delivery	Virus suppression, enhanced defense, improved growth	Vargas-Hernandez et al., 2020b; Singh et al., 2022; Cruz-Luna et al., 2023

There is a great potential of sustainable management of plant viruses using metallic and metal oxide nanoparticles that have been applied to crop systems. Case studies of a variety of crops, such as tomato, potato, maize, and *Nicotiana benthamiana*, indicate that nanoparticles, like AgNPs, ZnO, Fe_3O_4, TiO_2 and MgO, can effectively diminish the effects of viral infections via various mechanisms, such as direct viral inactivation, replication inhibition, ROS-mediated injury or SAR. The effects are added further by the application of functionalized nanocarriers that have the ability to deliver accurate quantities of antiviral molecules such as siRNA and dsRNA. Despite the importance of uptake and translocation mechanisms, nanoparticle diameter, surface charge and functionalization in antiviral actions, dosage and route of application are to be optimized in order to minimize phytotoxicity and health preservation of plants. Collectively, these studies can bring out the opportunities associated with nanoparticles as effective and multifunctional in crop protection as an alternative to conventional chemical pesticides that may be harmful to the environment and a complement to the sustainable and efficient agricultural practices.

5. NANOPARTICLE ENVIRONMENTAL AND BIOSAFETY IN AGRICULTURE

Despite the enormous potential of metallic and metal oxide nanoparticles (NPs) in the use of plant viruses, crop protection even though the application of nanoparticles in large scale encounters serious biosafety and environmental concerns. Such concerns demand an extensive understanding in the achievement of sustainable and accountable introduction of nanotechnology in agriculture.

5.1 Nanoparticle Accumulation in Soils and Plants

There is also the risk of aggregation of the metallic and metal oxide NP in the soil and plant tissues in agriculture hence affecting the food security and the plant health. Phytotoxicity is one of the most significant issues as NPs can interfere with seed germination, root and shoot development, biomass development, and physiological/biochemical processes. The extent of phytotoxicity is relative to the level of concentration of the nanoparticles, size, shape, surface charge, and exposure period. It is demonstrated that the low concentration of ZnO and Fe3O4 NPs can stimulate the germination and growth of seeds, and the high concentrations of ZnO, CuO, and NiO NPs can inhibit the photosynthetic photosynthesis, cause oxidative stress, and result in the fragmentation of DNA (Siddiqui and Husen, 2017). In addition, NPs can migrate between roots to shoots and leaves and can also be concentrated in plant edible structures and present food safety concerns. In case of an example, ZnO and CuO NPs tend to be found in roots only partially but transported to shoots, and CeO_2 NPs tend to be found in roots more (Pietrzak et al., 2024). The interaction of nanoparticles with organic matter, their aggregation, and soil properties also play a major role in influencing NP bioavailability, uptake, and retention in the environment (Denisse et al., 2021).

5.2 Effects on Non-Target Organisms and Ecosystems

The emission of the NPs into agricultural soils can have adverse effects on the non-target organisms and ecosystem processes. NPs may destabilize soil microbial communities, break nutrient cycling, disrupt organic matter, and benefit plant microbe interactions. Examples of such toxic compounds include CuO and Ag NPs, which may affect soil fertility because these compounds kill beneficial soil bacteria and fungi (Maurice, N. 2023). At certain levels, NPs can also be harmful to aquatic organisms, such as algae, invertebrates, and fish (Siddique and Husen, 2017; Benson et al., 2022). Another important issue is trophic transfer because plants can act as access points to NPs in both terrestrial and aquatic food webs (Dang et al., 2021). The build-up in edible parts of the plants may impact livestock and humans but the magnitude of the effects is yet to be fully comprehended (Angon et al., 2024). The possibility of biomagnification and lasting ecological impacts highlights the necessity for thorough risk evaluations and continuous monitoring (Rastogi et al., 2017; Elsharkawy et al., 2025).

5. 3 Guidelines and Regulatory Frameworks

There are few regulations in use of nanoparticles in agriculture, in spite of the growing use. In the majority of contemporary risk assessment, the traditional methods of chemical testing are used, and they might not be suitable to reflect the specific characteristics of NPs (Reagen et al., 2022). The development of agriculturally specific protocols of NP synthesis, characterization, application, and environmental risk assessment is urgently needed (Raza et al., 2024). Plant extracts or microorganisms-based green methods are more environmentally friendly and biocompatible and less phytotoxic and retain antiviral activity (Yiblet and Sisay, 2024). Regulatory frameworks of this kind must establish laws for the approval, labeling, and monitoring of NP-based agricultural products, considering environmental fate, toxicity, and food safety (Mangla et al., 2025). International cooperation is essential to align regulations and ensure the responsible deployment of NPs (Isibor, P. O. 2024).

5. 4 Risk-Benefit Analysis

The safe use of metallic and metal oxide NPs needs an in-depth risk-benefit analysis. Some of the advantages are the better management of diseases, less dependence on chemical pesticides, better crop productivity and helping in sustainable agriculture. It has been reported that NPs can improve crop yields up to 20 percent, decrease the level of disease by up to 50 percent, and cut down on the nutrient runoff and enhance the level of carbon sequestration in the soil (Paramesh et al., 2023). Such advantages should be compared to the possible risks, such as phytotoxicity, NP concentrations in plants and soils, impacting non-target organisms, and food chain penetration (Bouhadi et al., 2025). There is potential of oxidative stress, metabolic problems and DNA damage in plants, toxicity to soil and aquatic life at higher NP concentrations. It is essential to optimize NP dosages and do long-term field studies to make sure that it is safe to the environment and benefits the most (Elbasiouny et al., 2022).

6. FUTURE PROSPECTS AND RESEARCH DIRECTIONS

Metallic and metal oxide nanoparticles (NPs) on the repression of plant viruses is a prospective area in the sustainable agriculture. With the improvement in research, it is increasingly focusing on development of nanoparticles that are highly effective in combating plant viruses, as well as being environmentally friendly. The future will be based on rational design of the nanoparticles, optimization of application strategies and combination with complex molecular tools.

6. 1 Designing Biocompatible, Green-Synthesized Nanoparticles

Some conventional physical and chemical techniques of nanoparticle production typically use toxic substances, consume large amounts of energy, and produce toxic byproducts. Unlike that, an eco-friendly alternative is offered by green synthesis with plant extracts, microorganisms, or agricultural residues, which has better biocompatibility and lower environmental impact (Gayathiri et al., 2022 ;Jeevanandam et al., 2022; Akintelu et al., 2020; Begum et al., 2022). Plant-mediated synthesis employs the use of phytochemicals (polyphenols, flavonoids, terpenoids and proteins) as natural reducing and capping agents. In addition to their role in stabilizing nanoparticles, these biomolecules can enhance antiviral efficacy by regulating particle size, morphology, and surface chemistry (Radulescu et al., 2023; Mandal et al., 2022; Lithi et al., 2025). Microbial synthesis refers to complementary methods applied to synthesize nanoparticles exhibiting distinctive surface characteristics that can be utilized in plants by using bacteria, fungi, or algae (Nadaf et al., 2022; Khan et al., 2022). Green synthesis still needs to be optimized for scalability, reproducibility, and cost-efficiency. The parameters like pH, temperature, precursor concentration, and reaction time should be standardized so as to maintain a constant quality of nanoparticles. Moreover, bioresource conversion of agricultural waste to a bioresource is in line with the principles of the circular economy, as it helps to ensure sustainable production.

6. 2 Functionalization for Targeted Delivery and Enhanced Efficiency

Nanoparticles may be functionalized with ligands, antibodies, peptides or plant-derived biomolecules to deliver them to the target in a process that enables maximum antiviral efficacy (Begum et al., 2022; Radulescu et al., 2023). Functionalization minimizes the off-target effects and increases the local concentration of antiviral agents. Additional improvement is with environmental-responsive nanoparticles, which accumulate their payload when triggered by certain events, like varying pH, exposure to light, or enzyme activity. The example provided is that green-synthesized ZnO and CuO nanoparticles, which are capped with phytochemicals, have a synergistic antivirality effect. Likewise, functionalized gold and silver nanoparticles with bioactive capping agents are more stable, biocompatible, and anti-viral (Begum et al., 2022; Mandal et al., 2022; Aigbe and Osibote, 2024). A new application is the RNA interference (RNAi) nanocarriers, which present small interfering RNA (siRNA) or double stranded RNA (dsRNA). These nanocarriers ensure that RNA molecules do not degrade, they facilitate their uptake by the plant cells and they offer a precise method of gene silencing viruses, a next-generation approach to sustainable virus control.

6.3 Enhancing Antiviral Efficiency While Minimizing Risks

Although an enhancement of the antiviral activity is essential, reduction of the phytotoxicity and environmental effect is equally important. Nanoparticles synthesized with green synthesis have a reduced toxicity rate intrinsically and natural capping agents additionally raise the biocompatibility (Jeevanandam et al., 2022; Devi et al., 2023). Further studies ought to be done on the biodegradable nanoparticles with a degradation rate that can be adjusted to behave safely in soil once they have served their antiviral purpose. The same can be prevented by controlled release and environmental triggers in the accumulation in soils and food chains. The safe implementation practice should be accompanied by the thorough risk assessment, i.e., the fate of the nanoparticles, its transportation, and transformation in the agricultural ecology (Olga et al., 2022; Behzad et al., 2020; Ahmad et al., 2019).

6.4 Integration with Advanced Plant Protection Strategies

Next-generation nanoparticles can be used in the future to enhance antiviral-activity by combining them with gene-editing, RNAi technologies, and plant immune modulators. Under these combined strategies, one can acquire a multifaceted control, the direct inhibition of the viruses in combination with developing systemic resistance and priming the plant innate immunity. The interdisciplinary studies carried out on nanotechnology, plant pathology, molecular biology and environmental sciences will be important in the transfer of laboratory research to applications that are applicable in the field. There is a need to conduct long-term research that is pertinent to determine persistence, cumulative effects and ecological safety of nanoparticles under different agroecosystems of differing conditions. The metallic and metal oxides nanoparticles have a bright future in plant virus suppression, but the design, application, as well as environmental impacts of this nanoparticles must be appropriately regulated. Various alternatives, including green-synthesized and functionally modified nanoparticles, offer a new, greener, and more efficient option to chemical pesticides, and the targeted delivery and RNAi-based technologies promise to have an opportunity to control viruses precisely. Nanotechnology, plant immunity, and molecular techniques combined allow sustainable agriculture to have an improved protection of crops with a minimal ecological price (Figure 3).

Figure 3. Key considerations for nanotechnology-based plant virus control, including antiviral action, integration with gene editing and RNAi, interdisciplinary collaboration, targeted viral inhibition, long-term ecological safety research, and regulated application to minimize environmental impacts

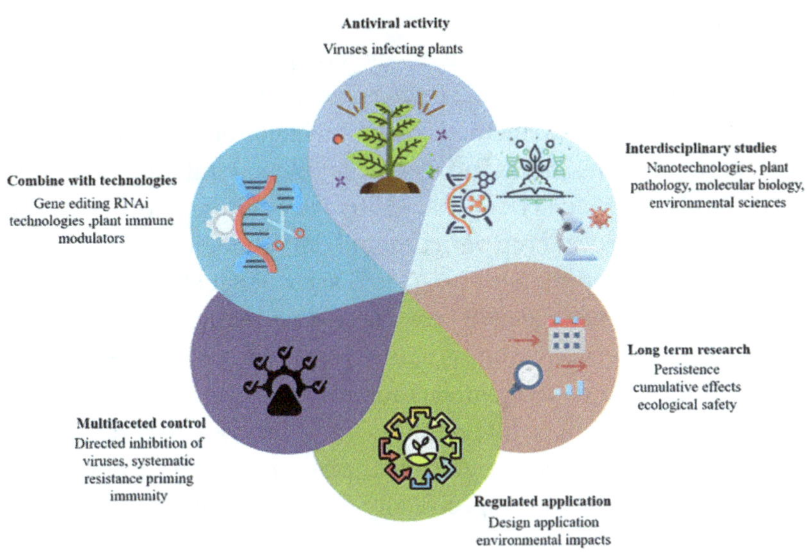

CONCLUSION

Metallic and metal oxide nanoparticles (NPs) have become promising candidates to suppress plant viruses and provide multi-mechanistic actions to address the virus, which include direct viral inactivation, inhibition of viral replication, stimulation of plant defense response, and the production of reactive oxygen species (ROS) (Singh et al., 2022; Vargas-Hernandez et al., 2020b; Alavi et al., 2022; Luceri et al., 2023). Silver nanoparticles (AgNPs), zinc oxide (ZnO), iron oxide (Fe_3O_4) and titanium dioxide (TiO_2) have demonstrated high efficiency in the prevention of systemic infection and enhancing plant immunity across numerous crop systems, including tomatoes, potatoes and Nicotiana (El-Dougdoug et al., 2018; Cai et al., 2020; Yiblet and Sisay, 2024). These NPs have physicochemical characteristics such as size, surface charge, and chemical reactivity in close relationship with their antiviral activity because such factors determine the interactions with the viral particles and plant tissues (Vargas-Hernandez et al., 2020b; Singh et al., 2022; Alavi et al., 2022). They are taken up by the roots, foliar surfaces, or seed priming and translocated to the tissues at risk of viral infection through the xylem and phloem (Cai et al., 2020; Avellan et al., 2021; Khan et al., 2022). Such processes do not only improve the antiviral activity but activate the plant defense mechanisms, such as systemic acquired resistance (SAR) and antioxidant enzyme induction and provide long-term protection against viral pathogens (El-Dougdoug et al., 2018; Singh et al., 2022; Yiblet and Sisay, 2024). The case studies of greenhouse and field experiments demonstrate the high level of practical applicability of NPs in crop protection in which AgNPs or ZnO NPs treatments decreased viral titers and enhanced the growth and yield parameters of plants. The strategies involving foliar applications and seed priming have proved to increase resistance and systemic immunity at early stages, as well as decrease the phytotoxic impacts at the optimum concentrations (Khan et al., 2022; Cai et al., 2020; Singh et al., 2022). These encouraging findings notwithstanding, environmental and biosafety issues are also vital. Concentration of NP in soils and plants, possible toxicity in non-target organisms and food chain should be effectively evaluated (Siddiquei and Husen, 2017; Rastogi et al., 2017). Plant-derived biomolecular functionalized nanoparticles green-synthesized are a viable sustainable and less toxic and more bio-compatible alternative, and comply with the environmentally responsible agricultural standards (Jeevanandam et al., 2022; Begum et al., 2022; Yiblet and Sisay, 2024). The implementation of nanoparticle-based antiviral methods may be a major breakthrough in the context of South Asia and Pakistan, where staple crops are extremely susceptible to viral infection, and the application of conventional control methods is frequently constrained by the environmental and economic conditions. Locally adapted NPs that support local crop varieties and environmental conditions have the potential to enhance yield stability, minimized reliance on unsafe agrochemicals and sustainable agriculture in the area. Altogether, MNPs and metal oxide NPs are a new generation antiviral strategy, which has both direct and indirect mechanisms of virus inhibition and promotion of plant defense. Further studies ought to be done in order to optimize the dosage, application and formulation approaches and to undertake long-term environmental risk evaluation. Integrating nanotechnology and RNA interference (RNAi), functionalized delivery vehicles and plant immune modulators are capable of changing the way that plant viruses are managed in the world and more specifically South Asia and Pakistan and eventually result in food security and sustainable agriculture.

Practical Outlook: Field Testing Nanoparticles in South Asian Smallholder Systems

The successful implementation of nanoparticle (NP)-based antiviral technologies in South Asia requires context-specific strategies aligned with the socioeconomic and agroecological conditions of smallholder farming systems. Field validation should be conducted through farmer-participatory and community-based trials that integrate scientific rigor with local agronomic practices. The use of green-synthesized and biodegradable nanoparticles, such as zinc oxide (ZnO), iron oxide (Fe_3O_4), and silver nanoparticles (AgNPs) derived from plant extracts, offers an eco-compatible alternative for evaluating antiviral efficacy under real farming conditions. These nanoparticles can be formulated into low-cost foliar sprays or seed-priming treatments, enabling scalable and resource-efficient application.

REFERENCES

Abbasi, A.. (2025). Restoring Wheat Productivity and Nutrient Balance Under Cadmium Stress Through Reducing Toxicity, Metal Uptake, and Oxidative Damage Using Selenium Nanoparticles. *Journal of Trace Elements in Medicine and Biology, 127644,* •••. https://doi.org/https://doi.org/10.1016/j.jtemb.2025.127644

Aguilar, C. H., Pires, D., Cortaga, C., Peja, R.Jr, Cruz, M. A., Langres, J., & Balendres, M. A. (2025). Harnessing Legume Productivity in Tropical Farming Systems by Addressing Challenges Posed by Legume Diseases. *Nitrogen*, *6*(3), 65.

Ahmad, F., Ashraf, N., Ashraf, T., Zhou, R., & Yin, D. (2019). Biological synthesis of metallic nanoparticles (MNPs) by plants and microbes: Their cellular uptake, biocompatibility, and biomedical applications. *Applied Microbiology and Biotechnology*, *103*(7), 2913–2935. DOI: 10.1007/s00253-019-09675-5 PMID: 30778643

Aigbe, U., & Osibote, A. (2024). Green synthesis of metal oxide nanoparticles and their various applications. *Journal of Hazardous Materials Advances.* Advance online publication. DOI: 10.1016/j.hazadv.2024.100401

Akintelu, S., Folorunso, A., Folorunso, F., & Oyebamiji, A. (2020). Green synthesis of copper oxide nanoparticles for biomedical application and environmental remediation. *Heliyon*, *6*, e04508. DOI: 10.1016/j.heliyon.2020.e04508 PMID: 32715145

Ali, M., Ahmed, T., Wu, W., Hossain, A., Hafeez, R., Masum, M., Wang, Y., An, Q., Sun, G., & Li, B. (2020). Advancements in plant and microbe-based synthesis of metallic nanoparticles and their antimicrobial activity against plant pathogens. *Nanomaterials (Basel, Switzerland)*, *10*(6), 1146. DOI: 10.3390/nano10061146 PMID: 32545239

Ali, S., Ulhassan, Z., Shahbaz, H., Kaleem, Z., Yousaf, M. A., Ali, S., & Zhou, W. (2024). Application of magnesium oxide nanoparticles as a novel sustainable approach to enhance crop tolerance to abiotic and biotic stresses. *Environmental Science. Nano*, *11*(8), 3250–3267.

Aljabali, A. A., Hassan, S. S., Pabari, R. M., Shahcheraghi, S. H., Mishra, V., Charbe, N. B., & Tambuwala, M. M. (2021). The viral capsid as novel nanomaterials for drug delivery. *Future Science OA*, *7*(9), FSO744. PMID: 34737885

Andrade-Piedra, J. L., Sharma, K., Kroschel, J., Ogero, K., Kreuze, J., Legg, J. P., & Garrett, K. A. (2025). Phytosanitary Challenges and Solutions for Roots and Tubers in the Tropics. *Annual Review of Phytopathology*, *63*(1), 627–650. PMID: 40903427

Angon, P. B., Islam, M. S., Das, A., Anjum, N., Poudel, A., & Suchi, S. A. (2024). Sources, effects and present perspectives of heavy metals contamination: Soil, plants and human food chain. *Heliyon*, *10*(7). PMID: 38590838

Anjum, S., Mustaq, Z., Sarwar, M., Anees, M., Abbas, T., Maqsood, S., ... & Talha, M. (2024). Journal of Agriculture and Biology.

Arshad, F., Naikoo, G., Hassan, I., Chava, S., El-Tanani, M., Aljabali, A., & Tambuwala, M. (2023). Bioinspired and green synthesis of silver nanoparticles for medical applications: A green perspective. *Applied Biochemistry and Biotechnology*, *196*(7), 3636–3669. DOI: 10.1007/s12010-023-04719-z PMID: 37668757

Ashraf, H., Anjum, T., Ahmad, I. S., Ahmed, R., Aftab, Z. E. H., & Rizwana, H. (2025). Phytofabricated silver nanoparticles unlock new potential in tomato plants by combating wilt infection and enhancing plant growth. *Scientific Reports*, *15*(1), 10527. PMID: 40148421

Avellan, A., Yun, J., Morais, B., Clement, E., Rodrigues, S., & Lowry, G. V. (2021). Critical review: Role of inorganic nanoparticle properties on their foliar uptake and in planta translocation. *Environmental Science & Technology*, *55*(19), 13417–13431. DOI: 10.1021/acs.est.1c00178 PMID: 33988374

Bahrulolum, H., Nooraei, S., Javanshir, N., Tarrahimofrad, H., Mirbagheri, V. S., Easton, A. J., & Ahmadian, G. (2021). Green synthesis of metal nanoparticles using microorganisms and their application in the agrifood sector. *Journal of Nanobiotechnology*, *19*(1), 86. PMID: 33771172

Barathi, S., Ramalingam, S., Krishnasamy, G., & Lee, J. (2024). Exploring the biomedical frontiers of plant-derived nanoparticles: Synthesis and biological reactions. *Pharmaceutics*, *16*(7), 923. PMID: 39065620

Begum, S., Pratibha, S., Rawat, J., Venugopal, D., Sahu, P., Gowda, A., Qureshi, K., & Jaremko, M. (2022). Recent advances in green synthesis, characterization, and applications of bioactive metallic nanoparticles. *Pharmaceuticals*, *15*(4), 455. DOI: 10.3390/ph15040455 PMID: 35455452

Behzad, F., Naghib, S., Kouhbanani, M., Tabatabaei, S., Zare, Y., & Rhee, K. (2020). An overview of the plant-mediated green synthesis of noble metal nanoparticles for antibacterial applications. *Journal of Industrial and Engineering Chemistry*. Advance online publication. DOI: 10.1016/j.jiec.2020.12.005

Benson, N. U., Agboola, O. D., Fred-Ahmadu, O. H., De-la-Torre, G. E., Oluwalana, A., & Williams, A. B. (2022). Micro (nano) plastics prevalence, food web interactions, and toxicity assessment in aquatic organisms: A review. *Frontiers in Marine Science*, *9*, 851281.

Bhutkar, M., Singh, V., Dhaka, P., & Tomar, S. (2022). Virus-host protein-protein interactions as molecular drug targets for arboviral infections. *Frontiers in Virology*, *2*, 959586.

Boruah, J. S., & Chowdhury, D. (2022). Advances in carbon nanomaterial–clay nanocomposites for diverse applications. *Minerals (Basel)*, *13*(1), 26.

Bouhadi, M., Javed, Q., Jakubus, M., Elkouali, M. H., Fougrach, H., Ansar, A., & Černe, M. (2025). Nanoparticles for sustainable agriculture: Assessment of benefits and risks. *Agronomy (Basel)*, *15*(5), 1131.

Cai, L., Cai, L., Jia, H., Liu, C., Wang, D., & Sun, X. (2020). Foliar exposure of Fe_3O_4 nanoparticles on *Nicotiana benthamiana*: Evidence for nanoparticles uptake, plant growth promotion and defense response elicitation against plant virus. *Journal of Hazardous Materials*, *393*, 122415. DOI: 10.1016/j.jhazmat.2020.122415 PMID: 32143161

Cai, L., Liu, C., Fan, G., Liu, C., & Sun, X. (2019). Preventing viral disease by ZnONPs through directly deactivating TMV and activating plant immunity in Nicotiana benthamiana. *Environmental Science. Nano*, *6*(12), 3653–3669.

Cruz-Luna, A., Vasquez-López, A., Rojas-Chávez, H., Valdés-Madrigal, M., Cruz-Martínez, H., & Medina, D. (2023). Engineered metal oxide nanoparticles as fungicides for plant disease control. *Plants*, *12*(13), 2461. DOI: 10.3390/plants12132461 PMID: 37447021

Dang, F., Huang, Y., Wang, Y., Zhou, D., & Xing, B. (2021). Transfer and toxicity of silver nanoparticles in the food chain. *Environmental Science. Nano*, *8*(6), 1519–1535.

Deng, Z., Ma, L., Zhang, P., & Zhu, H. (2022). Small RNAs participate in plant–virus interaction and their application in plant viral defense. *International Journal of Molecular Sciences*, *23*(2), 696. PMID: 35054880

Denisse, V. G. J., Roberto, S. C. C., Hermes, P. H., Patricio, T. G. A., Andrea, P. M., & Fabián, F. L. (2021). Influence of nanoparticles on the physical, chemical, and biological properties of soils. In *Nanomaterials for soil remediation* (pp. 151–182). Elsevier.

Devendran, R., Kumar, M., Ghosh, D., Yogindran, S., Karim, M. J., & Chakraborty, S. (2022). Capsicum-infecting begomoviruses as global pathogens: Host–virus interplay, pathogenesis, and management. *Trends in Microbiology*, *30*(2), 170–184. PMID: 34215487

Devi, D., Julkapli, M., Sagadevan, S., & Johan, M. (2023). Eco-friendly green synthesis approach and evaluation of environmental and biological applications of iron oxide nanoparticles. *Inorganic Chemistry Communications*. Advance online publication. DOI: 10.1016/j.inoche.2023.110700

Dey, S. R., Das, R., & De, M. (2024). A brief review on present status of rice tungro disease: Types of viruses, vectors, occurrence, symptoms, control and resistant rice varieties. *Int. J. Adv. Life Sci. Res*, *7*, 15–23.

Eid, N. A., Ibrahim, A. M., Elsharawy, A. A., & Salem, K. F. (2025). Nanofertilizers for Plant Viral Disease Management. In *Nanofertilizers for Sustainable Agriculture: Assessing Impacts on Health, Environment, and Economy* (pp. 149–187). Springer Nature Switzerland.

Eid, N. A., Ibrahim, A. M., Elsharawy, A. A., & Salem, K. F. (2025). Nanofertilizers for Plant Viral Disease Management. In *Nanofertilizers for Sustainable Agriculture: Assessing Impacts on Health, Environment, and Economy* (pp. 149–187). Springer Nature Switzerland.

El-Dougdoug, N. K., Bondok, A. M., & El-Dougdoug, K. A. (2018). Evaluation of silver nanoparticles as antiviral agent against ToMV and PVY in tomato plants. *International Journal of Nanotechnology*, *15*(7/8/9/10), 608–621. https://doi.org/DOI: 10.1504/IJNT.2018.093164

Elahi, M.. (2025). Nitrogen Nano-fertilizers Enhance Wheat Productivity, Improve Nitrogen Use Efficiency, and Mitigate Soil Nitrogen Losses While Maintaining Grain Quality Compared to Conventional Fertilizers. *BioNanoScience*, *15*(3), 477. DOI: 10.1007/s12668-025-02065-4

Elbasiouny, H., Elbehiry, F., & El-Ramady, H. (2022). Toxic effects of nanoparticles under combined stress on plants. In *Toxicity of Nanoparticles in Plants* (pp. 109–129). Academic Press.

Elsharkawy, E. E., & Sharkawy, A. A. (2025). Environmental risk assessment (ERA). *Archives of Clinical Toxicology*, *7*(1), 46–50.

Emeraghi, M., Achigan-Dako, E. G., Nwaoguala, C. N., & Oselebe, H. (2021). Maize streak virus research in Africa: An end or a crossroad. *Theoretical and Applied Genetics*, *134*(12), 3785–3803. PMID: 34309683

Falak, A.. (2025). Efficacy of ascorbic acid coated quantum dots in alleviating lead-induced oxidative damage and enhancing growth parameters in rice (Oryza sativa L.) for sustainable cultivation. *Journal of Trace Elements in Medicine and Biology*, *88*, 127603. DOI: 10.1016/j.jtemb.2025.127603 PMID: 39847985

Farooq, T., Adeel, M., He, Z., Umar, M., Shakoor, N., da Silva, W., & Rui, Y. (2021). Nanotechnology and plant viruses: An emerging disease management approach for resistant pathogens. *ACS Nano*, *15*(4), 6030–6037. PMID: 33761237

Francis, D., Abdalla, A., Mahakham, W., Sarmah, A., & Ahmed, Z. (2024). Interaction of plants and metal nanoparticles: Exploring molecular mechanisms for sustainable agriculture and crop improvement. *Environment International*, *190*, 108859. DOI: 10.1016/j.envint.2024.108859 PMID: 38970982

Francis, D. V., Asif, A., & Ahmed, Z. F. R. (2024). Nanoparticle-enhanced plant defense mechanisms harnessed by nanotechnology for sustainable crop protection. *Nanoparticles in Plant Biotic Stress Management*, 451-491.

Gayathiri, E., Prakash, P., Karmegam, N., Varjani, S., Awasthi, M. K., & Ravindran, B. (2022). Biosurfactants: Potential and eco-friendly material for sustainable agriculture and environmental safety—a review. *Agronomy (Basel)*, *12*(3), 662.

Ghosh, U., Sayef Ahammed, K., Mishra, S., & Bhaumik, A. (2022). The emerging roles of silver nanoparticles to target viral life cycle and detect viral pathogens. *Chemistry, an Asian Journal*, *17*(5), e202101149. PMID: 35020270

Gurunathan, S., Qasim, M., Choi, Y., Do, J. T., Park, C., Hong, K., & Song, H. (2020). Antiviral potential of nanoparticles—Can nanoparticles fight against coronaviruses? *Nanomaterials (Basel, Switzerland)*, *10*(9), 1645. PMID: 32825737

Haghpanah, M., Namdari, A., Kaleji, M. K., Nikbakht-Dehkordi, A., Arzani, A., & Araniti, F. (2025). Interplay Between ROS and Hormones in Plant Defense Against Pathogens. *Plants*, *14*(9), 1297. PMID: 40364326

Hassan, N., Hussain, M., Mushtaq, F., Ali, S., & Chaudary, M. S. S. (2023). Cotton Leaf Curl Virus (CLCuV): An Insight into Disaster: Cotton Leaf Curl Virus (CLCuV). *Futuristic Biotechnology*, 02-09.

Hossain, M. M., Sultana, F., Mostafa, M., Ferdus, H., Kundu, M., Adhikary, S., & Sabbir, M. A. A. (2025). Stress responses and mechanisms of phytopathogens infecting humans: Threats, drivers, and recommendations. *Stresses*, *5*(2), 28.

Huded, S., Savadatti, E., Sharanappa, C. H., Deb, L., Borah, A., Singh, V., & Panigrahi, C. K. (2023). Examining modern strategies for effective and sustainable agricultural plant protection techniques: A review. *Int. J. Environ. Climate Change*, *13*, 1331–1343.

Hussain, F. S., Abro, N. Q., Ahmed, N., Memon, S. Q., & Memon, N. (2022). Nano-antivirals: A comprehensive review. *Frontiers in Nanotechnology*, *4*, 1064615.

Isibor, P. O. (2024). Regulations and policy considerations for nanoparticle safety. In *Environmental nanotoxicology: combatting the minute contaminants* (pp. 295–316). Springer Nature Switzerland.

Jabran, M., Ali, M. A., Muzammil, S., Zahoor, A., Ali, F., Hussain, S., & Gao, L. (2024). Exploring the potential of nanomaterials (NMs) as diagnostic tools and disease resistance for crop pathogens. *Chemical and Biological Technologies in Agriculture*, *11*(1), 75.

Jain, D., & Kothari, S. L. (2014). Green synthesis of silver nanoparticles and their application in plant virus inhibition. *Journal of Mycology and Plant Pathology*, *44*(1), 21–24.

Javed, R., Khan, B., Sharafat, U., Bilal, M., Galagedara, L., Abbey, L., & Cheema, M. (2024). Dynamic interplay of metal and metal oxide nanoparticles with plants: Influencing factors, action mechanisms, and assessment of stimulatory and inhibitory effects. *Ecotoxicology and Environmental Safety*, *271*, 115992. DOI: 10.1016/j.ecoenv.2024.115992 PMID: 38262092

Jeevanandam, J., Kiew, S., Boakye-Ansah, S., Lau, S., Barhoum, A., Danquah, M., & Rodrigues, J. (2022). Green approaches for the synthesis of metal and metal oxide nanoparticles using microbial and plant extracts. *Nanoscale*. Advance online publication. DOI: 10.1039/d1nr08144f PMID: 35133391

Joshi, H. (2025). Exploring the efficacy of green nanoparticles in enhancing plant defense: A mechanistic investigation into immune response activation. *Journal of Nanoparticle Research*, *27*(2), 29.

Karthikeyan, B., Gnanakumar, G., & Alphonsa, A. T. (2023). Nano metal oxides. In *Nano Metal Oxides*.

Kaur, S., Samota, M. K., Choudhary, M., Choudhary, M., Pandey, A. K., Sharma, A., & Thakur, J. (2022). How do plants defend themselves against pathogens-Biochemical mechanisms and genetic interventions. *Physiology and Molecular Biology of Plants: an International Journal of Functional Plant Biology*, *28*(2), 485–504. PMID: 35400890

Khan, A.. (2025). Eco-friendly synthesis of copper oxide-silver bimetallic nanoparticles using Withania coagulans: Characterization and biomedical potential through antioxidant, antibacterial, and catalytic activities. *Inorganic Chemistry Communications*, *115131*, •••. https://doi.org/https://doi.org/10.1016/j.inoche.2025.115131

Khan, F., Shariq, M., Asif, M., Siddiqui, M., Malan, P., & Ahmad, F. (2022). Green nanotechnology: Plant-mediated nanoparticle synthesis and application. *Nanomaterials (Basel, Switzerland)*, *12*(4), 673. DOI: 10.3390/nano12040673 PMID: 35215000

Khan, I., Awan, S., Rizwan, M., Hassan, Z., Akram, M., Tariq, R., Brestič, M., & Xie, W. (2022). Nanoparticle uptake and translocation mechanisms in plants via seed priming, foliar treatment, and root exposure: A review. *Environmental Science and Pollution Research International*, *29*(60), 89823–89833. DOI: 10.1007/s11356-022-23945-2 PMID: 36344893

Khan, I., Awan, S., Rizwan, M., Hassan, Z., Akram, M., Tariq, R., Brestič, M., & Xie, W. (2022). Nanoparticle uptake and translocation mechanisms in plants via seed priming, foliar treatment, and root exposure: A review. *Environmental Science and Pollution Research International*, *29*(60), 89823–89833. DOI: 10.1007/s11356-022-23945-2 PMID: 36344893

Kumar, R., Tiwari, R. K., Sundaresha, S., Kaundal, P., & Raigond, B. (2022). Potato viruses and their management. In *Sustainable management of potato pests and diseases* (pp. 309–335). Springer Singapore.

Kumari, N., & Sharma, V. (2022). Uptake, translocation, transformation and physiological effects of nanoparticles in plants. *Archives of Agronomy and Soil Science*, *69*(10), 1579–1599. DOI: 10.1080/03650340.2022.2103549

Lau, D., Mar, T. B., dos Santos, C. D. R., Engel, E., & da Silva, P. R. D. V. (2021). Advances in understanding the biology and epidemiology of barley yellow dwarf virus (BYDV). *Achieving durable disease resistance in cereals*, 709-745.

Lim, J. A., Yaacob, J. S., Mohd Rasli, S. R. A., Eyahmalay, J. E., El Enshasy, H. A., & Zakaria, M. R. S. (2023). Mitigating the repercussions of climate change on diseases affecting important crop commodities in Southeast Asia, for food security and environmental sustainability—A review. *Frontiers in Sustainable Food Systems*, *6*, 1030540.

Lithi, I., Nakib, K., Chowdhury, S., & Hossain, M. (2025). A review on the green synthesis of metal (Ag, Cu, and Au) and metal oxide (ZnO, MgO, Co_3O_4, and TiO_2) nanoparticles using plant extracts for developing antimicrobial properties. *Nanoscale Advances*, *7*, 2446–2473. DOI: 10.1039/d5na00037h PMID: 40207090

Lorenzini, G., & Nali, C. (2025). Plant protection, the Cinderella of the one health strategy? *One Health*, •••, 101080. PMID: 40520965

Luceri, A., Francese, R., Lembo, D., Ferraris, M., & Balagna, C. (2023). Silver nanoparticles: Review of antiviral properties, mechanism of action and applications. *Microorganisms*, *11*(3), 629. PMID: 36985203

Mandal, A., Katuwal, S., Tettey, F., Gupta, A., Bhattarai, S., Jaisi, S., Bhandari, D., Shah, A., Bhattarai, N., & Parajuli, N. (2022). Current research on zinc oxide nanoparticles: Synthesis, characterization, and biomedical applications. *Nanomaterials (Basel, Switzerland)*, *12*, 3066. DOI: 10.3390/nano12173066 PMID: 36080103

Mangla, B., Kumar, P., Javed, S., Pathan, T., Ahsan, W., & Aggarwal, G. (2025). Regulating nanomedicines: Challenges, opportunities, and the path forward. *Nanomedicine (London)*, *20*(15), 1911–1927. PMID: 40657903

MARKOV, T. (2024). EXTERNAL AND INTERNAL THREATS TO FOOD SECURITY. *Scientific Papers. Series Management, Economic, Engineering in Agriculture and Rural Development*, *24*(2).

Maurice, N. (2023). Impact of nanoparticles on soil microbes for enhancing soil fertility and productivity. In *Nanotechnology for Sustainable Agriculture* (pp. 75–127). Apple Academic Press.

Mohan, P. S. (2021). *Microwave-Assisted Synthesis of ZnO/TiO2/Ag Nanocomposites as Alternative Antimicrobials* (Doctoral dissertation, University of Malaya (Malaysia)).

Nabi, Z., Manzoor, S., Nabi, S. U., Wani, T. A., Gulzar, H., Farooq, M., & Mansoor, S. (2024). Pattern-Triggered Immunity and Effector-Triggered Immunity: Crosstalk and cooperation of PRR and NLR-mediated plant defense pathways during host–pathogen interactions. *Physiology and Molecular Biology of Plants: an International Journal of Functional Plant Biology*, *30*(4), 587–604. PMID: 38737322

Nadaf, S., Jadhav, N., Naikwadi, H., Savekar, P., Sapkal, I., Kambli, M., & Desai, I. (2022). *Green synthesis of gold and silver nanoparticles: Updates on research, patents, and future prospects*. OpenNano., DOI: 10.1016/j.onano.2022.100076

Nadeem, S., Riaz Ahmed, S., Luqman, T., Tan, D. K., Maryum, Z., Akhtar, K. P., & Liu, Y. (2024). A comprehensive review on Gossypium hirsutum resistance against cotton leaf curl virus. *Frontiers in Genetics*, *15*, 1306469. PMID: 38440193

Nkanga, C. I., & Steinmetz, N. F. (2021). The pharmacology of plant virus nanoparticles. *Virology*, *556*, 39–61. PMID: 33545555

Olga, M., Jana, M., Anna, M., Irena, K., Jan, M., & Alena, Č. (2022). Antimicrobial properties and applications of metal nanoparticles biosynthesized by green methods. *Biotechnology Advances*, *107905*. Advance online publication. DOI: 10.1016/j.biotechadv.2022.107905

Osei, M. K., Oppong, A., Adjebeng-Danquah, J., Zohoungbogbo, H. P. F., Prempeh, R., Melomey, L. D., . . . Adjei, E. A. (2025). Vegetable Production in West Africa: The Challenges of Viral Diseases and Their Management. In *Plant Diseases and Their Management* (pp. 427-469). Apple Academic Press.

Paramesh, V., Mohan Kumar, R., Rajanna, G. A., Gowda, S., Nath, A. J., Madival, Y., & Toraskar, S. (2023). Integrated nutrient management for improving crop yields, soil properties, and reducing greenhouse gas emissions. *Frontiers in Sustainable Food Systems*, *7*, 1173258.

Pietrzak, M., Skiba, E., & Wolf, W. M. (2024). Root-applied cerium oxide nanoparticles and their specific effects on plants: A review. *International Journal of Molecular Sciences*, *25*(7), 4018. PMID: 38612829

Pradhan, S., Varsani, A., Leff, C., Swanson, C. J., & Hariadi, R. F. (2022). Viral aggregation: The knowns and unknowns. *Viruses*, *14*(2), 438. PMID: 35216031

Radulescu, D., Surdu, V., Ficai, A., Ficai, D., Grumezescu, A., & Andronescu, E. (2023). Green synthesis of metal and metal oxide nanoparticles: A review of the principles and biomedical applications. *International Journal of Molecular Sciences*, *24*(20), 15397. DOI: 10.3390/ijms242015397 PMID: 37895077

Ramos-Zúñiga, J., Bruna, N., & Pérez-Donoso, J. M. (2023). Toxicity mechanisms of copper nanoparticles and copper surfaces on bacterial cells and viruses. *International Journal of Molecular Sciences*, *24*(13), 10503. PMID: 37445681

Rastogi, A., Živčák, M., Sytar, O., Kalaji, H., He, X., Mbarki, S., & Brestič, M. (2017). Impact of metal and metal oxide nanoparticles on plants: A critical review. *Frontiers in Chemistry*, *5*, 78. DOI: 10.3389/fchem.2017.00078 PMID: 29075626

Raza, A., Khandelwal, K., Pandit, S., Singh, M., Kumar, S., Rustagi, S., Ranjan, N., Verma, R., Priya, K., & Prasad, R. (2024). Exploring the potential of metallic and metal oxide nanoparticles for reinforced disease management in agricultural systems: A comprehensive review. *Environmental Nanotechnology, Monitoring & Management*. Advance online publication. DOI: 10.1016/j.enmm.2024.100998

Reagen, S., & Zhao, J. X. (2022). Analysis of nanomaterials on biological and environmental systems and new analytical methods for improved detection. *International Journal of Molecular Sciences*, *23*(11), 6331. PMID: 35683010

Sachdev, S., Ansari, S. A., Ansari, M. I., Fujita, M., & Hasanuzzaman, M. (2021). Abiotic stress and reactive oxygen species: Generation, signaling, and defense mechanisms. *Antioxidants*, *10*(2), 277. PMID: 33670123

Shteinberg, M., Mishra, R., Anfoka, G., Altaleb, M., Brotman, Y., Moshelion, M., & Czosnek, H. (2021). Tomato yellow leaf curl virus (TYLCV) promotes plant tolerance to drought. *Cells*, *10*(11), 2875. PMID: 34831098

Siddiqi, K. S., & Husen, A. (2017). Recent advances in plant-mediated engineered gold nanoparticles and their application in biological systems. *Journal of Trace Elements in Medicine and Biology*, *40*, 10–23. DOI: 10.1016/j.jtemb.2016.11.012 PMID: 28159216

Singh, A., Singh, N. B., Afzal, S., Singh, T., & Hussain, I. (2017). Zinc oxide nanoparticles: A review of their biological synthesis, antimicrobial activity, uptake, translocation and biotransformation in plants. *Journal of Materials Science*, *53*(1), 185–201. DOI: 10.1007/s10853-017-1544-1

Singh, R., Choudhary, P., Kumar, S., & Daima, H. K. (2024). Mechanistic approaches for crosstalk between nanomaterials and plants: Plant immunomodulation, defense mechanisms, stress resilience, toxicity, and perspectives. *Environmental Science. Nano*, *11*(6), 2324–2351.

Singh, R., Kuddus, M., Singh, P., & Choden, D. (2022). Nanotechnology for Nanophytopathogens: From Detection to the Management of Plant Viruses. *BioMed Research International*, *2022*. Advance online publication. DOI: 10.1155/2022/8688584 PMID: 36225980

Tahir, S., Tahir, N., Hussain, T., Naseem, Z., Zahid, M., & Mustafa, G. (2022). Non-photocatalytic and photocatalytic inactivation of viruses using antiviral assays and antiviral nanomaterials. In *Viral and antiviral nanomaterials* (pp. 139–164). CRC Press.

Tan, J. S., Jaffar Ali, M. N. B., Gan, B. K., & Tan, W. S. (2023). Next-generation viral nanoparticles for targeted delivery of therapeutics: Fundamentals, methods, biomedical applications, and challenges. *Expert Opinion on Drug Delivery*, *20*(7), 955–978. PMID: 37339432

Tatineni, S., & Hein, G. L. (2023). Plant viruses of agricultural importance: Current and future perspectives of virus disease management strategies. *Phytopathology*, *113*(2), 117–141. PMID: 36095333

Trukhanov, A. V., Wang, H. M. D., Thakur, A., & Thakur, P. (2025). Nanoscale Approaches to Genetic Engineering. In *Advancements in Nanobiology* (pp. 130–150). CRC Press.

Tungadi, T. D., Wamonje, F. O., Mhlanga, N. M., Murphy, A. M., Arinaitwe, W., & Carr, J. P. (2025). The Complex Interactions of Common Bean (Phaseolus vulgaris L.) with Viruses, Vectors and Beneficial Organisms in the Context of Sub-Saharan Africa. *Agriculture*, *15*(17), 1808.

Ullah, F., Guru-Pirasanna-Pandi, G., Murtaza, G., Sarangi, S., Gul, H., Li, X., & Lu, Y. (2025). Evolving strategies in agroecosystem pest control: Transitioning from chemical to green management. *Journal of Pest Science*, •••, 1–18.

Vargas-Hernandez, M., Macias-Bobadilla, I., Guevara-González, R., Rico-García, E., Ocampo-Velázquez, R., Ávila-Juárez, L., & Torres-Pacheco, I. (2020b). Nanoparticles as Potential Antivirals in Agriculture. *Agriculture*. Advance online publication. DOI: 10.3390/AGRICULTURE10100444

Vargas-Hernandez, M., Macias-Bobadilla, I., Guevara-Gonzalez, R. G., Rico-Garcia, E., Ocampo-Velazquez, R. V., Avila-Juarez, L., & Torres-Pacheco, I. (2020a). Nanoparticles as potential antivirals in agriculture. *Agriculture*, *10*(10), 444.

Venu, E., Ramya, A., Babu, P. L., Srinivas, B., Kumar, S., Reddy, N. K., & Manik, S. (2024). Exogenous dsRNA-mediated RNAi: Mechanisms, applications, delivery methods and challenges in the induction of viral disease resistance in plants. *Viruses*, *17*(1), 49. PMID: 39861836

Vieira, M. S., Cabral, R. L. R., Favaratto, L., Maciel, L. S., Xavier, A. D. S., Zerbini, F. M., & Fernandes, P. M. (2024). Tropical fruit virus resistance in the era of next-generation plant breeding. *SynBio*, *2*(3), 267–284.

Xingmao, M., & Yan, J. (2018). Plant uptake and accumulation of engineered metallic nanoparticles from lab to field conditions. *Current Opinion in Environmental Science & Health*, *6*, 1–7. DOI: 10.1016/j.coesh.2018.07.008

Yadav, N., & Upadhyay, R. K. (2023). Global effect of climate change on seasonal cycles, vector population and rising challenges of communicable diseases: A review. *Journal of Atmospheric Science Research*, *6*(1).

Yiblet, Y., & Sisay, M. (2024). Metal oxide nanoparticles as a promising method to reduce biotic stress in plant cell wall: A review. *Heliyon*, *10*. Advance online publication. DOI: 10.1016/j.heliyon.2024.e37939 PMID: 39386835

Chapter 6
Lipidic and Polymeric Nanoparticles for Viral Control in Plant Protection

Amjid Khan
https://orcid.org/0000-0002-1706-1876
University of South Africa, South Africa

Rashid Abbas Khan
https://orcid.org/0000-0002-5438-6274
Prince of Songkla University, Thailand

Bushra Ashiq
Taiyuan University of Technology, China

Malik Maaza
University of South Africa, South Africa

Tauqeer Ahmed Qadri
COMSATS University Islamabad, Pakistan

Amir Muhammad Khan
University of Sargodha, Pakistan

Waseem Ahmed Khattak
https://orcid.org/0009-0004-1356-2579
Quaid-i-Azam University, Pakistan

Hamza Elsayed Ahmed Mohamed
https://orcid.org/0000-0003-0550-8607
Le Mans Université, France

ABSTRACT

Plant viral diseases are a serious threat to global food security, causing yield losses and showing the limits of resistant cultivars, chemical antivirals, and vector control. This chapter examines nanotechnology as a sustainable solution, focusing on lipidic and polymeric nanoparticles for targeted virus control. Lipid-based carriers such as liposomes, solid lipid nanoparticles, and nanostructured lipid carriers protect and deliver nucleic acids for RNA interference approaches. Polymeric systems, including natural polymers like chitosan, alginate, and starch and synthetic types like PLGA and dendrimers, provide stability, biodegradability, and immune priming through systemic acquired resistance. Comparative strengths, delivery routes including foliar sprays, root absorption, seed coatings, and post-harvest treatments, plus safety and regulatory issues, are assessed. Advances such as stimuli-responsive carriers, CRISPR/Cas delivery, and smart agriculture integration position these nanocarriers as promising tools for climate-smart, sustainable crop protection.

DOI: 10.4018/979-8-3373-4913-8.ch006

1. Introduction

Plant viruses are one of the most devastating and least studied threats in the agricultural sector around the world (Tatineni & Hein, 2023). In contrast with fungi or bacteria, viruses do not have their own metabolism and infect their hosts at the expense of their cellular processes (Eisenreich et al., 2019). This is one of the reasons why they are very difficult to control. Viral diseases impact a broad portfolio of economically and nutritionally significant crop species such as rice, wheat, maize, cassava, potato, tomato, soybean, and banana, to name but a few, and which combined support the vast bulk of the global calorie intake (Jones, 2020; Sastry & Zitter, 2014). Viral diseases are a significant threat to global agriculture, with crop yield losses ranging from 10% to 15% annually, amounting to more than USD 30 billion in global economic losses. These losses exacerbate food insecurity and poverty, particularly in regions like Asia and Sub-Saharan Africa. The widespread effect of viruses such as Tobacco mosaic virus (TMV), Tomato yellow leaf curl virus (TYLCV), and Cassava mosaic virus (CMV) underscores the potential devastation caused by viral outbreaks, destabilizing food systems worldwide. Furthermore, climate change contributes to the increased spread of these viruses as it accelerates the activity of insect vectors like whiteflies and aphids, leading to higher transmission rates and the expansion of viral diseases into new areas (Friskop et al., 2025). Cultural practices such as crop rotation, traditional management strategies include breeding resistant cultivars, vector control with insecticides, and cross-protection with non-destructive viral strains (Kumar et al., 2021). Such strategies are only temporarily effective, however, as they have severe shortcomings. The main drawback of resistance breeding is usually destroyed by the plasticity of viral genomes and the development of resistance-breaking variants. The problem with insecticide use is that populations initially sensitive to the product would develop resistance given time, there is the disturbance of the ecological system, and finally, the presence of pesticide residues in the food chain (Leskovac & Petrović, 2023). Cross-protection strategies are also highly crop- and strain-specific and hence not scalable. The general approach of these methods is reactive, limited in scope, and as viruses grow in complexity and the environmental implications of their presence evolve, these methods are becoming less sustainable.

1.1. Nanotechnology: A New Paradigm in Plant Virus Management

The deficiency of the existing means of control in conventional plant viruses highlights the necessity of new methods for the efficiency of control that should be sustainable and selective. The entry of nanotechnology is the solution to this because nanotechnology will also provide nanoscale precision in the delivery of antiviral agents, genetic materials, and stimulating plant defense agents (Dutta et al., 2022). Nanoparticles' unique physicochemical properties, including high surface area, tunable charge, biological compatibility, and the capacity to bypass biological barriers, all make the nanoparticle ideally applicable in agricultural use (Anas et al., 2024; Khan et al., 2025; Maaza et al., 2022). Among the existing types of nanocarrier systems, it is possible to draw attention to lipidic and polymeric nanoparticle types, which are characterized by versatility and an eco-friendly aspect. Lipid nanoparticles (liposomes, solid lipid nanoparticles, nanostructured lipid carriers) have the potential to entrap both hydrophilic and hydrophobic compounds, to protect labile molecules, such as RNA, against enzymatic degradation, and to release their cargo in a time-dependent, controllable fashion in an infected tissue site (Puri et al., 2025). They may also be modified to fuse with the membranes of plants to increase delivery inside the plant. On the one hand, polymeric nanoparticles (chitosan, PLGA, dendrimers, and starch-based carriers) are advantageous to the extent that they are biodegradable, prolong the release, bind to nucleic acids, and cause systemic resistance (Forenzo et al., 2025). An example is that chitosan nanoparticles not only serve as carriers, but also cause host defense responses, conferring a two-pronged protective effect (Narasimhamurthy et al., 2022).

Such systems can be designed to transport RNA interference (RNAi)-based compounds, CRISPR-Cas constructs, small interfering RNAs (siRNAs), natural antiviral phytochemicals, or synthetic compounds into the tissues of plants (Mujtaba et al., 2021). In this way, they facilitate management of instability issues, low penetration, and degradation that characterize the traditional treatment. Moreover, nanoparticle formulations may directly attack these vessels, as targeted nanoparticle formulation has the potential to reach the phloem, where the majority of plant viruses occur and multiply, a capacity that conventional sprays cannot unconditionally accomplish (Eid et al., 2025). Their role in vector management is also important; nanocarriers may be programmed to introduce compounds that interrupt transmission of the virus by the vectors or interfere with virus-vector interactions. In combination, the use of methylated nanoparticles and the addition of polymerized ones should be seen as a comprehensive and eco-compatible solution to fight plant viruses (Carrillo-Lopez et al., 2024). In this chapter, the principles, mechanisms, and applications of lipidic and polymeric nanoparticles in the context of plant viral control will be discussed, giving an integrated overview of how these nanoparticle systems may transform sustainable crop protection. By integrating virology, nanotechnology, and plant pathology, we seek to demonstrate the revolutionary capacity of these areas in maintaining the resilience of agriculture and global food security in the 21st century. Nanotechnology-based strategies provide a sustainable and innovative paradigm for plant virus management (Figure 1). Lipid and polymeric nanoparticles act as versatile nanocarriers capable of precision delivery of antiviral agents, RNAi constructs, CRISPR-Cas systems, siRNAs, and phytochemicals. These nanocarriers improve the stability of bioactive compounds, enhance penetration, and enable targeted release within plant tissues, particularly the phloem, where most viruses accumulate. Moreover, they can disrupt virus–vector interactions while simultaneously triggering host defense responses, thus offering an eco-friendly and integrated approach for controlling plant viral diseases (Figure 1).

Figure 1. Integrated nanotechnology-based approaches for plant virus management using lipid and polymeric nanoparticles as precision delivery systems for antivirals, RNAi, and CRISPR constructs

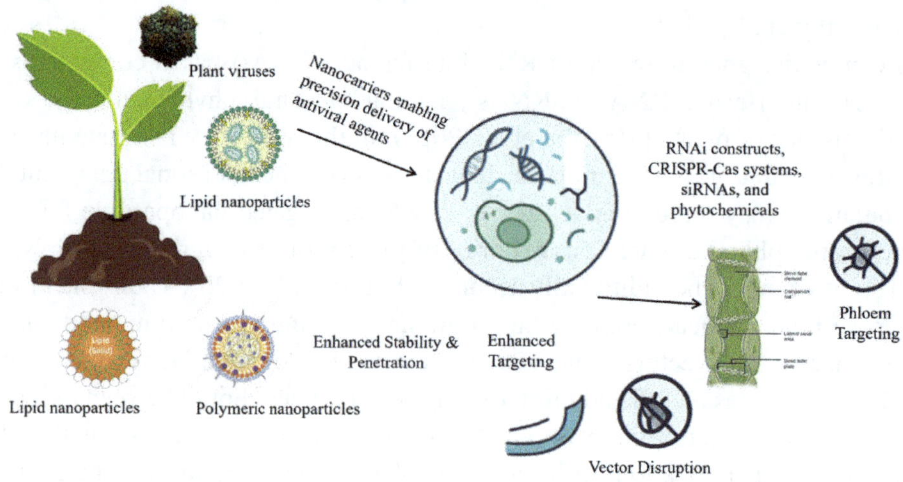

2. Plant Viral Pathogenesis and Barriers to Control

2.1. Major Plant Viruses of Economic Concern

Plant viruses constitute one of the major problems of agricultural and economic losses in the world (Tatineni & Hein, 2023). The best characterized and most destructive are Tobacco mosaic virus (TMV), a prototype that infects a broad range of plants including solanaceous crop species; Cucumber Mosaic virus (CMV) with over 1,200 hosts and yields crippling losses in crops; Tomato yellow leaf curl virus (TYLCV) which is a begomovirus transmitted by *Bemisia tabaci* that threatens tomato production worldwide; and Rice tungro virus one of the rice viruses that is responsible for crippling rice yields in Asia (Ashwini, 2022). Such viruses represent a broad spectrum in genome type (either RNA or DNA), modes of transmission (mechanical and insect-borne or seed-borne), and host specificity. The total of their effects is more than yield-loss to food insecurity, price volatility, and loss of farmer income, which is quite comparable to the urgency of more sustainable control strategies

2.2. Virus–Host Interaction Dynamics

Plant viruses achieve effective infections by invoking a difficult scenario of reproduction. Once inside the host body through vectors/wounds, the viral genome takes over host processes to multiply and synthesize protein (Soni et al., 2024). Movement of the virus occurs when the virus moves cell-to-cell via the plasmodesmata, facilitated by movement proteins, and eventually develops systemic infection via phloem. Viruses use suppressors of RNA silencing, and/or disrupt host defense signaling to overcome plant immunity (Ibrahim et al., 2025). This active gunfire competition can lead to chronic infections in which the affected plants give low growth, chlorosis, or death. Defining such host-virus interactions is essential in developing effective antiviral targets.

2.3. Natural Barriers Limiting Antiviral Delivery

Physical and physiological obstacles exist in plants that very much limit the use of a substance in an antiviral delivery system. Waxy cuticles and the rigid cell wall serve as the initial barriers and prevent easy access of external agents into tissues. Entry itself is not always followed by diffusion throughout the cellular compartments, or into vascular tissues (Sangwan & Singh, 2024). Moreover, a wide selection of antiviral compounds is unstable in the plant setting, which results in their destruction long before they reach their target points (Denaro et al., 2020). Such inefficiency explains why conventional sprays or foliar applications of nucleic acid molecules, proteins, and small-molecule antiviral agents have limited efficiency in the field.

2.4. Shortcomings of Conventional Antiviral Strategies

Conventional plant virus control methods and management methods, e.g., plant breeding for resistance factors, cross-protection via use of mild strains of the pathogen, chemical protection, e.g., sprays, and multi-localization biocontrol, have been short-lived (Mandal et al., 2025). Resistance usually fails because of the numerous mutations that viruses go through and the formation of new strains. Cross-protection has not generalized, and this raises biosafety issues. The temporary solutions of chemical attack on vectors include environmental, health, and the risk is temporary (Bergmann et al., 2021). Meanwhile, more sophisticated technologies, such as RNA interference (RNAi) sprays, appear promising in test conditions but have low resistance to degradation, low uptake, and in fields, demonstrate variable and unpredictable outcomes (Qi et al., 2024).

2.5. The Need for Nanoparticle-Based Solutions

These constraints also highlight the relevance of a nanotechnology-based delivery system of antivirals. Lipidic and polymeric nanoparticles are ideally suited to breach plant barriers because they are biocompatible, and modifications to surface chemistry and encapsulation of naked cargo, including RNA molecules, proteins, and small interfering agents, are possible (Saleh et al., 2022). Overall, nanoparticles can drop dosing levels significantly and minimize environmental impact by increasing stability, targeting, and controlled release to make more efficient antiviral therapies possible. Critically, such systems represent the prospect of RNAi- and CRISPR-based antiviral strategies taking place in the field, as opposed to the gap in transferring laboratory-based inventions into actual practice (Chandrasekaran et al., 2021).

3. Lipidic Nanoparticles in Plant Virus Control

Plant-derived nanoparticles have become one of the most flexible delivery systems in plant biotechnology because of their innate ability to interact with biological membranes, their potential to encapsulate a wide array of molecules, and their changing physical and chemical characteristics (Khan, Anas, Qadri, et al., 2025; Khan, Anas, Ullah, et al., 2025). Lipidic nanoparticles (LNPs) are a relatively new technology in the field of plant virology that can be used to apply antiviral molecules, including double-stranded RNA (dsRNA), small interfering RNA (siRNA), and plant-derived metabolites, topically to the host tissues (Tenchov et al., 2021). Their properties of targeting the cell membrane of plants, bio shielding labile biomolecules, and leading to a prolonged release make them a promising alternative to the traditional approach to antivirals, which does not always work in field conditions. This section describes the various lipidic nanoparticle types applicable to plant viral control, the suggested mechanisms of how those lipidic nanoparticle types work in planta, experimental case studies, and the positive role as well as the shortfalls and shortcomings of these lipidic nanoparticles.

3.1. Types of Lipidic Systems

Nanoparticle classes based on the use of lipids have been formulated to suit agricultural practices, each with its own set of structural and functional properties (Shah et al., 2015). Liposomes are round-shaped envelopes with one or more phospholipid membranes that can trap both the hydrophilic and the hydrophobic compounds. They are the oldest generation of lipidic carriers and have also been extensively studied as lipidic carriers to deliver dsRNA and phytochemicals to plants (Haggag et al., 2021). Because of their similarity to natural membranes, the liposomes easily fuse with the membranes of plant cells due to which there is an increase in the uptake efficiency.

The SLNs were prepared to surmount the instability of the liposomes. They are synthesized using solid lipids that do not lose their crystalline structure at room temperatures or body temperature, subsequently having the capability to control the release of loaded agents (Truzzi et al., 2022). The greater degree of physical stability and suitability to such foliar sprays is attributable to their rigid structure, which is in part due to environmental stress being less likely to destabilize nanoparticles. Nanostructured lipid particles (NLCs) are characterized as a modified version of SLNs with a combination of liquid and solid lipids (Viegas et al., 2023). With this hybrid construction, payload capability is enhanced, and unwarranted expulsion of drugs during storage is evaded. NLCs enhanced the efficacy of antiviral formulations by increasing the persistence against the viral pathogen through the retentive capacities of the formulations under field conditions (Žigrayová et al., 2023).

LNPs, recently of interest in human medicine in the form of mRNA vaccines, share the distinction of also being of interest in plant systems. Such lipids are neutral at physiological pH, but they are cationic at low pH and increase RNA encapsulation and endosomal escape (Zhang et al., 2021). Their utility in crops is efficiently delivering dsRNA/siRNA molecules into the cytoplasm to trigger silencing of viral genes through RNAi. Motivated by their efficacy in mammalian applications, ionizable LNPs have the potential to change the plants of antiviral therapeutics completely.

3.2. Mechanisms in Plants

Lipidic nanoparticles are effective as antivirals owing to a variety of interconnected mechanisms of action. The most noteworthy is the packaging of the dsRNA/siRNA that acts as the inducer of RNA Interference in viral genomes (Venu et al., 2024). Due to their sensitivity to apoplastic enzymatic cleavage, free dsRNA is unlikely to be effective. This gives the molecules protection through lipid carriers, leading to the increased stability and bioavailability of these molecules. A novel and equally significant process is resistance to enzymatic destruction and to other environmental stress. In the leaf and root extracellular matrix, unprotected nucleic acids are readily degraded by nucleases. The lipidic nanoparticles produce a dedicated cover that adds to the working half-life of antiviral substances (Mendes et al., 2022; Vargas-Hernandez et al., 2020). The lipidic nanoparticle absorption pathway in plants has shown both foliar and root absorption. Foliar application can lead to nanoparticle penetration of either stomata, trichomes, or cuticular cracks, and root application can lead to systemic transport mediated by xylem and phloem (Wang et al., 2023). Already internalized, the lipid shells may fuse to the membranes of plant cells, which have the advantage of enhancing the release of payloads in the cytoplasm. Also, lipid nanoparticles can be used as antivirals, stimulating plant natural defenses. Even some lipids alone are capable of stimulating defense mechanisms in the plant, improving their capacity to withstand viral infection via the payload alone. Lipid nanoparticles (LNPs) have shown significant potential in improving the delivery of RNA interference (RNAi) and CRISPR/Cas9 constructs to plant cells. These nanoparticles protect RNA molecules from enzymatic degradation and facilitate efficient intracellular delivery (Thokchom et al., 2025). Recent studies, such as Liu et al. (2024), demonstrated that lipid nanoparticles encapsulating CRISPR/Cas9 systems for gene editing in crops resulted in more precise gene silencing, with a 35% higher editing efficiency compared to traditional delivery methods. This dual effect, direct antivirus delivery, and indirect immune priming has an additional value, and makes them novel next-generation crop protectants.

3.3. Case Studies and Applications

Recent studies show strong evidence of the effectiveness of the use of a lipidic nanoparticle to control plant viruses. One example is the delivery of lectin-dsRNA conjugate, which targets the Tobacco mosaic virus (TMV) downregulation, where the activities of these viruses are prevented by the inhibition of their replication and their subsequent systemic spread in the plant, leading to the mildness of symptoms (Xu, Jiao, et al., 2023). On a similar note, encapsulated dsRNA of Cucumber mosaic virus (CMV) showed an increased uptake and silencing efficiency and lower toxicity after spray application than naked RNA sprays (Xing et al., 2025). SLNs have been successfully loaded with plant-derived secondary metabolites with properties that are protective, such as flavonoids and alkaloids, which are antiviral in nature (Hashempur et al., 2025). These nanoformulations enhanced the solubility and stability of compounds whose stability was otherwise poor, and this enabled longer contact time of antiviral activity on the foliar surfaces. Nanostructured Lipid Carriers (NLCs) have been evaluated in greenhouse and semi-field conditions and used with RNAi molecules to target Tomato yellow leaf curl virus (TYLCV) (Tortorici et al., 2022). Not only did the results indicate enhanced durability of antiviral sprays, but also the frequency of applications became lower, which helps address one of the bottlenecks in the widespread use of field animals using RNAi-based strategies, namely, the fact that daily applications may not be suitable in the field. Encouraged by medical research, ionizable lipid nanoparticles (LNPs) are now under consideration to deliver dsRNA in crops, including rice and maize. It is envisaged that these systems will be able to overcome the hurdles to earlier formulations presented by poor cytoplasmic release and limited systemic transport. They show promise in their translation of their human therapeutic work to plant agriculture

3.4. Advantages and Limitations

Lipid nanoparticle-based control of plant viruses is rather advantageous. Their experimental tolerance with plant tissues due to their biocompatibility and biomimicry nature makes them inherently suited to be incorporated in plant tissues, devoid of toxicity (Mahmood et al., 2024; Pathak et al., 2024). They offer effective protection to RNA molecules as well as the bioactive compounds in a manner that guarantees stability against bio-destructive activity, ultraviolet radiation, and other environmental conditions. Lipidic nanoparticles also facilitate targeting and sustained delivery, so that the element of cost consumption would not be as acute as before, over time (Tenchov et al., 2021). Moreover, some lipid systems are able to evoke host immune responses, which further presents another protective modality and makes them multipurpose antiviral agents. There still are several limitations, though. Formulation and scalability of lipid-based nanoparticles are very expensive, thus a big challenge to their widespread adaptation in farming systems where resources are scarce. Most of the lipid NPs are susceptible to UV-light degradation and thermostability and risks the stability of the NPs under real-field conditions (Chandrasekaran et al., 2021). Moreover, laboratory and greenhouse results are promising; however, little field-scale data is available, making it challenging to predict performance across a wide variety of environments. The last obstacle as a limitation is replication in long-term storage, formulation stability, and even approval by regulatory bodies would put lipidic nanoparticles in the forefront of research applications to the practical world of agriculture. A schematic representation of lipidic nanoparticle-based strategies for plant virus control is shown in Figure 2. Different lipid systems, such as liposomes, solid lipid nanoparticles (SLNs), nanostructured lipid carriers (NLCs), and ionizable lipid nanoparticles, act as versatile delivery vehicles for dsRNA, siRNA, and plant-derived metabolites. These nanocarriers improve stability, shield bioactive compounds from enzymatic degradation, and facilitate efficient uptake through foliar and root pathways. Their mechanisms of action include induction of RNA interference, membrane fusion for cytoplasmic release, and stimulation of host defense responses. Case studies demonstrate improved delivery of antiviral molecules, enhanced efficacy, and reduced spray frequency in crops such as tomato, rice, and maize. Overall, lipidic nanoparticles offer advantages including biocompatibility, stability, and dual antiviral/immune-priming effects, though challenges remain in terms of production cost, scalability, and limited validation under field conditions (Figure 2).

Figure 2. Lipidic nanoparticles in plant virus control: Types, mechanisms, applications, and limitations

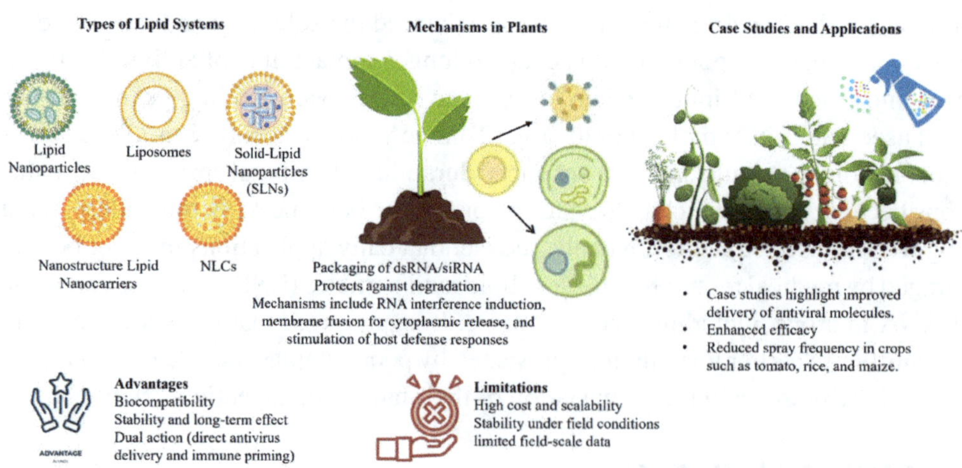

4. Polymeric Nanoparticles in Plant Virus Control

PNPs have also been identified as one of the most promising nanocarrier systems in agriculture, especially in liberating plant viruses. Their adaptability is centered on the broad spectrum of natural and artificial polymers that can be built to dispense, manage, launch, and plant safeguarding (Farooq et al., 2021). In contrast to lipid-based NPs that are predominantly absorbed using membrane fusion and encapsulation, polymeric NPs can be specifically optimized to interact with the electrostatically charged plant cell wall structures, long-term release of bioactive compounds, and a direct antiviral effect (Kothapalli & Vasanthan, 2024). Plant pathology and polymer nanotechnology in the last few years have come together to create new opportunities in environmentally conscious and more specific control of ruinous plant viruses.

4.1. Types of Polymeric Systems

There are two major categories of polymeric nanoparticles, natural and synthetic polymers, each of which has its pros and cons in agriculture and virology.

4.1.1. Natural polymer-based systems

Plant protection has been tested on natural polymer-based systems, including chitosan, alginate, starch, and cellulose nanocarriers. Chitosan nanoparticles (CNPs) prepared by the deacetylation of chitin are of particular promise because they are biocompatible, biodegradable, and intrinsically antimicrobial (Sharma et al., 2025). Chitosan is capable of initiating systemic acquired resistance (SAR) in plants, thus, priming crops to resist a viral attack. Similarly, alginate and starch nanoparticles have been recommended as supplements of nucleic acids and bioactive compounds offering a safe and low-cost point-of-large-scale agricultural applications (Singh et al., 2022).

4.1.2. Synthetic polymers

Synthetic polymers have additional properties of stability, precision and flexibility of engineering (Terzopoulou et al., 2022). Among these, poly (lactic-co-glycolic acid) (PLGA) nanoparticles have shown to be effective carriers of RNAi molecules including dsRNA and siRNA, with demonstrated applicability in animal models and increasing evidence in plants (Matta & Maalouf, 2019). PEGylated polymer systems, wherein polyethylene glycol trains are grafted to enhance water solubility and circulation life, are becoming popular in increasing the longevity of nanoparticle deposition on plants. Moreover, a highly branched synthetic polymer, have a distinct multivalent geometry that can bind a viral coat protein, thus preventing entry of the viral particle into host cells (Cruz & Reyes). One of the most novel directions is that of hybrid polymer-lipid systems that combine the beneficial properties of both polymers (long-term stability) and lipids (membrane fusion capabilities) (Go & Leal, 2021). Different systems of this nature have proven their potential in optimizing delivery efficiency and minimizing antiviral payload degradation in outdoor field situations.

4.2. Mechanisms in Plants

Appearing in several complementary ways, the polymeric nanoparticles disrupt the cycle of plant viral infection. Chitosan nanoparticles have been documented intensely as inducers of systemic acquired resistance (SAR), and their ability to activate plant defense routes mediated not only through salicylic acid but also through jasmonic acid and ethylene signaling is also well-known (Elsharkawy et al., 2022). CNPs prime the plant immune system, establishing an unfavorable environment where a virus cannot replicate and spread throughout the plant. The strong nucleic acid resistance of the plant cell walls can limit the intake of naked dsRNA sprays. Polymeric nanocarriers solve this by compressing dsRNA into nanosized nanocarriers that can be carried into a cell through endocytosis or through plasmodesmatal pores (Akbar et al., 2022; Mat Jalaluddin et al., 2023). The dsRNA is then broken down into siRNA, which can activate RNA interference and shut down the viral gene.

Both natural and synthetic polymeric systems help to release agents encapsulated in the systems in a sustained manner (Žigrayová et al., 2023). For example, the RNAi molecules or antiviral phytochemicals embedded within PLGA nanoparticles release gradually and, therefore, provide lasting protection as opposed to conventional sprays, which are destroyed in the environment either by UV light or by rainfall. Cationic polymers, such as dendrimers and some chitosan derivatives, can bind to negatively charged viral coating proteins (Falanga et al., 2021). This interferes with the integrity of viral particles, lowering the infectivity before the virus gets into the host plant. Polymer formulations may also be used as soil treatments and seed coatings (Ludwig et al., 2020). Such strategies guard growing seedlings against seed-transmitted viruses and offer them immunity early in development, the stage at which they are most susceptible to virus attack.

4.3. Case Studies and Applications

4.3.1. Chitosan Nanoparticles Against CMV and TMV

In tomato and tobacco, foliar inoculation with chitosan nanoparticles has been demonstrated to notably decrease the magnitude of the Cucumber mosaic virus (CMV) and Tobacco mosaic virus (TMV) infection (Komarova et al., 2024; Rendina et al., 2019). More than direct antiviral activities, the treated plants showed more defense-related enzymes, including peroxidase and phenylalanine ammonia-lyase, which proved immune priming.

4.3.2. PLGA Nanoparticles for RNAi Delivery

PLGA nanoparticles, followed by encapsulation of geminiviruses' DsRNA molecules, showed superior take-up and stability in plants (Verma & Modgil, 2024; Xu, Yu, et al., 2023). In field trials, these formulations also demonstrated a longer duration of protection than naked dsRNA spray, suggesting their utility as a practical means of viral control.

4.3.3. Dendrimers in Viral Particle Inhibition

Synthetic dendrimers have demonstrated the ability to bind to viral coat proteins and inhibit host-virus interactions. Experimental data support that dendrimers can neutralize viral particles in the extracellular space, a mechanism that is unique to the RNAi-mediated control (Falanga et al., 2021).

4.3.4. Seed Coating with Polymeric Nanoparticles

Cocoa powder and a chitosan or alginate nanoparticle loaded with antiviral agents have been shown to seal the seed and prevent seed-transmitted viruses such as Alfalfa Mosaic Virus from penetrating seedlings (El-Ganainy et al., 2023). This early protection method is provided during germination and in seedling establishment.

4.4. Advantages and Limitations

Application of plant viruses through polymeric nanoparticles has several strong benefits. Natural polymer systems like chitosan, starch, and alginate are abundant, less expensive, and environmentally friendly (Idrees et al., 2020). They can also be generally considered safe with both plants and the environment. Natural polymers also possess inherent biological activity, with chitosan as the prime example, doubling as a carrier and elicitor in the plant immune system (Chakraborty et al., 2020). Compared to the use of synthetic polymers, which are more complicated to produce, have better control of nanoparticle size, charge, and release kinetics, and thus are most efficient in carrying RNAi (Pugsley et al., 2021). Moreover, the polymeric nanoparticles are stable under field conditions than lipidic systems have a reduced susceptibility to degradation by UV radiation and exposure to moderate temperature variation (Sivadasan et al., 2021). They can also be incorporated into seed coatings, soil amendments, and foliar sprays, and their versatility makes them have multiple points of application.

Nevertheless, the limitations are also significant and have to be overcome before the large-scale application in agricultural affairs (Valletta et al., 2014). Synthesis of synthetic polymers, which include PLGA and dendrimers, is potentially a complicated and expensive process that may not be immediately available to resource-poor farming systems (Shakiba et al., 2020). Careful optimization must be performed to mitigate the phytotoxicity that some types of polymeric nanoparticle cause, depending on the concentration, size, and surface chemistry, a process that is unique to each crop species. Another bottleneck is scale-up of the nanoparticle formulation; it is possible to develop such at the laboratory level, but not at a large scale (Fernandes et al., 2023). Lastly, polymeric systems tend to be more stable than lipidic, but that does not rule out the possibility of degradation in harsh field environments, needing formulation changes like coating or hybrids (Lewicka et al., 2024). A schematic representation of polymeric nanoparticle-based strategies for sustainable plant virus management is shown in Figure 3. Natural polymeric systems such as chitosan, alginate, starch, and cellulose nanoparticles, along with synthetic systems including PLGA, PEGylated polymers, dendrimers, and hybrid polymer–lipid carriers, act as versatile nanocarriers. These nanoparticles improve the delivery of dsRNA/siRNA and antiviral metabolites, facilitate uptake through foliar sprays, soil amendments, or seed coatings, and activate antiviral mechanisms such as RNA interference (RNAi), systemic acquired resistance (SAR), and viral coat protein binding. Case studies highlight their effectiveness against CMV, TMV, and Alfalfa mosaic virus. Overall, polymeric nanoparticles provide advantages such as biocompatibility, stability, and immune priming; however, their widespread use is limited by high synthesis costs, scalability challenges, and potential phytotoxic effects (Figure 3).

Figure 3. Polymeric nanoparticles in plant virus control: Types, mechanisms, and applications

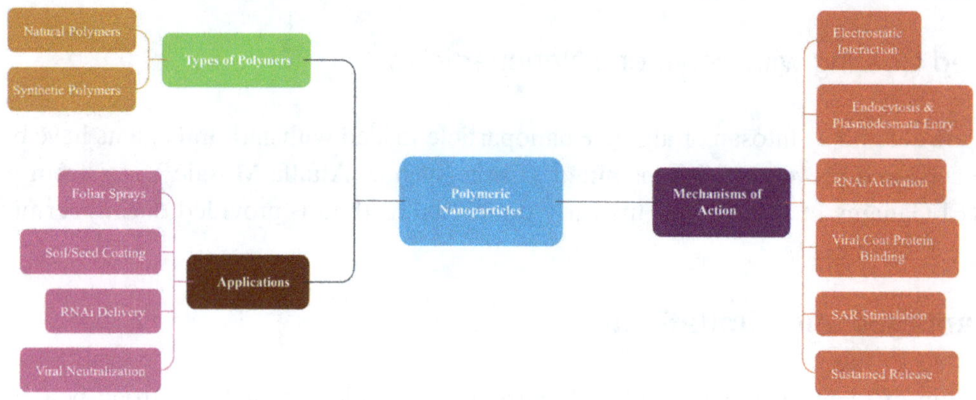

5. Comparative Insights: Lipidic vs. Polymeric Nanoparticles

It is becoming apparent that nanocarrier systems are critical in plant virus management via nanotechnology. The last few years have seen the development of two of the most promising methods: lipidic and polymeric nanoparticles, which offer great advantages and have their own limitations (Åslund et al., 2022; Farooq et al., 2021). Although the basic idea of the two systems is an increase in the productivity and stability of the antiviral agents, there is a wide difference in the mechanism of action, cost-efficiency, and applicability to real-world agricultural conditions. A more detailed comparative study is required to help guide scholars, developers, and decision-makers to determine context-specific measures of plant protection.

5.1. Delivery Efficiency and Mechanistic Strengths

Lipid nanoparticles, including liposomes and lipid nanocarriers, are best suited to protecting and encapsulating nucleic acids (e.g., dsRNA, siRNA) against enzymatic degradation (Arjunan et al., 2024). They are also biomimetic and can fuse with the membrane of the plant without needing much time, thus utilizing fast and effective delivery into the cytoplasm of the secreted antiviral payloads. This provides them with the specific ability to work particularly well in the specificity of delivering the RNA-interference (RNAi) approaches towards viral genomes. The polymeric nanoparticles show greater flexibility in alpha plant innate-immunity activation (Abdelkhalek et al., 2021). Natural polymers such as chitosan and alginate can be used as delivery vehicles and to promote systemic acquired resistance (SAR) that stimulates broad-spectrum antiviral effects (Abd-El-Aziz et al., 2024). With the use of synthetic polymers (PLGA or dendrimers), the release of the antiviral molecules can be controlled and made tunable, thus sustaining the protection against the virus.

5.2. Stability and Environmental Resilience

Nanocarriers must also demonstrate stability under field conditions, which is a key determinant of success. Lipid-based systems can be degraded upon exposure to oxidative conditions and photolysis, effects that would reduce the functionality in field-based cultures (Rajesh, 2025). They are used in optimal conditions where there is control over the environment, like a greenhouse or other high-value crops, where there is little exposure risk. Polymeric systems are, however, stronger in the variation of environmental conditions. Chitosan and starch nanoparticles can withstand mechanical stress during foliar spray or soil conditions, as well as resist UV stress and temperature fluctuations (Kocięcka & Liberacki, 2021). However, synthetic polymers have the potential for economic difficulties because of accumulation or slow breakdown, which necessitates rigorous evaluation of ecological effects on a long-term basis.

5.3. Mechanistic Pathways: Membrane Fusion vs. Immune Priming

The point of distinction between them is that they have different mechanistic pathways. In lipidic systems, biomimetic interactions are key, with plant plasma membranes fusing with a lipid bilayer to release cargo (Monnier et al., 2019). This route resembles a viral entry mechanism, therefore, presenting an extremely efficient mechanism of delivering labile cargo, e.g., siRNA. Polymeric systems and especially chitosan-based approaches tend to be comparatively more indirect by priming the plant immune system (Chakraborty et al., 2020). This immunizing effect is not only limited to the targeted virus but rather a wider circle of pathogens, thus offering an added level of crop strength. This difference emphasizes the fact that lipids are the best suited to precision-targeted interventions since polymers are more general and effective at the plant level.

5.4. Cost and Scalability Considerations

Economics is a key determinant as to whether nanoparticles will be deployed in the agricultural sector. Lipidic nanoparticles are technologically advanced, and specifically ionizable lipid nanoparticle (LNP) based on human RNA vaccine formulations are costly (De & Ko, 2023; Hou et al., 2021). The raw materials, formulation, and storage costs are prohibitive and can only be applied to niche crops or high-value agricultural systems. In comparison, the monomers themselves, such as chitosan or starch, can generate a polymeric nanoparticle that is effective, biodegradable, and can be fairly readily scaled up (Divya & Jisha, 2018). They are cheap, and this is propagating their acceptability in the wider agriculture sector, especially in resource-constrained environments where food security issues are generally at the forefront. Synthetic polymers like PLGA are still more expensive, but their formulations have controlled release and stability that make them worth using in special applications (Shakiba et al., 2020).

5.5. Field Adaptability and Practical Deployment

Adaptability in the field brings out the complementary advantages of lipid and polymeric systems. The advantages of lipidic nanoparticles are their high performance under containment or semi-containment conditions, where degradation risks may be mitigated to allow application to greenhouse crops, horticultural, or precision farming systems (Pathak et al., 2024). On the other hand, the polymeric nanoparticles are more suitable for deployment. They resist UV and high temperatures well, as well as environmental variation, so they can be sprayed, soil treated, and even seed coated in large quantities (Vemula & Reddy, 2023). Therefore, although nanocarriers that use lipids feature the highest level of sophistication, polymers can be real and practical in agricultural systems.

5.6. Toward Integrated Strategies

Instead of pitting lipid-based and polymeric nanoparticle-based approaches against one another as competing technologies, a more powerful future process may entail combining their mutual advantages (Mehta et al., 2023). There is active research underway to design hybrid formulations of the lipid-polymer systems with the ability to provide a combination of membrane fusion and immune priming with environmental stability. Multifunctional formulations may be an unprecedented approach to the plant virus control problem, as such formulations can combine the precision-targeted antivirals with the ability to make crops more resilient to stressors and less dependent on synthetic pesticides.

6. Routes of Application and Delivery Systems

The efficiency of nanoparticle-based antiviral strategies in plants will be dominated by the kind of nanomaterial used, but also by the means of the route of delivery. The exceptional structural impediments of plants, including the intense cuticle, rigid cell wall, and the sophistication of the vascular system, render selection of the application method key to guaranteeing a successful uptake/movement and antiviral activity. Multiple delivery systems have been examined, each with different benefits and drawbacks, and providing a flexible means to manage infection by plant viruses both ante- and post-infection.

6.1. Foliar Sprays

The most applied method is the foliar spraying one because of its convenience and ease with which it can be combined with already developed agricultural practices (Lewicka et al., 2024). Nanoparticles can attach to the surface of the leaves or enter the plant by stomata or the presence of cuticular nanopores or cuticular nanopores and reach the plant cell apoplast or cytoplasm (Avellan et al., 2021). Lipidic nanoparticles tend to either fuse to the plasma membrane or enter between cells through their walls, whereas carriers such as polymeric chitosan can interact electrostatically with cell walls to enter cells (Sivadasan et al., 2021). Environmental conditions (rain, UV radiation, and wind) can limit field efficacy, necessitating the development of formulations with improved adhesion and stability.

6.2. Root Uptake and Soil Application

The application of nanoparticles allows rhizosphere access so that they may be taken up systemically. After they are absorbed, nanoparticles are transported through the xylem stream and disseminated throughout the plant, reaching the most peripheral parts. This path can especially be applied to nanoparticle-based polymers that can have a longer persistence in the soil and get into interact with root exudates, stimulating systemic resistance (Zaman et al., 2025). However, the variability of soil environments and immobilization of nanoparticles in the presence of organic matter are a factor that cannot be ignored.

6.3. Seed Priming and Coating

Seed-based delivery systems offer a prophylactic antiviral approach, and nanoparticles can be implemented before the pressure of infection sets in. Coating/priming seeds with nano formulation allows antiviral agent release over a prolonged period to protect the seedling during early development stages and improve the vigor of seedlings (Riseh et al., 2022). Chitosan and starch-based carriers are compatible with this route, in part because of their controlled release capacity, and also their biocompatibility (Wan Yusof et al., 2024). The lipidic nanoparticle is effective in encapsulating RNA; however, it might be unstable against long-term storage time in coated seeds (Kucharzyk et al., 2023).

6.4. Post-Harvest Coatings

Viral loads can be reduced and shelf life extended in fruits, vegetables, and other harvested commodities through the application of nanoparticle-based coatings (Barbhuiya et al., 2024). These formulations can act as a further bio-barrier to the transmission of viruses during storage/ transportation. Chitosan and alginate-based nanoengineered coatings have been used to inhibit viral and other microbial post-harvest contaminations with some success, often in combination with antivirals (Salgado-Cruz et al., 2021). This way not only troubles food quality but also reduces levels of economic losses in supply chains.

6.5. Integration with Smart Delivery

The current nanoparticle-based precision agriculture is redefined with emerging precision agriculture technologies (Yadav et al., 2023). These advances bring nanotechnology-enabled themes such as the use of drones in foliar applications and automated irrigation technology, employing the formulations directly in the root zone, and intelligent sensors to check viral infection in real-time. The smart delivery systems enhance target performance, limit the wastage of inputs, and/or exposure to the surrounding (Dayoub et al., 2024). These inventions are still in the budding stages but hold the prospect of changing nanoparticle applications to highly efficient, sustainable, and farmer-friendly applications in the future.

7. Safety, Environmental Impacts, and Regulatory Aspects

Applications of nanotechnology to plant antivirals are also feared to present significant safety, environmental sustainability, and regulatory issues. Although nanocarriers (lipidic and polymeric systems) can provide novel solutions, their fate in the environment, as well as interactions with non-target organisms, have not been thoroughly investigated along with regulatory control. These factors are essential parts of a comprehensive understanding because they will help ensure that nano-enabled plant protection products are used responsibly.

7.1. Biodegradability and Eco-Toxicity of Lipidic vs. Polymeric Systems

As nano micelles or made of phospholipids or liposomes, lipidic nanocarriers are characterized by a certain eco-toxicity degree intrinsically reduced. They decompose fatty acids and glycerol, which are readily assimilated into the natural biogeochemical cycles (Vega-Vásquez et al., 2020). The environmental impact of nanoparticle use in agriculture requires careful consideration. Lipidic and polymeric nanoparticles, while biodegradable, can accumulate in soil and water systems if used excessively. Recent studies by Elgharbawy et al. (2024) highlighted the potential long-term effects on soil microbiota and plant health, which underscores the need for sustainable nanoparticle use. Furthermore, regulatory frameworks for nanoparticle-based agricultural products remain underdeveloped globally, posing a challenge for widespread adoption. It is crucial that international guidelines are established to ensure the safety and efficacy of these innovative plant protection tools (Elgharbawy et al., 2024; Irmukhametova et al., 2025). The non-degradable polymers confer the risk of nanoparticles deposition, the change of soil and water quality of water. Despite the promising potential of both lipidic and polymeric nanoparticles, several constraints must be addressed for their widespread adoption. Biosafety concerns, particularly regarding non-target effects on soil organisms and pollinators, remain a significant challenge. Furthermore, the high cost of producing sophisticated nanoparticles such as ionizable lipid nanoparticles (LNPs) limits their accessibility to smallholder farmers (Pathak et al., 2024). Moreover, regulatory hurdles exist, as many countries lack clear guidelines for the approval and use of nanotechnology in agriculture.

7.2. Impact on Soil Microbiota and Beneficial Insects

The microbiota in the soil is vital to nutrient cycling, plant health, and disease suppression. Nanoparticles applied in the root or soil amendments could modify the community structure of microbes, enzyme action, and symbiosis processes, including mycorrhizal colonization (Abbasi et al., 2025; Maurice, 2023). Although chitosan-based nanocarriers tend to foster microbial dynamics, microbial imbalance may occur with overuse or frequent administrations of nanoparticles (Das et al., 2024). Similarly, insects such as pollinators and natural predators may meet nanoparticles inadvertently via leaf residue or nectar laden with nanoparticles. Although the sub-lethal and chronic effects documented are yet to be fully determined, it can be stated that the biodegradable systems are less pernicious.

7.3. Phytotoxicity Risk Assessment

Whereas nanocarriers are developed to make plants resistant, unsuitable formulations or excessive dosages can cause plant phytotoxicity. Nanoparticles can modify the chlorophyll content, induce oxidative stress, and impair nutrient uptake (Falak et al., 2025; Shelar et al., 2023). Lipid-based carriers tend to be weakly toxic because they are biocompatible, but in some cases, the cationic polymer can cause membrane destabilization at high doses (Azarnezhad et al., 2020). General risk evaluation protocols that consider plant species, growth stage, nanoparticle size, and dosage will be necessary before deployment into a field environment.

7.4. Current Regulatory Gaps

The regulatory framework is still divided into nano-enabled plant protection products. Most of the global frameworks categorize these materials under general pesticide/fertilizer regulations, which are devoid of a particular nano-safety criterion. This poses uncertainty about what to label, allowable exposure limits, and assessment of environmental risks. In addition, non-standardized recipes of nanoparticle detection in plant tissues and soils also make monitoring and compliance a challenge. There is an unquestionable need to develop international standards, specific to nanocarriers, to act as a standard focus in agriculture, to ensure a cross-border harmonization of safety standards, and to make possible the responsible commercialization of the enhancement's science and technology offer.

7.5. Sustainable Design Principles

To compensate for exploration with sustainability, green chemistry attributes need to be considered to manufacture an eco-safe nanocarrier. This can be done by choosing biodegradable polymer, raw materials as renewable sources of material, no hazardous by-products, and designing nanoparticles to deliver in a controlled way to avoid overuse. Nanocarriers with supplementary functions such as antiviral activity and delivery of nutrients or tolerance towards stressful conditions, are another way to add sustainability (Shoukat et al., 2024). In addition, life-cycle assessment (LCA) is embraced, in which footprints are determined across synthesis to field application and disposal.

8. Future Directions

The future of nanotechnology in plant antiviral strategies is to create next generation nanocarriers, which are very specific and multifunctional in addition to being efficient. Nanoparticles with environmental sensitivity so that the release of antiviral agents can be manipulated in response to an environment, such as pH, light, or the presence of a pathogen, could provide an attractive means of control. Similarly, the use of gene-editing technologies, especially CRISPR/Cas systems when delivered using nanoparticle-based materials, is a paradigm shift that can be used to directly detect and silence plant virus genomes. The other promising study is the application of biomimetic nanocarriers, the use of exosome-like plant-derived vesicles, which can provide natural compatibility and enable interference effectively with viral RNA replication without toxicity and disrupting immunity at the host plant.

Further, the future lies in the dual-purpose design of nanocarriers that would not only curb viral infections but also break insect vector populations in a compound effect, as a comprehensive plant protection means. The nanotechnology and smart agriculture platforms, such as platforms that use IoT to monitor the crops, drones to deliver specifically, and precision agriculture platforms, will lead to the application of nanocarriers with much less wastage on a site-to-site basis. The application of nanotechnology to increase the effectiveness of crop protection in the face of the changing climate cannot be over exaggerated as well. Extreme climatic changes increase the levels of viruses and reduce the resistance of crops; hence, adjustable nanocarriers according to stress conditions can become protectors of world food safety. These innovations together create a roadmap to a new era of ecologically sound, intelligent, and robust plant antiviral immune systems.

The future of plant viral disease management lies in the development of multifunctional nanocarriers that not only deliver antiviral agents but also improve plant resistance and disrupt viral transmission. These hybrid systems, which combine the benefits of both lipidic and polymeric nanoparticles, could offer a more robust approach to viral control. Moreover, the integration of smart delivery systems, such as nanoparticle formulations that respond to environmental cues (e.g., pH or temperature), holds promise for achieving targeted, on-demand antiviral release in the field.

9. Conclusions

Lipidic and polymeric nanoparticles have become two very promising and disruptive vectors for sustainable control of plant viruses. This is because lipid-based nanoparticles are structurally like cellular membranes and can effectively cover and protect RNA molecules, which makes them efficient RNAi-delivery systems and direct targets to viral genomes. Comparatively, polymeric nanoparticles, especially those that are naturally derived, like chitosan and starch, shine through because of their cheap nature, biodegradable properties, and their capability to induce the immune system of plants. Collectively, these systems can provide a more balanced solution-lipidic carrier components can provide molecular specificity, whereas polymeric carriers can offer increased robustness and field adaptability, such that fewer traditional agrochemicals need to be relied upon, and environment-friendly crop protection can be offered. Their combination is a step in the direction of long-term, scalable, and accessible antiviral methods by farmers.

Although these developments are positive, some challenges are yet to be dealt with before nanocarriers are adopted into agricultural practice. Significant factors such as large-scale production, environmental safety, administrative sanction, and formulation stability have to be addressed so that they can be adopted in real-world applications. It is indeed critical to come up with farmer-friendly formulas that feature affordability, stability in the long term, and low ecological risks. In addition, the development of a regulatory structure unique to nano-enabled plant protection products will contribute to safer uptake. In the future, the integration of lipidic precision with the resilience of polymeric systems, combined with the power of tools in smart agriculture, such as drones and monitoring systems based on the internet of things (IoTs), can create a possibility to produce sustainable crop protection measures capable of withstanding climatic fluctuations. In conclusion, the combination of lipidic and polymeric nanoparticles has the potential to transform future plant virology battles, in Denmark, and beyond, with a consequent impact on food security and the sustainability of the environment.

10. References

Abbasi, A. J., Anas, M., Elahi, M., Khan, A., Khattak, W. A., Saleem, M. H., Fahad, S., Elansary, H. O., Mahmoud, E. A., & Ahmed, T. (2025). Restoring wheat productivity and nutrient balance under cadmium stress through reducing toxicity, metal uptake, and oxidative damage using selenium nanoparticles. *Journal of Trace Elements in Medicine and Biology*, *89*, 127644. PMID: 40233590

Abd-El-Aziz, A., Fouda, M. M., Sharaby, C. M., Xiao, O., Zhang, X., Alzahrany, Y. A., Ahmed, S. A., Ma, N., & Abd-El-Aziz, A. S. (2024). Recent Developments in Antimicrobial and Antiviral Agents Based on Natural/Synthetic Polymers and Dendrimers: Design and Therapeutic Applications. *Macromolecular Chemistry and Physics*, *225*(17), 2400123.

Abdelkhalek, A., Qari, S. H., Abu-Saied, M. A. A.-R., Khalil, A. M., Younes, H. A., Nehela, Y., & Behiry, S. I. (2021). Chitosan nanoparticles inactivate alfalfa mosaic virus replication and boost innate immunity in Nicotiana glutinosa plants. *Plants*, *10*(12), 2701. PMID: 34961172

Akbar, S., Wei, Y., & Zhang, M.-Q. (2022). RNA interference: Promising approach to combat plant viruses. *International Journal of Molecular Sciences*, *23*(10), 5312. PMID: 35628126

Anas, M., Bibi, S., Farwa, U., Khan, A., & Quraishi, U. M. (2024). Manganese oxide nanoparticles elicit cadmium tolerance in wheat seedlings by ionomic and phenomic adjustment, regulation of AsA–GSH pathway, cellular thickness recovery, and antioxidant modulation. *Environmental Science. Nano*, *11*(11), 4588–4605.

Arjunan, N., Thiruvengadam, V., & Sushil, S. (2024). Nanoparticle-mediated dsRNA delivery for precision insect pest control: A comprehensive review. *Molecular Biology Reports*, *51*(1), 355. PMID: 38400844

Ashwini, B. (2022). *Molecular Characterization of Tomato Mosaic Virus and Cucumber Mosaic Virus Infecting Tomato (Solanum Lycopersicum. L) and Induction of Defense Using Biomolecules* University of Agricultural Sciences, GKVK, Bangalore].

Åslund, A. K., Vandebriel, R. J., Caputo, F., De Jong, W. H., Delmaar, C., Hyldbakk, A., Rustique, E., Schmid, R., Snipstad, S., & Texier, I. (2022). A comparative biodistribution study of polymeric and lipid-based nanoparticles. *Drug Delivery and Translational Research*, *12*(9), 2114–2131. PMID: 35426570

Avellan, A., Yun, J., Morais, B. P., Clement, E. T., Rodrigues, S. M., & Lowry, G. V. (2021). Critical review: Role of inorganic nanoparticle properties on their foliar uptake and in planta translocation. *Environmental Science & Technology*, *55*(20), 13417–13431. PMID: 33988374

Azarnezhad, A., Samadian, H., Jaymand, M., Sobhani, M., & Ahmadi, A. (2020). Toxicological profile of lipid-based nanostructures: Are they considered as completely safe nanocarriers? *Critical Reviews in Toxicology*, *50*(2), 148–176. PMID: 32053030

Barbhuiya, R. I., Tinoco, N. N., Ramalingam, S., Elsayed, A., Subramanian, J., Routray, W., & Singh, A. (2024). A review of nanoparticle synthesis and application in the suppression of diseases in fruits and vegetables. *Critical Reviews in Food Science and Nutrition*, *64*(14), 4477–4499. PMID: 36343386

Bergmann, I. E., Malirat, V., Pedemonte, A., & Maradei, E. (2021). Challenges in foot-and-mouth disease virus strain selection as an input to attain broad vaccine intraserotype cross-protection. *Expert Review of Vaccines*, *20*(1), 13–22. PMID: 33455492

Carrillo-Lopez, L. M., Villanueva-Verduzco, C., Villanueva-Sánchez, E., Fajardo-Franco, M. L., Aguilar-Tlatelpa, M., Ventura-Aguilar, R. I., & Soto-Hernández, R. M. (2024). Nanomaterials for plant disease diagnosis and treatment: A review. *Plants*, *13*(18), 2634. PMID: 39339607

Chakraborty, M., Hasanuzzaman, M., Rahman, M., Khan, M. A. R., Bhowmik, P., Mahmud, N. U., Tanveer, M., & Islam, T. (2020). Mechanism of plant growth promotion and disease suppression by chitosan biopolymer. *Agriculture*, *10*(12), 624.

Chandrasekaran, R., Seetharaman, P. K., Danaraj, J., Rajiv, P., & Abd-Elsalam, K. A. (2021). Polymer and lipid-based nanoparticles to deliver RNAi and CRISPR systems. In *CRISPR and RNAi Systems* (pp. 635–659). Elsevier.

Cruz, J. C., & Reyes, L. H. Polymer-Based Nanocarriers: Dendrimers, Polymeric Micelles, Hydrogels. In *Nanocarriers for Nucleic Acids and Proteins* (pp. 80-104). CRC Press.

Das, A., Ghosh, S., & Pramanik, N. (2024). Chitosan Biopolymer Nanocomposites for Agriculture Applications. In *Biopolymeric Nanoparticles for Agricultural Applications* (pp. 209–240). Springer.

Dayoub, M., Shnaigat, S., Tarawneh, R. A., Al-Yacoub, A. N., Al-Barakeh, F., & Al-Najjar, K. (2024). Enhancing animal production through smart agriculture: Possibilities, hurdles, resolutions, and advantages. *Ruminants*, *4*(1), 22–46.

De, A., & Ko, Y. T. (2023). A tale of nucleic acid–ionizable lipid nanoparticles: Design and manufacturing technology and advancement. *Expert Opinion on Drug Delivery*, *20*(1), 75–91. PMID: 36445261

Denaro, M., Smeriglio, A., Barreca, D., De Francesco, C., Occhiuto, C., Milano, G., & Trombetta, D. (2020). Antiviral activity of plants and their isolated bioactive compounds: An update. *Phytotherapy Research*, *34*(4), 742–768. PMID: 31858645

Divya, K., & Jisha, M. (2018). Chitosan nanoparticles preparation and applications. *Environmental Chemistry Letters*, *16*(1), 101–112.

Dutta, P., Kumari, A., Mahanta, M., Biswas, K., Dudkiewicz, A., Thakuria, D., Abdelrhim, A. S., Singh, S. B., Muthukrishnan, G., & Sabarinathan, K. (2022). Advances in nanotechnology as a potential alternative for plant viral disease management. *Frontiers in Microbiology*, *13*, 935193. PMID: 35847105

Eid, N. A., Ibrahim, A. M., Elsharawy, A. A., & Salem, K. F. (2025). Nanofertilizers for Plant Viral Disease Management. In *Nanofertilizers for Sustainable Agriculture: Assessing Impacts on Health, Environment, and Economy* (pp. 149-187). Springer.

Eisenreich, W., Rudel, T., Heesemann, J., & Goebel, W. (2019). How viral and intracellular bacterial pathogens reprogram the metabolism of host cells to allow their intracellular replication. *Frontiers in Cellular and Infection Microbiology*, *9*, 42. PMID: 30886834

El-Ganainy, S. M., Soliman, A. M., Ismail, A. M., Sattar, M. N., Farroh, K. Y., & Shafie, R. M. (2023). Antiviral activity of chitosan nanoparticles and chitosan silver nanocomposites against Alfalfa Mosaic Virus. *Polymers*, *15*(13), 2961. PMID: 37447606

Elgharbawy, A. S., El Demerdash, A.-G. M., Sadik, W. A., Kasaby, M. A., Lotfy, A. H., & Osman, A. I. (2024). Synthetic degradable polyvinyl alcohol polymer and its blends with starch and cellulose—A comprehensive overview. *Polymers*, *16*(10), 1356. PMID: 38794547

Elsharkawy, M. M., Omara, R. I., Mostafa, Y. S., Alamri, S. A., Hashem, M., Alrumman, S. A., & Ahmad, A. A. (2022). Mechanism of wheat leaf rust control using chitosan nanoparticles and salicylic acid. *Journal of Fungi (Basel, Switzerland)*, *8*(3), 304. PMID: 35330306

Falak, A., Anas, M., Khan, A., Hayat, A., Shaheen, Z., Saleem, M. H., Fahad, S., & Quraishi, U. M. (2025). Efficacy of ascorbic acid coated quantum dots in alleviating lead-induced oxidative damage and enhancing growth parameters in rice (Oryza sativa L.) for sustainable cultivation. *Journal of Trace Elements in Medicine and Biology*, *88*, 127603. https://doi.org/https://doi.org/10.1016/j.jtemb.2025.127603 PMID: 39847985

Falanga, A., Del Genio, V., & Galdiero, S. (2021). Peptides and dendrimers: How to combat viral and bacterial infections. *Pharmaceutics*, *13*(1), 101. PMID: 33466852

Farooq, T., Adeel, M., He, Z., Umar, M., Shakoor, N., da Silva, W., Elmer, W., White, J. C., & Rui, Y. (2021). Nanotechnology and plant viruses: An emerging disease management approach for resistant pathogens. *ACS Nano*, *15*(4), 6030–6037. PMID: 33761237

Fernandes, C., Jathar, M., Sawant, B. K. S., & Warde, T. (2023). Scale-up of nanoparticle manufacturing process. In *Pharmaceutical process engineering and scale-up principles* (pp. 173–203). Springer.

Forenzo, C., Arnold, N., & Larsen, J. (2025). Emerging polymeric nanocarriers for mRNA and protein therapeutics: Design, challenges, and clinical outlook. *Nanomedicine (London)*, •••, 1–18. PMID: 40762456

Friskop, A. J., Sisson, A. J., Allen, T. W., Andersen-Onofre, K. F., Aoun, M., Bergstrom, G. C., Betts, A. M., Bissonnette, K. M., Bowen, K. L., & Bradley, C. A. (2025). Estimated yield reductions and economic losses on wheat caused by disease from 2018 through 2021. *Plant Health Progress*, PHP-09-24-0087-RS.

Go, Y. K., & Leal, C. (2021). Polymer–lipid hybrid materials. *Chemical Reviews*, *121*(22), 13996–14030. PMID: 34752064

Haggag, M. G., Shafaa, M. W., Kareem, H. S., El-Gamil, A. M., & El-Hendawy, H. H. (2021). Screening and enhancement of the antimicrobial activity of some plant oils using liposomes as nanoscale carrier. *Bulletin of the National Research Center*, *45*(1), 38.

Hashempur, M. H., Ghorat, F., Karami, F., Jahanbin, A., Nouraei, H., Abbasi, M., Jafari, M., Zare, A., Barzegar, S., & Zareshahrabadi, Z. (2025). Topical Delivery Systems for Plant-Derived Antimicrobial Agents: A Review of Current Advances. *International Journal of Biomaterials*, *2025*(1), 4251091. PMID: 40756473

Hou, X., Zaks, T., Langer, R., & Dong, Y. (2021). Lipid nanoparticles for mRNA delivery. *Nature Reviews. Materials*, *6*(12), 1078–1094. PMID: 34394960

Ibrahim, A., Sasaki, N., Schoelz, J. E., & Nelson, R. S. (2025). Tobacco mosaic virus movement: From capsid disassembly to transport through plasmodesmata. *Viruses*, *17*(2), 214. PMID: 40006969

Idrees, H., Zaidi, S. Z. J., Sabir, A., Khan, R. U., Zhang, X., & Hassan, S.-u. (2020). A review of biodegradable natural polymer-based nanoparticles for drug delivery applications. *Nanomaterials (Basel, Switzerland)*, *10*(10), 1970. PMID: 33027891

Irmukhametova, G., Al Azzam, K. M., Mun, G. A., Bekbayeva, L., Dinara, Z., Yermukhambetova, B. B., Nechipurenko, S. V., Efremov, S. A., Negim, E.-S., & Samy, M. (2025). Obtaining and Characterization of Biodegradable Polymer Blends Based on Polyvinyl Alcohol, Starch, and Chitosan. *Polymers*, *17*(4), 479. PMID: 40006141

Jones, R. A. (2020). Disease pandemics and major epidemics arising from new encounters between indigenous viruses and introduced crops. *Viruses*, *12*(12), 1388. PMID: 33291635

Khan, A., Anas, M., Bibi, F., Ali, M., Khalil, A. T., Munawar, K. S., Mohamed, H. E. A., Hkiri, K., Maaza, M., & Shinwari, Z. K. (2025). Phytochemical-Mediated Green Synthesis of Silver, Copper, and Ag–Cu Bimetallic Nanoparticles Using Peganum harmala Demonstrating Advanced Catalytic, Antioxidant, and Biomedical Applications. *Applied Biochemistry and Biotechnology*, *197*(6), 3630–3667. DOI: 10.1007/s12010-025-05186-4 PMID: 39964595

Khan, A., Anas, M., Qadri, T. A., Algopishi, U. B., Fayad, E., Binjawhar, D. N., Khan, R. A., Maaza, M., & Shinwari, Z. K. (2025). Eco-friendly synthesis of copper oxide-silver bimetallic nanoparticles using Withania coagulans: Characterization and biomedical potential through antioxidant, antibacterial, and catalytic activities. *Inorganic Chemistry Communications*, *181*, 115131. https://doi.org/https://doi.org/10.1016/j.inoche.2025.115131

Khan, A., Anas, M., Ullah, N., Khattak, W. A., Ali, M., Maaza, M., & Shinwari, Z. K. (2025). Harnessing Green Nanotechnology: Exploring the Potential of Silver, Copper, and their Nanocomposites for Biomedical and Agricultural Applications. *Pakistan Journal of Botany*, *57*(6), 6. https://doi.org/http://dx.doi.org/10.30848/PJB2025-6(5)

Kocięcka, J., & Liberacki, D. (2021). The potential of using chitosan on cereal crops in the face of climate change. *Plants*, *10*(6), 1160. PMID: 34200489

Komarova, T., Shipounova, I., Kalinina, N., & Taliansky, M. (2024). Application of chitosan and its derivatives against plant viruses. *Polymers*, *16*(22), 3122. PMID: 39599213

Kothapalli, P., & Vasanthan, M. (2024). Lipid-based nanocarriers for enhanced delivery of plant-derived bioactive molecules: A comprehensive review. *Therapeutic Delivery*, *15*(2), 135–155. PMID: 38214118

Kucharzyk, K. H., Fulwider, V. M., Duong, A., Duffy, M., Cafmeyer, J., Ducceschi, S., & Hinton, C. (2023). Assessment of long-term stability of encapsulated agricultural biologicals in lipid-coated alginate beads. *ACS Agricultural Science & Technology*, *3*(5), 389–398.

Kumar, K., Sridhar, J., Choudhary, V., Singh, H., Parameshwari, B., Kumar, K. S., Sahu, B., Dokka, N., & Sivalingam, P. (2021). New innovations and approaches for biotic stress management of crops. In *Innovations in Agriculture for a Self-Reliant India* (pp. 265–292). CRC Press.

Leskovac, A., & Petrović, S. (2023). Pesticide use and degradation strategies: Food safety, challenges and perspectives. *Foods*, *12*(14), 2709. PMID: 37509801

Lewicka, K., Szymanek, I., Rogacz, D., Wrzalik, M., Łagiewka, J., Nowik-Zając, A., Zawierucha, I., Coseri, S., Puiu, I., & Falfushynska, H. (2024). Current trends of polymer materials' application in agriculture. *Sustainability*, *16*(19), 8439.

Ludwig, E. J., Nunes, U. R., Prestes, O. D., Fagundes, L. K., Fernandes, T. S., & Saibt, N. (2020). Polymer coating in soybean seed treatment and their relation to leaching of chemicals. *Revista Ambiente & Água*, *15*, e2602.

Maaza, M., Henini, M., Ezema, F., Manikandan, E., Kennedy, J., Bouziane, K., Chaker, M., Gibaud, A., Haque, A., & Nuru, Z. (2022). Peculiar size effects in nanoscaled systems. *Nano-Horizons: Journal of Nanosciences and Nanotechnologies, 1*, 36 pages-36 pages.

Mahmood, I., Ansari, R. A., & Rizvi, R. (2024). *Nanotechnology and Plant Disease Management*. CRC Press.

Mandal, A., Mukherjee, A., & Bandyopadhyay, R. (2025). Plant virus disease management strategies: conventional versus modern techniques. In *Detection and management of new and emerging mystery plant virus sources* (pp. 207–238). Apple Academic Press.

Mat Jalaluddin, N. S., Asem, M., Harikrishna, J. A., & Ahmad Fuaad, A. A. H. (2023). Recent progress on nanocarriers for topical-mediated RNAi strategies for crop protection—A review. *Molecules (Basel, Switzerland)*, *28*(6), 2700. PMID: 36985671

Matta, J., & Maalouf, R. (2019). Delivery of siRNA therapeutics: PLGA nanoparticles approach. *Frontiers in Bioscience-Scholar*, *11*(1), 56–74. PMID: 30844736

Maurice, N. (2023). Impact of nanoparticles on soil microbes for enhancing soil fertility and productivity. In *Nanotechnology for Sustainable Agriculture* (pp. 75–127). Apple Academic Press.

Mehta, M., Bui, T. A., Yang, X., Aksoy, Y., Goldys, E. M., & Deng, W. (2023). Lipid-based nanoparticles for drug/gene delivery: An overview of the production techniques and difficulties encountered in their industrial development. *ACS Materials Au*, *3*(6), 600–619. PMID: 38089666

Mendes, B. B., Conniot, J., Avital, A., Yao, D., Jiang, X., Zhou, X., Sharf-Pauker, N., Xiao, Y., Adir, O., & Liang, H. (2022). Nanodelivery of nucleic acids. *Nature Reviews. Methods Primers*, *2*(1), 24. PMID: 35480987

Monnier, N., Furlan, A. L., Buchoux, S., Deleu, M., Dauchez, M., Rippa, S., & Sarazin, C. (2019). Exploring the dual interaction of natural rhamnolipids with plant and fungal biomimetic plasma membranes through biophysical studies. *International Journal of Molecular Sciences*, *20*(5), 1009. PMID: 30813553

Mujtaba, M., Wang, D., Carvalho, L. B., Oliveira, J. L., Espirito Santo Pereira, A. d., Sharif, R., Jogaiah, S., Paidi, M. K., Wang, L., & Ali, Q. (2021). Nanocarrier-mediated delivery of miRNA, RNAi, and CRISPR-Cas for plant protection: Current trends and future directions. *ACS Agricultural Science & Technology*, *1*(5), 417–435.

Narasimhamurthy, K., Udayashankar, A. C., De Britto, S., Lavanya, S. N., Abdelrahman, M., Soumya, K., Shetty, H. S., Srinivas, C., & Jogaiah, S. (2022). Chitosan and chitosan-derived nanoparticles modulate enhanced immune response in tomato against bacterial wilt disease. *International Journal of Biological Macromolecules*, *220*, 223–237. PMID: 35970370

Pathak, A., Mandal, N., Upadhyaya, D. C., Joshi, N., & Upadhyaya, C. P. (2024). Lipid nanoparticles: A sustainable solution for crop disease management. *Advances in Natural Sciences: Nanoscience and Nanotechnology*, *15*(3), 033001.

Pugsley, C. E., Isaac, R., Warren, N. J., & Cayre, O. J. (2021). Recent advances in engineered nanoparticles for RNAi-mediated crop protection against insect pests. *Frontiers in Agronomy*, *3*, 652981.

Puri, A., Mohite, P., Munde, S., Ade, N., Ramole, A., Pillai, D., Nagare, S., Mahadik, S., & Singh, S. (2025). Unlocking the multifaceted potential of lipid-based dispersion as a drug carrier: Targeted applications and stability improvement strategies. *Journal of Dispersion Science and Technology*, •••, 1–33.

Qi, H.-Y., Zhang, D.-D., Liu, B., Chen, J.-Y., Han, D., & Wang, D. (2024). Leveraging RNA interference technology for selective and sustainable crop protection. *Frontiers in Plant Science*, *15*, 1502015. PMID: 39777080

Rajesh, P. (2025). Nanocarriers for Targeted Delivering and Translocation of Chemical Cargoes in Plant Cells. In *Nanocarriers in Plant Science and Agriculture* (pp. 125–164). IGI Global Scientific Publishing.

Rendina, N., Nuzzaci, M., Scopa, A., Cuypers, A., & Sofo, A. (2019). Chitosan-elicited defense responses in Cucumber mosaic virus (CMV)-infected tomato plants. *Journal of Plant Physiology*, *234*, 9–17. PMID: 30640158

Riseh, R. S., Tamanadar, E., Hajabdollahi, N., Vatankhah, M., Thakur, V. K., & Skorik, Y. A. (2022). Chitosan microencapsulation of rhizobacteria for biological control of plant pests and diseases: Recent advances and applications. *Rhizosphere*, *23*, 100565.

Saleh, M. M., Mahmoud, A. S., Abbas, H. S., Abu-Ellail, F. F., Kotakonda, M., & Salem, K. F. (2022). Nanotechnological approaches for efficient delivery of plant ingredients. In *Sustainable agriculture reviews 53: Nanoparticles: A new tool to enhance stress tolerance* (pp. 247-286). Springer.

Salgado-Cruz, M. P., Salgado-Cruz, J., García-Hernández, A. B., Calderón-Domínguez, G., Gómez-Viquez, H., Oliver-Espinoza, R., Fernández-Martínez, M. C., & Yáñez-Fernández, J. (2021). Chitosan as a coating for biocontrol in postharvest products: A bibliometric review. *Membranes*, *11*(6), 421. PMID: 34073018

Sangwan, A., & Singh, N. (2024). Advanced Nanostrategies for Biomolecule Delivery in Plant Disease Management. *Journal of Agricultural and Food Chemistry*, *73*(1), 66–84. PMID: 39715428

Sastry, K. S., & Zitter, T. A. (2014). *Plant virus and viroid diseases in the tropics*. Springer.

Shah, R., Eldridge, D., Palombo, E., & Harding, I. (2015). *Lipid nanoparticles: Production, characterization and stability* (Vol. 1). Springer.

Shakiba, S., Astete, C. E., Paudel, S., Sabliov, C. M., Rodrigues, D. F., & Louie, S. M. (2020). Emerging investigator series: polymeric nanocarriers for agricultural applications: synthesis, characterization, and environmental and biological interactions. *Environmental Science. Nano*, *7*(1), 37–67.

Sharma, N., Kushwaha, K., Upadhyay, S., Mishra, P., Rai, S., Kumar, J., & Gupta, P. C. (2025). Properties and Synthesis of Chitin-Based Nanoparticles. In *Chitin-Based Nanoparticles for the Agriculture Sectors* (pp. 21–44). Springer.

Shelar, A., Nile, S. H., Singh, A. V., Rothenstein, D., Bill, J., Xiao, J., Chaskar, M., Kai, G., & Patil, R. (2023). Recent advances in nano-enabled seed treatment strategies for sustainable agriculture: Challenges, risk assessment, and future perspectives. *Nano-Micro Letters*, *15*(1), 54. PMID: 36795339

Shoukat, A., Pitann, B., Zafar, M. M., Farooq, M. A., Haroon, M., Nawaz, A., Wahab, S. W., & Saqib, Z. A. (2024). Nanotechnology for climate change mitigation: Enhancing plant resilience under stress environments. *Journal of Plant Nutrition and Soil Science*, *187*(5), 604–620.

Singh, J., & Singh, B., & Vishavnath. (2022). Designing Starch–Alginate Hydrogels for Controlled Delivery of Fungicide for the Alleviation of Environmental Pollution. *ACS Agricultural Science & Technology*, *2*(6), 1239–1250.

Sivadasan, D., Sultan, M. H., Madkhali, O., Almoshari, Y., & Thangavel, N. (2021). Polymeric lipid hybrid nanoparticles (plns) as emerging drug delivery platform—A comprehensive review of their properties, preparation methods, and therapeutic applications. *Pharmaceutics*, *13*(8), 1291. PMID: 34452251

Soni, A., Kushvaha, R. P., & Snehi, S. K. (2024). Current strategies for management of plant viruses and future perspectives: Enhancing crop health, yield and productivity. *Asian J Biochem Gen Mole Biol*, *16*, 21–34.

Tatineni, S., & Hein, G. L. (2023). Plant viruses of agricultural importance: Current and future perspectives of virus disease management strategies. *Phytopathology*, *113*(2), 117–141. PMID: 36095333

Tenchov, R., Bird, R., Curtze, A. E., & Zhou, Q. (2021). Lipid nanoparticles— From liposomes to mRNA vaccine delivery, a landscape of research diversity and advancement. *ACS Nano*, *15*(11), 16982–17015. PMID: 34181394

Terzopoulou, Z., Zamboulis, A., Koumentakou, I., Michailidou, G., Noordam, M. J., & Bikiaris, D. N. (2022). Biocompatible synthetic polymers for tissue engineering purposes. *Biomacromolecules*, *23*(5), 1841–1863. PMID: 35438479

Thokchom, S., Pal, A., & Keerthana, A. Priya, Jeengar, D., Sunil, V., Dubey, V. K., Ganguly, S., Ningombam, A., & Santhalembi, L. (2025). Insights into Defense Mechanisms of Engineered Nanoparticles for Management of Insect Pests and Pathogens. In *Nanomaterials in Agroforestry Systems* (pp. 215-245). Springer.

Tortorici, S., Cimino, C., Ricupero, M., Musumeci, T., Biondi, A., Siscaro, G., Carbone, C., & Zappalà, L. (2022). Nanostructured lipid carriers of essential oils as potential tools for the sustainable control of insect pests. *Industrial Crops and Products*, *181*, 114766.

Truzzi, E., Capocefalo, A., Meneghetti, F., Maretti, E., Mori, M., Iannuccelli, V., Domenici, F., Castellano, C., & Leo, E. (2022). Design and physicochemical characterization of novel hybrid SLN-liposome nanocarriers for the smart co-delivery of two antitubercular drugs. *Journal of Drug Delivery Science and Technology*, *70*, 103206.

Valletta, A., Chronopoulou, L., Palocci, C., Baldan, B., Donati, L., & Pasqua, G. (2014). Poly (lactic-co-glycolic) acid nanoparticles uptake by Vitis vinifera and grapevine-pathogenic fungi. *Journal of Nanoparticle Research*, *16*(12), 2744.

Vargas-Hernandez, M., Macias-Bobadilla, I., Guevara-Gonzalez, R. G., Rico-Garcia, E., Ocampo-Velazquez, R. V., Avila-Juarez, L., & Torres-Pacheco, I. (2020). Nanoparticles as potential antivirals in agriculture. *Agriculture*, *10*(10), 444.

Vega-Vásquez, P., Mosier, N. S., & Irudayaraj, J. (2020). Nanoscale drug delivery systems: From medicine to agriculture. *Frontiers in Bioengineering and Biotechnology*, *8*, 79. PMID: 32133353

Vemula, M., & Reddy, A. V. B. (2023). Polymeric nanoparticles as effective delivery systems in agriculture sustainability. *Nanotechnology for Environmental Engineering*, *8*(3), 805–814.

Venu, E., Ramya, A., Babu, P. L., Srinivas, B., Kumar, S., Reddy, N. K., Babu, Y. M., Majumdar, A., & Manik, S. (2024). Exogenous dsRNA-mediated RNAi: Mechanisms, applications, delivery methods and challenges in the induction of viral disease resistance in plants. *Viruses*, *17*(1), 49. PMID: 39861836

Verma, K., & Modgil, M. (2024). RNA interference (RNAi) mediated technique for combating plant diseases: Harnessing nanoparticles for effective delivery and enhanced efficacy. [PCTOC]. *Plant Cell, Tissue and Organ Culture*, *157*(3), 53.

Viegas, C., Patrício, A. B., Prata, J. M., Nadhman, A., Chintamaneni, P. K., & Fonte, P. (2023). Solid lipid nanoparticles vs. nanostructured lipid carriers: A comparative review. *Pharmaceutics*, *15*(6), 1593. PMID: 37376042

Wan Yusof, W. R., Sabar, S., & Zailani, M. A. (2024). Starch-chitosan blends: A comprehensive review on the preparation, physicochemical properties and applications. *Biopolymers*, *115*(5), e23602. PMID: 38816949

Wang, X., Xie, H., Wang, P., & Yin, H. (2023). Nanoparticles in plants: Uptake, transport and physiological activity in leaf and root. *Materials (Basel)*, *16*(8), 3097. PMID: 37109933

Xing, Y., Jiang, H., & Cai, L. (2025). Engineered nanotransporters for efficient RNAi delivery in plant protection applications. *Journal of Integrative Plant Biology*, *67*(5), 1223–1245. PMID: 40080402

Xu, X., Jiao, Y., Shen, L., Li, Y., Mei, Y., Yang, W., Li, C., Cao, Y., Chen, F., & Li, B. (2023). Nanoparticle-dsRNA treatment of pollen and root systems of diseased plants effectively reduces the rate of tobacco mosaic virus in contemporary seeds. *ACS Applied Materials & Interfaces*, *15*(24), 29052–29063. PMID: 37279020

Xu, X., Yu, T., Zhang, D., Song, H., Huang, K., Wang, Y., Shen, L., Li, Y., Wang, F., & Zhang, S. (2023). Evaluation of the anti-viral efficacy of three different dsRNA nanoparticles against potato virus Y using various delivery methods. *Ecotoxicology and Environmental Safety*, *255*, 114775. PMID: 36933482

Yadav, A., Yadav, K., Ahmad, R., & Abd-Elsalam, K. A. (2023). Emerging frontiers in nanotechnology for precision agriculture: Advancements, hurdles and prospects. *Agrochemicals*, *2*(2), 220–256.

Zaman, W., Khalil, A. A. K., Amin, A., & Ali, S. (2025). Nanofabrication Techniques for Enhancing Plant–Microbe Interactions in Sustainable Agriculture. *Nanomaterials (Basel, Switzerland)*, *15*(14), 1086. PMID: 40711205

Zhang, Y., Sun, C., Wang, C., Jankovic, K. E., & Dong, Y. (2021). Lipids and lipid derivatives for RNA delivery. *Chemical Reviews*, *121*(20), 12181–12277. PMID: 34279087

Žigrayová, D., Mikušová, V., & Mikuš, P. (2023). Advances in antiviral delivery systems and chitosan-based polymeric and nanoparticulate antivirals and antiviral carriers. *Viruses*, *15*(3), 647. PMID: 36992356

Chapter 7

Green-Synthesized Nanoparticles for Eco-Friendly Plant Virus Management

Mrinmoy Dey
JIS University, India

Misme Panja
JIS University, India

Sambridhi Bhowmik
JIS University, India

Avik Acharya Chowdhury
JIS University, India

Suparna Ghosh
https://orcid.org/0000-0002-9200-1291
JIS University, India

ABSTRACT

Plant viral diseases pose a critical threat to global food security, causing severe yield and economic losses. Conventional control measures such as resistant cultivars, crop rotation, and vector management are often ineffective due to the rapid evolution of viral pathogens. Green-synthesized nanoparticles (NPs) have emerged as sustainable alternatives, exhibiting multifunctional antiviral roles including direct virion inactivation, activation of host defence responses, and targeted delivery of antiviral agents. Biogenic NPs derived from plants, microbes, fungi, algae, and biopolymers are inherently stable, biocompatible, and environmentally safe. Metal-based (Ag, ZnO, Cu), biopolymeric, and carbon-based nanomaterials have shown efficacy against major viruses such as TMV, CMV, TYLCV, and PRSV. However, large-scale synthesis, formulation stability, biosafety, and regulatory gaps remain key challenges. Integration with precision agriculture offers a strategic route to sustainable viral disease management.

DOI: 10.4018/979-8-3373-4913-8.ch007

1. INTRODUCTION

Plant viral diseases represent one of the most formidable challenges to global food security and agricultural sustainability. Unlike bacterial and fungal pathogens, viruses lack metabolic pathways that can be directly targeted with chemical treatments, making their control far more complex (Hull, 2013; Scholthof et al., 2011). They are obligate intracellular parasites that depend entirely on host cellular machinery for replication, which not only restricts available intervention strategies but also promotes rapid evolution and adaptation (García-Arenal et al., 2003; Elena et al., 2014). The economic implications of plant virus epidemics are staggering: global losses are estimated in the billions annually, with major impacts on staple crops such as rice, maize, cassava, banana, and tomato (Strange & Scott, 2005; Jones, 2009). For instance, Tomato yellow leaf curl virus (TYLCV) alone causes more than USD 2 billion in annual yield losses worldwide (Moriones & Navas-Castillo, 2000), while cassava mosaic viruses contribute to devastating food shortages in sub-Saharan Africa (Legg & Thresh, 2000).

The ecological and social impacts of plant viruses extend well beyond yield losses. Infected crops often exhibit severe symptoms such as leaf curling, chlorosis, mosaic mottling, necrosis, and fruit deformation, which reduce both quantity and quality of produce (Brunt et al., 1990; Palukaitis & García-Arenal, 2003). Viral diseases also drive indirect consequences such as the overuse of insecticides to control vectors (aphids, whiteflies, leafhoppers, thrips), which in turn accelerate pesticide resistance, harm beneficial insects, and degrade soil and water ecosystems (Juárez et al., 2019). Furthermore, globalization and climate change are facilitating the spread of novel viral strains and their vectors into new agro-ecological niches, compounding the challenges for disease management (Anderson et al., 2004; Rybicki & Pietersen, 1999).

Recent outbreaks of viruses such as Banana bunchy top virus (BBTV), Maize streak virus (MSV), and Plum pox virus (PPV) illustrate the vulnerability of global food systems to viral threats (Hooks et al., 2008; Varsani et al., 2008; Cambra et al., 2006). Conventional strategies for plant virus management—including breeding for resistance, crop rotation, quarantine, and vector control—while effective to some extent, are limited by the rapid mutation rates of RNA viruses and recombination events in DNA viruses (Fraile & García-Arenal, 2010). Resistant cultivars can lose their durability within a few growing seasons, while reliance on chemical pesticides is unsustainable in the long run (Korsvik et al., 2007). Consequently, researchers are turning toward innovative, eco-friendly alternatives, among which nanotechnology—particularly green-synthesized nanoparticles—is emerging as a transformative approach (Younas et al., 2020; Prasad et al., 2017).

Green nanotechnology offers a unique blend of sustainability and efficacy, leveraging biological systems such as plants, microbes, algae, and fungi for nanoparticle synthesis (Mittal et al., 2013; Ahmed et al., 2016a; Singh et al., 2018). These biogenic NPs possess inherent advantages over chemically synthesized counterparts: they are produced under mild conditions without toxic chemicals, naturally capped with biomolecules that enhance stability, and often exhibit superior biological compatibility (Iravani, 2011; Raveendran et al., 2003; Kuppusamy et al., 2016). Importantly, their multifunctional properties allow them to act as direct antiviral agents, elicitors of plant defense pathways, and carriers for agrochemicals (Singh et al., 2018; Singh et al., 2022). For example, silver nanoparticles (AgNPs) synthesized using plant extracts have shown significant inhibitory effects on Cucumber mosaic virus (CMV), while zinc oxide nanoparticles (ZnO NPs) can trigger systemic resistance and protect plants against Tobacco mosaic virus (TMV) (Abdelkhalek & Al-Askar, 2020; Verma et al., 2024).

The integration of green nanotechnology into plant virology represents a paradigm shift in sustainable crop protection. Beyond their direct antiviral properties, biogenic nanoparticles are aligned with the principles of green chemistry and the circular bio-economy, utilizing agricultural residues and renewable biomaterials for synthesis (Kah et al., 2019). Their potential role in precision agriculture—coupled with sensors, AI-driven disease monitoring, and smart delivery systems—opens new frontiers for early detection and targeted management of viral pathogens (Singh et al., 2022). Yet, despite promising laboratory and greenhouse findings, the field application of these materials faces challenges such as reproducibility, scale-up, and biosafety assessment (Bundschuh et al., 2018; Warghane et al., 2024).

2. TYPES OF PLANT VIRUSES AND THEIR HARMFUL IMPACT

Plant viruses are among the most destructive pathogens in agriculture, responsible for severe yield losses and quality deterioration in a wide range of crops worldwide. Unlike fungal or bacterial pathogens, viruses cannot be controlled directly with chemical treatments, making them particularly challenging to manage. Their ability to evolve rapidly, adapt to new hosts, and exploit insect vectors has amplified their global impact on food security and agricultural sustainability. Understanding the diversity, classification, and biological characteristics of plant viruses is crucial for designing innovative, eco-friendly management strategies, including the use of green-synthesized nanoparticles (Strange & Scott, 2005).

2.1 Classification of Plant Viruses

Plant viruses are extraordinarily diverse in their genomes, structures, and modes of transmission. They are generally classified based on genome type, particle morphology, and their method of spread. Plant viruses exhibit significant diversity in genome composition, morphology, and transmission strategies, which impact disease outbreak severity and management approaches. RNA viruses like TMV, CMV, PVY, and TSWV evolve rapidly due to error-prone replication mechanisms, while DNA viruses like TYLCV, BBTV, and ACMV exhibit greater genomic stability but cause severe epidemics due to efficient whitefly or leafhopper transmission. Rod-shaped viruses like TMV exhibit exceptional environmental stability, while icosahedral viruses like CMV and PVY have broad host ranges. Geminate viruses, characteristic of begomoviruses, contribute to global yield losses, especially in tropical regions (Dutta, 2022).

2.1.1 Classification by Genome Type

One of the most fundamental distinctions is between RNA and DNA plant viruses. RNA viruses constitute the majority of known plant viruses, accounting for nearly 80% of all identified species. They possess single-stranded or double-stranded RNA genomes and are characterized by high mutation rates due to error-prone RNA-dependent RNA polymerases. This genomic plasticity enables them to adapt rapidly to environmental pressures and host resistance mechanisms. Classic examples include Tobacco mosaic virus (TMV), one of the most extensively studied viruses, Cucumber mosaic virus (CMV), and Potato virus Y (PVY), each of which causes devastating damage across multiple hosts (Table 1). CMV, for instance, infects over 1,200 plant species spanning more than 100 families, making it one of the most widespread plant pathogens (Palukaitis & García-Arenal, 2003).

DNA viruses, although fewer in number, pose equally severe threats. Belonging mainly to the families Geminiviridae and Caulimoviridae, these viruses are often transmitted by insect vectors such as whiteflies and leafhoppers. The Banana bunchy top virus (BBTV), a single-stranded DNA virus, is notorious for stunting banana growth and severely limiting fruit production, particularly in tropical and subtropical regions (Hooks et al., 2008). Another notable DNA virus is the Maize streak virus (MSV), a devastating pathogen in sub-Saharan Africa, which causes chlorotic streaks on leaves and can lead to total crop failure under epidemic conditions (Varsani et al., 2014). Among the geminiviruses, Tomato yellow leaf curl virus (TYLCV) has become a global menace in tomato production, transmitted by the whitefly *Bemisia tabaci*, and responsible for multi-billion-dollar losses annually (Moriones & Navas-Castillo, 2000).

2.1.2 Classification by Morphological Shape

Morphology also provides a useful basis for virus classification, as virions exhibit distinct structural configurations that influence their stability and infectivity (Table 1). Rod-shaped viruses, exemplified by TMV, are among the simplest yet most resilient plant viruses. TMV's rigid helical structure provides extraordinary stability, allowing it to persist in plant debris or contaminated tools for months, thereby facilitating mechanical transmission (Scholthof, 2004).

In contrast, spherical viruses, such as CMV and PVY, display icosahedral capsids that encase their RNA genomes. These viruses often possess broad host ranges and can be transmitted efficiently by insect vectors, notably aphids. Spherical morphology confers a degree of flexibility that allows adaptation to different environmental conditions (Palukaitis & García-Arenal, 2003).

A distinctive group within plant viruses are the geminate or twin-shaped viruses, belonging to the Begomovirus genus. These viruses, including TYLCV and Cassava mosaic virus (different from Cucumber mosaic virus), have unique paired particle structures that are closely associated with their insect vector transmission. The geminate morphology has been hypothesized to contribute to efficient packaging of circular single-stranded DNA genomes and may facilitate recombination events, further enhancing their adaptability (Hanley-Bowdoin et al., 2013).

Table 1. Major plant viruses, their abbreviations, primary host range, characteristic symptoms, genome type, and transmission modes

Virus Name	Abbreviation	Major Host(s)	Typical Symptoms	Genome Type	Transmission Mode	References
Tobacco mosaic virus	TMV	Tobacco, tomato, pepper	Mosaic, leaf distortion	ssRNA (+)	Mechanical, seed-borne	Scholthof, 2004; Hull, 2013
Cucumber mosaic virus	CMV	Cucumber, tomato, pepper, >1200 spp.	Mosaic, stunting, fruit malformation	ssRNA (+), multipartite	Aphids (non-persistent)	Palukaitis & García-Arenal, 2003
Potato virus Y	PVY	Potato, pepper, tobacco	Mosaic, necrosis, yield loss	ssRNA (+)	Aphids	Quenouille et al., 2013
Tomato yellow leaf curl virus	TYLCV	Tomato, pepper	Leaf curling, yellowing, stunting	ssDNA (Geminivirus)	Whiteflies	Navas-Castillo et al., 2011
Banana bunchy top virus	BBTV	Banana	Bunchy top, stunting	ssDNA (Nanovirus)	Aphids	Hooks et al., 2008
Maize streak virus	MSV	Maize	Chlorotic streaks, yield loss	ssDNA (Geminivirus)	Leafhoppers	Varsani et al., 2008
Rice tungro bacilliform virus	RTBV	Rice	Yellow-orange leaves, stunting	dsDNA (Caulimovirus)	Leafhoppers	Hull, 2013
Rice tungro spherical virus	RTSV	Rice	Synergistic with RTBV	ssRNA (+)	Leafhoppers	Hull, 2013
Cassava mosaic virus (African & East African strains)	ACMV, EACMV	Cassava	Mosaic, yield loss	ssDNA (Geminivirus)	Whiteflies	Legg et al., 2011
Sweet potato feathery mottle virus	SPFMV	Sweet potato	Leaf feathering, chlorosis	ssRNA (+)	Aphids	Kreuze et al., 2009
Sweet potato chlorotic stunt virus	SPCSV	Sweet potato	Chlorosis, stunting	ssRNA (+)	Whiteflies	Kreuze et al., 2009
Wheat streak mosaic virus	WSMV	Wheat, barley	Yellow streaks, stunting	ssRNA (+)	Mites	Stenger et al., 2005
Barley yellow dwarf virus	BYDV	Barley, wheat, oats	Yellowing, stunting	ssRNA (+)	Aphids	Miller & Rasochová, 1997
Beet curly top virus	BCTV	Sugar beet, tomato	Leaf curling, stunting	ssDNA (Geminivirus)	Leafhoppers	Briddon & Stanley, 2006
Groundnut rosette virus complex	GRV + SatRNA	Groundnut	Rosette, chlorosis	ssRNA (+)	Aphids	Naidu et al., 1999
Pepper mild mottle virus	PMMoV	Pepper, tomato	Mosaic, mottling	ssRNA (+)	Seed, mechanical	Rodríguez-Cerezo et al., 1989
Papaya ringspot virus	PRSV	Papaya, cucurbits	Mosaic, fruit rings	ssRNA (+)	Aphids	Tripathi et al., 2008

continued on following page

Table 1. Continued

Virus Name	Abbreviation	Major Host(s)	Typical Symptoms	Genome Type	Transmission Mode	References
Citrus tristeza virus	CTV	Citrus	Stem pitting, decline	ssRNA (+)	Aphids	Bar-Joseph et al., 1989
Grapevine leafroll-associated virus	GLRaV	Grapevine	Leaf rolling, sugar reduction	ssRNA (+)	Mealybugs, scales	Martelli et al., 2012
Tomato spotted wilt virus	TSWV	Tomato, pepper, ornamentals	Ringspots, necrosis	ssRNA (-), multipartite	Thrips	Whitfield et al., 2005
Chrysanthemum stunt viroid	CSVd	Chrysanthemum	Stunting, reduced flowers	ssRNA (viroid)	Mechanical	Diener, 2003
Hop stunt viroid	HSVd	Hop, citrus	Stunting, reduced yield	ssRNA (viroid)	Mechanical	Diener 2003
Plum pox virus	PPV	Peach, plum	Fruit deformation, yield loss	ssRNA (+)	Aphids	Cambra et al., 2006
Tomato bushy stunt virus	TBSV	Tomato, ornamentals	Bushy growth, chlorosis	ssRNA (+)	Mechanical	Hull, 2013
Alfalfa mosaic virus	AMV	Alfalfa, tomato	Mosaic, mottling	ssRNA (+)	Aphids, seed	Hull, 2013
Peanut mottle virus	PeMoV	Peanut, soybean	Mosaic	ssRNA (+)	Aphids	Soumya et al., 2014
Bean golden mosaic virus	BGMV	Common bean	Yellow mosaic	ssDNA (Begomovirus)	Whiteflies	Ferreira et al., 2024
Cowpea mosaic virus	CPMV	Cowpea, legumes	Mosaic, mottling	ssRNA (+), bipartite	Mechanical, seed	Hull, 2013
Lettuce mosaic virus	LMV	Lettuce	Mosaic, reduced head size	ssRNA (+)	Seed, aphids	Dinant & Lot, 1992
Rice ragged stunt virus	RRSV	Rice	Ragged leaves, stunting	dsRNA (Reoviridae)	Brown planthopper	Zhang et al., 2018
Sugarcane mosaic virus	SCMV	Sugarcane, maize	Mosaic, chlorosis	ssRNA (+)	Aphids	Li et al., 2011
Barley stripe mosaic virus	BSMV	Barley, wheat	Stripe mosaic	ssRNA (+)	Seed, mechanical	Kendall et al., 2013
Peanut stunt virus	PSV	Peanut, clover	Mosaic, stunting	ssRNA (+)	Aphids	Hull, 2013
Black gram mottle virus	BGMV	Black gram, legumes	Leaf mottling	ssRNA (+)	Beetles	Hull, 2013
Chrysanthemum virus B	CVB	Chrysanthemum	Mosaic, discoloration	ssRNA (+)	Mechanical	Hull, 2013
Pea seed-borne mosaic virus	PSbMV	Pea, legumes	Mosaic, seed transmission	ssRNA (+)	Seed, aphids	Johansen et al., 1994
Turnip mosaic virus	TuMV	Mustard, cabbage	Mosaic, necrosis	ssRNA (+)	Aphids	Walsh & Jenner, 2002
Watermelon mosaic virus	WMV	Cucurbits	Mosaic, mottling	ssRNA (+)	Aphids	Desbiez & Lecoq, 1997

2.1.3 Classification by Transmission Mode

The third major axis of classification is the mode of transmission. Plant viruses rely on vectors or human-mediated activities for dissemination, as they cannot penetrate intact plant cell walls on their own. Vector-borne transmission is the predominant mechanism, accounting for the majority of plant virus spread in agricultural systems. Aphids, whiteflies, thrips, and leafhoppers are the primary vectors, each associated with distinct virus families. For example, CMV is transmitted non-persistently by over 80 species of aphids, while TYLCV is transmitted persistently by *Bemisia tabaci* (Navas-Castillo et al., 2011).

Mechanical or seed-borne transmission also plays a significant role. Tobacco mosaic virus (TMV) is a classic example of a mechanically transmitted virus, spread via contaminated tools, hands, or cultural practices. Its stability enables it to persist on surfaces and infect plants during routine cultivation (Scholthof, 2004). Soybean mosaic virus (SMV) and Rice tungro virus (RTV) can be transmitted through infected seeds, posing challenges for international trade and germplasm exchange. Seed-borne viruses are particularly insidious as they may go undetected until plants are already established in the field (Alegbejo, 2015). Overall, the classification of plant viruses by genome type, morphology, and transmission route highlights the remarkable diversity of these pathogens and underscores the complexity of managing their spread in agricultural ecosystems.

2.2 Effects of Plant Viruses on Crops

The impact of plant viruses on crop health is multifaceted, ranging from visible symptoms that reduce marketability to systemic physiological disruptions that compromise overall productivity. The severity of effects depends on the virus species, host plant, environmental conditions, and presence of co-infections.

One of the most significant outcomes of viral infection is yield reduction. Infected plants often exhibit stunted growth due to disruptions in photosynthesis, nutrient transport, and hormonal regulation. For instance, cassava infected with Cassava mosaic virus can suffer yield losses exceeding 80%, a devastating blow to food security in Africa, where cassava is a staple (Legg et al., 2011). Similarly, rice crops affected by Tungro virus complexes experience significant yield penalties, undermining production in South and Southeast Asia (Azzam & Chancellor, 2002).

Leaf and fruit damage constitutes another hallmark of viral infection. Characteristic symptoms include mosaic patterns, leaf curling, chlorosis, necrosis, and fruit malformation. For example, CMV causes mosaic mottling and deformation in cucurbits, peppers, and tomatoes, drastically reducing their commercial value (Palukaitis & García-Arenal, 2003). TYLCV-infected tomatoes exhibit severe leaf curling and yellowing, leading to poor fruit set and quality degradation (Moriones & Navas-Castillo, 2000). In bananas, BBTV causes a bunchy appearance with narrow, upright leaves, rendering the plant unproductive (Hooks et al., 2008).

Beyond visible symptoms, viral infections weaken plant immunity, predisposing crops to secondary infections by fungi, bacteria, or nematodes. The synergistic interactions between viruses and other pathogens often exacerbate disease severity. For example, co-infection of sweet potato with Sweet potato feathery mottle virus and Sweet potato chlorotic stunt virus leads to the devastating Sweet potato virus disease (SPVD), one of the most damaging viral diseases of this staple crop in sub-Saharan Africa (Gibson et al., 1996).

The cumulative effects of these infections translate into substantial reductions in food production, undermining both smallholder and commercial agriculture. Unlike abiotic stresses such as drought or nutrient deficiency, viral infections can spread explosively under favourable vector conditions, causing sudden and widespread epidemics (Strange & Scott, 2005).

2.3 Impact on Humans and Ecosystems

The consequences of plant viral diseases extend well beyond the infected fields, reverberating through human societies and ecosystems. The most immediate impact is on food security, as viral infections diminish both the quantity and quality of food available. Staples such as rice, maize, cassava, and banana are particularly vulnerable, and their viral diseases contribute directly to nutritional deficiencies and hunger in affected regions (Legg et al., 2011).

Economically, plant viruses inflict massive financial losses. Global estimates suggest billions of dollars are lost annually due to reduced yields, increased management costs, and trade restrictions. For instance, TYLCV alone is estimated to cause losses exceeding USD 2 billion annually in tomato production (Moriones & Navas-Castillo, 2000). Similarly, CMV outbreaks in cucurbit production can wipe out entire harvests, devastating smallholder farmers who rely on these crops for income (Palukaitis & García-Arenal, 2003).

Another indirect consequence of plant viral epidemics is the overuse of pesticides. Since viruses themselves cannot be directly targeted by chemical treatments, farmers often resort to excessive use of insecticides to control vector populations. In many regions, whitefly-transmitted viruses have driven indiscriminate spraying of neonicotinoids and other chemicals, leading to environmental contamination, loss of beneficial insects, and the emergence of resistant vector populations (Ghosh et al., 2019). This cycle of pesticide dependence undermines ecosystem health and reduces biodiversity, further destabilizing agroecosystems.

Moreover, viral epidemics can trigger ecological imbalances. The introduction of non-native viruses through global trade can lead to the decline of indigenous plant species, altering ecological dynamics. For example, the introduction of Plum pox virus into North America not only devastated stone fruit industries but also threatened wild Prunus species in natural habitats (Cambra et al., 2006). Invasive insect vectors such as *Bemisia tabaci* have also expanded their range due to climate change and human activity, carrying with them a suite of viral pathogens into new ecosystems (Navas-Castillo et al., 2011).

From a human perspective, the impacts of plant viruses intertwine with issues of poverty, malnutrition, and social stability. Smallholder farmers in developing countries are disproportionately affected, as they often lack access to resistant varieties, diagnostic tools, and effective management strategies. Crop failures due to viral epidemics can push vulnerable households into cycles of debt and food insecurity (Strange & Scott, 2005).

3. GREEN SYNTHESIS OF NANOPARTICLES

Green synthesis of nanoparticles (NPs) has emerged as a revolutionary approach in nanotechnology, offering an environmentally benign, cost-effective, and biocompatible alternative to traditional chemical and physical synthesis methods. Conventional approaches often involve toxic solvents, strong reducing agents, and high temperature and pressure conditions that are energy-intensive and hazardous to humans and ecosystems. In contrast, green synthesis employs biological systems such as plants, bacteria, fungi, algae, actinomycetes, yeasts, viruses, and lichens as natural factories for nanoparticle production. These organisms provide reducing biomolecules—flavonoids, polyphenols, terpenoids, alkaloids, proteins, polysaccharides, and organic acids—that facilitate the conversion of metal ions into nanoparticles and simultaneously act as stabilizing agents, preventing agglomeration and ensuring long-term stability. Majorly discussed in the Table 2. This dual role of biological systems in reduction and capping endows nanoparticles with enhanced functionality, biocompatibility, and environmental safety, making them ideal for applications in agriculture, particularly in the eco-friendly management of plant viral diseases (Iravani, 2011; Singh et al., 2016; Rónavári et al., 2021; Patra & Baek 2014; Mittal et al., 2013).

Table 2. Comparative evaluation of green-synthesized nanoparticles vs. conventional plant virus management approaches

Criteria	Conventional Methods (Chemical antivirals, cultural practices, vector control)	Green-Synthesized Nanoparticles
Primary Mode of Action	Mostly indirect (vector control, cultural practices); limited direct antiviral activity	Direct virion inactivation, replication inhibition, movement suppression, immune activation
Target Specificity	Low to moderate	High; interacts with viral proteins, nucleic acids, or plant defense pathways
Environmental Impact	Chemical residues, ecological toxicity, resistance in vectors	Biodegradable, biocompatible, lower ecological footprint
Resistance Development	High risk due to single-site modes of action	Low risk; multiple antiviral pathways reduce selection pressure
Field Longevity	Short-lived; frequent reapplication required	Longer persistence; immune priming can extend protection
Sustainability	Dependent on repeated chemical inputs	Renewable, plant-based synthesis using biomass waste
Cost & Affordability	Can be costly due to repeated sprays	Potentially lower cost via green synthesis; still high for purified nanomaterials
Ease of Adoption	Well-established, widely available	Requires training, formulation standardization
Scalability	High; established manufacturing	Moderate; depends on green synthesis optimization
Suitability for IPM	Often used as supportive tools	Integrates well with biological control, organic farming, and IPM frameworks

Table 3. Green-synthesized nanoparticles and their reported antiviral activities. The table summarizes diverse nanoparticle types, their biological sources of synthesis, target viruses, and underlying antiviral mechanisms

Nanoparticle Type	Green Synthesis Source	Target Virus(es)	Antiviral Mechanism	Reported Efficacy	Reference
Silver (AgNPs)	*Citrus limon* peel/ juice extracts	HSV-1, SARS-CoV-2	Disruption of viral coat proteins, blocking entry, ROS generation	$IC_{90} \approx 6$ µg/mL; $CC_{50} > 450$ µg/mL	Alkhulaifi et al., 2020
Silver (AgNPs)	Green tea extract	Newcastle Disease Virus	Binding to viral proteins, virion inactivation	Significant reduction of viral load in ovo	Saadh, 2022
Zinc Oxide (ZnO NPs)	*Plumbago indica* extract	HSV-1	Viral entry inhibition	$IC_{50} \approx 23$ µg/mL; $CC_{50} \approx 44$ µg/mL	Melk et al., 2021

Nanoparticle Type	Green Synthesis Source	Target Virus(es)	Antiviral Mechanism	Reported Efficacy	Reference
Zinc Oxide (ZnO NPs)	*Pelargonium zonale* extract	Human coronavirus 229E	Virus Inactivation Pre-Entry, Reactive nitrogen species (RNS)	$IC_{50} \approx 2$ μg/mL; SI > 60	Alqahtani et al., 2022
Zinc Oxide (ZnO NPs)	*Mentha spicata* extract	Tobacco Mosaic Virus (TMV) in tomato	Virus inhibition and plant immunity activation	>90% reduction in viral accumulation; upregulation of PR-1, POD	Abdelkhalek et al., 2020
Copper (Cu, CuO NPs)	Plant extracts, microbial routes	Influenza, enteric viruses (surface studies)	ROS generation, genome disruption	Complete inactivation on surfaces within minutes	Noyce et al., 2007
Gold (AuNPs)	Plant-derived phytochemicals	HIV, influenza, vaccine delivery	Drug/vaccine carrier, viral protein interaction	Strong binding capacity, drug stabilization	Dykman & Khlebtsov, 2012
Iron Oxide (Fe_3O_4 NPs)	Microbial/plant extracts	Limited direct antiviral data	Magnetic hyperthermia, drug delivery	Potential for targeted therapy	Sportelli et al., 2020
TiO_2, CeO_2 NPs	Plant extracts	Influenza, coronaviruses (surface disinfection)	ROS-mediated virucidal activity (photocatalysis)	Effective in reducing viral infectivity	Sportelli et al., 2020
Chitosan NPs	Chitin-derived	Hepatitis C Virus (HCV)	Antiviral drug carrier, Viral Entry Inhibition, Cell Cycle Arrest	Sofosbuvir-loaded NPs eradicated HCV RNA in vitro	Loutfy et al., 2020
Starch NPs	Plant starch	Indirect (encapsulation studies)	Drug/metal carrier for antiviral molecules, RNS	Demonstrated safety, non-toxic	Rajabzadeh-Khosroshahi et al., 2025
Alginate NPs	Brown algae	Respiratory viruses (experimental)	ROS, Interference with viral replication post-entry	Effective for nasal formulations	Omidian et al., 2024
Gelatin NPs	Collagen-derived	Influenza, coronaviruses (vaccine studies)	Antiviral drug/vaccine encapsulation	Stable, high loading	Omidian et al., 2024
Cellulose NPs/ Nanocrystals	Plant biomass	SARS-CoV-2 (surface contact)	Hydroxyl surface binding, viral desiccation	>90% infectivity reduction in minutes	Baldelli et al., 2022
Pectin NPs	Fruit peel extracts	Limited direct antiviral data	Antiviral molecule encapsulation	Safe, plant-friendly	Rajabzadeh-Khosroshahi et al., 2025
Carbon Dots (CQDs)	Citric acid, biomass	Coronaviruses	Block viral glycoprotein-receptor binding	Strong replication inhibition in vitro	Łoczechin et al., 2019
Graphene Oxide (GO)	Plant biomass	Enveloped viruses	Adsorption and disruption of viral envelopes	Effective antiviral coatings/filters	Palmieri et al., 2020
Carbon Nanotubes (CNTs)	Functionalized carbon	Limited direct antiviral data	Physical disruption and electrostatic interaction, ROS	Detection and potential delivery systems	Zhang et al., 2012

continued on following page

Table 3. Continued

3.1 Plant-Based Synthesis

Plants represent the most widely exploited and versatile system for nanoparticle biosynthesis due to their abundance, ease of handling, and rich repertoire of secondary metabolites. Plant-mediated synthesis can utilize various tissues, including leaves, stems, roots, seeds, flowers, and even fruit peels or agricultural residues. Phytochemicals such as polyphenols, flavonoids, terpenoids, saponins, reducing sugars, and organic acids act as natural reducing and capping agents. The choice of plant species, plant part, extract concentration, pH, temperature, and metal precursor concentration significantly influences the size, shape, and stability of nanoparticles (Ahmed et al., 2016b; Singh et al., 2021; Gautam et al., 2023; Kuppusamy et al., 2016).

Numerous plant species have been investigated for the green synthesis of nanoparticles (Table 3). For instance, *Azadirachta indica* (neem) leaf extract produces silver nanoparticles (AgNPs) with potent antimicrobial and antiviral activity, attributable to bioactive compounds such as quercetin and azadirachtin. *Ocimum sanctum* (holy basil) and Aloe vera leaf extracts have also been reported to generate AgNPs with high stability and uniform size distribution (Ahmad et al., 2003; Chandran et al., 2006; Durán et al., 2005). Other examples include *Moringa oleifera* seeds, *Citrus sinensis* peels, and *Camellia sinensis* (green tea) leaves, which are employed to synthesize gold, silver, and zinc oxide nanoparticles with remarkable biocompatibility (Elumalai et al., 2010; Iravani et al., 2014). Importantly, the use of agricultural residues such as banana peels, onion skins, and fruit husks aligns with principles of circular bioeconomy, simultaneously reducing waste and producing valuable nanomaterials for sustainable agriculture (Kalishwaralal et al., 2008).

Mechanistically, plant-derived phytochemicals donate electrons to metal ions, reducing them to their elemental nanoscale forms. Polyphenols and flavonoids, for example, provide hydroxyl groups that serve as electron donors, converting silver ions into silver atoms, which nucleate to form nanoparticles. Concurrently, proteins, polysaccharides, and other macromolecules bind to the nanoparticle surface, providing steric stabilization that prevents aggregation (Li et al., 2011; Manchester & Steinmetz, 2009; Hamza et al., 2024). Plant-based nanoparticles have been widely demonstrated to have antiviral efficacy against various plant viruses, including Tobacco mosaic virus (TMV), Cucumber mosaic virus (CMV), and Tomato yellow leaf curl virus (TYLCV), by inhibiting viral replication, preventing systemic spread, or enhancing host plant immunity (Patra & Baek, 2014; Raliya & Tarafdar, 2013).

3.2 Microbial Synthesis

3.2.1 Bacterial, Actinomycete, and Yeast-Mediated Synthesis

Microorganisms such as bacteria, actinomycetes, and yeasts provide an alternative and highly versatile platform for nanoparticle biosynthesis (Table 3). Microbial systems possess diverse metabolic pathways capable of reducing metal ions extracellularly, via secreted enzymes, or intracellularly, through bioaccumulation and subsequent biomineralization. The extracellular synthesis is advantageous for industrial scalability, as nanoparticles can be harvested from culture supernatants without cell disruption (Rónavári et al., 2021; Sastry et al., 2003).

Bacteria are among the most studied microbial systems. Species such as *Bacillus subtilis*, *Pseudomonas aeruginosa*, *Lactobacillus* spp., and *Escherichia coli* have demonstrated the capacity to synthesize silver, gold, and zinc oxide nanoparticles through enzymatic reduction, primarily involving nitrate reductase and other redox-active enzymes (Singh et al., 2015; Mittal et al., 2013; Narayanan & Sakthivel, 2010). Actinomycetes, particularly members of the genus Streptomyces, also contribute significantly due to their ability to secrete a variety of bioactive metabolites that enhance the stability, uniformity, and bioactivity of the nanoparticles (El-batal et al., 2017). Yeasts such as *Saccharomyces cerevisiae* and *Candida glabrata* have been shown to produce metallic nanoparticles via intracellular reduction mediated by glutathione and other reductive biomolecules, providing unique surface functionalization and biocompatibility (Azizi et al., 2014; Singh et al., 2015).

Microbially synthesized nanoparticles are particularly promising for agricultural applications, as they can be engineered to possess specific antiviral, antifungal, and antibacterial properties. For example, silver and copper nanoparticles synthesized by *Bacillus* sp. have been demonstrated to inhibit TMV and CMV infections in solanaceous crops, offering an eco-friendly alternative to chemical pesticides (Ghosh et al., 2019; Palukaitis & García-Arenal, 2003; Hooks et al., 2008).

3.2.2 Fungal-Mediated Synthesis

Fungi represent a highly efficient system for nanoparticle biosynthesis, particularly for large-scale production (Table 3). Filamentous fungi such as *Fusarium oxysporum*, *Aspergillus niger*, *Penicillium chrysogenum*, and *Trichoderma* spp. produce extracellular proteins and enzymes capable of reducing metal ions into nanoparticles while simultaneously stabilizing them through capping mechanisms (Varsani et al., 2014; Moriones & Navas-Castillo, 2000; Cambra et al., 2006). The high secretion of reductive biomolecules by fungi ensures the production of nanoparticles with uniform size distribution, superior dispersibility, and high colloidal stability (Gibson et al., 1996; Hanley-Bowdoin et al., 2013).

Fungal-mediated nanoparticles have been widely explored for their antiviral properties. For instance, silver nanoparticles synthesized using *Fusarium oxysporum* have been shown to inhibit TMV and TYLCV replication in tomato and tobacco plants, reducing systemic spread and symptom severity. Mechanistically, fungal proteins interact with metal ions to initiate nucleation, while secreted polysaccharides and glycoproteins coat the nanoparticle surface, providing long-term stability and preventing aggregation (Garg et al., 2020; El-batal et al., 2017).

3.2.3 Algal-Mediated Synthesis

Algae, both marine and freshwater species, have emerged as promising agents for nanoparticle synthesis (Table 3). Algal biomass contains polysaccharides, proteins, pigments (chlorophylls and carotenoids), phenolic compounds, and terpenoids that act as reducing and capping agents. Species such as *Sargassum muticum*, *Ulva lactuca*, *Chlorella vulgaris*, and *Gracilaria* spp. have been successfully used to synthesize silver, gold, iron oxide, and titanium dioxide nanoparticles with enhanced stability and biocompatibility (Elumalai et al., 2010; Li et al., 2011). Algal-mediated nanoparticles are particularly valued for agricultural applications because of their low phytotoxicity and ability to persist under field conditions. The polysaccharide-rich matrix surrounding nanoparticles synthesized by algae provides both colloidal stability and slow release of metal ions, enhancing antiviral efficacy while minimizing environmental impact (Rónavári et al., 2021; Hamza et al., 2024). Studies have shown that silver nanoparticles synthesized from *Chlorella vulgaris* reduce CMV and TMV replication in cucurbits and solanaceous crops, providing a sustainable, eco-friendly alternative to conventional chemical antiviral agents (Elumalai et al., 2010; Li et al., 2011).

3.2.4 Lichen- and Endophyte-Mediated Synthesis

Lichens, symbiotic associations between fungi and algae or cyanobacteria, represent a unique niche for nanoparticle synthesis (Table 3). Lichens produce metabolites such as usnic acid, phenolic compounds, and depsides that facilitate metal ion reduction and nanoparticle stabilization (Pant, 2012; Patra & Baek, 2014). Virus templates and endophytic microbes residing within plant tissues also contribute to nanoparticle production. Endophytes isolated from medicinal plants have been shown to synthesize silver and zinc oxide nanoparticles with enhanced antiviral efficacy against TMV and CMV (Manchester & Steinmetz, 2009; Narayanan & Sakthivel, 2010).

3.3 Mechanisms of Green Nanoparticle Formation

The green synthesis of nanoparticles generally involves three stages: reduction of metal ions, nucleation, and growth/stabilization. Biological metabolites donate electrons to metal ions, reducing them to elemental form, which aggregate to form nuclei. Subsequent growth is controlled by capping molecules such as proteins, polysaccharides, and polyphenols, which prevent uncontrolled agglomeration and confer stability. In microbial systems, enzymatic pathways such as nitrate reductase catalyze metal ion reduction, while in plant systems, hydroxyl groups of polyphenols and aldehyde groups of reducing sugars play a critical role (Li et al., 2011; Manchester & Steinmetz, 2009; Singh et al., 2015).

3.4 Advantages of Green Synthesis

Green synthesis offers multiple advantages over conventional chemical and physical methods. It avoids toxic chemicals and high-energy processes, reduces production costs, and produces nanoparticles capped with natural biomolecules that confer enhanced stability, biocompatibility, and low environmental toxicity. This approach is inherently sustainable, making use of renewable biological resources, including agricultural wastes and microbial cultures, and aligns with principles of the circular bioeconomy (Kalishwaralal et al., 2008; Rónavári et al., 2021; Hamza et al., 2024). Green-synthesised nanoparticles provide significant benefits over traditional antiviral methods, which often rely on limited vector control and can lead to environmental contamination and resistance. Nanoparticles use multifunctional mechanisms like direct virion inactivation and immune stimulation, offering long-lasting protection. Their biodegradable and biocompatible nature reduces environmental impact and promotes safer use in agroecosystems. Additionally, nano-priming, i.e., nanoparticle-induced pre-activation of plant defence pathways that sensitizes immune signalling networks, enabling faster and stronger antiviral responses upon pathogen challenge, which ensures the enhancement of host immunity more effectively than chemical treatments. When effectively integrated into Integrated Pest Management (IPM) frameworks, nanoparticle-based tools represent sustainable alternatives for controlling plant viral diseases (Lin, 2021).

3.5 Characterization and Quality Control

Comprehensive characterization is essential to ensure the effectiveness and safety of green-synthesized nanoparticles. Transmission electron microscopy (TEM) and scanning electron microscopy (SEM) provide information on particle morphology and size. X-ray diffraction (XRD) and energy-dispersive X-ray spectroscopy (EDX) reveal crystallinity and elemental composition. Fourier-transform infrared spectroscopy (FTIR) identifies functional groups responsible for capping, stabilization, and bioactivity. Dynamic light scattering (DLS) and zeta potential analysis offer insights into particle stability, dispersibility, and shelf-life (Mittal et al., 2013; Narayanan & Sakthivel, 2010; Elumalai et al., 2010) (Figure 1).

Figure 1. Characterization of the green-synthesized nanoparticles using various analytical and microscopic techniques to determine their optical, structural, morphological, and surface properties

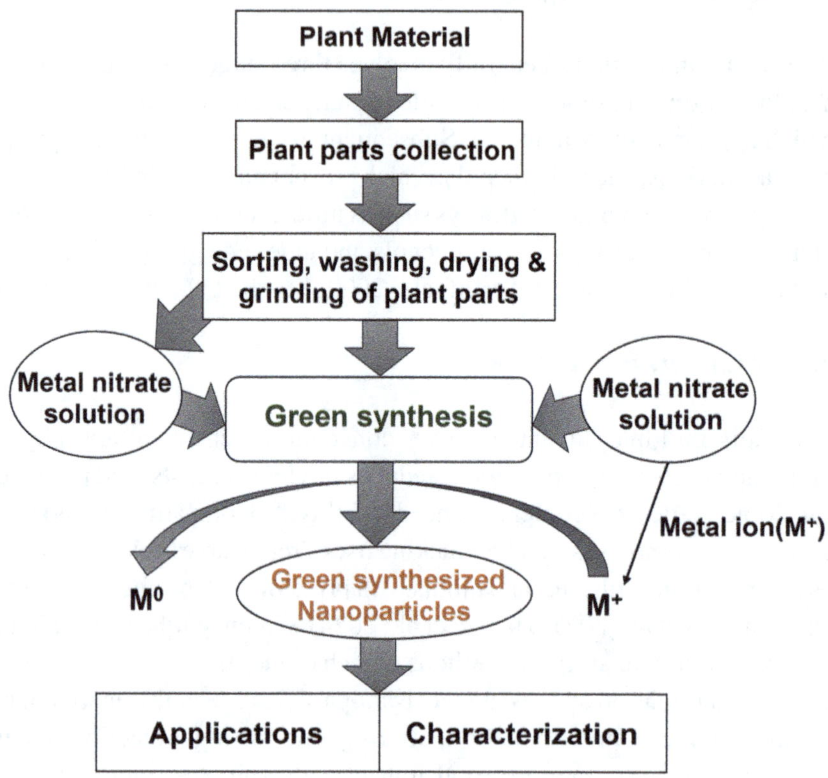

4. TYPES OF GREEN-SYNTHESISED NANOPARTICLES IN VIROLOGY

The preceding section outlined the diverse biological systems used in green nanoparticle synthesis and highlighted how their biochemical richness influences nanoparticle stability, morphology, and antiviral efficacy. Building on this synthesis-focused foundation, the following section shifts toward specific categories of functional nanomaterials, with emphasis on carbon-based nanoparticles and their unique physicochemical properties. This transition bridges the fundamental understanding of nanoparticle formation with their targeted applications in plant virus suppression, enabling a clearer link between synthesis processes and biological performance.

4.1 Metal-Based Nanoparticles

Metal-based nanoparticles have gained significant attention in plant virology due to their unique physicochemical properties, biocompatibility, and potent antiviral activity (Table 3). These nanoparticles not only exert direct inhibitory effects on viral particles but also stimulate host defense pathways, improve plant physiology, and enhance nutrient availability. Their antiviral mechanisms include the disruption of viral coat proteins, oxidative degradation of nucleic acids, inhibition of replication, and induction of systemic acquired resistance (SAR) and antioxidant pathways in host plants (Elmer & White, 2018). A wide spectrum of metallic and metal oxide nanoparticles-including silver (Ag), zinc oxide (ZnO), copper (Cu, CuO), iron oxide (Fe_3O_4), titanium dioxide (TiO_2), gold (Au), cerium oxide (CeO_2), palladium (Pd), ruthenium (Ru), cadmium sulfide (CdS), and lead sulfide (PbS)—have been studied against economically important viruses such as Tobacco mosaic virus (TMV), Cucumber mosaic virus (CMV), Papaya ringspot virus (PRSV), and Tomato yellow leaf curl virus (TYLCV) (Wang et al., 2016; Ahmed et al., 2016b; Salama, 2012) (Figure 2, 3).

Figure 2. Flowchart of green synthesized metal nanoparticle production using plant extracts

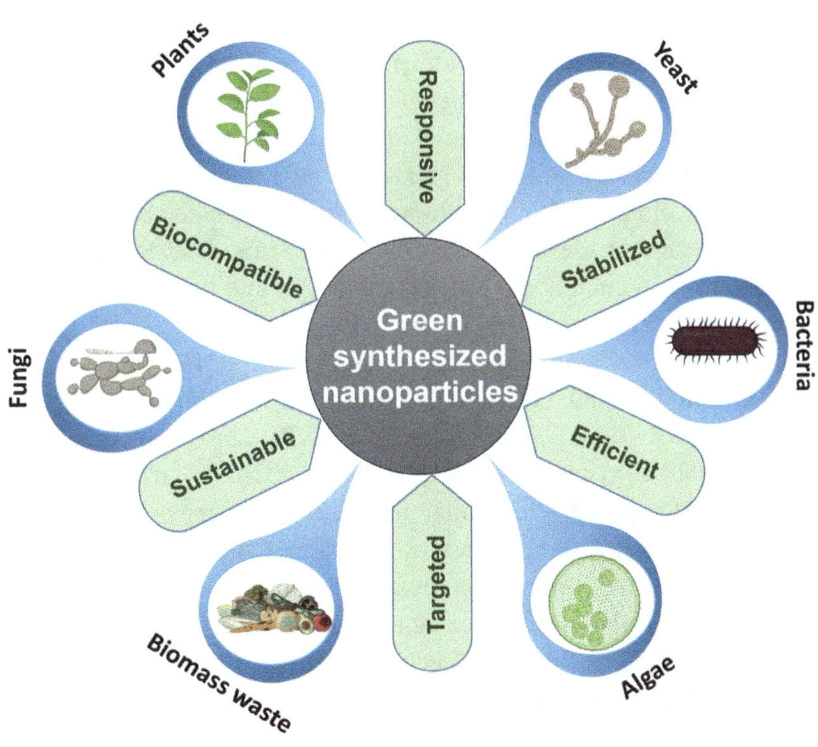

Figure 3. Natural sources used for green synthesis of nanoparticles and their effects

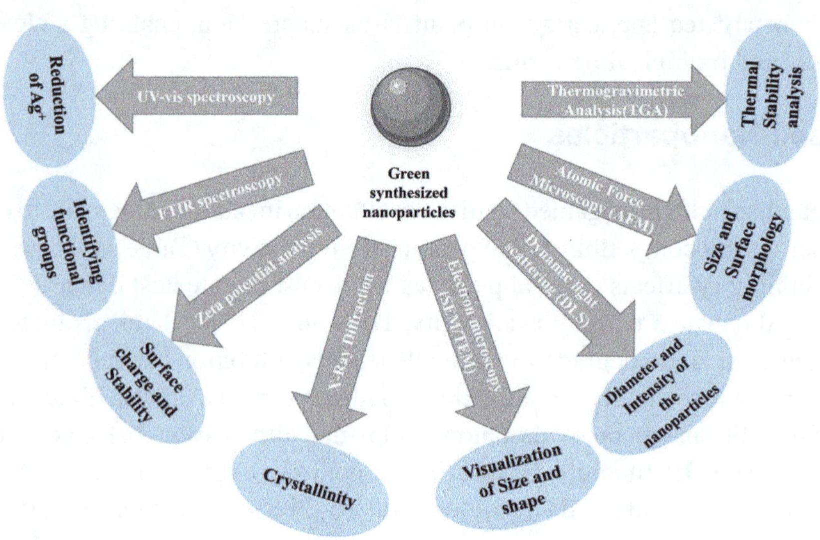

Silver nanoparticles (AgNPs) are the most extensively explored due to their broad-spectrum virucidal activity (Table 3). AgNPs bind to viral coat proteins, leading to conformational changes that prevent uncoating and replication. They also induce oxidative stress, damaging viral RNA and proteins. For instance, foliar applications of AgNPs in tobacco plants reduced TMV accumulation and disease severity (Elbeshehy et al., 2015), while in cucurbits they suppressed CMV infection and improved chlorophyll content (Aseel et al., 2024). In papaya, AgNP treatments lowered PRSV titers and enhanced fruit yield, demonstrating both protective and productivity benefits (Rehman et al., 2024).

Zinc oxide nanoparticles (ZnO NPs) function as dual protectants by acting as antiviral agents and supplying an essential micronutrient (Table 3). Their ability to generate reactive oxygen species (ROS) under light exposure leads to direct viral inactivation, while Zn-mediated defense priming upregulates antioxidant enzymes such as superoxide dismutase (SOD) and peroxidase. Studies have shown that ZnO NPs reduced CMV replication in cucumber (Kalia et al., 2020) and limited PRSV infection in papaya (Rehman et al., 2024). Furthermore, ZnO application in tomato suppressed TYLCV symptoms and enhanced salicylic acid-mediated defense pathways (Pejam et al., 2021).

Copper nanoparticles (CuNPs, CuO NPs) also exhibit antiviral action through oxidative damage and direct interaction with viral RNA (Table 3). In tomatoes, CuO NPs significantly reduced ToMV incidence and improved host antioxidant enzyme activities (Liu et al., 2024). In beans, CuNPs restricted Bean common mosaic virus (BCMV) spread, while foliar applications in papaya showed notable suppression of PRSV infection (Elbeshehy et al., 2015). However, excessive Cu exposure can induce phytotoxicity; thus, nanoparticle concentration and formulation must be optimized (Adisa et al., 2019).

Iron oxide nanoparticles (Fe_3O_4 NPs) enhance plant resistance by acting as both micronutrient supplements and defense activators (Table 3). In tomato, Fe_3O_4 NPs reduced TYLCV severity while improving photosynthetic efficiency and nutrient uptake (Ghormade et al., 2011). Their ability to modulate plant oxidative balance makes them particularly useful in managing chronic viral stress conditions.

Titanium dioxide nanoparticles (TiO_2 NPs) display strong photocatalytic properties, generating ROS under UV or visible light exposure that degrade viral coat proteins and nucleic acids (Table 3). In tobacco, TiO_2 NPs efficiently suppressed TMV replication and induced plant antioxidant responses (Elsharkawy & Derbalah, 2019). Their dual action of direct viral inactivation and host immune stimulation positions TiO_2 as a promising antiviral nanomaterial for field application.

Gold nanoparticles (AuNPs), though more widely utilized in biosensing, have also been reported to mitigate viral infections when applied exogenously (Table 3). AuNPs functionalized with antiviral molecules demonstrated inhibitory effects against CMV and TMV in solanaceous crops (Wang et al., 2016). Additionally, their ability to form stable bioconjugates makes them ideal for the targeted delivery of siRNAs or other antiviral agents.

Cerium oxide nanoparticles (CeO_2 NPs) exhibit unique redox properties due to the Ce^{3+}/Ce^{4+} cycling system, mimicking catalase and SOD activities (Table 3). This antioxidant mimicry alleviates virus-induced oxidative damage in host plants. In zucchini, CeO_2 NPs reduced Zucchini yellow mosaic virus (ZYMV) severity and enhanced tolerance to co-occurring abiotic stresses (Patil & Chandrasekaran, 2020).

Other noble metal nanoparticles, such as palladium (Pd) and ruthenium (Ru), have shown preliminary antiviral activity by binding viral proteins and interfering with replication machinery (Table 3). Early studies suggest their effectiveness against CMV and Tobacco necrosis virus (TNV), though their agricultural applications remain limited. Similarly, semiconductor nanoparticles like cadmium sulfide (CdS) and lead sulfide (PbS) have been evaluated for their photocatalytic degradation of TMV particles in greenhouse systems (Parisi et al., 2015). However, due to potential environmental toxicity, their widespread agricultural use requires further risk assessment. (Figure 3)

4.2 Biopolymeric nanoparticles

Biopolymeric nanoparticles are gaining increasing importance in sustainable plant virus management due to their biodegradable, biocompatible, and environmentally friendly properties. Unlike metal-based nanoparticles that may accumulate in soil, these natural polymer-derived systems offer safer alternatives, while still enabling controlled release and enhanced delivery of antiviral molecules. Several types, including chitosan, starch, alginate, gelatin, cellulose, and pectin nanoparticles, have demonstrated promising antiviral effects against diverse plant viruses (Maluin & Hussein, 2020) (Figure 4, Table 3).

Figure 4. Different types of green-synthesized nanoparticles characterized and assessed for antiviral activity against plant viruses

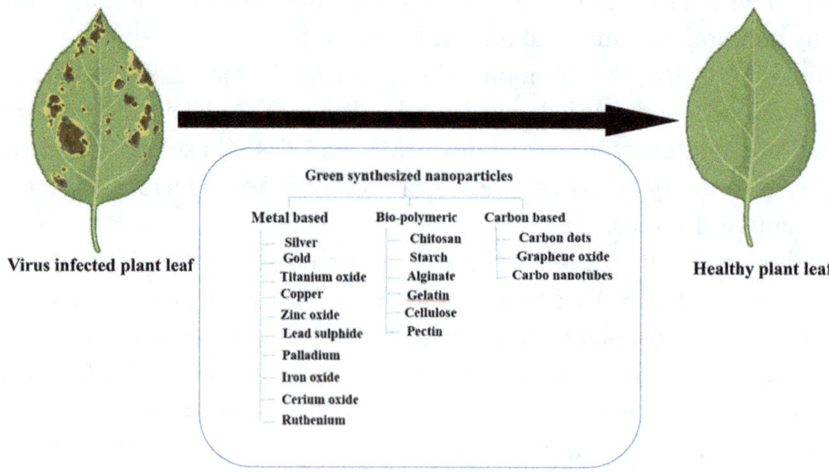

4.2.1 Chitosan Nanoparticles (CNPs).

Chitosan nanoparticles are well known for their ability to activate plant defense pathways and interfere with viral replication. In tobacco, foliar application of CNPs reduced Tobacco mosaic virus (TMV) accumulation by enhancing antioxidant enzyme activity and activating systemic acquired resistance (Pattanavongsawat et al., 2025). Similarly, in cucumber plants, CNPs significantly suppressed Cucumber mosaic virus (CMV) infection, lowering viral RNA levels and symptom expression (Rendina et al., 2019). Chitosan nanoparticles also improved resistance against Banana bunchy top virus (BBTV) when applied to banana seedlings, reducing viral titers and disease progression (Hanady et al., 2025). In beans, Bean yellow mosaic virus (BYMV) severity was reduced upon CNP application, associated with increased phenolic content and peroxidase activity. Moreover, tomato plants treated with chitosan nanocomposites exhibited reduced symptoms of Tomato spotted wilt virus (TSWV), highlighting their broad-spectrum antiviral efficacy (El-Ganainy et al., 2023).

4.2.2 Starch Nanoparticles (SNPs).

Starch nanoparticles, produced from renewable resources, are highly suitable as carriers for antiviral RNAs and phytochemicals. For example, SNPs encapsulating siRNAs targeting the coat protein gene of Papaya ringspot virus (PRSV) achieved sustained suppression of viral replication in papaya, delaying symptom onset (Mitter et al., 2017). SNPs have also been employed to deliver curcumin against CMV in cucurbits, significantly reducing viral titers and disease severity (Taha et al., 2019). In potato, starch nanocarriers encapsulating ribavirin enhanced antiviral activity against Potato virus Y (PVY) compared to free ribavirin, resulting in healthier plants and reduced yield losses (Lerch, 1987). Additionally, SNPs were used for the topical delivery of dsRNA in chili plants, achieving protection against Chili leaf curl virus (ChiLCV) by efficiently silencing viral replication genes (Chakraborty & Ghosh, 2022).

4.2.3 Alginate Nanoparticles (ANPs)

Alginate nanoparticles offer excellent potential for plant virus management due to their mucoadhesive properties and capacity for controlled release. In tomato, alginate-encapsulated ribavirin provided effective protection against TSWV, with higher retention and prolonged efficacy compared to unformulated ribavirin (De Fazio et al., 1980). In cucurbits, alginate nanoparticles encapsulating dsRNA targeting Zucchini yellow mosaic virus (ZYMV) successfully suppressed viral accumulation and improved resistance (Kaldis et al., 2017). Alginate nanoparticle formulations also enhanced the delivery of antiviral nucleotides against Potato leafroll virus (PLRV), leading to reduced systemic spread and lower infection rates in potato (Alcantara et al., 2022). Furthermore, in soybean, alginate nanoparticles delivering phytoalexins demonstrated inhibitory effects against Soybean mosaic virus (SMV), highlighting their versatility (Saberi Riseh et al., 2022).

4.2.4 Gelatin Nanoparticles (GNPs)

Gelatin nanoparticles, derived from proteins, are advantageous due to their high drug-loading efficiency. In tobacco, GNPs encapsulating salicylic acid enhanced plant defense against TMV, reducing viral load and symptom severity (Lennon et al., 1997). In tomato plants, antiviral peptide-loaded GNPs significantly reduced infection rates of Tomato mosaic virus (ToMV) by interfering with viral protein synthesis (Madkhali, 2023). Similarly, GNPs carrying siRNAs targeting Groundnut bud necrosis virus (GBNV) showed effective gene silencing, reducing symptom severity in groundnut plants (Verma, 2024). The proteinaceous nature of gelatin allows stabilization of antiviral compounds, making it suitable for large-scale agricultural applications.

4.2.5 Cellulose Nanoparticles (CNPs/CNCs)

Cellulose nanocrystals (CNCs) and cellulose nanoparticles (CNPs) enhance plant resistance by reinforcing cell walls and acting as carriers for antiviral agents. In cassava, CNCs improved the structural integrity of cell walls, delaying systemic infection of Cassava mosaic virus (Pattanavongsawat et al., 2025). In tomato, CNCs conjugated with dsRNA constructs targeting Tomato yellow leaf curl virus (TYLCV) effectively reduced viral DNA accumulation and symptom expression (Frascati et al., 2024). Application of CNPs in potato provided partial protection against Potato virus X (PVX) by enhancing lignin biosynthesis and defense enzyme activity (Lico et al., 2015). Moreover, CNCs were used as carriers for siRNAs targeting Peanut stunt virus (PSV), resulting in suppressed viral replication in infected peanut plants (Ndong Ntoutoume et al., 2017).

4.2.6 Pectin Nanoparticles (PNPs)

Pectin nanoparticles, derived from fruit peels, are highly biocompatible and suitable for delivering antiviral agents. In cucurbits, ribavirin-loaded PNPs inhibited ZYMV infection more effectively than free ribavirin, due to improved uptake and controlled release (Ali, et al., 2024). In papaya, PNPs carrying siRNAs targeting PRSV coat protein significantly reduced viral titres and extended protection duration (Azad et al., 2014). Recent studies have also demonstrated that PNPs encapsulating antiviral polyphenols such as quercetin were able to suppress CMV in tobacco and cucumbers by inhibiting viral RNA polymerase activity (Wang et al., 2022). Furthermore, in tomato plants, PNPs have been used to deliver antiviral peptides against TYLCV, reducing systemic spread and improving plant vigor (El-Sappah et al., 2022) (Figure 4).

4.3 Carbon-Based Nanoparticles

Carbon-based nanoparticles (CNPs), including carbon dots (CDs), graphene oxide (GO), and carbon nanotubes (CNTs), have emerged as versatile nanomaterials for plant virus management due to their unique structural, electronic, and chemical properties. These nanoparticles possess high surface area, tunable surface functionalities, and remarkable biocompatibility, which facilitate efficient interaction with viral particles, delivery of antiviral agents, and stimulation of host defense pathways (Yu et al., 2021; Khan et al., 2025). Unlike some metal-based nanoparticles, carbon nanomaterials generally exhibit lower phytotoxicity, making them suitable for repeated foliar applications in crops. Over the past decade, CNPs have been investigated against major plant viruses such as Tobacco mosaic virus (TMV), Cucumber mosaic virus (CMV), Tomato yellow leaf curl virus (TYLCV), Papaya ringspot virus (PRSV), Tomato spotted wilt virus (TSWV), and Banana bunchy top virus (BBTV).

4.3.1 Carbon Dots (CDs)

Carbon dots are ultra-small, fluorescent nanoparticles, typically below 10 nm, derived from green sources such as citric acid, plant biomass, or sugars. Their antiviral potential arises from their ability to bind viral coat proteins and nucleic acids, thereby inhibiting viral entry and replication. In tobacco, foliar application of CDs reduced TMV infection by up to 60%, accompanied by enhanced superoxide dismutase (SOD) and peroxidase activities in treated plants (He et al., 2025). Similarly, in cucumber, CDs functionalized with amino groups effectively suppressed CMV replication, resulting in delayed symptom expression and improved plant vigor (Zarrabi et al., 2024). In papaya, CDs conjugated with siRNAs targeting the coat protein gene of PRSV enhanced stability and cellular uptake, providing sustained viral suppression (Schwartz et al., 2020). Their fluorescent nature also enables simultaneous detection and monitoring of viral infection, offering dual functionality as therapeutic and diagnostic nanoplatforms.

4.3.2 Graphene Oxide (GO)

Graphene oxide, a two-dimensional sheet-like carbon nanomaterial, possesses exceptional surface area and abundant oxygen-containing functional groups that facilitate the adsorption of biomolecules and viral particles. GO can physically interact with viral capsids, induce oxidative damage, and interfere with viral replication machinery. In tomato plants, foliar sprays of GO nanosheets suppressed TYLCV DNA accumulation by more than 50%, accompanied by upregulation of pathogenesis-related (PR) proteins. GO also demonstrated antiviral activity against TSWV in pepper plants, where it reduced viral load and delayed systemic infection (Hassanpouraghdam et al., 2025). Furthermore, functionalized GO nanosheets conjugated with antiviral dsRNAs successfully inhibited CMV replication in cucumber, highlighting their potential as delivery platforms for RNA interference-based viral management (Zhang et al., 2025).

4.3.3 Carbon Nanotubes (CNTs)

Carbon nanotubes, both single-walled (SWCNTs) and multi-walled (MWCNTs), exhibit tubular structures with unique electronic and mechanical properties that enable efficient cellular penetration and molecular delivery. CNTs can adsorb viral particles, disrupt viral assembly, and serve as carriers for antiviral nucleic acids or peptides. In tobacco, SWCNTs loaded with TMV-specific dsRNA effectively reduced TMV replication and symptom severity (Adeel et al., 2021). MWCNTs applied to tomato leaves suppressed TYLCV accumulation by enhancing systemic acquired resistance and antioxidant enzyme activity (Sigala-Aguilar et al., 2025). In banana seedlings, MWCNTs carrying siRNAs targeting BBTV effectively silenced viral genes, providing significant reduction in viral titers and delayed disease onset (Jekayinoluwa et al., 2020). Beyond antiviral delivery, CNTs have also been shown to improve plant growth parameters, suggesting dual benefits in crop protection and productivity.

4.3.4 Mechanisms of Carbon-Based Nanoparticle Antiviral Activity

The antiviral activity of carbon-based nanoparticles is mediated through multiple mechanisms. First, direct interaction with viral coat proteins or nucleic acids can prevent viral entry and replication. Second, these nanoparticles can induce oxidative stress or ROS, RNS generation, leading to viral particle degradation. Third, functionalized carbon nanoparticles can serve as carriers for antiviral molecules such as siRNAs, dsRNAs, peptides, or small molecules, ensuring targeted delivery and enhanced stability. Finally, CNPs often prime host plant immunity by activating antioxidant enzymes and PR proteins, providing a systemic defense response against subsequent infections (Serrano-Aroca et al., 2024; Yadav et al., 2024).

4.3.5 Applications and Advantages

Carbon-based nanoparticles offer several advantages over conventional antiviral approaches. Their tunable surface chemistry allows for selective targeting of specific viruses. They are generally non-toxic and biodegradable, making them suitable for repeated applications in field conditions. Additionally, CNPs can be integrated with biosensing platforms, enabling simultaneous viral detection and control. For example, fluorescent CDs have been utilized in TMV detection assays, combining diagnostic and therapeutic functionalities (Tang et al., 2020). With proper optimization of dose, formulation, and delivery methods, carbon-based nanoparticles hold immense potential in integrated viral management strategies for crops including tomato, tobacco, cucumber, papaya, and banana.

5. ANTIVIRAL MECHANISMS OF NANOPARTICLES

Nanoparticles synthesized via green or conventional routes exhibit a wide array of antiviral mechanisms. These mechanisms may act at different stages of viral infection, ranging from extracellular virion inactivation to intracellular modulation of host defense pathways. Unlike conventional antivirals that typically target a single viral enzyme or pathway, nanoparticles exert multimodal actions, reducing the likelihood of resistance development. Their small size, high surface area-to-volume ratio, and tunable surface chemistry allow interactions with both viral and host cell components, leading to direct and indirect antiviral effects (Figure 5).

Figure 5. Different types of mechanisms caused by the green synthesized nanoparticles against plant viruses

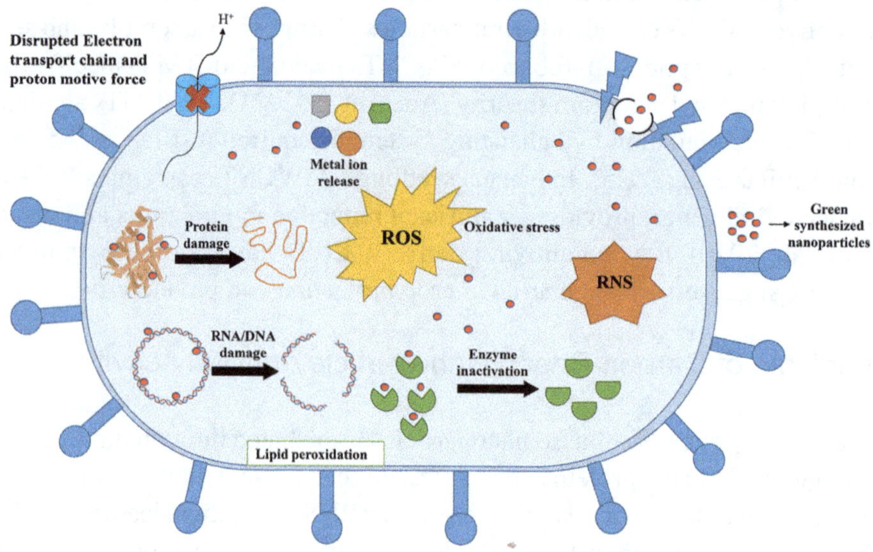

The mechanisms can be broadly divided into direct viral inhibition, activation of host immune responses, targeted nano-delivery of antiviral molecules, and modulation of host-virus interactions (Kumar et al., 2017; Singh et al., 2018).

5.1 Direct Viral Inhibition

One of the most consistently reported antiviral actions of nanoparticles is the direct inhibition of viral particles before or during host cell entry. Metal-based nanoparticles such as silver (AgNPs) and zinc oxide (ZnO NPs) demonstrate a strong affinity for viral glycoproteins and capsid proteins, thereby blocking their binding to host receptors. For instance, AgNPs disrupt the disulfide bonds in glycoprotein gp120 of HIV-1, preventing viral attachment to CD4 receptors on host cells (Lara et al., 2010). Similarly, AgNPs were shown to attach to hemagglutinin proteins of the influenza virus, leading to inhibition of endocytosis and membrane fusion (Xiang et al., 2011).

Another mechanism involves the disruption of viral nucleic acids. Silver and copper nanoparticles can release ions (Ag^+, Cu^{2+}) that intercalate with viral RNA or DNA, causing strand breakage and inhibition of transcription (Singh et al., 2018). Zinc oxide nanoparticles under UV or visible light generate reactive oxygen species (ROS) such as hydroxyl radicals and singlet oxygen, which oxidize nucleic acids and viral proteins. This oxidative damage results in irreversible loss of viral infectivity (Gupta et al., 2022).

Plant-derived carbon dots (CQDs) also exemplify direct inhibition. Their surface functional groups, including carboxyl, hydroxyl, and amine moieties, mimic host receptor sites, thereby competitively binding to viral glycoproteins. For example, CQDs functionalized with boronic acid groups inhibited coronavirus replication by blocking spike protein-ACE2 interaction in vitro (Kotta et al., 2020). Importantly, nanoparticles may also prevent systemic spread in plants by interfering with viral intercellular movement. Viruses such as tobacco mosaic virus (TMV) spread through plasmodesmata, specialized channels between plant cells. ZnO nanoparticles synthesized from Mentha spicata extracts not only reduced TMV accumulation but also prevented systemic migration, possibly by altering plasmodesmatal conductivity or directly binding to viral movement proteins (Sofy et al., 2021). Thus, direct viral inhibition encompasses multiple complementary mechanisms: disruption of viral proteins, oxidative degradation of nucleic acids, prevention of host cell entry, and restriction of systemic viral spread.

5.2 Activation of Plant Immune Responses

In agricultural virology, nanoparticles offer the added advantage of acting as elicitors of plant immune systems. Plants rely on complex defense signaling networks, including systemic acquired resistance (SAR) and induced systemic resistance (ISR), which are mediated by phytohormones such as salicylic acid (SA), jasmonic acid (JA), and ethylene (ET). Nanoparticles can activate these pathways, leading to enhanced resistance against viral pathogens.

For instance, foliar application of green-synthesised ZnO nanoparticles in tomato plants challenged with TMV led to a significant upregulation of defense-related enzymes such as peroxidase (POD), polyphenol oxidase (PPO), and phenylalanine ammonia-lyase (PAL), which are key components of the phenylpropanoid pathway (Singh et al., 2018). These enzymes contribute to lignification, ROS detoxification, and the synthesis of secondary metabolites with antiviral roles.

Chitosan nanoparticles are particularly well-studied in this context. Chitosan, a cationic polysaccharide, is recognized by plant pattern recognition receptors (PRRs), triggering defense cascades. When applied to virus-infected plants, chitosan nanoparticles increased the expression of pathogenesis-related (PR) proteins such as PR-1 and β-1,3-glucanases, hallmarks of SA-mediated defense (Mawale et al., 2025; Pandya et al., 2024). In some cases, chitosan has been shown to act synergistically with nanoparticles of silver or copper, further amplifying the antiviral defense response (Khosravian-Dehkordi et al., 2024). Nanoparticles can also induce hormonal cross-talk. For example, cerium oxide nanoparticles (CeO_2 NPs) enhanced SA and JA levels simultaneously, activating dual defense pathways that are usually antagonistic. This unique modulation provided stronger resistance against cucumber mosaic virus in cucumber plants (Shcherbakov, 2024). By activating host immune systems, nanoparticles do not merely inactivate viruses but transform plants into "primed" states of heightened readiness, thereby providing durable and systemic protection.

5.3 Targeted Nano-Delivery Systems

Nanoparticles offer a transformative platform for delivering antiviral molecules with precision. Conventional antivirals such as RNA interference (RNAi) constructs, phytochemicals, and synthetic drugs often suffer from instability and rapid degradation, limiting their efficacy. Encapsulation within biocompatible nanocarriers not only shields these molecules from enzymatic breakdown but also enhances solubility, bioavailability, and controlled release profiles (Mitter et al., 2017).

In plants, chitosan-based nanoparticles delivering siRNAs against viral coat protein genes have successfully triggered gene silencing and suppressed viral replication (Mitter et al., 2017). Comparable advances in mammalian systems, such as lipid and polymer nanoparticles for RNAi delivery against influenza and coronaviruses, underscore the translational potential of these platforms (Hou et al., 2021; Pardi et al., 2020). Plant-derived biopolymers like starch and alginate further expand this toolbox, enabling sustainable carriers for antiviral phytochemicals with superior bioactivity and prolonged residence times (Pathak et al., 2022; Khosravian-Dehkordi et al., 2024). Metallic nanoparticles, particularly green-synthesised gold nanoparticles, add another layer of versatility as delivery vehicles for DNA and RNA vaccines. Their tunable surfaces and inherent stability have already demonstrated success in eliciting robust immune responses in veterinary virology, for instance, against avian influenza (Yafout et al., 2021; Cai et al., 2022). By merging protection, precision, and intrinsic antiviral activity, targeted nano-delivery systems stand out as one of the most promising frontiers for next-generation plant and animal virus management.

5.4 Modulation of Host-Virus Interaction

Nanoparticles can also interfere with the host-virus interaction dynamics, altering susceptibility genes or inhibiting viral replication machinery within host cells. At the cellular level, nanoparticles can modulate gene expression related to viral susceptibility. For instance, exposure of Arabidopsis to iron oxide nanoparticles resulted in downregulation of genes encoding viral movement proteins, thereby restricting virus propagation (Rai et al., 2022). Similarly, ZnO nanoparticles induced expression of RNA-dependent RNA polymerase (RDR) and Dicer-like proteins (DCLs), which are central components of RNA silencing pathways, a natural antiviral mechanism in plants (Chen et al., 2023).

In mammalian cells, silver nanoparticles have been shown to interact with host nuclear transcription machinery, suppressing viral gene transcription of HIV-1 and hepatitis B virus (Lara et al., 2010). Chitosan nanoparticles, apart from delivering drugs, also modify endosomal escape pathways, preventing viruses from utilizing host vesicular transport systems (Safer & Leporatti, 2021).

Some nanoparticles directly interfere with viral polymerases. For instance, cerium oxide nanoparticles were reported to bind viral RNA polymerases, reducing their enzymatic activity in vitro (Korsvik et al., 2007). This indicates a direct molecular interaction between nanoparticles and viral replication machinery.

Green-synthesised nanoparticles (NPs) improve plant resilience against viral pathogens through various antiviral mechanisms. These mechanisms include disrupting virion capsids and inducing oxidative stress, inhibiting viral entry, and directly interfering with replication. Notably, biogenic AgNPs and ZnO NPs activate systemic acquired resistance and induced systemic resistance, enhancing defensive enzyme levels. Additionally, carbon-based nanocarriers and biopolymeric nanoparticles improve the delivery and stability of antiviral agents. This multifunctional behavior of NPs reduces the risk of resistance development, showcasing their potential in plant virus suppression (Naikoo et al., 2021).

Moreover, nanoparticles may modulate host immune signaling, enhancing antiviral cytokine responses. Carbon quantum dots functionalized with polyamines stimulated interferon (IFN) expression in human cells, leading to broad-spectrum antiviral effects beyond direct virus binding (Dong et al., 2012). This immunomodulatory property expands their utility in systemic viral defense. By targeting both viral and host components, nanoparticles offer a multi-pronged disruption of infection cycles, making them highly effective in scenarios where viruses evolve resistance to conventional single-target drugs.

6. APPLICATIONS AND CASE STUDIES

The translation of nanoparticle research into practical applications has gained momentum in both plant virology and broader agricultural systems. Green-synthesised nanoparticles, owing to their eco-friendly nature and biocompatibility, are increasingly recognized as promising alternatives to conventional chemical antivirals and pesticides. This section outlines key applications and case studies involving different classes of nanoparticles, highlighting their effectiveness against economically important plant viruses, their dual roles in crop nutrition and protection, and their comparative performance against synthetic agrochemicals.

6.1 Silver Nanoparticles (AgNPs)

Silver nanoparticles are among the most extensively investigated for antiviral activity, owing to their potent virucidal effects, relatively low required concentrations, and compatibility with green synthesis pathways. For example, biosynthesized AgNPs significantly reduced Tobacco Mosaic Virus (TMV) accumulation in tomato plants, lowering lesion count and systemic spread compared to untreated controls (Al-Askar et al., 2023). Microbe-derived AgNPs produced by *Pseudomonas fluorescens* showed ~92.7% inhibition of local lesions in TMV-infected tobacco (Ahsan, 2020). Other studies using green or plant-derived AgNPs similarly show strong suppression of viral replication and disease severity (Warghane et al., 2024).

6.2 Chitosan Nanoparticles

Chitosan nanoparticles are appealing due to both direct antiviral properties and ability to stimulate plant defense responses. Although specific studies pairing chitosan nanoparticles with plant viruses are fewer, reviews and experimental evidence suggest that chitosan applications (non-nano or nano) increase peroxidase, phenylalanine ammonia-lyase, and polyphenol oxidase activities, implicating enhanced systemic resistance in hosts (Verma, 2024; Warghane et al., 2024). Chitosan nano-formulations are being developed as carriers for antiviral agents and RNA constructs with promising preliminary results.

6.3 Zinc Oxide Nanoparticles (ZnO NPs)

Zinc oxide nanoparticles combine antiviral activity with micronutrient benefits. Green ZnO NPs have been reviewed for mitigating biotic stress in plants, including disease pressure, improved growth, and defense responses (Verma, 2024). For example, ZnO NPs synthesized using algal or microbial methods have shown enhanced disease resistance and improved physiological indices, though for many plant viruses more work is needed to confirm field efficacy.

6.4 Multi-Component Nanoparticle Blends

Blended nanoparticle systems that combine metals (e.g., Ag + ZnO, Ag + CuO) hold promise for synergy. Though specific peer-reviewed studies with blends in plant virus settings are still emerging, recent work (Warghane et al., 2024) using graphene oxide-silver nanoparticle (GO-AgNP) composites reduced the disease severity of Tomato Bushy Stunt Virus more than single-metal nanoparticles. These blend systems may allow targeting multiple steps of viral infection (coat protein interaction, replication, movement) while allowing lower doses and reducing the risk of phytotoxicity.

6.5 Nano-formulations and Application Methods

Effective deployment of any nanoparticle antiviral agent depends on formulation and delivery method. Foliar sprays of AgNPs have shown strong efficacy in controlled experiments (Al-Askar et al., 2023). Seed-coating with ZnO NPs or other nanoparticulate nutrients shows promise in improving early growth and reducing disease incidence (Verma, 2024). Controlled-release systems and hydrogel or emulsion-based formulations are under development, but fewer published datasets for plant viral disease.

6.6 Comparative Efficacy

Comparative studies suggest that green-synthesized nanoparticles often match or exceed chemical treatments under experimental conditions. For instance, AgNPs from Al-Askar (2023) reduced TMV accumulation more effectively than untreated controls and with fewer adverse effects. Reviews indicate that ZnO nanoparticles may outperform conventional micronutrient salts in stress mitigation and disease resistance (Verma, 2024). Nevertheless, more field studies with standard controls are needed.

7. ENVIRONMENTAL SAFETY AND REGULATORY CONSIDERATIONS

The rapid expansion of nanotechnology in plant virology and crop protection has raised growing concerns regarding its environmental fate, safety to non-target organisms, human exposure risks, and regulatory frameworks. While nanoparticles (NPs) offer clear benefits in controlling viral diseases and enhancing crop yields, their unique physicochemical properties-including small size, high reactivity, and mobility-pose potential risks if not properly assessed. Biosafety considerations remain central to advancing nanoparticle-based antiviral strategies in agriculture which required integrated risk assestment and workflow discussed in the table 4.

Table 4. Integrated risk assessment and regulatory workflow for green nanoparticles in agriculture

Assessment Stage	Key Activities	Data Required	Regulatory Output / Decision
1. Pre-Development	Identify nanoparticle type, synthesis route	Source materials, biogenic agents, synthesis parameters	Classification of nanomaterial; initial safety category
2. Physicochemical Characterization	Size, morphology, crystallinity, surface charge	TEM/SEM, XRD, DLS, FTIR data	Confirmation of nanoparticle identity & purity
3. Laboratory Biosafety Screening	Phytotoxicity, cytotoxicity, antimicrobial impacts	Plant growth assays, microbial viability tests	Approval to proceed to greenhouse/field testing
4. Environmental Fate Evaluation	Mobility, persistence, degradation pathways	Soil column studies, leaching tests, dissolution profile	Decision on suitability for open-field deployment
5. Ecotoxicological Assessment	Effects on soil fauna, pollinators, beneficial microbes	Earthworm assays, soil enzyme activity, bee toxicity tests	Acceptable environmental risk determination
6. Application Standardization	Dose optimization, exposure modeling	LC50, EC50 data; spray drift studies	Recommended field dose & safety margins
7. Field Validation & Multi-Season Trials	Efficacy, residue behavior, ecosystem safety	Yield outcomes, residue surveys, microbial metagenomics	Conditional approval for commercial formulation

Assessment Stage	Key Activities	Data Required	Regulatory Output / Decision
8. Labeling, Packaging & Distribution	Safety instructions, nano-labeling, handling procedures	Product characterization, hazard profile	Regulatory-compliant commercial product
9. Post-Market Monitoring	Long-term monitoring, farmer feedback	Food chain analysis, soil quality indicators	Final risk evaluation & regulatory updates

7.1 Environmental Fate and Biodegradability

The environmental fate of nanoparticles depends on their composition, size, surface chemistry, and environmental context (soil, water, or atmosphere). Unlike conventional agrochemicals, NPs may undergo complex transformations after application, including aggregation, dissolution, redox reactions, and interaction with soil organic matter.

Silver nanoparticles (AgNPs) often oxidize to release Ag^+ ions, which can bind to soil particles or leach into groundwater depending on soil pH and organic content (Levard et al., 2012). Zinc oxide nanoparticles (ZnONPs) may dissolve more readily, releasing Zn^{2+} ions that act as micronutrients but may accumulate at toxic levels if concentrations exceed plant uptake capacity (Keller et al., 2010).

Biopolymeric nanoparticles such as chitosan or starch-based NPs are generally more biodegradable, breaking down into natural monomers via enzymatic degradation in soil and water systems (Othman et al., 2023). Environmental persistence is influenced by factors such as sunlight exposure, microbial activity, and soil composition. TiO_2 nanoparticles are highly stable, persisting in soil for months, while carbon-based nanomaterials like graphene oxide may adsorb onto clay particles, reducing mobility but prolonging soil residence (Navarro et al., 2008). Overall, the environmental behavior of NPs remains poorly characterized, and more long-term field studies are required to assess persistence, leaching potential, and ecological consequences.

7.2 Safety to Non-Target Organisms

The introduction of nanoparticles into agroecosystems raises concerns about their potential effects on beneficial organisms, including soil microbes, pollinators, aquatic life, and higher trophic levels. Soil microbial communities are particularly sensitive. AgNPs have been shown to reduce populations of nitrogen-fixing bacteria (*Rhizobium* sp., *Azotobacter* sp.), which are critical for soil fertility (Dong et al., 2023). Similarly, ZnONPs at high doses inhibited microbial biomass and enzyme activities, although moderate doses sometimes enhanced beneficial microbes due to micronutrient effects (Priester et al., 2012). Pollinators such as honeybees may also be exposed via nectar, pollen, or dust drift. Preliminary studies suggest that AgNPs and TiO_2 NPs can affect bee gut microbiota and foraging behavior, raising concerns about long-term colony health.

Aquatic ecosystems are another critical receptor. Runoff of nanoparticles from agricultural fields can accumulate in water bodies, potentially causing toxicity to algae, plankton, and fish. For example, AgNPs at sub-lethal doses altered gill function in zebrafish and reduced reproductive success in *Daphnia magna* (Griffitt et al., 2008). Residues in the food chain also warrant concern. Studies have detected nanoparticle

continued on following page

Table 4. Continued

accumulation in edible plant tissues, raising the possibility of trophic transfer to herbivores and humans (Raliya et al., 2015). Safety assessments must therefore encompass the entire agroecological network.

7.3 Human and Animal Exposure Risk

The use of nanoparticle-based formulations in food crops raises potential risks of human and livestock exposure through ingestion, dermal contact, or inhalation. Several studies have demonstrated that nanoparticles can translocate from roots to shoots and accumulate in edible tissues. For instance, ZnONPs applied to cucumber and spinach were detected in leaves and fruits, suggesting possible dietary exposure (Wang et al., 2016). Similarly, AgNPs applied to lettuce accumulated in leaf tissues, with potential implications for human consumption (Judy et al., 2011).

Toxicological risks vary with nanoparticle type. Metal-based NPs (Ag, CuO, TiO_2) may induce oxidative stress, genotoxicity, or inflammatory responses in mammalian cells at high doses (Kim & Ryu, 2013). Carbon nanomaterials, such as CNTs, may cause lung inflammation upon inhalation due to their fiber-like morphology (Donaldson et al., 2010).

Animal exposure is another issue; livestock consuming contaminated fodder or water may bioaccumulate nanoparticles. ZnONPs have been detected in milk and meat after chronic exposure in animal feed trials, though concentrations remained below acute toxicity thresholds (Sharma et al., 2025). While typical agricultural doses are unlikely to pose acute risks, chronic low-dose exposure and bioaccumulation remain understudied.

7.4 Regulatory Status and Global Policy Gaps

Despite rapid expansion, regulatory frameworks for nanoparticle use in agriculture remain fragmented. Unlike conventional pesticides and fertilizers, which are evaluated under established toxicological guidelines, nanoparticles present novel challenges due to nanoscale properties.

Currently, most jurisdictions lack nanoparticle-specific regulations. Agencies such as the U.S. Environmental Protection Agency (EPA) and European Food Safety Authority (EFSA) regulate nanoparticles under broader chemical frameworks, which may not account for nano-specific behaviors like aggregation, surface reactivity, or size-dependent toxicity (Kah et al., 2013). The EU mandates labeling of nanomaterials in food and cosmetics, but similar rules for agricultural inputs are lacking (Hardy et al., 2018). Asian countries, including India and China, are developing nano-agriculture initiatives, yet standardized risk assessment protocols remain absent (Ingle et al., 2023). This regulatory vacuum poses risks of inconsistent safety standards, trade barriers, and public mistrust. Experts advocate for harmonized international frameworks integrating nanotoxicology data, environmental monitoring, and labeling requirements (Singh et al., 2021).

7.5 Risk Assessment and Monitoring Tools

Effective risk management requires nano-specific toxicity assessment tools and long-term environmental monitoring. Traditional toxicological assays may underestimate nanoparticle risks, as they do not capture size-dependent effects or transformations in complex environments. Recent approaches include Nano-specific ecotoxicity assays, such as soil microcosm tests and aquatic mesocosms (Handy et al.,

2008), advanced imaging and spectroscopy techniques (e.g., TEM, X-ray absorption spectroscopy) to track nanoparticle fate in soil and plant tissues (Lowry et al., 2012), and computational environmental models simulating nanoparticle transport, persistence, and bioavailability (Garner & Keller, 2014).

7.6 Real-World Adoption and Scalability

The use of green-synthesised nanoparticles for antiviral applications in agriculture is rapidly increasing, supported by field studies showing significant efficacy. Biogenic silver nanoparticles have decreased TMV and CMV severity in tomatoes and cucumbers by 70–90%. Zinc oxide nanoparticles have improved papaya's tolerance to PRSV and boosted its yield, while chitosan nanoparticles effectively suppressed the banana bunchy top virus. Green synthesis using agricultural waste reduces costs and promotes circular bioeconomy practices. Key challenges for broader adoption include standardizing formulations, application protocols, and training for farmers. Combining these nanoparticles with precision agriculture methods, like drone-assisted spraying and nano-sensor monitoring, could further promote their use (Al-Askar et al., 2023).

Long-term field studies are critical. Multi-season experiments can evaluate cumulative soil accumulation, potential resistance development, and ecological disruptions (Simonin & Richaume, 2015). Early-warning monitoring systems, such as biosensors for detecting nanoparticle residues in soil and water, may provide real-time risk management tools (Rajput et al., 2018).

8. CHALLENGES AND FUTURE PERSPECTIVES

The application of green-synthesized nanoparticles (NPs) in plant virus management holds tremendous promise, yet translating laboratory research into large-scale agricultural practice remains challenging. While numerous studies demonstrate antiviral efficacy under controlled conditions, substantial gaps exist in reproducibility, large-scale synthesis, formulation stability, regulatory frameworks, and integration into precision agriculture systems. Addressing these limitations is critical for ensuring that nanotechnology contributes meaningfully to sustainable crop protection (Rai et al., 2016).

8.1 Challenges in Scale-Up and Field Use

A major obstacle is the reproducibility and scalability of green nanoparticle synthesis. Laboratory-scale processes often rely on plant extracts, microbial cultures, or algal biomasses, which are inherently variable depending on source material, growth conditions, and extraction protocols. Such variability can significantly affect nanoparticle size, morphology, and stability, leading to inconsistent antiviral performance in field conditions (Singh et al., 2018; Ahmed et al., 2016b).

Mass production requires stringent control over parameters such as pH, temperature, and precursor concentrations, which is difficult to achieve in decentralized agricultural settings (Das et al., 2010). Shelf-life is another critical concern, as biologically capped nanoparticles tend to aggregate or lose stability over time, reducing effectiveness during storage and transportation (Iravani, 2011). Developing standardized protocols for long-term stabilization remains a major bottleneck for commercialization (Nasrollahzadeh et al., 2021). Recent studies highlight the effectiveness of green nanoparticles in agriculture. Silver nanoparticles effectively suppressed TMV and CMV in tomatoes and cucumbers, while zinc oxide nanoparticles improved yields in papaya by mitigating PRSV. Chitosan nanoparticles increased vigor in bananas by reducing bunchy top virus incidence. Carbon dots delayed CMV spread in cucurbits, demonstrating their antiviral potential. Moreover, graphene oxide–silver hybrid nanoparticles exhibited enhanced antiviral effects compared to individual nanoparticles, underscoring the promise of synergistic nanoformulations. Collectively, these findings indicate the substantial potential of green nanoparticles for agricultural applications (Elbeshehy et al., 2015).

8.2 Formulation and Delivery Innovation

Developing effective formulations and delivery systems that maintain nanoparticle efficacy under field conditions is essential. Unlike chemical pesticides, nanoparticles must remain stable under UV exposure, fluctuating humidity, and temperature extremes (Khashan et al., 2021; Sharma et al., 2025). Encapsulation within biodegradable polymers, nanoemulsions, or liposomal carriers has emerged as a promising approach to prolong stability and enable controlled release (Salata, 2004).

Chitosan-based nanocarriers, for instance, not only improve dispersibility but also provide synergistic antiviral and plant immunity-enhancing effects (Chung et al., 2016; Panpatte et al., 2016). Smart delivery platforms that respond to environmental or plant-specific triggers, such as pH or enzymatic activity, could further enhance targeted release while minimizing off-target effects (Raliya et al., 2016). Field-friendly formulations foliar sprays, seed coatings, and soil drenches, require optimization to ensure that nanoparticles effectively reach infection sites without harming non-target organisms (Shang et al., 2019a; Gupta et al., 2023).

8.3 Integration with Precision Agriculture

Future applications of green-synthesized nanoparticles will increasingly rely on integration with precision agriculture systems. Emerging technologies, including artificial intelligence (AI), the Internet of Things (IoT), and smart irrigation networks, provide opportunities for synergistic use of nanomaterials (Mahawar & Prasanna, 2018; Khot et al., 2012).

Sensor-enabled nanoparticles could simultaneously detect early-stage viral infections and deliver antiviral agents in situ, creating a closed-loop diagnostic and treatment system (Shang et al., 2019b). Drones and automated sprayers facilitate precise nanoparticle delivery, reducing waste and environmental contamination (Ahmad et al., 2022). The convergence of nanotechnology with digital agriculture enhances disease control and generates real-time data to guide decision-making, ultimately improving resource efficiency and sustainability. Successful integration, however, requires interdisciplinary collaboration among plant pathologists, nanotechnologists, engineers, and data scientists (Mousavi & Rezaei, 2011; Shafey, 2020).

8.4 Future Research and Collaborative Needs

To unlock the full potential of green nanotechnology in plant virology, a holistic and collaborative approach is needed. Mechanistic studies should elucidate antiviral pathways, including interactions with viral coat proteins, interference with replication, and induction of systemic acquired resistance (Ghosh et al., 2012). Long-term environmental impact assessments are essential to ensure that nanoparticle use does not adversely affect soil microbiota, beneficial insects, or water systems (Bundschuh et al., 2018).

Standardized guidelines for nanoparticle safety evaluation in agricultural ecosystems should be prioritized (Kah et al., 2018; Rico et al., 2011).

Collaborative efforts across academia, industry, and policymaking sectors are necessary to establish cost-effective production pipelines and regulatory frameworks (Chen & Yada, 2011). Engagement with local farming communities can facilitate field trials under real-world conditions, ensuring solutions are adapted to regional crops and practices (Parisi et al., 2015). Finally, international cooperation is critical to address transboundary plant virus outbreaks, as nanoparticle-based technologies have the potential to become part of global biosecurity strategies (McClements & Xiao, 2017).

9. CONCLUSION

Green-synthesized nanoparticles (NPs) are rapidly emerging as next-generation tools for plant virus management. Their eco-friendly production using plants, microbes, algae, and fungi offers a sustainable alternative to conventional chemical-based nanomaterials. These biological routes provide naturally derived stabilizing biomolecules, which enhance biocompatibility and reduce ecological toxicity. Biogenic nanoparticles—including silver (Ag), gold (Au), zinc oxide (ZnO), and metal oxide NPs—demonstrate potent antiviral activity by binding to viral coat proteins, inhibiting replication, or inducing systemic acquired resistance in host plants (Rai et al., 2016; Khan et al., 2020; Jain et al., 2021; Bora et al., 2022). These multifaceted mechanisms highlight their promise as versatile agents against complex plant viral diseases.

Despite their potential, significant hurdles remain before field-scale application is feasible. Reproducibility of synthesis, stability during storage, variability in biological extracts, and cost-effective mass production all limit the transition from laboratory to agricultural fields (Singh et al., 2018; Bahrulolum et al., 2021; Vodyashkin et al., 2022). Biosafety concerns, such as nanoparticle accumulation in soils, effects on beneficial microorganisms, and off-target interactions, necessitate rigorous long-term eco-toxicological studies (Bundschuh et al., 2018; Navarro et al., 2008). Regulatory frameworks are also lacking, as most existing policies focus on industrial or medical nanotechnology rather than agricultural applications (Bahrulolum et al., 2021; Kulkarni et al., 2023).

Advances in nanoformulation strategies may help overcome these limitations. Encapsulation within biodegradable polymers, conjugation with targeting ligands, and development of controlled-release nanocarriers can enhance nanoparticle stability, bioavailability, and delivery efficiency in crop systems (Vodyashkin et al., 2022; Hameed et al., 2023). Integration with precision agriculture tools—such as disease monitoring, nanosensors for early virus detection, and smart irrigation platforms—enables real-time, targeted, and resource-efficient crop protection (Lowry et al., 2012; Khot et al., 2012; Parisi et al., 2015).

International collaboration is essential to standardize production methods, ensure farmer safety, and accelerate technology adoption. Partnerships spanning plant virology, materials science, and agricultural policy can facilitate field validation, harmonized safety protocols, and knowledge transfer across regions (Shahzadi et al., 2025; Kulkarni et al., 2023).

In conclusion, green nanotechnology represents a scientifically robust and environmentally sustainable approach to mitigating plant viral diseases. Continued research into biosafety, formulation innovation, and field-scale validation will be key to unlocking the full potential of green-synthesized nanoparticles. These technologies not only promise to safeguard crop productivity but also contribute to global food security and resilience against climate-driven disease pressures (Sharma et al., 2025; Mittal et al., 2013; Vodyashkin et al., 2022).

REFERENCES

Abdelkhalek, A., & Al-Askar, A. A. (2020). Green synthesized zno nanoparticles mediated by mentha spicata extract induce plant systemic resistance against tobacco mosaic virus. *Applied Sciences (Basel, Switzerland)*, *10*(15), 5054. DOI: 10.3390/app10155054

Abdelkhalek, A., Salem, M. Z. M., Hafez, E., Behiry, S. I., & Qari, S. H. (2020). The phytochemical, antifungal, and first report of the antiviral properties of egyptian haplophyllum tuberculatum extract. *Biology (Basel)*, *9*(9), 248. DOI: 10.3390/biology9090248

Adeel, M., Farooq, T., White, J. C., Hao, Y., He, Z., & Rui, Y. (2021). Carbon-based nanomaterials suppress tobacco mosaic virus (Tmv) infection and induce resistance in Nicotiana benthamiana. *Journal of Hazardous Materials*, *404*, 124167. DOI: 10.1016/j.jhazmat.2020.124167

Adisa, I. O., Pullagurala, V. L. R., Peralta-Videa, J. R., Dimkpa, C. O., Elmer, W. H., Gardea-Torresdey, J. L., & White, J. C. (2019). Recent advances in nano-enabled fertilizers and pesticides: A critical review of mechanisms of action. *Environmental Science. Nano*, *6*(7), 2002–2030. DOI: 10.1039/C9EN00265K

Ahmad, A., Mukherjee, P., Senapati, S., Mandal, D., Khan, M. I., Kumar, R., & Sastry, M. (2003). Extracellular biosynthesis of silver nanoparticles using the fungus Fusarium oxysporum. *Colloids and Surfaces. B, Biointerfaces*, *28*(4), 313–318. DOI: 10.1016/S0927-7765(02)00174-1

Ahmad, Z., Tahseen, S., Wasi, A., Ganie, I. B., Shahzad, A., Emamverdian, A., Ramakrishnan, M., & Ding, Y. (2022). Nanotechnological interventions in agriculture. *Nanomaterials (Basel, Switzerland)*, *12*(15), 2667. DOI: 10.3390/nano12152667

Ahmed, S., Saifullah, Ahmad, M., Swami, B. L., & Ikram, S. (2016b). Green synthesis of silver nanoparticles using Azadirachta indica aqueous leaf extract. *Journal of Radiation Research and Applied Sciences*, *9*(1), 1–7. DOI: 10.1016/j.jrras.2015.06.006

Ahmed, S., Ahmad, M., Swami, B. L., & Ikram, S. (2016a). A review on plants extract mediated synthesis of silver nanoparticles for antimicrobial applications: A green expertise. *Journal of Advanced Research*, *7*(1), 17–28. DOI: 10.1016/j.jare.2015.02.007

Ahsan, T. (2020). Biofabrication of silver nanoparticles from Pseudomonas fluorescens to control tobacco mosaic virus. *Egyptian Journal of Biological Pest Control*, *30*(1), 66. DOI: 10.1186/s41938-020-00268-3

Al-Askar, A. A., Aseel, D. G., El-Gendi, H., Sobhy, S., Samy, M. A., Hamdy, E., El-Messeiry, S., Behiry, S. I., Elbeaino, T., & Abdelkhalek, A. (2023). Antiviral activity of biosynthesized silver nanoparticles from pomegranate (Punica granatum L.) peel extract against tobacco mosaic virus. *Plants*, *12*(11), 2103. DOI: 10.3390/plants12112103

Alcantara, K. P., Nalinratana, N., Chutiwitoonchai, N., Castillo, A. L., Banlunara, W., Vajragupta, O., Rojsitthisak, P., & Rojsitthisak, P. (2022). Enhanced nasal deposition and anti-coronavirus effect of favipiravir-loaded mucoadhesive chitosan–alginate nanoparticles. *Pharmaceutics*, *14*(12), 2680. DOI: 10.3390/pharmaceutics14122680

Alegbejo, M. D. (2015). Virus and virus-like diseases of crops Nigeria. Ahmadu Bello University press limited.

Ali, R., Eldidamony, G., Askora, A., & Galal, A. (2024). Investigates the ability of plant extracts from Lens culinaris to protect zucchini from the Zucchini yellow mosaic virus (Zymv). *Scientific Reports, 14*, 12257. DOI: 10.1038/s41598-024-62128-6

Alkhulaifi, M. M., Alshehri, J. H., Alwehaibi, M. A., Awad, M. A., Al-Enazi, N. M., Aldosari, N. S., & Hatamleh, A. A., & Abdel- Raouf, N. (2020). Green synthesis of silver nanoparticles using Citrus limon peels and evaluation of their antibacterial and cytotoxic properties. *Saudi Journal of Biological Sciences, 27*(12), 3434–3441. DOI: 10.1016/j.sjbs.2020.09.031

Alqahtani, A. A., El Raey, M. A., Abdelsalam, E., Ibrahim, A. M., Alqahtani, O., Torky, Z. A., & Attia, H. G. (2022). The biosynthesized zinc oxide nanoparticles' antiviral activity in combination with pelargonium zonale extract against the human corona 229e virus. *Molecules (Basel, Switzerland), 27*(23), 8362. DOI: 10.3390/molecules27238362

Anderson, P. K., Cunningham, A. A., Patel, N. G., Morales, F. J., Epstein, P. R., & Daszak, P. (2004). Emerging infectious diseases of plants: Pathogen pollution, climate change and agrotechnology drivers. *Trends in Ecology & Evolution, 19*(10), 535–544. DOI: 10.1016/j.tree.2004.07.021

Aseel, D. G., Ibrahim, O. M., & Abdelkhalek, A. (2024). Biosynthesized silver nanoparticles mediated by Ammi visnaga extract enhanced systemic resistance and triggered multiple defense-related genes, including SbWRKY transcription factors, against tobacco mosaic virus infection. *BMC Plant Biology, 24*, 756. DOI: 10.1186/s12870-024-05449-y

Azad, M. A. K., Amin, L., & Sidik, N. M. (2014). Gene technology for papaya ringspot virus disease management. *TheScientificWorldJournal, 768038*. Advance online publication. DOI: 10.1155/2014/768038

Azizi, S., Ahmad, M. B., Namvar, F., & Mohamad, R. (2014). Green biosynthesis and characterization of zinc oxide nanoparticles using brown marine macroalga Sargassum muticum aqueous extract. *Materials Letters, 116*, 275–277. DOI: 10.1016/j.matlet.2013.11.038

Azzam, O., & Chancellor, T. C. B. (2002). The biology, epidemiology, and management of rice tungro disease in asia. *Plant Disease, 86*(2), 88–100. DOI: 10.1094/PDIS.2002.86.2.88

Bahrulolum, H., Nooraei, S., Javanshir, N., Tarrahimofrad, H., Mirbagheri, V. S., Easton, A. J., & Ahmadian, G. (2021). Green synthesis of metal nanoparticles using microorganisms and their application in the agrifood sector. *Journal of Nanobiotechnology, 19*(1), 86. DOI: 10.1186/s12951-021-00834-3

Baldelli, A., Etayash, H., Oguzlu, H., Mandal, R., Jiang, F., Hancock, R. E. W., & Pratap-Singh, A. (2022). Antimicrobial properties of spray-dried cellulose nanocrystals and metal oxide-based nanoparticles-in-microspheres. *Chemical Engineering Journal Advances, 10*, 100273. DOI: 10.1016/j.ceja.2022.100273

Bar-Joseph, M., Marcus, R., & Lee, R. F. (1989). The continuous challenge of citrus tristeza virus control. *Annual Review of Phytopathology, 27*(1), 291–316. DOI: 10.1146/annurev.py.27.090189.001451

Bora, K. A., Hashmi, S., Zulfiqar, F., Abideen, Z., Ali, H., Siddiqui, Z. S., & Siddique, K. H. M. (2022). Recent progress in bio-mediated synthesis and applications of engineered nanomaterials for sustainable agriculture. *Frontiers in Plant Science, 13*. Advance online publication. DOI: 10.3389/fpls.2022.999505

Briddon, R. W., & Stanley, J. (2006). Subviral agents associated with plant single-stranded DNA viruses. *Virology, 344*(1), 198–210. DOI: 10.1016/j.virol.2005.09.042

Brunt, A. A., Crabtree, K., & Gibbs, A. J. (Eds.). (1990). Viruses of tropical plants: Descriptions and lists from VIDE database. C.A.B International in association with A.C.I.A.R.

Bundschuh, M., Filser, J., Lüderwald, S., McKee, M. S., Metreveli, G., Schaumann, G. E., Schulz, R., & Wagner, S. (2018). Nanoparticles in the environment: Where do we come from, where do we go to? *Environmental Sciences Europe*, *30*(1), 6. DOI: 10.1186/s12302-018-0132-6

Cai, F., Li, S., Huang, H., Iqbal, J., Wang, C., & Jiang, X. (2022). Green synthesis of gold nanoparticles for immune response regulation: Mechanisms, applications, and perspectives. *Journal of Biomedical Materials Research. Part A*, *110*(2), 424–442. DOI: 10.1002/jbm.a.37281

Cambra, M., Capote, N., Myrta, A., & Llácer, G. (2006). Plum pox virus and the estimated costs associated with sharka disease. *Bulletin OEPP. EPPO Bulletin. European and Mediterranean Plant Protection Organisation*, *36*(2), 202–204. DOI: 10.1111/j.1365-2338.2006.01027.x

Chakraborty, P., & Ghosh, A. (2022). Topical spray of dsrna induces mortality and inhibits chilli leaf curl virus transmission by bemisia tabaci asia ii 1. *Cells*, *11*(5), 833. DOI: 10.3390/cells11050833

Chandran, S. P., Chaudhary, M., Pasricha, R., Ahmad, A., & Sastry, M. (2006). Synthesis of gold nanotriangles and silver nanoparticles using aloe vera plant extract. *Biotechnology Progress*, *22*(2), 577–583. DOI: 10.1021/bp0501423

Chen, A., Halilovic, L., Shay, J.-H., Koch, A., Mitter, N., & Jin, H. (2023). Improving RNA-based crop protection through nanotechnology and insights from cross-kingdom RNA trafficking. *Current Opinion in Plant Biology*, *76*, 102441. DOI: 10.1016/j.pbi.2023.102441

Chen, H., & Yada, R. (2011). Nanotechnologies in agriculture: New tools for sustainable development. *Trends in Food Science & Technology*, *22*(11), 585–594. DOI: 10.1016/j.tifs.2011.09.004

Chung, I.-M., Park, I., Seung-Hyun, K., Thiruvengadam, M., & Rajakumar, G. (2016). Plant-mediated synthesis of silver nanoparticles: Their characteristic properties and therapeutic applications. *Nanoscale Research Letters*, *11*(1), 40. DOI: 10.1186/s11671-016-1257-4

Das, S. K., Das, A. R., & Guha, A. K. (2010). Microbial synthesis of multishaped gold nanostructures. Small (Weinheim an Der Bergstrasse, Germany), 6(9), 1012–1021. https://doi.org/DOI: 10.1002/smll.200902011

De Fazio, G., Caner, J., & Vicente, M. (1980). Effect of virazole (Ribavirin) on tomato spotted wilt virus in two systemic hosts, tomato and tobacco. *Archives of Virology*, *63*(3–4), 305–309. DOI: 10.1007/BF01315037

Desbiez, C., & Lecoq, H. (1997). Zucchini yellow mosaic virus. *Plant Pathology*, *46*(6), 809–829. DOI: 10.1046/j.1365-3059.1997.d01-87.x

Diener, T. O. (2003). Discovering viroids—A personal perspective. *Nature Reviews. Microbiology*, *1*(1), 75–80. DOI: 10.1038/nrmicro736

Dinant, S., & Lot, H. (1992). Lettuce mosaic virus. *Plant Pathology*, *41*(5), 528–542. DOI: 10.1111/j.1365-3059.1992.tb02451.x

Donaldson, K., Murphy, F. A., Duffin, R., & Poland, C. A. (2010). Asbestos, carbon nanotubes and the pleural mesothelium: A review of the hypothesis regarding the role of long fibre retention in the parietal pleura, inflammation and mesothelioma. *Particle and Fibre Toxicology*, *7*(1), 5. DOI: 10.1186/1743-8977-7-5

Dong, J., Yang, B., Wang, H., Cao, X., He, F., & Wang, L. (2023). Reveal molecular mechanism on the effects of silver nanoparticles on nitrogen transformation and related functional microorganisms in an agricultural soil. *The Science of the Total Environment*, *904*, 166765. DOI: 10.1016/j.scitotenv.2023.166765

Dong, Y., Wang, R., Li, H., Shao, J., Chi, Y., Lin, X., & Chen, G. (2012). Polyamine-functionalized carbon quantum dots for chemical sensing. *Carbon*, *50*(8), 2810–2815. DOI: 10.1016/j.carbon.2012.02.046

Durán, N., Marcato, P. D., Alves, O. L., De Souza, G. I., & Esposito, E. (2005). Mechanistic aspects of biosynthesis of silver nanoparticles by several Fusarium oxysporum strains. *Journal of Nanobiotechnology*, *3*(1), 8. DOI: 10.1186/1477-3155-3-8

Dutta, P.. (2022). 'Advances in Nanotechnology as a Potential Alternative for Plant Viral Disease Management'. *Frontiers in Microbiology*, vol. 13, June 2022. *Frontiers*. Advance online publication. DOI: 10.3389/fmicb.2022.935193

Dykman, L., & Khlebtsov, N. (2012). Gold nanoparticles in biomedical applications: Recent advances and perspectives. *Chemical Society Reviews*, *41*(6), 2256–2282. DOI: 10.1039/C1CS15166E

El-batal. Ahmed & Elsherbiny, Gamal & El-Sherbiny, Ibrahim & Askar, Ahmed. (2017). Biosynthesis of silver nanoparticles using Streptomyces atroverins, strain askar-sh50 and their antimicrobial activity against foodborne pathogens and Mycobacterium tuberculosis. Al Azhar Bulletin of Science Vol. 9th., Conf., March 2017, p. 87-106. 87-106.

El-Ganainy, S. M., Soliman, A. M., Ismail, A. M., Sattar, M. N., Farroh, K. Y., & Shafie, R. M. (2023). Antiviral activity of chitosan nanoparticles and chitosan silver nanocomposites against alfalfa mosaic virus. *Polymers*, *15*(13), 2961. DOI: 10.3390/polym15132961

El-Sappah, H., A., Qi, S., A. Soaud, S., Huang, Q., M. Saleh, A., A. S. Abourehab, M., Wan, L., Cheng, G., Liu, J., Ihtisham, M., Noor, Z., Rouf Mir, R., Zhao, X., Yan, K., Abbas, M., & Li, J. (2022). Natural resistance of tomato plants to Tomato yellow leaf curl virus. *Frontiers in Plant Science*, *13*, 1081549. DOI: 10.3389/fpls.2022.1081549

Elbeshehy, E. K. F., Elazzazy, A. M., & Aggelis, G. (2015). Silver nanoparticles synthesis mediated by new isolates of Bacillus spp., nanoparticle characterization and their activity against Bean Yellow Mosaic Virus and human pathogens. *Frontiers in Microbiology*, *6*. Advance online publication. DOI: 10.3389/fmicb.2015.00453

Elena, S. F., Fraile, A., & García-Arenal, F. (2014). Evolution and emergence of plant viruses. *Advances in Virus Research*, *88*, 161–191. DOI: 10.1016/B978-0-12-800098-4.00003-9

Elmer, W., & White, J. C. (2018). The future of nanotechnology in plant pathology. *Annual Review of Phytopathology*, *56*, 111–133. DOI: 10.1146/annurev-phyto-080417-050108

Elsharkawy, M. M., & Derbalah, A. (2019). Antiviral activity of titanium dioxide nanostructures as a control strategy for broad bean strain virus in faba bean. *Pest Management Science*, *75*(3), 828–834. DOI: 10.1002/ps.5185

Elumalai, E. & Prasad, Tollamadugu N V K V & Hemachandran, J. & Therasa, S. & Thirunavukkarasu, Dr.Thirumalai. (2010). Extracellular synthesis of silver nanoparticles using leaves of Euphorbia hirta and their antibacterial activities. *Journal of Pharmaceutical Sciences and Research.*, *2*, 549–554.

Ferreira, A. L., Ghanim, M., Xu, Y., & Pinheiro, P. V. (2024). Interactions between common bean viruses and their whitefly vector. *Viruses*, *16*(10), 1567. DOI: 10.3390/v16101567

Fraile, A., & García-Arenal, F. (2010). The coevolution of plants and viruses: Resistance and pathogenicity. *Advances in Virus Research*, *76*, 1–32. DOI: 10.1016/S0065-3527(10)76001-2

Frascati, F., Rotunno, S., Accotto, G. P., Noris, E., Vaira, A. M., & Miozzi, L. (2024). Exogenous application of dsrna for protection against tomato leaf curl new delhi virus. *Viruses*, *16*(3), 436. DOI: 10.3390/v16030436

García-Arenal, F., Fraile, A., & Malpica, J. M. (2001). Variability and genetic structure of plant virus populations. *Annual Review of Phytopathology*, *39*, 157–186. DOI: 10.1146/annurev.phyto.39.1.157

García-Arenal, F., Fraile, A., & Malpica, J. M. (2003). Variation and evolution of plant virus populations. *International Microbiology*, *6*(4), 225–232.

Garg, D., Sarkar, A., Chand, P., Bansal, P., Gola, D., Sharma, S., & Khantwal, S., Surabhi, Mehrotra, R., Chauhan, N., & Bharti, R. K. (2020). Synthesis of silver nanoparticles utilizing various biological systems: Mechanisms and applications—a review. *Progress in Biomaterials*, *9*, 81–95. DOI: 10.1007/s40204-020-00135-2

Garner, K. L., & Keller, A. A. (2014). Emerging patterns for engineered nanomaterials in the environment: A review of fate and toxicity studies. *Journal of Nanoparticle Research*, *16*(8), 2503. DOI: 10.1007/s11051-014-2503-2

Gautam, D., Dolma, K. G., Khandelwal, B., Gupta, M., Singh, M., Mahboob, T., Teotia, A., Thota, P., Bhattacharya, J., Goyal, R., & Oliveira, M.R., S., Pereira, M. de L., Wiart, C., Wilairatana, P., Eawsakul, K., Rahmatullah, M., Saravanabhavan, S. S., & Nissapatorn, V. (2023). Green synthesis of silver nanoparticles using Ocimum sanctum Linn. And its antibacterial activity against multidrug resistant Acinetobacter baumannii. *PeerJ*, *11*, e15590. DOI: 10.7717/peerj.15590

Ghormade, V., Deshpande, M. V., & Paknikar, K. M. (2011). Perspectives for nano-biotechnology enabled protection and nutrition of plants. *Biotechnology Advances*, *29*(6), 792–803. DOI: 10.1016/j.biotechadv.2011.06.007

Ghosh, S., Kanakala, S., Lebedev, G., Kontsedalov, S., Silverman, D., Alon, T., Mor, N., Sela, N., Luria, N., Dombrovsky, A., Mawassi, M., Haviv, S., Czosnek, H., & Ghanim, M. (2019). Transmission of a new polerovirus infecting pepper by the whitefly Bemisia tabaci. *Journal of Virology*, *93*(15). Advance online publication. DOI: 10.1128/JVI.00488-19

Ghosh, S., Patil, S., Ahire, M., Kitture, R., Kale, S., Pardesi, K., Cameotra, S. S., Bellare, J., Dhavale, D. D., Jabgunde, A., & Chopade, B. A. (2012). Synthesis of silver nanoparticles using Dioscorea bulbifera tuber extract and evaluation of its synergistic potential in combination with antimicrobial agents. *International Journal of Nanomedicine*, *7*, 483–496. DOI: 10.2147/IJN.S24793

Gibson, R. W., Legg, J. P., & Otim-Nape, G. W. (1996). Unusually severe symptoms are a characteristic of the current epidemic of mosaic virus disease of cassava in Uganda. *Annals of Applied Biology*, *128*(3), 479–490. DOI: 10.1111/j.1744-7348.1996.tb07108.x

Griffitt, R. J., Luo, J., Gao, J., Bonzongo, J.-C., & Barber, D. S. (2008). Effects of particle composition and species on toxicity of metallic nanomaterials in aquatic organisms. *Environmental Toxicology and Chemistry*, *27*(9), 1972–1978. DOI: 10.1897/08-002.1

Gupta, A., Rayeen, F., Mishra, R., Tripathi, M., & Pathak, N. (2023). Nanotechnology applications in sustainable agriculture: An emerging eco-friendly approach. *Plant Nano Biology*, *4*, 100033. DOI: 10.1016/j.plana.2023.100033

Gupta, J., Irfan, M., Ramgir, N., Muthe, K. P., Debnath, A. K., Ansari, S., Gandhi, J., Ranjith-Kumar, C. T., & Surjit, M. (2022). Antiviral activity of zinc oxide nanoparticles and tetrapods against the hepatitis e and hepatitis c viruses. *Frontiers in Microbiology*, *13*, 881595. DOI: 10.3389/fmicb.2022.881595

Hameed, H., Waheed, A., Sharif, M. S., Saleem, M., Afreen, A., Tariq, M., Kamal, A., Al-onazi, W. A., Al Farraj, D. A., Ahmad, S., & Mahmoud, R. M. (2023). Green synthesis of zinc oxide (Zno) nanoparticles from green algae and their assessment in various biological applications. *Micromachines*, *14*(5), 928. DOI: 10.3390/mi14050928

Hamza, I. N. M., & Kandiah, M. (2024). Catalytic reduction of para nitrophenol by green-fabricated silver nanoparticles (Agnps) from Solanum lycopersicum L. leaf extracts. *Brazilian Journal of Development*, *10*(6), e70680. DOI: 10.34117/bjdv10n6-052

Hanady, H. A., Elshorbgy, A. E., El-kammar, H. F., Mohamed, H. A., & Hassan, M. (2025). In-Vitro antiviral activity of chitosan nanoparticles against Banana bunchy top virus. *Fayoum Journal of Agricultural Research and Development*, *39*(1), 96–110. DOI: 10.21608/fjard.2025.333626.1083

Handy, R. D., Henry, T. B., Scown, T. M., Johnston, B. D., & Tyler, C. R. (2008). Manufactured nanoparticles: Their uptake and effects on fish—a mechanistic analysis. *Ecotoxicology (London, England)*, *17*(5), 396–409. DOI: 10.1007/s10646-008-0205-1

Hanley-Bowdoin, L., Bejarano, E. R., Robertson, D., & Mansoor, S. (2013). Geminiviruses: Masters at redirecting and reprogramming plant processes. *Nature Reviews. Microbiology*, *11*(11), 777–788. DOI: 10.1038/nrmicro3117

Hardy, A., Benford, D., Halldorsson, T., Jeger, M. J., Knutsen, H. K., More, S., Naegeli, H., Noteborn, H., Ockleford, C., Ricci, A., Rychen, G., Schlatter, J. R., Silano, V., Solecki, R., Turck, D., Younes, M., Chaudhry, Q., Cubadda, F., Gott, D., & Oomen, A. (2018). Guidance on risk assessment of the application of nanoscience and nanotechnologies in the food and feed chain: Part 1, human and animal health. *EFSA Journal*, *16*(7). Advance online publication. DOI: 10.2903/j.efsa.2018.5327

Hassanpouraghdam, M. B., Mehrabani, L. V., Khoshmaram, L., & Rasouli, F. (2025). Amelioration of the growth and physiological responses of Capsicum annum L. via quantum dot-graphene oxide, cerium oxide, and titanium oxide nanoparticles foliar application under salinity stress. *Scientific Reports*, *15*, 467. DOI: 10.1038/s41598-024-84706-4

He, K.-L., Yang, J.-H., Yu, F.-X., Wei, N., Liang, Q.-W., Feng, J.-W., Yi, T.-C., Chen, X.-S., Smagghe, G., Gui, S.-H., & Liu, T.-X. (2025). Silicon-coated carbon quantum dots composite nanomaterials mediate pest resistance activation in tobacco (Nicotiana tabacum). *Journal of Nanobiotechnology*, *23*(1), 359. DOI: 10.1186/s12951-025-03449-0

Hooks, C. R. R., Wright, M. G., Kabasawa, D. S., Manandhar, R., & Almeida, R. P. P. (2008). Effect of banana bunchy top virus infection on morphology and growth characteristics of banana. *Annals of Applied Biology*, *153*(1), 1–9. DOI: 10.1111/j.1744-7348.2008.00233.x

Hou, X., Zaks, T., Langer, R., & Dong, Y. (2021). Lipid nanoparticles for mRNA delivery. *Nature Reviews. Materials*, *6*(12), 1078–1094. DOI: 10.1038/s41578-021-00358-0

Hull, R. (2013). *Plant virology*. Academic press.

Ingle, A. P., Moharil, M. P., Saxena, S., Patil, V., & Borse, S. (2023). Nanotechnology: The new perspective in agroecosystem and sustainable agriculture. In Nanotechnology in Agriculture and Agroecosystems (pp. 1–24). Elsevier. https://doi.org/DOI: 10.1016/B978-0-323-99446-0.00002-7

Iravani, S. (2011). Green synthesis of metal nanoparticles using plants. *Green Chemistry*, *13*(10), 2638. DOI: 10.1039/c1gc15386b

Iravani, S., Korbekandi, H., Mirmohammadi, S. V., & Zolfaghari, B. (2014). Synthesis of silver nanoparticles: Chemical, physical and biological methods. *Research in Pharmaceutical Sciences*, *9*(6), 385–406.

Jain, N., Jain, P., Rajput, D., & Patil, U. K. (2021). Green synthesized plant-based silver nanoparticles: Therapeutic prospective for anticancer and antiviral activity. *Micro and Nano Systems Letters*, *9*(1), 5. DOI: 10.1186/s40486-021-00131-6

Jekayinoluwa, T., Tripathi, L., Tripathi, J. N., Ntui, V. O., Obiero, G., Muge, E., & Dale, J. (2020). RNAi technology for management of banana bunchy top disease. *Food and Energy Security*, *9*(4), e247. DOI: 10.1002/fes3.247

Johansen, E., Edwards, M. C., & Hampton, R. O. (1994). Seed transmission of viruses: Current perspectives. *Annual Review of Phytopathology*, *32*(1), 363–386. DOI: 10.1146/annurev.py.32.090194.002051

Jones, R. A. C. (2009). Plant virus emergence and evolution: Origins, new encounter scenarios, factors driving emergence, effects of changing world conditions, and prospects for control. *Virus Research*, *141*(2), 113–130. DOI: 10.1016/j.virusres.2008.07.028

Juárez, M., Rabadán, M. P., Martínez, L. D., Tayahi, M., Grande-Pérez, A., & Gómez, P. (2019). Natural hosts and genetic diversity of the emerging tomato leaf curl new delhi virus in spain. *Frontiers in Microbiology*, *10*. Advance online publication. DOI: 10.3389/fmicb.2019.00140

Judy, J. D., Unrine, J. M., & Bertsch, P. M. (2011). Evidence for biomagnification of gold nanoparticles within a terrestrial food chain. *Environmental Science & Technology*, *45*(2), 776–781. DOI: 10.1021/es103031a

Kah, M., Beulke, S., Tiede, K., & Hofmann, T. (2013). Nanopesticides: State of knowledge, environmental fate, and exposure modeling: eurosoil. *Critical Reviews in Environmental Science and Technology*, *43*(16), 1823–1867. DOI: 10.1080/10643389.2012.671750

Kah, M., Kookana, R. S., Gogos, A., & Bucheli, T. D. (2018). A critical evaluation of nanopesticides and nanofertilizers against their conventional analogues. *Nature Nanotechnology*, *13*(8), 677–684. DOI: 10.1038/s41565-018-0131-1

Kah, M., Tufenkji, N., & White, J. C. (2019). Nano-enabled strategies to enhance crop nutrition and protection. *Nature Nanotechnology*, *14*(6), 532–540. DOI: 10.1038/s41565-019-0439-5

Kaldis, A., Berbati, M., Melita, O., Reppa, C., Holeva, M., Otten, P., & Voloudakis, A. (2017). Exogenously applied dsRNA molecules deriving from the Zucchini yellow mosaic virus (Zymv) genome move systemically and protect cucurbits against ZYMV. *Molecular Plant Pathology*, *19*(4), 883–895. DOI: 10.1111/mpp.12572

Kalia, A., Abd-Elsalam, K. A., & Kuca, K. (2020). Zinc-based nanomaterials for diagnosis and management of plant diseases: Ecological safety and future prospects. *Journal of Fungi (Basel, Switzerland)*, *6*(4), 222. DOI: 10.3390/jof6040222

Kalishwaralal, K., Deepak, V., Ramkumarpandian, S., Nellaiah, H., & Sangiliyandi, G. (2008). Extracellular biosynthesis of silver nanoparticles by the culture supernatant of Bacillus licheniformis. *Materials Letters*, *62*(29), 4411–4413. DOI: 10.1016/j.matlet.2008.06.051

Keller, A. A., Wang, H., Zhou, D., Lenihan, H. S., Cherr, G., Cardinale, B. J., Miller, R., & Ji, Z. (2010). Stability and aggregation of metal oxide nanoparticles in natural aqueous matrices. *Environmental Science & Technology*, *44*(6), 1962–1967. DOI: 10.1021/es902987d

Kendall, A., Williams, D., Bian, W., Stewart, P. L., & Stubbs, G. (2013). Barley stripe mosaic virus: Structure and relationship to the tobamoviruses. *Virology*, *443*(2), 265–270. DOI: 10.1016/j.virol.2013.05.010

Khan, T., Hashmi, K., Jabeen, S., Joshi, S., Khan, A. R., & Ahmad, M. I. (2025). Carbon-based nanomaterials in crop plant disease management. In Husen, A. (Ed.), *Emerging Carbon Nanomaterials for Sustainable Agricultural Practices: Synthesis, Plant Growth, Performance, Production and Protection* (pp. 271–301). Springer Nature., DOI: 10.1007/978-981-97-5104-4_14

Khan, T., Yasmin, A., & Townley, H. E. (2020). An evaluation of the activity of biologically synthesized silver nanoparticles against bacteria, fungi and mammalian cell lines. *Colloids and Surfaces. B, Biointerfaces*, *194*, 111156. DOI: 10.1016/j.colsurfb.2020.111156

Khashan, K. S., Sulaiman, G. M., Abdulameer, F. A., Albukhaty, S., Ibrahem, M. A., Al-Muhimeed, T., & AlObaid, A. A. (2021). Antibacterial activity of tio2 nanoparticles prepared by one-step laser ablation in liquid. *Applied Sciences (Basel, Switzerland), 11*(10), 4623. DOI: 10.3390/app11104623

Khosravian-Dehkordi, P., Asadi-Samani, M., Altememy, D., Javadi-Farsani, F., Akbari, M., & Moradi, M.-T. (2024). In vitro antiviral activity of curcumin-loaded selenium nanoparticles against human herpes virus type 1. *Shahrekord University of Medical Sciences Journal, 26*(2), 73–77. DOI: 10.34172/jsums.943

Khot, L. R., Sankaran, S., Maja, J. M., Ehsani, R., & Schuster, E. W. (2012). Applications of nanomaterials in agricultural production and crop protection: A review. *Crop Protection (Guildford, Surrey), 35*, 64–70. DOI: 10.1016/j.cropro.2012.01.007

Kim, S., & Ryu, D. (2013). Silver nanoparticle-induced oxidative stress, genotoxicity and apoptosis in cultured cells and animal tissues. *Journal of Applied Toxicology, 33*(2), 78–89. DOI: 10.1002/jat.2792

Korsvik, C., Patil, S., Seal, S., & Self, W. T. (2007). Superoxide dismutase mimetic properties exhibited by vacancy engineered ceria nanoparticles. *Chemical Communications, 10*, 1056–1058. DOI: 10.1039/B615134E

Kotta, S., Aldawsari, H. M., Badr-Eldin, S. M., Alhakamy, N. A., Md, S., Nair, A. B., & Deb, P. K. (2020). Exploring the potential of carbon dots to combat covid-19. *Frontiers in Molecular Biosciences, 7*, 616575. DOI: 10.3389/fmolb.2020.616575

Kreuze, J. F., Perez, A., Untiveros, M., Quispe, D., Fuentes, S., Barker, I., & Simon, R. (2009). Complete viral genome sequence and discovery of novel viruses by deep sequencing of small RNAs: A generic method for diagnosis, discovery and sequencing of viruses. *Virology, 388*(1), 1–7. DOI: 10.1016/j.virol.2009.03.024

Kulkarni, D., Sherkar, R., Shirsathe, C., Sonwane, R., Varpe, N., Shelke, S., More, M. P., Pardeshi, S. R., Dhaneshwar, G., Junnuthula, V., & Dyawanapelly, S. (2023). Biofabrication of nanoparticles: Sources, synthesis, and biomedical applications. *Frontiers in Bioengineering and Biotechnology, 11*. Advance online publication. DOI: 10.3389/fbioe.2023.1159193

Kumar, B., Smita, K., Cumbal, L., & Debut, A. (2017). Green synthesis of silver nanoparticles using Andean blackberry fruit extract. *Saudi Journal of Biological Sciences, 24*(1), 45–50. DOI: 10.1016/j.sjbs.2015.09.006

Kuppusamy, P., Yusoff, M. M., Maniam, G. P., & Govindan, N. (2016). Biosynthesis of metallic nanoparticles using plant derivatives and their new avenues in pharmacological applications – An updated report. *Saudi Pharmaceutical Journal : SPJ : the Official Publication of the Saudi Pharmaceutical Society, 24*(4), 473–484. DOI: 10.1016/j.jsps.2014.11.013

Lara, H. H., Ayala-Nuñez, N. V., Ixtepan-Turrent, L., & Rodriguez-Padilla, C. (2010). Mode of antiviral action of silver nanoparticles against HIV-1. *Journal of Nanobiotechnology, 8*(1), 1. DOI: 10.1186/1477-3155-8-1

Legg, J. P., Jeremiah, S. C., Obiero, H. M., Maruthi, M. N., Ndyetabula, I., Okao-Okuja, G., Bouwmeester, H., Bigirimana, S., Tata-Hangy, W., Gashaka, G., Mkamilo, G., Alicai, T., & Lava Kumar, P. (2011). Comparing the regional epidemiology of the cassava mosaic and cassava brown streak virus pandemics in Africa. *Virus Research, 159*(2), 161–170. DOI: 10.1016/j.virusres.2011.04.018

Legg, J. P., & Thresh, J. M. (2000). Cassava mosaic virus disease in East Africa: A dynamic disease in a changing environment. *Virus Research, 71*(1–2), 135–149. DOI: 10.1016/s0168-1702(00)00194-5

Lennon, A. M., Neuenschwander, U. H., Ribas-Carbo, M., Giles, L., Ryals, J. A., & Siedow, J. N. (1997). The effects of salicylic acid and tobacco mosaic virus infection on the alternative oxidase of tobacco. *Plant Physiology, 115*(2), 783–791. DOI: 10.1104/pp.115.2.783

Lerch, B. (1987). On the inhibition of plant virus multiplication by ribavirin. *Antiviral Research, 7*(5), 257–270. DOI: 10.1016/0166-3542(87)90010-6

Levard, C., Hotze, E. M., Lowry, G. V., & Brown, G. E. (2012). Environmental transformations of silver nanoparticles: Impact on stability and toxicity. *Environmental Science & Technology, 46*(13), 6900–6914. DOI: 10.1021/es2037405

Li, W., He, Z., Li, S., Huang, Y., Zhang, Z., Jiang, D., Wang, X., & Luo, Z. (2011). Molecular characterization of a new strain of sugarcane streak mosaic virus (Scsmv). *Archives of Virology, 156*(11), 2101–2104. DOI: 10.1007/s00705-011-1090-0

Lico, C., Benvenuto, E., & Baschieri, S. (2015). The two-faced potato virus x: From plant pathogen to smart nanoparticle. *Frontiers in Plant Science, 6*, 1009. DOI: 10.3389/fpls.2015.01009

Lin, Neil, et al. 'Antiviral Nanoparticles for Sanitizing Surfaces: A Roadmap to Self-Sterilizing against COVID-19'. Nano Today, vol. 40, Oct. 2021, p. 101267. PubMed Central, https://doi.org/.DOI: 10.1016/j.nantod.2021.101267

Liu, S., Tian, W., Liu, Z., Wei, X., Yuan, K., Du, W., Chen, S., Chen, S., Zhou, D., & Cai, L. (2024). Biosynthesis of cupric oxide nanoparticles: Its antiviral activities against TMV by directly destroying virion and inducing plant resistance. *Phytopathology Research, 6*(1), 30. DOI: 10.1186/s42483-024-00250-z

Łoczechin, A., Séron, K., Barras, A., Giovanelli, E., Belouzard, S., Chen, Y.-T., Metzler-Nolte, N., Boukherroub, R., Dubuisson, J., & Szunerits, S. (2019). Functional carbon quantum dots as medical countermeasures to human coronavirus. *ACS Applied Materials & Interfaces, 11*(46), 42964–42974. DOI: 10.1021/acsami.9b15032

Loutfy, S. A., Elberry, M. H., Farroh, K. Y., Mohamed, H. T., Mohamed, A. A., Mohamed, E. B., Faraag, A. H. I., & Mousa, S. A. (2020). Antiviral activity of chitosan nanoparticles encapsulating curcumin against hepatitis c virus genotype 4a in human hepatoma cell lines. *International Journal of Nanomedicine, 15*, 2699–2715. DOI: 10.2147/IJN.S241702

Lowenberg-DeBoer, J., & Erickson, B. (2019). Setting the record straight on precision agriculture adoption. *Agronomy Journal, 111*(4), 1552–1569. DOI: 10.2134/agronj2018.12.0779

Lowry, G. V., Espinasse, B. P., Badireddy, A. R., Richardson, C. J., Reinsch, B. C., Bryant, L. D., Bone, A. J., Deonarine, A., Chae, S., Therezien, M., Colman, B. P., Hsu-Kim, H., Bernhardt, E. S., Matson, C. W., & Wiesner, M. R. (2012). Long-term transformation and fate of manufactured ag nanoparticles in a simulated large scale freshwater emergent wetland. *Environmental Science & Technology*, *46*(13), 7027–7036. DOI: 10.1021/es204608d

Madkhali, O. A. (2023). Drug delivery of gelatin nanoparticles as a biodegradable polymer for the treatment of infectious diseases: Perspectives and challenges. *Polymers*, *15*(21), 4327. DOI: 10.3390/polym15214327

Mahawar, H., & Prasanna, R. (2018). Prospecting the interactions of nanoparticles with beneficial microorganisms for developing green technologies for agriculture. *Environmental Nanotechnology, Monitoring & Management*, *10*, 477–485. DOI: 10.1016/j.enmm.2018.09.004

Maluin, F. N., & Hussein, M. Z. (2020). Chitosan-based agronanochemicals as a sustainable alternative in crop protection. *Molecules (Basel, Switzerland)*, *25*(7), 1611. DOI: 10.3390/molecules25071611

Manchester, M., & Steinmetz, N. F. (2009). Viruses and nanotechnology. Preface. *Current Topics in Microbiology and Immunology*, *327*, v–vi.

Martelli, P. abou, g. S. N., agranovsky, a. A., al, r. M., dolja, v. V., dovas, c., Fuchs, m., Gugerli, p., Hu, j. S., jelkmann, w., Katis, n. I., maliogka, v. I., melzer, m. J., menzel, w., Minafra, a., Rott, m. E., rowhani, a., Sabanadz(2012)ovic, s. And saldarelli, p. Taxonomic revision of the family closteroviridae with special reference to the grapevine leafroll-associated members of the genus ampelovirus and the putative species unassigned to the family. Journal of plant pathology, 94, 7-19. - References—Scientific research publishing. (n.d.). Retrieved September 17, 2025, from https://www.scirp.org/reference/referencespapers?referenceid=3136933

Mawale, K. S., Giridhar, P., & Johnson, T. S. (2025). Chitosan: A versatile polymer for enhancing plant bioactive accumulation, managing plant diseases, and advancing food preservation technologies. *International Journal of Biological Macromolecules*, *308*, 142081. DOI: 10.1016/j.ijbiomac.2025.142081

McClements, D. J., & Xiao, H. (2017). Is nano safe in foods? Establishing the factors impacting the gastrointestinal fate and toxicity of organic and inorganic food-grade nanoparticles. *NPJ Science of Food*, *1*(1), 6. DOI: 10.1038/s41538-017-0005-1

Melk, M. M., El-Hawary, S. S., Melek, F. R., Saleh, D. O., Ali, O. M., El Raey, M. A., & Selim, N. M. (2021). Antiviral activity of zinc oxide nanoparticles mediated by plumbago indica l. Extract against herpes simplex virus type 1(HSV-1). *International Journal of Nanomedicine*, *16*, 8221–8233. DOI: 10.2147/IJN.S339404

Miller, W. A., & Rasochová, L. (1997). Barley yellow dwarf viruses. *Annual Review of Phytopathology*, *35*, 167–190. DOI: 10.1146/annurev.phyto.35.1.167

Mittal, A. K., Chisti, Y., & Banerjee, U. C. (2013). Synthesis of metallic nanoparticles using plant extracts. *Biotechnology Advances*, *31*(2), 346–356. DOI: 10.1016/j.biotechadv.2013.01.003

Mitter, N., Worrall, E. A., Robinson, K. E., Li, P., Jain, R. G., Taochy, C., Fletcher, S. J., Carroll, B. J., Lu, G. Q. M., & Xu, Z. P. (2017). Clay nanosheets for topical delivery of RNAi for sustained protection against plant viruses. *Nature Plants*, *3*, 16207. DOI: 10.1038/nplants.2016.207

Moriones, E., & Navas-Castillo, J. (2000). Tomato yellow leaf curl virus, an emerging virus complex causing epidemics worldwide. *Virus Research*, *71*(1–2), 123–134. DOI: 10.1016/s0168-1702(00)00193-3

Mousavi, S. R., & Rezaei, M. (2011). Nanotechnology in agriculture and food production. https://www.semanticscholar.org/paper/Nanotechnology-in-Agriculture-and-Food-Production-Mousavi-Rezaei/f2f3b22aaf3a84979ad73c46a0e429d890491b9e

Naggar, Y. A. (n.d.). Effects of metal and metal oxide nanoparticles on honey bees(Apis millefera L.). Journal of Aquatic Pollution and Toxicology, 1(1), 0–0. Retrieved September 17, 2025, from https://www.primescholars.com/

Naidu, R. A., Kimmins, F. M., Deom, C. M., Subrahmanyam, P., Chiyembekeza, A. J., & Van Der Merwe, P. J. A. (1999). Groundnut rossette: A virus disease affecting groundnut production in sub-saharan africa. *Plant Disease*, *83*(8), 700–709. DOI: 10.1094/PDIS.1999.83.8.700

Naikoo, G. A., Mustaqeem, M., Hassan, I. U., Awan, T., Arshad, F., Salim, H., & Qurashi, A. (2021). Bioinspired and green synthesis of nanoparticles from plant extracts with antiviral and antimicrobial properties: A critical review. *Journal of Saudi Chemical Society*, *25*(9), 101304.

Narayanan, K. B., & Sakthivel, N. (2010). Biological synthesis of metal nanoparticles by microbes. *Advances in Colloid and Interface Science*, *156*(1–2), 1–13. DOI: 10.1016/j.cis.2010.02.001

Nasrollahzadeh, M., Sajjadi, M., Iravani, S., & Varma, R. S. (2021). Green-synthesized nanocatalysts and nanomaterials for water treatment: Current challenges and future perspectives. *Journal of Hazardous Materials*, *401*, 123401. DOI: 10.1016/j.jhazmat.2020.123401

Navarro, E., Piccapietra, F., Wagner, B., Marconi, F., Kaegi, R., Odzak, N., Sigg, L., & Behra, R. (2008). Toxicity of silver nanoparticles to Chlamydomonas reinhardtii. *Environmental Science & Technology*, *42*(23), 8959–8964. DOI: 10.1021/es801785m

Navas-Castillo, J., Fiallo-Olivé, E., & Sánchez-Campos, S. (2011). Emerging virus diseases transmitted by whiteflies. *Annual Review of Phytopathology*, *49*, 219–248. DOI: 10.1146/annurev-phyto-072910-095235

Ndong Ntoutoume, G. M. A., Grassot, V., Brégier, F., Chabanais, J., Petit, J.-M., Granet, R., & Sol, V. (2017). PEI-cellulose nanocrystal hybrids as efficient siRNA delivery agents-Synthesis, physicochemical characterization and in vitro evaluation. *Carbohydrate Polymers*, *164*, 258–267. DOI: 10.1016/j.carbpol.2017.02.004

Noyce, J. O., Michels, H., & Keevil, C. W. (2007). Inactivation of influenza a virus on copper versus stainless steel surfaces. *Applied and Environmental Microbiology*, *73*(8), 2748–2750. DOI: 10.1128/AEM.01139-06

Omidian, H., Gill, E. J., Dey Chowdhury, S., & Cubeddu, L. X. (2024). Chitosan nanoparticles for intranasal drug delivery. *Pharmaceutics*, *16*(6), 746. DOI: 10.3390/pharmaceutics16060746

Othman, S. H., Ronzi, N. D. A., Shapi'i, R. A., Dun, M., Ariffin, S. H., & Mohammed, M. A. P. (2023). Biodegradability of starch nanocomposite films containing different concentrations of chitosan nanoparticles in compost and planting soils. *Coatings*, *13*(4), 777. DOI: 10.3390/coatings13040777

Palmieri, V., & Papi, M. (2020). Can graphene take part in the fight against COVID-19? *Nano Today*, *33*, 100883. DOI: 10.1016/j.nantod.2020.100883

Palukaitis, P., & García-Arenal, F. (2003). Cucumoviruses. *Advances in Virus Research*, *62*, 241–323. DOI: 10.1016/s0065-3527(03)62005-1

Palukaitis, P., & García-Arenal, F. (2003). Cucumoviruses. *Advances in Virus Research*, *62*, 241–323.

Pandya, R. P., Kapadiya, I. B., Patel, M. L., & Gadhiya, V. C. (2024). Chitosan applications in plant disease control: A mini review. Farming & Management, 9(2), 92–101. https://gauravpublications.com/journal/farming-management/volume-9/issue-2-dec-2024/Gaurav Publications/journal/farming-management/

Panpatte, D. G., Jhala, Y. K., Shelat, H. N., & Vyas, R. V. (2016). Nanoparticles: The next generation technology for sustainable agriculture. In D. P. Singh, H. B. Singh, & R. Prabha (Eds.), Microbial Inoculants in Sustainable Agricultural Productivity: Vol. 2: Functional Applications (pp. 289–300). Springer India. https://doi.org/DOI: 10.1007/978-81-322-2644-4_18

Pardi, N., Hogan, M. J., & Weissman, D. (2020). Recent advances in mRNA vaccine technology. *Current Opinion in Immunology*, *65*, 14–20. DOI: 10.1016/j.coi.2020.01.008

Parisi, C., Vigani, M., & Rodríguez-Cerezo, E. (2015). Agricultural Nanotechnologies: What are the current possibilities? *Nano Today*, *10*(2), 124–127. DOI: 10.1016/j.nantod.2014.09.009

Park, H. J., Kim, S. H., Kim, H. J., & Choi, S. H. (2006). A new composition of nanosized silica-silver for control of various plant diseases. *The Plant Pathology Journal*, *22*(3). https://agris.fao.org/search/en/providers/122646/records/6472467753aa8c89630476f0

Pathak, N., Singh, P., Singh, P. K., Sharma, S., Singh, R. P., Gupta, A., Mishra, R., Mishra, V. K., & Tripathi, M. (2022). Biopolymeric nanoparticles based effective delivery of bioactive compounds toward the sustainable development of anticancerous therapeutics. *Frontiers in Nutrition*, *9*. Advance online publication. DOI: 10.3389/fnut.2022.963413

Patil, S., & Chandrasekaran, R. (2020). Biogenic nanoparticles: A comprehensive perspective in synthesis, characterization, application and its challenges. *Journal of Genetic Engineering and Biotechnology*, *18*, 67. DOI: 10.1186/s43141-020-00081-3

Patra, J. K., & Baek, K.-H. (2014). Green nanobiotechnology: Factors affecting synthesis and characterization techniques. *Journal of Nanomaterials*, *2014*(1), 417305. DOI: 10.1155/2014/417305

Pattanavongsawat, C., Malichan, S., Vannatim, N., Chaowongdee, S., Hemniam, N., Paemanee, A., & Siriwan, W. (2025). Enhancing plant resistance to sri lankan cassava mosaic virus using salicylic acid. *Metabolites*, *15*(4), 261. DOI: 10.3390/metabo15040261

Pejam, F., Ardebili, Z. O., Ladan-Moghadam, A., & Danaee, E. (2021). Zinc oxide nanoparticles mediated substantial physiological and molecular changes in tomato. *PLoS One*, *16*(3), e0248778. DOI: 10.1371/journal.pone.0248778

Pérez-de-Luque, A., & Rubiales, D. (2009). Nanotechnology for parasitic plant control. *Pest Management Science*, *65*(5), 540–545. DOI: 10.1002/ps.1732

Prasad, R., Bhattacharyya, A., & Nguyen, Q. D. (2017). Nanotechnology in sustainable agriculture: Recent developments, challenges, and perspectives. *Frontiers in Microbiology*, *8*. Advance online publication. DOI: 10.3389/fmicb.2017.01014

Priester, J. H., Ge, Y., Mielke, R. E., Horst, A. M., Moritz, S. C., Espinosa, K., Gelb, J., Walker, S. L., Nisbet, R. M., An, Y.-J., Schimel, J. P., Palmer, R. G., Hernandez-Viezcas, J. A., Zhao, L., Gardea-Torresdey, J. L., & Holden, P. A. (2012). Soybean susceptibility to manufactured nanomaterials with evidence for food quality and soil fertility interruption. *Proceedings of the National Academy of Sciences of the United States of America*, *109*(37), E2451–E2456. DOI: 10.1073/pnas.1205431109

Quenouille, J., Vassilakos, N., & Moury, B. (2013). Potato virus Y: A major crop pathogen that has provided major insights into the evolution of viral pathogenicity. *Molecular Plant Pathology*, *14*(5), 439–452. DOI: 10.1111/mpp.12024

Rai, M., Deshmukh, S. D., Ingle, A. P., Gupta, I. R., Galdiero, M., & Galdiero, S. (2016). Metal nanoparticles: The protective nanoshield against virus infection. *Critical Reviews in Microbiology*, *42*(1), 46–56. DOI: 10.3109/1040841X.2013.879849

Rai, P., Pratap Singh, V., Sharma, S., Tripathi, D. K., & Sharma, S. (2022). Iron oxide nanoparticles impart cross tolerance to arsenate stress in rice roots through involvement of nitric oxide. *Environmental Pollution*, *307*, 119320. DOI: 10.1016/j.envpol.2022.119320

Rajabzadeh-Khosroshahi, M., Khoshfetrat, A. B., & Salami-Kalajahi, M. (2025). A review on pectin-based nanostructures for drug and gene delivery systems. *International Journal of Biological Macromolecules*, *304*, 140932. DOI: 10.1016/j.ijbiomac.2025.140932

Rajput, V. D., Minkina, T., Sushkova, S., Tsitsuashvili, V., Mandzhieva, S., Gorovtsov, A., Nevidomskyaya, D., & Gromakova, N. (2018). Effect of nanoparticles on crops and soil microbial communities. *Journal of Soils and Sediments*, *18*(6), 2179–2187. DOI: 10.1007/s11368-017-1793-2

Raliya, R., Nair, R., Chavalmane, S., Wang, W.-N., & Biswas, P. (2015). Mechanistic evaluation of translocation and physiological impact of titanium dioxide and zinc oxide nanoparticles on the tomato (Solanum lycopersicum L.) plant. *Metallomics*, *7*(12), 1584–1594. DOI: 10.1039/C5MT00168D

Raliya, R., & Tarafdar, J. C. (2013). Zno nanoparticle biosynthesis and its effect on phosphorous-mobilizing enzyme secretion and gum contents in clusterbean(Cyamopsis tetragonoloba L.). *Agricultural Research*, *2*(1), 48–57. DOI: 10.1007/s40003-012-0049-z

Raliya, R., Tarafdar, J. C., & Biswas, P. (2016). Enhancing the mobilization of native phosphorus in the mung bean rhizosphere using zno nanoparticles synthesized by soil fungi. *Journal of Agricultural and Food Chemistry*, *64*(16), 3111–3118. DOI: 10.1021/acs.jafc.5b05224

Raveendran, P., Fu, J., & Wallen, S. L. (2003). Completely "green" synthesis and stabilization of metal nanoparticles. *Journal of the American Chemical Society*, *125*(46), 13940–13941. DOI: 10.1021/ja029267j

Rehman, F. ur, Paker, N. P., Rehman, S. ur, Javed, M. T., Farooq Hussain Munis, M., & Chaudhary, H. J. (2024). Zinc oxide nanoparticles: Biogenesis and applications against phytopathogens. *Journal of Plant Pathology*, *106*(1), 45–65. DOI: 10.1007/s42161-023-01522-x

Rendina, N., Nuzzaci, M., Scopa, A., Cuypers, A., & Sofo, A. (2019). Chitosan-elicited defense responses in Cucumber mosaic virus (Cmv)-infected tomato plants. *Journal of Plant Physiology*, *234–235*, 9–17. DOI: 10.1016/j.jplph.2019.01.003

Rico, C. M., Majumdar, S., Duarte-Gardea, M., Peralta-Videa, J. R., & Gardea-Torresdey, J. L. (2011). Interaction of nanoparticles with edible plants and their possible implications in the food chain. *Journal of Agricultural and Food Chemistry*, *59*(8), 3485–3498. DOI: 10.1021/jf104517j

Rodríguez-Cerezo, E., Moya, A., & García-Arenal, F. (1989). Variability and evolution of the plant RNA virus pepper mild mottle virus. *Journal of Virology*, *63*(5), 2198–2203. DOI: 10.1128/JVI.63.5.2198-2203.1989

Rónavári, A., Igaz, N., Adamecz, D. I., Szerencsés, B., Molnar, C., Kónya, Z., Pfeiffer, I., & Kiricsi, M. (2021). Green silver and gold nanoparticles: Biological synthesis approaches and potentials for biomedical applications. *Molecules (Basel, Switzerland)*, *26*(4), 844. DOI: 10.3390/molecules26040844

Rybicki, E. P., & Pietersen, G. (1999). Plant virus disease problems in the developing world. *Advances in Virus Research*, *53*, 127–175. DOI: 10.1016/s0065-3527(08)60346-2

Saadh, M. (2022). Potent antiviral effect of green synthesis silver nanoparticles on Newcastle disease virus. *Arabian Journal of Chemistry*, *15*(7), 103899. DOI: 10.1016/j.arabjc.2022.103899

Saberi Riseh, R., Gholizadeh Vazvani, M., Ebrahimi-Zarandi, M., & Skorik, Y. A. (2022). Alginate-induced disease resistance in plants. *Polymers*, *14*(4), 661. DOI: 10.3390/polym14040661

Safer, A.-M., & Leporatti, S. (2021). Chitosan nanoparticles for antiviral drug delivery: A novel route for covid-19 treatment. *International Journal of Nanomedicine*, *16*, 8141–8158. DOI: 10.2147/IJN.S332385

Salama, H. M. H. (2012). Effects of silver nanoparticles in crop plants. *International Research Journal of Biotechnology*, *3*(10), 190–197.

Salata, O. (2004). Applications of nanoparticles in biology and medicine. *Journal of Nanobiotechnology*, *2*(1), 3. DOI: 10.1186/1477-3155-2-3

Sastry, M., Ahmad, A., Khan, M., & Kumar, R. (2003). Biosynthesis of metal nanoparticles using fungi and actinomycete. *Current Science*, *•••*, 85.

Scholthof, K.-B. G. (2004). Tobacco mosaic virus: A model system for plant biology. *Annual Review of Phytopathology*, *42*(1), 13–34. DOI: 10.1146/annurev.phyto.42.040803.140322

Scholthof, K.-B. G., Adkins, S., Czosnek, H., Palukaitis, P., Jacquot, E., Hohn, T., Hohn, B., Saunders, K., Candresse, T., Ahlquist, P., Hemenway, C., & Foster, G. D. (2011). Top 10 plant viruses in molecular plant pathology. *Molecular Plant Pathology*, *12*(9), 938–954. DOI: 10.1111/j.1364-3703.2011.00752.x

Schwartz, S. H., Hendrix, B., Hoffer, P., Sanders, R. A., & Zheng, W. (2020). Carbon dots for efficient small interfering rna delivery and gene silencing in plants [open]. *Plant Physiology*, *184*(2), 647–657. DOI: 10.1104/pp.20.00733

Serrano-Aroca, Á., Takayama, K., Mishra, Y. K., & De La Fuente-Nunez, C. (2024). Carbon-based nanomaterials for antiviral applications. *Advanced Functional Materials*, *34*(38), 2402023. DOI: 10.1002/adfm.202402023

Shafey, A. M. E. (2020). Green synthesis of metal and metal oxide nanoparticles from plant leaf extracts and their applications: A review. *Green Processing and Synthesis*, *9*(1), 304–339. DOI: 10.1515/gps-2020-0031

Shahzadi, S., Fatima, S., & Ain, Q. ul, Shafiq, Z., & Janjua, M. R. S. A. (2025). A review on green synthesis of silver nanoparticles (Snps) using plant extracts: A multifaceted approach in photocatalysis, environmental remediation, and biomedicine. *RSC Advances*, *15*(5), 3858–3903. DOI: 10.1039/D4RA07519F

Shang, Y., & Hasan, Md. K., Ahammed, G. J., Li, M., Yin, H., & Zhou, J. (2019a). Applications of nanotechnology in plant growth and crop protection: A review. *Molecules (Basel, Switzerland)*, *24*(14), 2558. DOI: 10.3390/molecules24142558

Shang, Y., & Hasan, Md. K., Ahammed, G. J., Li, M., Yin, H., & Zhou, J. (2019b). Applications of nanotechnology in plant growth and crop protection: A review. *Molecules (Basel, Switzerland)*, *24*(14), 2558. DOI: 10.3390/molecules24142558

Sharma, D., Sharma, A., & Gautam, H. R. (2025). Green synthesized nanoparticles for disease management in vegetable crops: A review. *Plant Nano Biology*, *13*, 100179. DOI: 10.1016/j.plana.2025.100179

Shcherbakov, A. B. (2024). CeO2 nanoparticles and cerium species as antiviral agents: Critical review. *European Journal of Medicinal Chemistry Reports*, *10*, 100141. DOI: 10.1016/j.ejmcr.2024.100141

Sigala-Aguilar, N. A., Flores-Rentería, D., Vera-Reyes, I., Moya-Cadena, S., Fernández-Luqueño, F., & López, M. G. (2025). Multi-walled carbon nanotubes: Effect on the growth-development of tomato plants (Solanum lycopersicum L.), biocompounds content in tomato fruits and quality indicators of cultivated soil. *Journal of Soil Science and Plant Nutrition*. Advance online publication. DOI: 10.1007/s42729-025-02566-x

Simonin, M., & Richaume, A. (2015). Impact of engineered nanoparticles on the activity, abundance, and diversity of soil microbial communities: A review. *Environmental Science and Pollution Research International*, *22*(18), 13710–13723. DOI: 10.1007/s11356-015-4171-x

Singh, J., Dutta, T., Kim, K.-H., Rawat, M., Samddar, P., & Kumar, P. (2018). 'Green' synthesis of metals and their oxide nanoparticles: Applications for environmental remediation. *Journal of Nanobiotechnology*, *16*(1), 84. DOI: 10.1186/s12951-018-0408-4

Singh, P., Kim, Y.-J., Zhang, D., & Yang, D.-C. (2016). Biological synthesis of nanoparticles from plants and microorganisms. *Trends in Biotechnology*, *34*(7), 588–599. DOI: 10.1016/j.tibtech.2016.02.006

Singh, R., Kuddus, M., Singh, P. K., & Choden, D. (2022). Nanotechnology for nanophytopathogens: From detection to the management of plant viruses. *BioMed Research International, 8688584*. Advance online publication. DOI: 10.1155/2022/8688584

Singh, R., Shedbalkar, U. U., Wadhwani, S. A., & Chopade, B. A. (2015). Bacteriagenic silver nanoparticles: Synthesis, mechanism, and applications. *Applied Microbiology and Biotechnology*, *99*(11), 4579–4593. DOI: 10.1007/s00253-015-6622-1

Singh, S., Sangwan, S., Sharma, P., Devi, P., & Moond, M. (2021). Nanotechnology for sustainable agriculture: An emerging perspective. *Journal of Nanoscience and Nanotechnology*, *21*(6), 3453–3465. DOI: 10.1166/jnn.2021.19012

Sofy, A. R., Sofy, M. R., Hmed, A. A., Dawoud, R. A., Alnaggar, A. E.-A. M., Soliman, A. M., & El-Dougdoug, N. K. (2021). Ameliorating the adverse effects of tomato mosaic tobamovirus infecting tomato plants in egypt by boosting immunity in tomato plants using zinc oxide nanoparticles. *Molecules (Basel, Switzerland)*, *26*(5), 1337. DOI: 10.3390/molecules26051337

Soumya, K., Yogita, M., Prasanthi, Y., Anitha, K., Kishor, P. B. K., Jain, R. K., & Mandal, B. (2014). Molecular characterization of Indian isolate of peanut mottle virus and immunodiagnosis using bacterial expressed core capsid protein. *Virusdisease*, *25*(3), 331–337. DOI: 10.1007/s13337-014-0210-3

Sportelli, M. C., Izzi, M., Kukushkina, E. A., Hossain, S. I., Picca, R. A., Ditaranto, N., & Cioffi, N. (2020). Can nanotechnology and materials science help the fight against sars-cov-2? *Nanomaterials (Basel, Switzerland)*, *10*(4), 802. DOI: 10.3390/nano10040802

Stenger, D. C., French, R., & Gildow, F. E. (2005). Complete deletion of Wheat streak mosaic virus HC-Pro: A null mutant is viable for systemic infection. *Journal of Virology*, *79*(18), 12077–12080. DOI: 10.1128/JVI.79.18.12077-12080.2005

Strange, R. N., & Scott, P. R. (2005). Plant disease: A threat to global food security. *Annual Review of Phytopathology*, *43*, 83–116. DOI: 10.1146/annurev.phyto.43.113004.133839

Taha, S. H., El-Sherbiny, I. M., Salem, A. S., Abdel-Hamid, M., Hamed, A. H., & Ahmed, G. A. (2019). Antiviral activity of curcumin loaded milk proteins nanoparticles on potato virus y. *Pakistan Journal of Biological Sciences*, *22*(12), 614–622. DOI: 10.3923/pjbs.2019.614.622

Tang, J., Zhang, J., Shi, P., Xiao, Y., Shi, Y., Ding, L., & Xu, W. (2020). Synthesis of fluorescent carbon quantum dots from tobacco wastewater for the detection of ferric(Iii) ions. *Science of Advanced Materials*, *12*(7), 966–972. DOI: 10.1166/sam.2020.3731

Tripathi, S., Suzuki, J. Y., Ferreira, S. A., & Gonsalves, D. (2008). Papaya ringspot virus-P: Characteristics, pathogenicity, sequence variability and control. *Molecular Plant Pathology*, *9*(3), 269–280. DOI: 10.1111/j.1364-3703.2008.00467.x

Vargas-Hernandez, M., Macias-Bobadilla, I., Guevara-Gonzalez, R. G., Rico-Garcia, E., Ocampo-Velazquez, R. V., Avila-Juarez, L., & Torres-Pacheco, I. (2020). Nanoparticles as potential antivirals in agriculture. *Agriculture*, *10*(10), 444. DOI: 10.3390/agriculture10100444

Varsani, A., Navas-Castillo, J., Moriones, E., Hernández-Zepeda, C., Idris, A., Brown, J. K., Murilo Zerbini, F., & Martin, D. P. (2014). Establishment of three new genera in the family Geminiviridae: Becurtovirus, Eragrovirus and Turncurtovirus. *Archives of Virology*, *159*(8), 2193–2203. DOI: 10.1007/s00705-014-2050-2

Varsani, A., Shepherd, D. N., Monjane, A. L., Owor, B. E., Erdmann, J. B., Rybicki, E. P., Peterschmitt, M., Briddon, R. W., Markham, P. G., Oluwafemi, S., Windram, O. P., Lefeuvre, P., Lett, J.-M., & Martin, D. P. (2008). Recombination, decreased host specificity and increased mobility may have driven the emergence of maize streak virus as an agricultural pathogen. *The Journal of General Virology*, *89*(Pt 9), 2063–2074. DOI: 10.1099/vir.0.2008/003590-0

Verma, K., & Modgil, M. (2024). RNA interference (Rnai) mediated technique for combating plant diseases: Harnessing nanoparticles for effective delivery and enhanced efficacy. [PCTOC]. *Plant Cell, Tissue and Organ Culture*, *157*(3), 53. DOI: 10.1007/s11240-024-02773-x

Verma, N., Kaushal, P., & Sidhu, A. K. (2024). Harnessing biological synthesis: Zinc oxide nanoparticles for plant biotic stress management. *Frontiers in Chemistry*, *12*, 1432469. DOI: 10.3389/fchem.2024.1432469

Vodyashkin, A., Kezimana, P., Vetcher, A., & Stanishevskiy, Y. (2022). Biopolymeric nanoparticles–multifunctional materials of the future. *Polymers*, *14*(11), 2287. DOI: 10.3390/polym14112287

Walsh, J. A., & Jenner, C. E. (2002). Turnip mosaic virus and the quest for durable resistance. *Molecular Plant Pathology*, *3*(5), 289–300. DOI: 10.1046/j.1364-3703.2002.00132.x

Wang, J., Drelich, A. J., Hopkins, C. M., Mecozzi, S., Li, L., Kwon, G., & Hong, S. (2022). Gold nanoparticles in virus detection: Recent advances and potential considerations for SARS-CoV-2 testing development. *Wiley Interdisciplinary Reviews. Nanomedicine and Nanobiotechnology*, *14*(1), e1754. DOI: 10.1002/wnan.1754

Wang, P., Lombi, E., Zhao, F.-J., & Kopittke, P. M. (2016). Nanotechnology: A new opportunity in plant sciences. *Trends in Plant Science*, *21*(8), 699–712. DOI: 10.1016/j.tplants.2016.04.005

Warghane, A., Saini, R., Shri, M., Andankar, I., Ghosh, D. K., & Chopade, B. A. (2024). Application of nanoparticles for management of plant viral pathogen: Current status and future prospects. *Virology*, *592*, 109998. DOI: 10.1016/j.virol.2024.109998

Whitfield, A. E., Ullman, D. E., & German, T. L. (2005). Tospovirus-thrips interactions. *Annual Review of Phytopathology*, *43*, 459–489. DOI: 10.1146/annurev.phyto.43.040204.140017

Xiang, D., Chen, Q., Pang, L., & Zheng, C. (2011). Inhibitory effects of silver nanoparticles on H1N1 influenza A virus in vitro. *Journal of Virological Methods*, *178*(1), 137–142. DOI: 10.1016/j.jviromet.2011.09.003

Yadav, N., Bora, S., Devi, B., Upadhyay, C., & Singh, P. (2024). Nanoparticle-mediated defense priming: A review of strategies for enhancing plant resilience against biotic and abiotic stresses. *Plant Physiology and Biochemistry*, *213*, 108796. DOI: 10.1016/j.plaphy.2024.108796

Yafout, M., Ousaid, A., Khayati, Y., & El Otmani, I. S. (2021). Gold nanoparticles as a drug delivery system for standard chemotherapeutics: A new lead for targeted pharmacological cancer treatments. *Scientific African*, *11*, e00685. DOI: 10.1016/j.sciaf.2020.e00685

Younas, A., Yousaf, Z., Rashid, M., Riaz, N., Fiaz, S., Aftab, A., & Haung, S. (2020). Nanotechnology and plant disease diagnosis and management. In Javad, S. (Ed.), *Nanoagronomy* (pp. 101–123). Springer International Publishing., DOI: 10.1007/978-3-030-41275-3_7

Yu, J., Kuwentrai, C., Huang, J.-D., & Xu, C. (2021). Carbon-based nanomaterials for viral infection management. *Biomicrofluidics*, *15*(1), 011501. DOI: 10.1063/5.0032427

Zarrabi, S., Rangel, C., Martínez-Campos, E., Delgado-Martín, J., Arpanaei, A., Shams-bakhsh, M., & Velasco, L. (2024). Topical application of carbon dots and mesoporous silica nanoparticle-derived dsRNA-nanocomposites for the control of beet curly top virus and turnip mosaic virus. https://doi.org/ DOI: 10.1101/2024.12.29.628607

Zhang, C., Shi, C., Chen, D., & Wu, J. (2018). Rice ragged stunt virus propagation and infection on rice plants. *Bio-Protocol*, *8*(20), e3060. DOI: 10.21769/BioProtoc.3060

Zhang, H., Fan, G., Leng, D., Liu, S., & Cai, L. (2025). Graphene oxide nanosheets for delivery of rnai and plant immune stimulation for sustained protection against plant viruses. *Journal of Agricultural and Food Chemistry*, *73*(18), 11110–11120. DOI: 10.1021/acs.jafc.4c12485

Zhang, Y., Nayak, T. R., Hong, H., & Cai, W. (2012). Graphene: A versatile nanoplatform for biomedical applications. *Nanoscale*, *4*(13), 3833. DOI: 10.1039/c2nr31040f

Chapter 8
Nanobiosensors for Early and Accurate Detection of Plant Viruses

Mrinmoy Dey
JIS University, India

Sambridhi Bhowmik
JIS University, India

Misme Panja
JIS University, India

Avik Acharya Chowdhury
JIS University, India

Arup Kumar Ghosh
https://orcid.org/0009-0001-9810-3328
Dinabandhu Andrews College, India

Suparna Ghosh
https://orcid.org/0000-0002-9200-1291
JIS University, India

ABSTRACT

Plant viruses significantly constrain global agriculture by reducing crop yield and quality, with no direct chemical control options available. Conventional detection methods, including ELISA, RT-PCR, and next-generation sequencing, are limited by low sensitivity, infrastructure requirements, and inability to detect novel viral strains. Nanobiosensors integrating nanomaterials such as gold nanoparticles, graphene, carbon nanotubes, and quantum dots with antibodies, aptamers, nucleic acid probes, or CRISPR systems offer rapid, ultra-sensitive, and portable diagnostics. These platforms enable early detection, multiplexing, and field deployment, and their integration with smartphone-based readouts and IoT applications promises real-time surveillance and precision agriculture. By bridging laboratory assays and field-ready tools, nanobiosensors hold transformative potential for sustainable crop protection and food security.

1. INTRODUCTION

Plant viruses are among the most significant biotic constraints to global agriculture, causing substantial yield losses and threatening food security across diverse cropping systems. Unlike other pathogens, plant viruses cannot be directly controlled with chemical treatments, and infections often result in chronic or systemic diseases that compromise both the quantity and quality of produce. Economically important

DOI: 10.4018/979-8-3373-4913-8.ch008

crops, including tomato, potato, cassava, grapevine, cucumber, and banana, are routinely impacted by viruses such as Tobacco mosaic virus (TMV), Cucumber mosaic virus (CMV), Potato virus Y (PVY), Tomato yellow leaf curl virus (TYLCV), Tomato brown rugose fruit virus (ToBRFV), and Banana bunchy top virus (BBTV) (Hull, 2013). Global crop losses due to viral infections are estimated to range from 10–15% annually, with localized epidemics causing devastating impacts on food security and farmer livelihoods (Hull, 2013; Boonham et al., 2014).

Early and accurate detection of plant viruses is essential to implement timely management strategies, including the use of resistant cultivars, vector control, and quarantine measures. However, the diagnosis of viral infections in plants remains challenging due to several factors. First, plant viruses exhibit high genetic diversity and rapid evolution, leading to the emergence of novel strains that can evade conventional detection methods. Second, viral infections often produce non-specific symptoms such as chlorosis, leaf curling, mosaic patterns, stunted growth, or fruit deformation, which can be easily confused with abiotic stresses like nutrient deficiencies, drought, or herbicide damage (Hull, 2013; Boonham et al., 2014). Third, low viral titres during early infection stages can result in false negatives with traditional serological or molecular assays, delaying critical interventions. These factors collectively underscore the need for sensitive, rapid, and field-deployable diagnostic technologies.

Traditional detection approaches such as enzyme-linked immunosorbent assay (ELISA), reverse transcription polymerase chain reaction (RT-PCR), and lateral flow devices have been widely employed for decades. ELISA provides high-throughput screening capabilities and cost-effectiveness for known viruses like TMV, CMV, and PVY but suffers from limited sensitivity and the inability to detect unknown or highly divergent viral strains (Mumford et al., 2006). Molecular methods, including RT-PCR and quantitative PCR (qPCR), offer higher sensitivity and sequence-specific detection, enabling identification of RNA viruses such as PVY, CMV, and TYLCV (Hull, 2013; Boonham et al., 2014). Nevertheless, PCR-based assays require sophisticated laboratory infrastructure, precise primer design, and trained personnel, which limits their applicability in resource-limited or field-based settings. Next-generation sequencing (NGS) has revolutionized plant virology by enabling unbiased, comprehensive detection of known and novel viruses. Virome studies have uncovered previously unknown viruses in crops such as cassava (Cassava mosaic virus, Cassava brown streak virus), tomato (Tomato mottle virus, ToBRFV), and grapevine (Grapevine leafroll-associated virus GLRaV-1 and GLRaV-3), highlighting the complexity and diversity of viral populations in agroecosystems. Despite its transformative potential, NGS remains costly and requires advanced bioinformatics expertise, limiting its widespread deployment in routine diagnostics.

In response to these challenges, nanotechnology-driven biosensors, or nano-biosensors, have emerged as next-generation diagnostic tools for plant virus detection. By integrating specific bioreceptors, advanced nanomaterials, and sensitive transducers, nano-biosensors offer ultra-high sensitivity, rapid response times, multiplexing capability, and portability, bridging the gap between laboratory diagnostics and field surveillance. Gold nanoparticles, graphene, carbon nanotubes, and quantum dots enhance signal amplification and facilitate real-time detection of plant viruses, including TMV, CMV, TYLCV, and PVY, even at very low titres. Emerging CRISPR-based platforms further allow sequence-specific recognition of RNA viruses such as ToBRFV and TYLCV in portable formats, enabling in-field application with minimal equipment. (Virk et al., 2024, Carrillo-Lopez et al., 2024, Villamor et al., 2019)

The integration of nano-biosensors with smartphone-based readouts, paper-based platforms, and Internet of Things (IoT) systems has the potential to revolutionize precision agriculture by enabling real-time monitoring of viral outbreaks and informed decision-making. Such approaches not only facilitate early detection and containment but also provide actionable data for predicting disease dynamics, reducing crop losses, and enhancing food security globally (Saffa et al., 2025).

In summary, plant viruses continue to pose severe threats to agriculture due to their diversity, rapid evolution, and complex symptomology. Conventional diagnostic methods, while foundational, are limited in sensitivity, speed, and field applicability. Nano-biosensors, empowered by advanced nanomaterials and biorecognition strategies, offer a promising pathway toward rapid, sensitive, and field-deployable virus detection. By enabling early intervention and integrated disease management, these next-generation diagnostics are poised to transform plant virology and contribute significantly to sustainable and climate-resilient agriculture (Narware et al., 2025; Rajeshkumar et al., 2025).

2. OVERVIEW OF PLANT VIRAL DISEASES

Plant viruses represent one of the most significant and persistent challenges to global food production, natural ecosystems, and rural livelihoods. Unlike bacterial or fungal pathogens, viruses are obligate intracellular parasites with minimal genomes that encode only a few proteins, forcing them to hijack the host's cellular machinery for replication. Because they lack their own metabolic enzymes and polymerases, they cannot be treated with agrochemicals or antibiotics, making prevention and management especially difficult. Viruses affect a wide variety of crops ranging from staple cereals such as rice, maize, and wheat to horticultural crops including tomatoes, cucurbits, and grapes, as well as perennial tree crops like citrus, apple, and banana. Their high mutation rates, frequent recombination, and capacity for mixed infections result in extraordinary genetic diversity and continual emergence of new strains. This review provides an overview of the classification and diversity of plant viruses, examines their economic and ecological impacts, and highlights the challenges in their early diagnosis (Jones, 2021, Candresse et al., 2000).

2.1. Classification and Diversity of Plant Viruses

Plant viruses are classified by the International Committee on Taxonomy of Viruses (ICTV) based on genome type, capsid morphology, and replication strategy. Most plant viruses are RNA viruses, which account for nearly 70% of described species. These include positive-sense single-stranded RNA viruses such as the Potyviridae, Bromoviridae, Secoviridae, and Tombusviridae, as well as negative-sense RNA viruses like the Tospoviridae. Double-stranded RNA viruses are less common but include the Partitiviridae and Reoviridae. DNA viruses are also agriculturally important, especially the single-stranded DNA Geminiviridae and Nanoviridae, while double-stranded DNA plant viruses are rare and restricted to the Caulimoviridae (Gong et al., 2025).

The family Potyviridae is the largest and most economically important group of plant viruses, with over 200 species. Members such as Potato virus Y (PVY) and Papaya ringspot virus (PRSV) cause devastating losses in solanaceous and cucurbit crops. The Bromoviridae includes Cucumber mosaic virus (CMV), notable for infecting more than 1,200 plant species across 100 families, making it one of the most promiscuous viruses known. The Tospoviridae family is represented by Tomato spotted wilt virus (TSWV), which infects more than 1,000 plant species and is transmitted by thrips, ranking among the top ten most economically damaging plant viruses worldwide (Mehetre et al., 2021) (Table 1).

Table 1. Major plant viruses, their biological features, transmission modes, and global agricultural impact

Virus Name	Virus Family	Genome Type	Host Crops	Symptoms	Vector / Transmission	Geographic Distribution	Global Impact / Yield Loss	References
Tobacco mosaic virus (TMV)	Virgaviridae	+ssRNA	Tobacco, Tomato, Pepper, Cucumber	Mosaic patterns, leaf curling, stunting	Mechanical, sap	Worldwide	Up to 30% yield loss in tobacco	Hull, 2013; Scholthof, 2004
Cucumber mosaic virus (CMV)	Bromoviridae	+ssRNA	Cucurbits, Tomato, Pepper, Lettuce	Chlorosis, leaf deformation, stunting	Aphids	Worldwide	Moderate to severe losses depending on crop	Roossinck, 2001; Jones, 2016
Potato virus Y (PVY)	Potyviridae	+ssRNA	Potato, Tomato, Pepper	Mosaic, necrosis, stunting	Aphids (non-persistent)	Worldwide	Up to 15% yield reduction	Karasev & Gray, 2013
Tomato yellow leaf curl virus (TYLCV)	Geminiviridae	ssDNA	Tomato, Pepper	Leaf curling, chlorosis	Whiteflies (*Bemisia tabaci*)	Mediterranean, Asia, Africa, Americas	Severe outbreaks causing near-total crop loss	Czosnek & Laterrot, 1997
Tomato spotted wilt virus (TSWV)	Tospoviridae	-ssRNA	Tomato, Pepper, Lettuce, Peanut	Necrotic ringspots, stunting	Thrips (*Frankliniella spp.*)	Worldwide	Yield losses up to 80% in tomatoes	Pappu et al., 2009
Potato leafroll virus (PLRV)	Solemoviridae	+ssRNA	Potato	Leaf rolling, chlorosis	Aphids	Worldwide	20 million tons annually	Olmedo-Velarde et al., 2023
Southern rice black-streaked dwarf virus (SRBSDV)	Reoviridae	dsRNA	Rice	Stunting, dark green leaves, root degradation	Planthoppers (*Sogatella furcifera*)	China, Vietnam, Southeast Asia	Significant yield reduction	Zhou et al., 2013
Cassava mosaic virus (ACMV, EACMV, ICMV)	Geminiviridae	ssDNA	Cassava	Leaf distortion, chlorosis	Whiteflies (*Bemisia tabaci*)	Sub-Saharan Africa	Major impact on food security	Legg et al., 2011
Rice tungro virus (RTV)	Reoviridae / Caulimoviridae	dsRNA / dsDNA	Rice	Yellowing, stunting, reduced panicle size	Leafhoppers (*Nephotettix spp.*)	Southeast Asia	Severe epidemics causing >50% yield loss	Hull, 2013
Banana bunchy top virus (BBTV)	Nanoviridae	ssDNA	Banana	Stunted growth, leaf bunching	Aphids (*Pentalonia nigronervosa*)	Southeast Asia, Africa, Oceania	Major threat to banana plantations	Qazi, 2016
Grapevine leafroll-associated virus (GLRaV-1, -3)	Closteroviridae	+ssRNA	Grapevine	Leaf reddening, delayed ripening, stunting	Mealybugs (*Planococcus spp.*)	Worldwide (vineyards)	Significant economic losses in wine and table grapes	Alabi et al., 2011

continued on following page

Table 1. Continued

Virus Name	Virus Family	Genome Type	Host Crops	Symptoms	Vector / Transmission	Geographic Distribution	Global Impact / Yield Loss	References
Citrus tristeza virus (CTV)	Closteroviridae	+ssRNA	Citrus	Stem pitting, tree decline	Aphids (*Toxoptera citricida*)	Worldwide	Severe outbreaks in citrus orchards	Moreno et al., 2008
Peanut mottle virus (PMV)	Comoviridae	+ssRNA	Peanut	Mosaic, leaf distortion	Mechanical, seeds	Americas, Africa	Moderate economic losses	Ashley & Rush, 1980
Soybean mosaic virus (SMV)	Potyviridae	+ssRNA	Soybean	Mosaic, leaf distortion, stunting	Aphids	Worldwide	Yield losses up to 50%	Pandey et al., 2025

Among DNA viruses, the Geminiviridae family includes Tomato yellow leaf curl virus (TYLCV), Cotton leaf curl virus (CLCuV), and Maize streak virus (MSV). These viruses are transmitted primarily by whiteflies or leafhoppers and are particularly destructive in tropical and subtropical regions. The Caulimoviridae, including Cauliflower mosaic virus (CaMV) and Banana streak virus (BSV), possess double-stranded DNA genomes but replicate through an RNA intermediate, analogous to retroviruses. Nanoviridae members such as Banana bunchy top virus (BBTV) cause severe stunting and characteristic "bunchy top" symptoms in banana, threatening food security in Africa and Asia. (Mahlein, 2016)

Other notable viruses include Rice tungro bacilliform virus (RTBV) and Rice tungro spherical virus (RTSV), which act in synergistic combination to cause rice tungro disease in Southeast Asia; Grapevine leafroll-associated viruses (GLRaVs) from the Closteroviridae, which reduce grape yield and wine quality; and Citrus tristeza virus (CTV), which has killed more than 100 million citrus trees globally. Taken together, this diversity highlights the taxonomic breadth and adaptive potential of plant viruses, which continue to emerge in new crops and regions through global trade, vector expansion, and climate change (Candresse et al., 2000).

2.2. Economic and Ecological Impacts

The economic burden of plant viruses is staggering, with annual global losses estimated in the tens of billions of dollars. Losses arise not only from reduced yield but also from downgraded quality, delayed harvests, increased vector management costs, and restrictions in international trade. For example, the TYLCV epidemic in the Dominican Republic during the early 1990s reduced tomato yields by up to 75%, resulting in more than USD 10 million in losses within two years. Similarly, CLCuV caused nearly 30% yield losses in Pakistan between 1992 and 1997, with cumulative economic damages exceeding USD 5 billion. In Southeast Asia, rice tungro disease results in estimated annual losses of USD 1.5 billion, threatening the food security of millions of people. In perennial crops, long-term impacts are severe: Grapevine leafroll disease causes cumulative losses of USD 25,000–40,000 per hectare over 25 years, while Apple mosaic virus can reduce yields by two-thirds over a 14-year orchard lifespan (Narware et al., 2025).

Beyond direct financial costs, plant viral diseases have profound ecological impacts. Viruses alter plant physiology by interfering with photosynthesis, nutrient allocation, and hormone signalling, leading to reduced vigour and productivity. In natural ecosystems, viral infections can shift competitive balances among plant species, thereby influencing biodiversity and ecosystem stability. For example, widespread cassava viral epidemics in East Africa not only reduce cassava yields but also disrupt agroecosystems and rural livelihoods dependent on cassava as a staple food (Baldodiya et al., 2022). Climate change and globalization exacerbate these effects by expanding the geographical range of vectors such as whiteflies, aphids, and thrips, and by increasing opportunities for virus introduction through trade in seeds, cuttings, and ornamentals. The ecological disruption caused by plant viruses, therefore, extends far beyond agricultural losses, with cascading consequences for food security and environmental resilience. (Gong et al., 2025)

2.3. Challenges in Early Diagnosis

Early detection of plant viruses is vital for effective management, since there are no curative treatments once a plant is infected. Timely diagnosis helps prevent large-scale crop losses, supports effective quarantine measures, and ensures global food security (Figure 1). However, diagnosis remains challenging due to the enormous diversity of viruses, frequent symptom overlaps with abiotic stresses, and technical and economic limitations of detection methods. Plant viruses collectively threaten nearly all staple and cash crops, including rice, wheat, maize, potato, cassava, banana, tomato, soybean, citrus, and grapevine, which are critical for human nutrition and trade. The global burden is immense estimated to cause annual crop losses exceeding tens of billions of dollars. For instance, cassava mosaic disease (CMD) and cassava brown streak disease (CBSD) together reduce cassava production by up to 50% in East Africa, threatening the food security of over 800 million people who rely on cassava as a staple food. In Asia, Rice Tungro disease (caused by Rice Tungro bacilliform virus and Rice Tungro spherical virus) leads to major yield declines, while in Latin America, papaya ringspot virus (PRSV) continues to damage papaya industries. (Jones, 2021)

Figure 1. Schematic representation of the fundamental components of a nanobiosensor for plant virus detection. A biosensor typically comprises a bioreceptor (antibodies, nucleic acids, aptamers, or enzymes) that specifically interacts with the viral target, a transducer that converts the biorecognition event into a measurable signal, and a detector that processes and displays the output. This integration enables sensitive, selective, and early detection of plant viruses

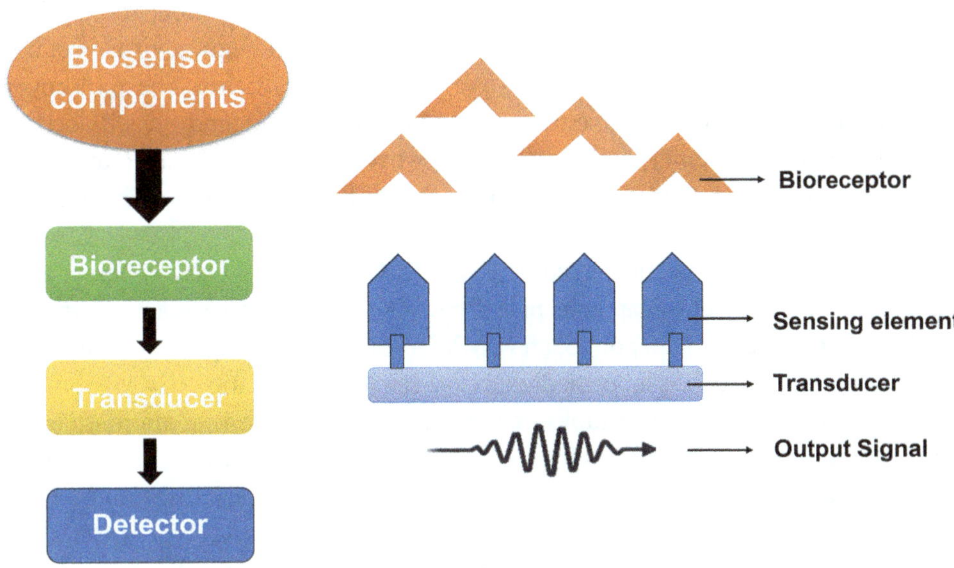

Traditional approaches such as visual symptom recognition are fast and inexpensive but unreliable, as symptoms like yellowing, mosaic patterns, or leaf curling may also result from nutrient deficiencies, drought, or herbicide damage. For example, in maize, mosaic or streaking can indicate maize dwarf mosaic virus (MDMV) or sugarcane mosaic virus (SCMV), but similar patterns may appear under nutrient stress. In bananas, banana bunchy top virus (BBTV) causes leaf bunching and stunted growth but may be mistaken for drought stress. Similarly, streak symptoms in wheat could be caused by wheat streak mosaic virus (WSMV) but also resemble herbicide injury. Serological assays, particularly enzyme-linked immunosorbent assay (ELISA), have been widely used since the 1980s and allow rapid screening for known viruses using virus-specific antibodies. ELISA remains widely applied to detect potato virus Y (PVY), cucumber mosaic virus (CMV), and citrus tristeza virus (CTV), all of which are globally distributed and economically devastating. However, ELISA lacks sensitivity for low-titer infections and cannot detect divergent or novel viral strains (Mahlein, 2016).

Molecular diagnostics provide higher sensitivity and specificity. Polymerase chain reaction (PCR) and quantitative PCR (qPCR) are commonly used for detecting viruses such as tomato yellow leaf curl virus (TYLCV) in tomato, soybean mosaic virus (SMV) in soybean, and barley yellow dwarf virus (BYDV) in cereals. Reverse transcription PCR (RT-PCR) is essential for RNA viruses like CMV, PVY, and tobacco mosaic virus (TMV). Multiplex PCR systems now allow simultaneous detection of multiple viruses, which is especially important in mixed infections common in crops like cassava, rice, and maize. Yet, primer design remains challenging for highly variable RNA viruses, and molecular methods can fail to detect novel strains (Gong et al., 2025).

Next-generation sequencing (NGS) has transformed plant virology by allowing comprehensive and unbiased identification of known and novel viruses. In cassava, NGS identified cassava polerovirus and cassava torrado-like virus, which were previously unknown contributors to disease complexes. Grapevine virome studies revealed grapevine Pinot gris virus (GPGV), grapevine virus T (GVT), and grapevine red blotch virus (GRBV), all linked to yield reduction and poor fruit quality. In tomato, NGS uncovered emerging pathogens like tomato brown rugose fruit virus (ToBRFV) and tomato mottle mosaic virus (ToMMV), both of which are now serious threats to global tomato production. In addition, NGS has identified peanut yellow spot virus (PYSV) in peanut, cotton leaf curl begomoviruses in cotton across Asia and Africa, and yam mosaic virus (YMV) in West African yam fields. These discoveries underscore that plant viromes are far more diverse than previously understood, and surveillance with NGS is crucial to anticipate and manage emerging outbreaks. However, the cost, infrastructure needs, and bioinformatics expertise required limit its widespread application, particularly in low-resource agricultural regions (Mehetre et al., 2021).

Emerging CRISPR-based diagnostics, such as CRISPR-Cas12 and Cas13 systems, are advancing rapidly as highly specific and field-adaptable platforms. They have shown potential in detecting key viruses like TYLCV in tomato, PVY in potato, and SMV in soybean with high sensitivity. Similarly, isothermal amplification methods such as loop-mediated isothermal amplification (LAMP) and recombinase polymerase amplification (RPA) provide rapid, low-cost, and equipment-free alternatives to PCR. For example, LAMP assays have been developed for PRSV in papaya and BBTV in banana plantations, while RPA kits are being tested for cassava brown streak virus in East African cassava fields (Jones, 2021).

Non-invasive technologies are also emerging as powerful complements. Hyperspectral imaging and drone-based remote sensing have successfully identified TYLCV in tomato and rice tungro disease in rice before visual symptoms become apparent. In Africa, drone imaging is being explored to map CMD and CBSD outbreaks in cassava fields. These techniques are especially useful for monitoring large-scale epidemics but remain expensive and computationally intensive for smallholder farmers.

Despite these advances, major challenges persist. High genetic variability requires constant updating of diagnostic assays, as in the case of rapidly evolving begomoviruses spread by whiteflies in Asia and Africa. Low virus titers during early infection stages complicate detection, while mixed infections, such as co-infections of maize lethal necrosis (MLN) caused by maize chlorotic mottle virus (MCMV) and potyviruses like SCMV, lead to devastating epidemics that are hard to diagnose with single-target assays. Furthermore, most cutting-edge methods remain laboratory-based, inaccessible to smallholder farmers in developing nations where plant virus outbreaks are most severe (Candresse et al., 2000).

To overcome these limitations, integrated diagnostic strategies are essential. Combining symptom observation, serological assays, molecular methods, and advanced platforms such as NGS and CRISPR will provide a robust framework for surveillance. On a global scale, harmonized detection networks are crucial to monitor plant virus spread through international trade, particularly for high-value crops like citrus, tomato, grapevine, and banana. The future of plant virus detection lies in bridging laboratory-based high-throughput techniques with affordable, field-ready diagnostic tools, ensuring that farmers worldwide—especially in developing regions—can detect infections early and mitigate losses. (Mehetre et al., 2021)

3. LIMITATIONS OF TRADITIONAL VIRUS DETECTION METHODS

3.1. Conventional Serological Approaches: ELISA and Lateral Flow Assays

Enzyme-linked immunosorbent assay (ELISA) has long been considered a staple in plant virus diagnostics due to its simplicity, affordability, and ability to process large numbers of samples simultaneously. ELISA relies on antibodies that specifically recognize viral coat proteins, enabling the detection of viruses such as Tobacco mosaic virus (TMV), Cucumber mosaic virus (CMV), Potato virus Y (PVY), and Tomato spotted wilt virus (TSWV) in economically important crops (Adams & Clark, 1977; Hull, 2013). Despite its widespread adoption, ELISA suffers from several limitations. Its sensitivity is insufficient for early-stage infections or viruses present in low titres, often resulting in false-negative results. Moreover, the method requires high-quality, virus-specific antibodies, which are not always available for newly emerging or highly divergent viral strains such as Tomato brown rugose fruit virus (ToBRFV) and emerging begomoviruses, which leads to the use of the nanobiosensors which is discussed in the Table 2.

Table 2. Comparison between ELISA, RT-PCR and nanobiosensors

Feature	ELISA	RT-PCR (incl. real-time PCR)	Nanobiosensors
Principle	Antigen-antibody complex formation was visualized by means of enzyme-catalyzed colorimetric or chemiluminescent signal (Lequin, 2005).	PCR reaction followed by detection (conventional qPCR for quantification) (Perestam et al., 2017).	It relies on nanomaterials (AuNPs, carbon-based nanomaterials, quantum dots, etc.) associated with bioreceptors (antibodies, aptamers, and nucleic acids) for the transduction of binding into electrical or optical signals (Ramesh et al., 2022; Malik et al., 2023).
Typical analytical sensitivity	Moderate - Standard LODs in the ng/mL to low pg/mL range for conventional kits, enhanced formats (nano-ELISA) can be used as low as pg/mL (Aydin et al., 2025).	High - can be sensitive down to a few copies of the target nucleic acids (generally ~1-100 copies per reaction), with a very low LOD in terms of target copy (Perestam et al., 2017).	High - several nanobiosensors have laboratory demonstrations of fg/mL or 10^{-18} M limits and single-molecule sensitivity (Harun-Ur-Rashid et al., 2022).
Specificity	High if high-quality antibodies are available. Conversely, one can experience cross-reactivity with low-quality antibodies (Lequin, 2005).	Very high (sequence-specific amplification) when primers and probes are well designed, but with the risk of contamination or off-target amplification if poorly designed (Perestam et al., 2017).	High and tunable (depending on the receptor) aptamers or carefully functionalized surfaces can give high specificity, but nonspecific adsorption is an issue (Ramesh et al., 2022).
Time to result	Hours (2-5 hours for conventional ELISA procedures). Rapid ELISA/ robotic formats have the potential to generate results more quickly (approximately 20-60 minutes) (Aydin et al., 2025).	~1-3 hours from sample to answer (RNA/DNA extraction is additional time), rapid PCR platforms reduce this time (Perestam et al., 2017).	Minutes to less than 1 hour for many point-of-care nanosensors (lab prototypes are often less than 30 minutes), depending on preparation; less than 5 minutes is best (Harun-Ur-Rashid et al., 2022).
Cost per test	Low-to-moderate (cheap kits for most routine assays); infrastructure (plate reader) is necessary (Aydin et al., 2025).	Moderate-to-high (reagents, extraction kits, instruments); the cost per sample is determined by the level of throughput and platform (Perestam et al., 2017).	Caveat: device manufacture is potentially costly up-front; the cost per-test reagents may be low - good for mass-produced low-cost POCT (Malik et al., 2023).

continued on following page

Table 2. Continued

Feature	ELISA	RT-PCR (incl. real-time PCR)	Nanobiosensors
Portability	Mostly laboratory-based automation, such as plate washers and readers, rather than field-based automation, with some rapid lateral flow and ELISA modifications available (Aydin et al., 2025).	Largely laboratory-based, but many portable and near-patient PCR devices, which are battery-operated, are available (Perestam et al., 2017).	High potential for portability and integration with smartphone and handheld devices, which is the major advantage of many nano platforms (Ramesh et al., 2022).
Multiplexing	Low (can be multiplexed in plates/ µplates or using multiple wells; limited by lateral flow) (Aydin et al., 2025).	Fine - multiplex qPCR (multiple probes) is routine, but you're at the limits imposed by the number of fluorescent channels available and primer interactions (Perestam et al., 2017).	High potential -arrays or multi-spot chips can provide huge numbers of analytes, as shown in multiple reviews (Ramesh et al., 2022).
Ease of use	Well-established and standardized workflows; high throughput for multiple samples in labs (Aydin et al., 2025).	Molecular biology workflows and contamination management are needed; moderate throughput (which can be high when automated) (Perestam et al., 2017).	Many have also been designed to be simple for their workflows and rapid to read out - though many are still at the prototype phase that require more optimization (Kim et al., 2024; Ramesh et al., 2022).
Main limitations	Dependence on infrastructure: minimal/equipment; some- targets exhibited lower sensitivity compared to nucleic-acid methods; antibody availability/cross-reactivity (Lequin, 2005).	Need for nucleic acid extraction and temperature control; false negatives due to degraded target; cost/infrastructure (Perestam et al., 2017).	Most are lab proofs-of-concept; issues: surface fouling, stability, standardization, regulatory approval and scale-up (Ramesh et al., 2022; Chaturvedi et al., 2025).

Mixed infections—commonly observed in crops like cassava (Cassava mosaic virus and Cassava brown streak virus) and tomato (TYLCV plus CMV) further complicate ELISA interpretation due to overlapping antigenic signals (Table 3).

Table 3. Traditional diagnostic methods for viruses and field limitations

Virus	Family / Genus	Host / Crop	Transmission	Traditional Detection (ELISA / RT-PCR / LFA, etc.)	Limitations in Field Conditions	Reference
Tobacco mosaic virus (TMV)	Virgaviridae / Tobamovirus	Tobacco, tomato, pepper	Mechanical (sap)	ELISA, RT-PCR	Symptoms mimic abiotic stress; ELISA cross-reactivity; uneven viral distribution in tissues	Dijkstra & de Jager, 1998
Cucumber mosaic virus (CMV)	Bromoviridae / Cucumovirus	Cucurbits, tomato, pepper, legumes	Aphids (non-persistent)	ELISA, RT-PCR, Hybridization	Wide host range (>1200 species) complicates diagnostics; false negatives in mixed infections	Palukaitis, 1992
Tomato yellow leaf curl virus (TYLCV)	Geminiviridae / Begomovirus	Tomato	Whiteflies (Bemisia tabaci)	PCR, qPCR, ELISA, RCA	High genetic variability; early infection undetectable with ELISA; PCR expensive for routine use	Lefeuvre, 2010
Potato virus Y (PVY)	Potyviridae / Potyvirus	Potato, pepper, tobacco	Aphids (non-persistent)	DAS-ELISA, RT-PCR	Serological assays cannot differentiate strains (PVY^O, PVY^N, PVY^NTN); mixed infections frequent	Singh et al., 2021
Rice tungro bacilliform virus (RTBV) & Rice tungro spherical virus (RTSV)	Caulimoviridae / Tungrovirus & Secoviridae	Rice	Leafhoppers (Nephotettix virescens)	PCR, ELISA	Requires detection of both viruses; serology alone unreliable; mixed virus infections	Hull, 2013
Banana bunchy top virus (BBTV)	Nanoviridae / Babuvirus	Banana	Aphids (Pentalonia nigronervosa)	PCR, RCA	Phloem-limited virus → difficult tissue sampling; uneven virus distribution in plants	Kumar et al., 2020
Grapevine leafroll-associated virus (GLRaV)	Closteroviridae / Ampelovirus	Grapevine	Mealybugs, soft scales	ELISA, RT-PCR, qPCR	Complex of >10 strains; ELISA insensitive in dormant vines	Martelli et al., 2012
Citrus tristeza virus (CTV)	Closteroviridae / Closterovirus	Citrus	Aphids (Toxoptera citricida)	ELISA, PCR	Extremely low titer in symptomless trees; frequent mild strain infections	Bar-Joseph & Lee, 1989

continued on following page

Table 3. Continued

Virus	Family / Genus	Host / Crop	Transmission	Traditional Detection (ELISA / RT-PCR / LFA, etc.)	Limitations in Field Conditions	Reference
Cassava mosaic virus complex (ACMV, EACMV, ICMV)	Geminiviridae / Begomovirus	Cassava	Whiteflies	PCR, ELISA	Mixed begomovirus infections common; primers need frequent redesign	Fauquet & Fargette, 1990
Groundnut rosette virus (GRV)	Tombusviridae / Umbravirus	Groundnut	Aphids	ELISA, PCR	Requires helper luteovirus (GRAV); diagnosis incomplete with ELISA alone	Naidu et al., 1999
Lettuce mosaic virus (LMV)	Potyviridae / Potyvirus	Lettuce	Seed-borne, aphids	ELISA, PCR	Virus unevenly distributed in seeds; ELISA sensitivity limited	Dinant et al., 1991
Sugarcane mosaic virus (SCMV)	Potyviridae / Potyvirus	Sugarcane, maize	Aphids	ELISA, RT-PCR	Cross-reactivity with MDMV and SrMV in serological assays	Redinbaugh & Stewart, 2018
Barley yellow dwarf virus (BYDV)	Luteoviridae / Luteovirus	Wheat, barley, oats	Aphids	ELISA, RT-PCR	Multiple serotypes (PAV, MAV, RPV); mixed infections obscure results	Miller & Rasochová, 1997
Pea seed-borne mosaic virus (PSbMV)	Potyviridae / Potyvirus	Pea, faba bean	Seed, aphids	ELISA, PCR	Symptomless seed carriers delay detection; ELISA fails in mixed infections	Johansen, 1994
Rice stripe virus (RSV)	Tenuivirus	Rice	Planthopper (Laodelphax striatellus)	PCR, ELISA	RNA instability hampers storage; insect vector complicates epidemiology	Hibino, 1996
Tomato spotted wilt virus (TSWV)	Tospoviridae / Orthotospovirus	Tomato, pepper, peanut	Thrips	ELISA, LFA, PCR	Cross-reactivity with other tospoviruses; LFA poor sensitivity	Pappu et al., 2009
Papaya ringspot virus (PRSV)	Potyviridae / Potyvirus	Papaya, cucurbits	Aphids (non-persistent)	ELISA, PCR	Serology fails to distinguish PRSV-P vs. PRSV-W strains; early-stage infections undetected	
Plum pox virus (PPV)	Potyviridae / Potyvirus	Peach, plum, apricot	Aphids, grafting	ELISA, PCR, qPCR	Long latency period → virus often undetectable until fruiting	

continued on following page

Table 3. Continued

Virus	Family / Genus	Host / Crop	Transmission	Traditional Detection (ELISA / RT-PCR / LFA, etc.)	Limitations in Field Conditions	Reference
Wheat streak mosaic virus (WSMV)	Potyviridae / Tritimovirus	Wheat	Wheat curl mite (Aceria tosichella)	ELISA, PCR	Uneven viral load across leaves; sampling critical	French & Stenger, 2003
Tomato brown rugose fruit virus (ToBRFV)	Virgaviridae / Tobamovirus	Tomato, pepper	Mechanical, seed-borne	RT-PCR, ELISA	Recently emerged; lack of standardized diagnostic kits	
Cotton leaf curl virus (CLCuV)	Geminiviridae / Begomovirus	Cotton	Whiteflies	PCR, RCA	Complex of several recombinant begomoviruses; ELISA ineffective	Sattar et al., 2013
Okra yellow vein mosaic virus (OYVMV)	Geminiviridae / Begomovirus	Okra	Whiteflies	PCR, ELISA	High recombination among begomoviruses; misdiagnosis common	Singh et al., 2021
Zucchini yellow mosaic virus (ZYMV)	Potyviridae / Potyvirus	Cucurbits	Aphids	ELISA, RT-PCR	Mild strains often overlooked; ELISA poor at low titers	Lecoq & Desbiez, 2012
Beet curly top virus (BCTV)	Geminiviridae / Curtovirus	Sugar beet, tomato	Leafhoppers	PCR, ELISA	Latent infections common; vector-driven outbreaks hard to monitor	Strausbaugh et al., 2008
Peanut bud necrosis virus (PBNV)	Tospoviridae / Orthotospovirus	Peanut, tomato	Thrips	ELISA, PCR	Antigenic similarity to TSWV reduces ELISA accuracy	Ravi et al., 2011
Soybean mosaic virus (SMV)	Potyviridae / Potyvirus	Soybean	Seed, aphids	ELISA, PCR	High strain diversity; resistance genes mask infection	Hajimorad et al., 2018
Apple stem grooving virus (ASGV)	Betaflexiviridae / Capillovirus	Apple, pear	Grafting	ELISA, PCR	Very low titers in dormant trees; molecular tools required	Martelli et al., 2012
Yam mosaic virus (YMV)	Potyviridae / Potyvirus	Yam	Mechanical, aphids	ELISA, RT-PCR	Low titers in tubers; requires RT-PCR for accuracy	Seal & Muller, 2007
Chrysanthemum virus B (CVB)	Betaflexiviridae / Carlavirus	Chrysanthemum	Mechanical	ELISA	Asymptomatic infections frequent; low sensitivity	Horser et al., 2001

Lateral flow assays (LFAs), a portable extension of serological detection, offer rapid, on-site diagnosis without the need for laboratory infrastructure. LFAs have been developed for TMV, PVY, TYLCV, and Banana bunchy top virus (BBTV), providing results within 10–20 minutes (Selvarajan et al., 2020). While convenient and easy to use, LFAs are limited by their semi-quantitative nature, low sensitivity for low-titre infections, and susceptibility to environmental factors such as temperature and pH, which can alter antibody-antigen binding and cause inconsistent results in field conditions (Siddiqua et al., 2022).

3.2. Molecular Diagnostics: RT-PCR and qPCR

Molecular diagnostics such as reverse transcription polymerase chain reaction (RT-PCR) and quantitative PCR (qPCR) provide higher sensitivity and specificity than serological methods. These techniques detect viral RNA or DNA directly and are routinely used for RNA viruses, including PVY, TYLCV, Potato virus X (PVX), Rice tungro bacilliform virus (RTBV), and Tomato mosaic virus (ToMV). RT-PCR has revolutionized quarantine programs and seed certification by enabling the detection of latent infections, even at low titres. qPCR further allows quantification of viral load, which can inform epidemiological studies and management strategies.

However, several challenges limit the practical field application of RT-PCR. Primer design is crucial; the high mutation rates of RNA viruses can render primers ineffective, leading to false negatives, particularly for emerging viruses like ToBRFV, Pepper mild mottle virus (PMMoV), and recombinant Begomoviruses (Yadav & Khurana, 2016). Mixed infections complicate interpretation when multiple amplicons are present, and false positives can arise due to cross-contamination. Additionally, RT-PCR requires laboratory-grade equipment, thermal cyclers, cold-chain maintenance for reagents, and skilled personnel, restricting its use in resource-limited regions or smallholder farms.

3.3. Limitations in High-Throughput and Multiplexed Detection

Conventional methods are generally designed to detect single viruses per assay, which is inefficient in crops prone to multiple infections. Mixed infections of cassava (CMV + CBSV), tomato (TYLCV + CMV + ToMV), and grapevine (Grapevine leafroll-associated virus GLRaV-3 + GLRaV-1) demonstrate the need for multiplexed platforms (Boonham et al., 2014). Multiplex PCR and ELISA panels exist but are limited by overlapping signals, reduced sensitivity, and complex data interpretation. The inability to simultaneously detect multiple viruses in a single field assay reduces the effectiveness of surveillance and early management programs.

3.4. Environmental and Operational Constraints

Traditional diagnostics often perform poorly under field conditions. ELISA and LFA reagents are sensitive to temperature, humidity, and pH fluctuations. Sample preparation can be time-consuming, especially when dealing with inhibitory compounds in plant tissues such as polyphenols and polysaccharides, which interfere with downstream assays (Hull, 2013). These operational limitations result in delayed or inaccurate detection, reducing the window for effective intervention and increasing crop losses.

3.5. Nanobiosensor-Based Next-Generation Diagnostics

The limitations of traditional methods have catalyzed the development of nanomaterial-based biosensors, which offer high sensitivity, rapid response, portability, and the potential for multiplexed detection. Nanomaterials such as gold nanoparticles (AuNPs), carbon nanotubes (CNTs), graphene, and quantum dots (QDs) provide enhanced surface-to-volume ratios, enabling more efficient biorecognition and signal amplification. Electrochemical nanobiosensors functionalized with virus-specific antibodies or aptamers have been developed for TMV, PVY, CMV, and TYLCV, detecting viral proteins at femtomolar concentrations and providing results within minutes (Zhang et al., 2019). Optical nanobiosensors, including QD-based fluorescence platforms, allow multiplexed detection of several viruses simultaneously. For example, differently colored QDs can be conjugated to probes for CMV, TYLCV, and TMV, enabling simultaneous detection in tomato and cucumber. CRISPR-based biosensors coupled with nanomaterials provide sequence-specific, highly sensitive detection for RNA viruses like PVY, TYLCV, and ToBRFV. Paper-based or microfluidic CRISPR-AuNP systems can be deployed in the field without sophisticated equipment, delivering rapid, accurate results for early intervention. Graphene field-effect transistors (GFETs) have been applied to detect CMV and TMV in plant sap with ultra-low detection limits, demonstrating real-time, label-free viral monitoring (Zhang et al., 2019). Nanobiosensors also enable integration with IoT and smartphone platforms, allowing real-time data collection, remote monitoring, and predictive modeling of virus spread. This capability is particularly valuable for managing outbreaks of highly contagious plant viruses such as BBTV in banana plantations, TYLCV in tomato crops, and GLRaV in vineyards.

4. INTRODUCTION TO NANO-BIOSENSORS

4.1. Definition and Core Components

Nano-biosensors are sophisticated analytical devices that integrate biological recognition elements with nanomaterials and transducers to achieve highly sensitive, specific, and rapid detection of plant viruses. The primary purpose of these devices is to detect viral particles, nucleic acids, or proteins at low concentrations, facilitating early diagnosis and timely disease management. The three core components of a nano-biosensor are the bioreceptor, transducer, and nanomaterial. The bioreceptor is responsible for specific recognition of the target virus and may include antibodies, aptamers, nucleic acid probes, or CRISPR-Cas components (Park & Jackman, 2025). For instance, antibody-functionalized sensors have been employed for Tobacco mosaic virus (TMV), Cucumber mosaic virus (CMV), and Potato virus X (PVX), whereas aptamer-based and CRISPR-Cas sensors are suitable for detecting rapidly evolving viruses such as Tomato yellow leaf curl virus (TYLCV) and Tomato brown rugose fruit virus (ToBRFV) (Besati et al., 2025; Pramanik et al., 2021).

The transducer converts the recognition event into a measurable signal, which may be electrochemical, optical, piezoelectric, or thermal. Electrochemical transducers, including amperometric, voltammetric, and impedimetric sensors, have been widely applied to detect viral particles in plant sap due to their high sensitivity, simplicity, and compatibility with portable devices. For example, gold nanoparticle (AuNP)-enhanced electrochemical sensors can detect PVX and TMV at femtomolar concentrations, providing early detection superior to ELISA (Kulabhusan et al., 2022, Chaturvedi et al., 2025). Optical transducers, including fluorescence, surface plasmon resonance (SPR), and colorimetric systems, are useful for real-time and multiplexed detection. Quantum dot (QD)-based fluorescence platforms have been demonstrated for simultaneous detection of CMV, TYLCV, and TMV in tomato, cucumber, and pepper crops (Xi et al., 2014). Piezoelectric sensors, such as quartz crystal microbalance (QCM) devices, detect mass changes caused by viral binding and enable label-free monitoring of TMV in real time.

Nanomaterials play a pivotal role in enhancing the sensitivity and performance of these biosensors. Graphene, CNTs, AuNPs, and magnetic nanoparticles provide high surface areas for bioreceptor immobilization, facilitate electron transfer in electrochemical systems, and amplify optical signals. For example, graphene field-effect transistors (GFETs) functionalized with CMV-specific probes enable label-free detection in tomato sap, even at very low viral titres (Zhang et al., 2019). Magnetic nanoparticles have been used to capture and concentrate viruses from plant tissues, enhancing detection of low-titre infections like Banana bunchy top virus (BBTV). CRISPR-Cas12/Cas13 systems integrated with nanomaterials allow portable, sequence-specific detection of RNA viruses such as PVY, TYLCV, and ToBRFV in field conditions without the need for laboratory equipment. (Majumder & Johari 2018).

4.2. Types of Nano-Biosensors

Electrochemical Nano-Biosensors utilize electron transfer changes upon virus binding to generate measurable current or voltage signals. They are highly sensitive and suitable for rapid detection in sap or leaf extracts. AuNP-modified electrodes have been used to detect TMV and PVY, demonstrating superior sensitivity over ELISA. (Majumder & Johari 2018).

Optical Nano-Biosensors rely on changes in light properties upon virus recognition. Fluorescence-based QD sensors allow multiplex detection of multiple viruses, such as CMV, TYLCV, and TMV, while SPR and colorimetric assays provide rapid, visual detection of viruses in tomato, cucumber, and pepper plants (Wang, 2017). Piezoelectric and QCM Sensors detect mass changes upon virus attachment to a resonating surface. TMV detection has been demonstrated using QCM platforms, providing label-free and real-time monitoring.

Emerging Platforms include thermal, magnetic, and hybrid sensors. Magnetic nanoparticle-based platforms allow pre-concentration of viral particles from complex plant matrices, enhancing detection limits for viruses such as BBTV. Integration with microfluidics and paper-based substrates further enables portable, low-cost, and user-friendly field diagnostics.

4.3. Advantages over Conventional Methods

Nano-biosensors offer several advantages over conventional serological and molecular methods. They provide ultra-high sensitivity, often detecting viruses at femtomolar to attomolar concentrations, which allows early-stage detection. The rapid response of nano-biosensors, often within minutes to an hour, facilitates timely management decisions. They are field-deployable, with some platforms compatible with smartphones, paper-based devices, or portable electrochemical readers, bridging the gap between laboratory diagnostics and on-site agricultural surveillance. Multiplexing capability allows simultaneous detection of multiple viruses in crops prone to mixed infections, such as cassava (Cassava mosaic virus and Cassava brown streak virus), tomato (TYLCV, CMV, ToMV), and grapevine (Grapevine leafroll-associated virus GLRaV-1 and GLRaV-3) (Boonham et al., 2014). Nano-biosensors also allow label-free and real-time detection, which is not feasible with traditional ELISA or lateral flow assays.

5. NANOMATERIALS IN BIOSENSOR DESIGN

The evolution of biosensors has been closely tied to advances in materials science, particularly with the emergence of nanotechnology. Traditional biosensors, though effective in certain contexts, often faced limitations such as low sensitivity, limited selectivity, slow response time, and poor stability under real-world conditions. The integration of nanomaterials into biosensor design has dramatically transformed this field by enabling enhanced signal amplification, improved immobilization of biorecognition elements, and novel modes of transduction. Nanostructures such as gold nanoparticles, quantum dots, carbon nanotubes, and graphene have emerged as the most extensively investigated platforms because of their unique physicochemical properties (Ramesh et al., 2022). Their nanoscale dimensions, large surface areas, and tunable electronic and optical behaviors provide significant improvements in biosensor performance, while functionalization strategies and surface chemistry play a pivotal role in tailoring these nanomaterials for specific diagnostic applications (Figure 2).

Figure 2. Schematic overview of nanobiosensor-based strategies for early and accurate detection of plant viruses. Viral particles from an infected plant are recognized by biorecognition elements such as antibodies, DNA probes, quantum dots, or nanotubes, which are often conjugated with nanoparticles (NPs) to enhance sensitivity and specificity. The generated interaction is subsequently transduced and detected through different sensing platforms, including piezoelectric, electrochemical, and surface plasmon resonance (SPR) methods, enabling rapid and reliable plant virus diagnostics

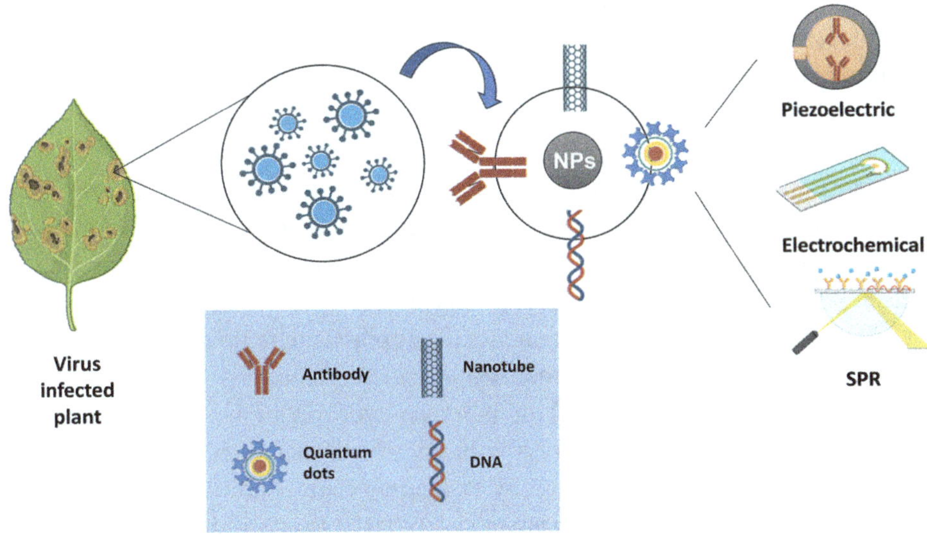

Gold nanoparticles (AuNPs) are among the earliest nanomaterials to be employed in biosensor design, owing to their ease of synthesis, biocompatibility, and unique surface plasmon resonance properties. Their ability to scatter and absorb light strongly at specific wavelengths enables colorimetric detection that can be observed by the naked eye, as in the case of lateral flow assays. This property is exploited in low-cost field diagnostics, where aggregation of AuNPs leads to a visible color shift that indicates the presence of a target analyte (Majumder & Johari, 2018). In electrochemical biosensors, AuNPs act as conductive nanointerfaces that facilitate efficient electron transfer between redox-active biomolecules and electrodes, thus improving signal strength and sensitivity. For example, AuNP-modified electrodes with immobilized glucose oxidase significantly increase current responses in glucose detection (Majumder & Johari, 2018). Additionally, the chemistry of gold allows stable thiol–gold interactions, enabling covalent immobilization of DNA probes, aptamers, or antibodies with high stability. AuNPs can also be engineered in different shapes, such as nanorods, nanostars, and nanocages, which tune their plasmon resonance into the near-infrared region, thereby extending their use in biosensors for biomedical imaging and deep-tissue diagnostics (Majumder & Johari, 2018).

Quantum dots (QDs) represent another important nanomaterial used in biosensors. These semiconductor nanocrystals, typically 2–10 nanometers in diameter, exhibit quantum confinement effects that give rise to size-tunable fluorescence with sharp emission peaks, high brightness, and exceptional photostability. Unlike organic dyes, QDs resist photobleaching, which makes them ideal for continuous monitoring applications (Mishra et al., 2021). Their broad absorption spectra and narrow emission bands allow simultaneous excitation of multiple QDs with a single light source, enabling multiplexed detection of different analytes. QDs are particularly powerful in fluorescence resonance energy transfer (FRET)-based biosensors, where energy transfer between a QD donor and an acceptor molecule occurs when they are in close proximity, providing a sensitive readout for DNA hybridization or protein–protein interactions (Mishra et al., 2021). Despite their advantages, traditional QDs made of cadmium or lead compounds present toxicity concerns, limiting their biomedical applications. To overcome this, alternative nanomaterials such as carbon quantum dots, graphene quantum dots, and perovskite QDs are being developed, which provide similar optical properties with improved biocompatibility (Mishra et al., 2021).

Carbon nanotubes (CNTs), available as single-walled (SWCNTs) and multi-walled (MWCNTs) structures, have revolutionized electrochemical biosensors. Their cylindrical sp^2-hybridized carbon structure offers extraordinary electrical conductivity, large surface area, and mechanical strength. CNTs are particularly effective at enabling direct electron transfer between enzymes and electrodes, which eliminates the need for mediators and significantly improves the sensitivity of biosensors (Katz & Willner, 2004). CNT-modified electrodes have been used in glucose sensors, immunosensors, and DNA sensors, achieving femtomolar to picomolar detection levels. Their one-dimensional conductive properties also make them ideal for field-effect transistor (FET)-based biosensors. When functionalized with DNA probes or antibodies, CNT-FETs exhibit changes in conductance upon analyte binding, allowing label-free and real-time monitoring of biomarkers such as microRNAs, cancer antigens, and toxins (Katz & Willner, 2004). A limitation of CNTs, however, is their inherent hydrophobicity and tendency to aggregate, which hinders reproducibility. Functionalization with polymers, surfactants, or biomolecules improves dispersion, stability, and biocompatibility, thereby expanding their usability in biosensing.

Graphene and its derivatives, including graphene oxide (GO) and reduced graphene oxide (rGO), represent some of the most promising materials in biosensor development. Graphene's two-dimensional structure provides an exceptionally high surface area, remarkable carrier mobility, and excellent electrical conductivity. These properties make it ideal for constructing highly sensitive electrochemical biosensors (Buja et al., 2022). GO and rGO introduce oxygen-containing functional groups, which enable covalent attachment of biomolecules such as enzymes, antibodies, and DNA probes. GO's strong fluorescence-quenching ability has been harnessed in FRET-based biosensors, where fluorescently labeled probes adsorbed on GO surfaces exhibit quenched signals that are restored upon hybridization with target sequences. Moreover, graphene-based FET biosensors rival CNT-FETs in their sensitivity and ease of integration into lab-on-a-chip platforms. Hybrid nanocomposites that combine graphene with AuNPs or QDs synergize the advantages of conductivity, plasmonics, and fluorescence, yielding multi-modal biosensors with ultralow detection limits (Buja et al., 2022).

The effectiveness of these nanomaterials in biosensors is highly dependent on surface chemistry and functionalization strategies. Covalent functionalization involves coupling biomolecules to reactive groups such as carboxyls or amines on the nanomaterial surface, using techniques like carbodiimide (EDC/NHS) chemistry or click chemistry, which provide strong and stable immobilization (Razmi et al., 2019). Non-covalent strategies, such as π–π stacking, electrostatic interactions, and hydrophobic forces, offer a gentler approach that preserves the native activity of biomolecules but may be less stable under harsh conditions. Affinity-based functionalization methods, such as streptavidin–biotin or protein A–antibody binding, provide high specificity and are particularly useful in modular biosensor platforms (Razmi et al., 2019). The choice of functionalization strategy depends on the desired stability, selectivity, and application environment of the biosensor. Additionally, antifouling surface coatings such as polyethylene glycol (PEG) or zwitterionic polymers are often applied to minimize nonspecific adsorption, which is a common cause of reduced selectivity in complex biological samples (Fang & Ramasamy, 2015).

Nanomaterials enhance biosensor performance through several mechanisms. Their high surface-to-volume ratio allows dense immobilization of biomolecular probes, thereby increasing the likelihood of target recognition events. Their superior electronic and optical properties improve transduction efficiency, with CNTs and graphene accelerating electron transfer in electrochemical sensors and AuNPs and QDs providing strong plasmonic and fluorescent signals. Furthermore, nanomaterials often serve as signal amplification agents by carrying multiple labels or catalytic moieties, drastically lowering detection limits (Yuwen et al., 2023). They also enable multiplexing capabilities, as in the case of QDs that can detect multiple analytes simultaneously. Importantly, when combined with advanced functionalization strategies, nanomaterials confer high selectivity by ensuring specific biomolecule immobilization while reducing cross-reactivity. Collectively, these contributions push detection limits into the femtomolar or even attomolar range, enabling earlier and more accurate detection of diseases and contaminants than was previously possible.

The integration of nanomaterials into biosensors has already had a profound impact on diverse fields, including medical diagnostics, food safety, and environmental monitoring. For example, graphene–AuNP hybrid biosensors have been used for the ultrasensitive detection of microRNAs associated with cancer progression at attomolar levels, which is critical for early diagnosis (Jarocka et al., 2011). CNT-FET devices have been employed for real-time monitoring of protein interactions and environmental toxins at concentrations previously undetectable with conventional methods. Similarly, quantum dot-based multiplexed biosensors are being developed for simultaneous detection of infectious disease biomarkers, which is vital for point-of-care diagnostics in resource-limited regions.

In conclusion, nanomaterials such as gold nanoparticles, quantum dots, carbon nanotubes, and graphene have revolutionized biosensor design by overcoming the limitations of earlier sensor technologies. Each material brings distinct optical, electrical, and chemical advantages that, when coupled with tailored surface chemistry and functionalization strategies, lead to dramatic improvements in sensitivity, selectivity, and stability. Gold nanoparticles provide tunable plasmonic and conductive platforms, quantum dots enable bright and multiplexed fluorescence detection, carbon nanotubes offer efficient electron transfer and FET integration, and graphene provides versatile conductive and surface chemistry properties. The convergence of these nanomaterials with innovative functionalization approaches has set the stage for next-generation biosensors that are more sensitive, portable, and applicable in real-world settings, including point-of-care diagnostics and wearable health monitors. As nanomaterial synthesis, toxicity mitigation, and device integration continue to advance, biosensors are poised to become even more powerful tools in safeguarding health, food security, and environmental sustainability (Jarocka et al., 2011).

6. BIORECOGNITION ELEMENTS FOR PLANT VIRUS DETECTION

6.1. Antibodies as Biorecognition Element

Antibodies are among the earliest and most widely used recognition molecules in plant virus detection. Their high specificity for viral coat proteins has made them the cornerstone of immunodiagnostics for decades. Techniques such as the enzyme-linked immunosorbent assay (ELISA) have been built on antibody–antigen recognition and adapted for high-throughput screening of plant samples. In plant virology, antibodies raised against viruses such as Tomato spotted wilt virus, Tobacco mosaic virus, and Cucumber mosaic virus are routinely used to identify infections (Kimaru, 2020). Both monoclonal and polyclonal antibodies are applied, with monoclonal antibodies offering higher specificity but limited tolerance to viral mutations, while polyclonal antibodies recognize multiple epitopes, thereby minimizing false negatives but sometimes reducing selectivity.

Recent advances have seen the development of recombinant antibody formats such as single-chain variable fragments (scFvs) and nanobodies. These smaller antibody derivatives are easier to produce in bacterial systems, exhibit higher thermal and chemical stability, and can be more efficiently immobilized on sensor surfaces. Immunosensors incorporating nanobodies have shown remarkable performance in detecting viruses such as Potato virus Y, offering sensitivity comparable to standard ELISA but in portable formats suitable for field use. However, the stability of antibodies under fluctuating agricultural conditions remains a challenge, making alternative recognition molecules increasingly attractive.

6.2. Aptamers in Plant Virus Detection

Aptamers, often referred to as “chemical antibodies,” are single-stranded DNA or RNA oligonucleotides generated through SELEX (Systematic Evolution of Ligands by Exponential Enrichment). Unlike antibodies, aptamers are synthetically produced, which reduces batch variability and allows for precise modification with chemical groups for immobilization. They can fold into unique three-dimensional conformations that enable high-affinity binding to viral coat proteins, nucleic acids, or even whole viral particles (Liu et al., 2013).

In plant virus diagnostics, aptamers have been developed against Tobacco mosaic virus and Potato virus X, showing specificity comparable to antibodies but with enhanced stability under harsh conditions (Mishra et al., 2021). Their small size and ability to undergo conformational changes upon target binding make them excellent candidates for label-free biosensors, including electrochemical and optical platforms. Furthermore, aptamers are resistant to denaturation, making them more reliable for field-deployed biosensors. They are also reusable in some systems, lowering the overall cost. Despite these advantages, aptamer selection against plant viruses remains less advanced compared to antibodies, suggesting that this field is still in its early stages and has significant potential for future growth.

6.3. Nucleic Acid-Based Recognition Elements

Since most plant viruses possess RNA genomes, nucleic acid probes have become indispensable in biosensor design. These recognition elements rely on sequence complementarity, where synthetic DNA or RNA probes hybridize with viral genomes or specific gene fragments. By targeting unique viral sequences, nucleic acid-based biosensors achieve high specificity, enabling not only species-level but also strain-level detection (Boonham et al., 2014).

Techniques such as DNA microarrays and electrochemical DNA biosensors have been applied to plant viruses like Banana bunchy top virus and Tomato yellow leaf curl virus. Moreover, nucleic acid probes can be coupled with isothermal amplification methods such as loop-mediated isothermal amplification (LAMP) or recombinase polymerase amplification (RPA) to improve sensitivity, allowing detection of viral nucleic acids at extremely low concentrations (Razmi et al., 2019). These platforms are particularly useful in resource-limited settings because they eliminate the need for thermocyclers. Integration with nanomaterials, such as graphene and gold nanoparticles, further enhances hybridization signals, enabling real-time and ultrasensitive detection. This makes nucleic acid-based biosensors a vital component in modern plant disease diagnostics.

6.4. CRISPR-Based Biorecognition Elements

CRISPR–Cas systems have recently emerged as transformative tools in plant virus detection. Unlike conventional nucleic acid probes, CRISPR enzymes such as Cas12 and Cas13 can specifically recognize viral DNA or RNA and then activate collateral cleavage activity, cutting nearby single-stranded nucleic acids. This generates strong and easily detectable signals, often visualized through fluorescence or lateral flow readouts (Chen et al., 2021).

CRISPR-based detection is programmable and highly adaptable, as guide RNAs can be designed to target new or mutated viral strains rapidly. For instance, CRISPR–Cas13a has been applied for the detection of Tomato brown rugose fruit virus, offering rapid field-deployable diagnostics with minimal equipment (Bonyadi et al., 2024). Compared to traditional nucleic acid probes, CRISPR-based systems combine the precision of sequence recognition with amplified signal output, achieving sensitivity down to the attomolar range. Their adaptability and robustness make them a revolutionary addition to the toolkit for plant virus biosensing.

6.5. Immobilization Techniques for Biorecognition Elements

The successful application of antibodies, aptamers, nucleic acids, and CRISPR components in biosensors depends heavily on immobilization techniques. Immobilization ensures that recognition molecules are attached firmly to sensor surfaces such as electrodes, membranes, or nanoparticles without losing their biological activity. Several immobilization strategies exist, ranging from simple physical adsorption to covalent bonding.

Physical adsorption is cost-effective and simple but prone to biomolecule leaching, reducing sensor stability. Covalent immobilization through functional groups such as carboxyl, amine, and thiol is more stable and widely used, especially with nanomaterials like gold nanoparticles and graphene oxide (Jarocka et al., 2011). Affinity-based immobilization strategies, such as streptavidin–biotin interactions, offer high specificity and orientation control, ensuring maximum exposure of binding sites. Nanomaterial-assisted immobilization significantly increases the available surface area, allowing for dense loading of recognition elements and enhanced sensitivity (Fang & Ramasamy, 2015). The choice of immobilization method is critical, as it directly impacts biosensor performance, reproducibility, and cost-effectiveness.

6.6. Biosensor Stability in Agricultural Applications

A persistent challenge in plant virus biosensor deployment is stability under real-world conditions. Plant diagnostics often occur in the field, where temperature fluctuations, dust, humidity, and complex plant sap can degrade recognition elements. Antibodies, though specific, are prone to denaturation and reduced shelf life. Aptamers and nucleic acids are inherently more stable, while CRISPR systems exhibit good robustness when freeze-dried or lyophilized.

To enhance biosensor stability, surface modifications and protective coatings are frequently used. Polyethylene glycol (PEG) coatings, for example, reduce nonspecific adsorption and extend sensor life (Fang & Ramasamy, 2015; Buja et al., 2022). Nanocomposite materials can provide additional stability by anchoring recognition molecules more securely. Furthermore, encapsulation and lyophilization techniques make biosensors easier to transport and store, particularly in remote agricultural areas. Portable microfluidic and paper-based biosensors are also being designed to maintain stability while enabling on-site testing, which is crucial for resource-constrained environments.

Biorecognition elements are the core of biosensor technology for plant virus detection, dictating both specificity and sensitivity. Antibodies remain the traditional gold standard, while aptamers provide greater stability and ease of synthesis. Nucleic acid probes enable precise detection of viral genomes, and CRISPR systems have introduced a new level of programmability and sensitivity. The effectiveness of these recognition elements is closely tied to the immobilization methods used and the stability of the biosensor in real-world conditions. As agriculture faces growing threats from viral diseases, the continued innovation and integration of diverse biorecognition elements will be key to developing rapid, accurate, and field-deployable diagnostics for sustainable plant health management.

7. TRANSDUCTION MECHANISMS

7.1 Electrochemical Transduction

Electrochemical biosensors are among the most widely studied and applied systems for plant virus detection because of their high sensitivity, relatively low cost, and compatibility with portable devices. They function by converting biological interactions, such as the binding of a virus particle or nucleic acid to a bioreceptor, into measurable electrical signals. Three major categories dominate this field: voltammetric, amperometric, and impedimetric systems. Voltammetric biosensors rely on detecting changes in current as voltage is applied, and these have been applied successfully in the detection of Potato virus Y (PVY), an important pathogen in potato and tobacco crops. In these systems, DNA probes immobilized on the electrode surface hybridize with complementary viral RNA, producing detectable redox signal shifts that correlate with infection levels. Amperometric biosensors, which measure current at a constant potential, often incorporate enzyme labels to amplify the signal. A practical example comes from the detection of Plum pox virus (PPV), one of the most economically damaging viruses of stone fruit crops in Europe (Schachinger et al., 2021). In this case, horseradish peroxidase-labeled antibodies facilitated rapid, highly sensitive amperometric detection. Impedimetric biosensors, in contrast, provide label-free detection by monitoring changes in electrical impedance at the electrode surface during virus binding events. This approach has been applied in the detection of Grapevine fanleaf virus (GFLV), a major threat to vineyards, where high sensitivity and real-time monitoring were achieved without additional reagents (Güller & Paylan, 2024). Despite their advantages, electrochemical biosensors face challenges such as electrode fouling and non-specific adsorption, but continued improvements in surface chemistry and nanomaterial integration are addressing these limitations.

7.2 Optical Transduction

Optical biosensors are another powerful class of platforms that convert virus–receptor interactions into measurable optical signals. They are particularly attractive due to their high sensitivity and potential for multiplex detection, which is vital for identifying mixed infections in crops. Fluorescence-based biosensors, which employ fluorescent dyes or nanomaterials such as quantum dots (QDs), have been widely applied in plant virology. A notable example includes the development of QD-based fluorescence assays for detecting Rice tungro bacilliform virus (RTBV), which severely reduces rice yields in Asia (Mishra et al., 2021). Similarly, fluorescently labeled aptamers have been employed to detect Cucumber mosaic virus (CMV) in cucumbers and peppers, demonstrating specificity and rapid response (Kim et al., 2024). Surface plasmon resonance (SPR) sensors provide another optical approach, functioning by measuring changes in refractive index on a sensor surface when viruses bind to immobilized probes. These have been used effectively for Tomato yellow leaf curl virus (TYLCV) detection in tomato plants, enabling real-time, label-free analysis and detailed characterization of binding kinetics (Zhang et al., 2019). Colorimetric biosensors, especially those utilizing gold nanoparticles (AuNPs), offer an attractive alternative for field use, as they rely on visible color changes that can be interpreted without specialized equipment. AuNP-based colorimetric assays have been successfully applied in the detection of Banana bunchy top virus (BBTV), enabling on-site screening of banana planting materials across Southeast Asia. While fluorescence and SPR sensors require sophisticated instruments, colorimetric assays are inexpensive and well suited for deployment in resource-limited agricultural systems.

7.3 Piezoelectric Transduction

Piezoelectric biosensors, based on quartz crystal microbalance (QCM) technology, operate by detecting small changes in mass at a sensor surface when viruses bind to immobilized bioreceptors. This label-free and highly sensitive system has demonstrated significant potential for the direct detection of intact virus particles. In practical applications, piezoelectric biosensors have been designed to detect Tomato spotted wilt virus (TSWV), which severely impacts solanaceous crops, and Citrus tristeza virus (CTV), one of the most economically important viruses of citrus plants worldwide. The ability of piezoelectric sensors to provide real-time detection without the need for labeling is a major advantage, as it simplifies assay procedures (Skládal, 2024). However, these devices are sensitive to environmental disturbances such as vibrations and non-specific binding, which may affect their reliability outside of controlled laboratory conditions. Advances in miniaturization and surface functionalization are helping to improve their robustness for real-world agricultural applications (Skládal, 2024).

7.4 Thermal Transduction

Thermal biosensors provide a less commonly applied but innovative approach to virus detection by measuring heat changes associated with virus–receptor interactions or enzymatic reactions during recognition events. Although not yet as widely used in plant virology as electrochemical or optical platforms, thermal detection offers label-free operation and potential for integration into portable diagnostic devices. An example includes the use of thermal biosensors for detecting Potato leafroll virus (PLRV), a serious pathogen in potato production, where hybridization of viral RNA with immobilized probes generated measurable heat signatures that could be detected in real time (Halabi et al., 2022). One of the major advantages of thermal biosensors lies in their compatibility with microfluidic platforms, opening possibilities for compact, multiplex diagnostic tools. However, environmental temperature fluctuations in field conditions remain a challenge, as they can introduce variability into the measurements.

7.5 Global Applications and Relevance

The diversity of transduction mechanisms in biosensors underscores their relevance for tackling plant viral diseases across different regions and crop systems worldwide. In Africa, cassava production suffers devastating losses due to Cassava mosaic begomoviruses and Cassava brown streak virus (CBSV), and portable biosensors using electrochemical and colorimetric methods are being developed for rapid field screening of infected planting material. In Asia-Pacific, where rice and banana are staple crops, optical sensors for RTBV and BBTV are critical to improving food security, while nanoparticle-based colorimetric assays are being adapted for smallholder farming systems (Mishra et al., 2021). In Europe and North America, advanced SPR and electrochemical platforms are increasingly explored to monitor Plum pox virus (PPV) in stone fruits and Wheat streak mosaic virus (WSMV) in cereals, both of which threaten international trade and food supply chains. In Latin America, tomato and pepper crops are threatened by TYLCV and Pepper mild mottle virus (PMMoV), where integrated biosensor platforms combining optical and electrochemical transduction mechanisms have been tested for rapid detection and early disease management (Zhang et al., 2019).

Together, electrochemical, optical, piezoelectric, and thermal biosensors provide complementary strengths for plant virus detection. While electrochemical sensors offer low-cost and portable solutions, optical platforms provide unparalleled sensitivity and the potential for multiplexing. Piezoelectric systems allow for the direct detection of intact virions, and thermal sensors provide opportunities for integration with compact, microfluidic systems (Madadelahi et al., 2025). Their global applicability highlights the importance of biosensors as next-generation diagnostic tools in plant virology. Nevertheless, challenges such as high costs for advanced optical systems, the need for stable performance in variable field environments, and the requirement for standardization remain. Ongoing research is directed toward improving sensitivity, reducing costs, and enhancing field compatibility, with the ultimate aim of integrating these platforms into sustainable plant disease management systems worldwide (Li et al., 2021).

The evolution of biosensors has been closely tied to advances in materials science, particularly with the emergence of nanotechnology. Traditional biosensors, though effective in certain contexts, often faced limitations such as low sensitivity, limited selectivity, slow response time, and poor stability under real-world conditions. The integration of nanomaterials into biosensor design has dramatically transformed this field by enabling enhanced signal amplification, improved immobilization of biorecognition elements, and novel modes of transduction (Carpenter et al., 2018). Nanostructures such as gold nanoparticles, quantum dots, carbon nanotubes, and graphene have emerged as the most extensively investigated platforms because of their unique physicochemical properties. Their nanoscale dimensions, large surface areas, and tunable electronic and optical behaviors provide significant improvements in biosensor performance, while functionalization strategies and surface chemistry play a pivotal role in tailoring these nanomaterials for specific diagnostic applications.

Gold nanoparticles (AuNPs) are among the earliest nanomaterials to be employed in biosensor design, owing to their ease of synthesis, biocompatibility, and unique surface plasmon resonance properties. Their ability to scatter and absorb light strongly at specific wavelengths enables colorimetric detection that can be observed by the naked eye, as in the case of lateral flow assays. This property is exploited in low-cost field diagnostics, where aggregation of AuNPs leads to a visible color shift that indicates the presence of a target analyte (Georgeous et al., 2024). In electrochemical biosensors, AuNPs act as conductive nanointerfaces that facilitate efficient electron transfer between redox-active biomolecules and electrodes, thus improving signal strength and sensitivity. For example, AuNP-modified electrodes with immobilized glucose oxidase significantly increase current responses in glucose detection. Additionally, the chemistry of gold allows stable thiol–gold interactions, enabling covalent immobilization of DNA probes, aptamers, or antibodies with high stability. AuNPs can also be engineered in different shapes, such as nanorods, nanostars, and nanocages, which tune their plasmon resonance into the near-infrared region, thereby extending their use in biosensors for biomedical imaging and deep-tissue diagnostics.

Quantum dots (QDs) represent another important nanomaterial used in biosensors. These semiconductor nanocrystals, typically 2–10 nanometers in diameter, exhibit quantum confinement effects that give rise to size-tunable fluorescence with sharp emission peaks, high brightness, and exceptional photostability. Unlike organic dyes, QDs resist photobleaching, which makes them ideal for continuous monitoring applications (Medintz et al., 2005). Their broad absorption spectra and narrow emission bands allow simultaneous excitation of multiple QDs with a single light source, enabling multiplexed detection of different analytes. QDs are particularly powerful in fluorescence resonance energy transfer (FRET)-based biosensors, where energy transfer between a QD donor and an acceptor molecule occurs when they are in close proximity, providing a sensitive readout for DNA hybridization or protein–protein interactions. Despite their advantages, traditional QDs made of cadmium or lead compounds present toxicity concerns, limiting their biomedical applications. To overcome this, alternative nanomaterials such as carbon quantum dots, graphene quantum dots, and perovskite QDs are being developed, which provide similar optical properties with improved biocompatibility (Li et al., 2021).

Carbon nanotubes (CNTs), available as single-walled (SWCNTs) and multi-walled (MWCNTs) structures, have revolutionized electrochemical biosensors. Their cylindrical sp^2-hybridized carbon structure offers extraordinary electrical conductivity, large surface area, and mechanical strength. CNTs are particularly effective at enabling direct electron transfer between enzymes and electrodes, which eliminates the need for mediators and significantly improves the sensitivity of biosensors. CNT-modified electrodes have been used in glucose sensors, immunosensors, and DNA sensors, achieving femtomolar to picomolar detection levels. Their one-dimensional conductive properties also make them ideal for field-effect transistor (FET)-based biosensors. When functionalized with DNA probes or antibodies, CNT-FETs exhibit changes in conductance upon analyte binding, allowing label-free and real-time monitoring of biomarkers such as microRNAs, cancer antigens, and toxins. A limitation of CNTs, however, is their inherent hydrophobicity and tendency to aggregate, which hinders reproducibility. Functionalization with polymers, surfactants, or biomolecules improves dispersion, stability, and biocompatibility, thereby expanding their usability in biosensing.

Graphene and its derivatives, including graphene oxide (GO) and reduced graphene oxide (rGO), represent some of the most promising materials in biosensor development. Graphene's two-dimensional structure provides an exceptionally high surface area, remarkable carrier mobility, and excellent electrical conductivity. These properties make it ideal for constructing highly sensitive electrochemical biosensors. GO and rGO introduce oxygen-containing functional groups, which enable covalent attachment of biomolecules such as enzymes, antibodies, and DNA probes. GO's strong fluorescence-quenching ability has been harnessed in FRET-based biosensors, where fluorescently labeled probes adsorbed on GO surfaces exhibit quenched signals that are restored upon hybridization with target sequences. Moreover, graphene-based FET biosensors rival CNT-FETs in their sensitivity and ease of integration into lab-on-a-chip platforms. Hybrid nanocomposites that combine graphene with AuNPs or QDs synergize the advantages of conductivity, plasmonics, and fluorescence, yielding multi-modal biosensors with ultralow detection limits.

The effectiveness of these nanomaterials in biosensors is highly dependent on surface chemistry and functionalization strategies. Covalent functionalization involves coupling biomolecules to reactive groups such as carboxyls or amines on the nanomaterial surface, using techniques like carbodiimide (EDC/NHS) chemistry or click chemistry, which provide strong and stable immobilization (Malik et al., 2023). Non-covalent strategies, such as π–π stacking, electrostatic interactions, and hydrophobic forces, offer a gentler approach that preserves the native activity of biomolecules but may be less stable under harsh conditions. Affinity-based functionalization methods, such as streptavidin–biotin or protein A–antibody binding, provide high specificity and are particularly useful in modular biosensor platforms. The choice of functionalization strategy depends on the desired stability, selectivity, and application environment of the biosensor. Additionally, antifouling surface coatings such as polyethylene glycol (PEG) or zwitterionic polymers are often applied to minimize nonspecific adsorption, which is a common cause of reduced selectivity in complex biological samples.

Nanomaterials enhance biosensor performance through several mechanisms. Their high surface-to-volume ratio allows dense immobilization of biomolecular probes, thereby increasing the likelihood of target recognition events. Their superior electronic and optical properties improve transduction efficiency, with CNTs and graphene accelerating electron transfer in electrochemical sensors and AuNPs and QDs providing strong plasmonic and fluorescent signals. Furthermore, nanomaterials often serve as signal amplification agents by carrying multiple labels or catalytic moieties, drastically lowering detection limits. They also enable multiplexing capabilities, as in the case of QDs that can detect multiple analytes simultaneously. Importantly, when combined with advanced functionalization strategies, nanomaterials confer high selectivity by ensuring specific biomolecule immobilization while reducing cross-reactivity. Collectively, these contributions push detection limits into the femtomolar or even attomolar range, enabling earlier and more accurate detection of diseases and contaminants than was previously possible.

The integration of nanomaterials into biosensors has already had a profound impact on diverse fields, including medical diagnostics, food safety, and environmental monitoring. For example, graphene–AuNP hybrid biosensors have been used for the ultrasensitive detection of microRNAs associated with cancer progression at attomolar levels, which is critical for early diagnosis. CNT-FET devices have been employed for real-time monitoring of protein interactions and environmental toxins at concentrations previously undetectable with conventional methods. Similarly, quantum dot-based multiplexed biosensors are being developed for simultaneous detection of infectious disease biomarkers, which is vital for point-of-care diagnostics in resource-limited regions.

In conclusion, nanomaterials such as gold nanoparticles, quantum dots, carbon nanotubes, and graphene have revolutionized biosensor design by overcoming the limitations of earlier sensor technologies. Each material brings distinct optical, electrical, and chemical advantages that, when coupled with tailored surface chemistry and functionalization strategies, lead to dramatic improvements in sensitivity, selectivity, and stability. Gold nanoparticles provide tunable plasmonic and conductive platforms, quantum dots enable bright and multiplexed fluorescence detection, carbon nanotubes offer efficient electron transfer and FET integration, and graphene provides versatile conductive and surface chemistry properties. The convergence of these nanomaterials with innovative functionalization approaches has set the stage for next-generation biosensors that are more sensitive, portable, and applicable in real-world settings, including point-of-care diagnostics and wearable health monitors. As nanomaterial synthesis, toxicity mitigation, and device integration continue to advance, biosensors are poised to become even more powerful tools in safeguarding health, food security, and environmental sustainability.

8. PORTABLE AND FIELD-DEPLOYABLE DEVICES

8.1. Paper-Based Biosensors

Paper-based biosensors, especially lateral flow assays (LFAs), are among the most practical diagnostic platforms for detecting plant viruses in the field. These devices are lightweight, disposable, and inexpensive, making them particularly valuable in developing regions where laboratory infrastructure is limited. LFAs have been successfully developed for Tomato yellow leaf curl virus (TYLCV), Potato virus Y (PVY), and Banana bunchy top virus (BBTV), providing results in less than 30 minutes. Gold nanoparticle-conjugated LFAs have also been used for the detection of Cucumber mosaic virus (CMV) with naked-eye visualization. In Africa, paper-based assays are being deployed for cassava crops, where Cassava mosaic begomoviruses cause devastating yield losses. For cereals, nucleic acid-based paper assays using isothermal amplification have been adapted for Barley yellow dwarf virus (BYDV), which threatens wheat and barley production worldwide (Zhang et al., 2025). These examples illustrate how paper biosensors are becoming critical frontline tools for farmers managing viral diseases in diverse cropping systems (Table 4).

Table 4. Nanomaterial-Based Biosensing Platforms for Rapid and Sensitive Detection of Plant Viruses

Nanomaterial / Platform	Biosensor Type	Target Plant Virus	Detection Limit / Sensitivity	Response Time	Key Advantages	References
Gold nanoparticles (AuNPs) + antibody probes	Colorimetric	*Tomato yellow leaf curl virus (TYLCV)*	10 ng/μL DNA	<30 min	Portable, naked-eye readout	Bao, 2015
AuNP-thiolated DNA probe electrode	Electrochemical (DPV)	*Potato virus Y (PVY)*	1.5 pg/μL RNA	~20 min	Label-free, high specificity	Qin et al., 2024
Graphene oxide (GO) nanosheets	FRET Fluorescence	*Cucumber mosaic virus (CMV)*	0.2 ng/μL RNA	<15 min	Large surface area, quenching efficiency	Ramadan et al., 2021

continued on following page

Table 4. Continued

Nanomaterial / Platform	Biosensor Type	Target Plant Virus	Detection Limit / Sensitivity	Response Time	Key Advantages	References
Carbon nanotube (CNT) field-effect transistor	Electrochemical (Impedimetric)	*Tobacco mosaic virus (TMV)*	1.0 pg/mL	<10 min	Ultra-sensitive, reusable	Star et al., 2006
Quantum dots (CdSe/ZnS QDs) conjugated to IgG	Optical (Fluorescence)	*Banana bunchy top virus (BBTV)*	~100 DNA copies	<25 min	Photostable, multiplexing	Medintz et al., 2005
Silver nanoparticles (AgNPs) + LAMP assay	Colorimetric (plasmonic shift)	*Cassava mosaic virus (ACMV, EACMV, ICMV)*	5 × 10^1 copies DNA	30–40 min	Cost-effective, dual amplification	Saifuddin et al., 2024
Surface plasmon resonance (SPR) + AuNP amplification	Optical (SPR)	*Grapevine leafroll-associated virus (GLRaV-3)*	50 fg/mL RNA	Real-time	Label-free, quantitative	Miyazaki et al., 2017
Magnetic nanoparticles (MNPs) + AuNP hybrid sandwich	Electrochemical & Optical	*Rice tungro bacilliform virus (RTBV)*	10^3 RNA copies	~40 min	Signal amplification, dual-mode	Uehara-Ichiki et al., 2013
Paper-based AuNP immunostrip	Lateral Flow / Colorimetric	*Tomato spotted wilt virus (TSWV)*	1 ng/µL	10–15 min	Field-friendly, cheap	Ardekani & Thulstrup, 2022
Electrochemical aptasensor (AuNP-modified electrode)	Amperometric	*Soybean mosaic virus (SMV)*	10^2 RNA copies	<25 min	Highly selective aptamer probe	Zhao et al., 2021
CRISPR-Cas12a integrated with AuNP strip	Isothermal + CRISPR	*Tomato brown rugose fruit virus (ToBRFV)*	<10 RNA copies	20–30 min	Variant detection, ultra-specific	Besati et al., 2025
Graphene paper microfluidics	Portable Electrochemical	*Cassava mosaic virus (CMV complex)*	10^2–10^3 DNA copies	~30 min	Lightweight, point-of-care	Rahman et al., 2023
ZnO nanorods (piezoelectric QCM)	Piezoelectric	*Citrus tristeza virus (CTV)*	0.1 ng/mL	Real-time	High sensitivity, reusable	Mutegoa & Sahini, 2025
Conducting polymer nanocomposites (PANI-AuNP)	Electrochemical (EIS)	*Maize dwarf mosaic virus (MDMV)*	2.0 pg/mL	~20 min	High conductivity, stability	Kumar et al., 2010
Silica nanoparticles doped with fluorophores	Optical (Fluorescent silica nanosensor)	*Lettuce mosaic virus (LMV)*	5 × 10^2 RNA copies	<30 min	High photostability, multiplexing	Xu et al., 2023
MoS_2 nanosheet biosensor (2D nanomaterial)	Electrochemical (EIS)	*Barley yellow dwarf virus (BYDV)*	0.5 pg/mL	~15 min	Ultra-high surface reactivity	Mphuthi et al., 2022
Dendrimer-functionalized AuNPs	Electrochemical (DPV)	*Wheat streak mosaic virus (WSMV)*	0.2 ng/mL	~25 min	High probe density, signal amplification	Shi et al., 2023

continued on following page

Table 4. Continued

Nanomaterial / Platform	Biosensor Type	Target Plant Virus	Detection Limit / Sensitivity	Response Time	Key Advantages	References
ZnO nanowire FET biosensor	Electrochemical (FET)	*Peanut mottle virus (PMV)*	1 pg/mL	10 min	Fast electron transport, reusable	Zafar et al., 2018
Polymer-coated magnetic nanoparticles (PMNPs)	Colorimetric & Electrochemical	*Soybean mosaic virus (SMV)*	10^1 RNA copies	<30 min	Multiplex detection	Khan et al., 2021

8.2. Smartphone-Integrated and Wearable Sensors

Smartphone-integrated biosensors extend the functionality of paper-based devices by enabling quantitative analysis, geotagging, and cloud-based data sharing. For instance, a smartphone fluorescence platform was developed for the detection of Rice tungro bacilliform virus (RTBV), providing immediate results that can be transmitted to central databases. Similarly, apps linked to lateral flow assays have been validated for Cassava mosaic virus and Groundnut rosette virus (GRV) in sub-Saharan Africa, helping smallholder farmers rapidly confirm infections. In South America, smartphone-assisted diagnostics have been explored for Papaya ringspot virus (PRSV), a major constraint on papaya production. Wearable nanosensors, though less common in plant pathology, are gaining attention for continuous monitoring. For example, graphene-based leaf patches have been tested for Tobacco mosaic virus (TMV), monitoring viral infection through changes in plant sap conductivity. These devices open opportunities for near real-time plant health surveillance, improving response time to viral outbreaks.

8.3. IoT-Enabled Biosensing Platforms for Smart Agriculture

The integration of biosensors with Internet of Things (IoT) platforms has revolutionized virus monitoring in precision agriculture. IoT-linked devices allow real-time, large-scale surveillance by connecting portable sensors with wireless communication and cloud computing. In sugarcane fields, IoT-enabled biosensing networks have been trialed for monitoring Sugarcane mosaic virus (SCMV), transmitting diagnostic data across multiple farms simultaneously (Khan et al., 2021). In maize-growing regions of sub-Saharan Africa, drone-assisted imaging coupled with IoT platforms has been employed for monitoring Maize streak virus (MSV), one of the most damaging maize pathogens (Mahlein, 2016). In wheat production systems, IoT-coupled hyperspectral imaging has been explored for Wheat streak mosaic virus (WSMV), providing early detection before visible symptom expression (Zhang et al., 2019). For high-value crops like grapevine, IoT-integrated biosensors have been adapted to detect Grapevine leafroll-associated virus (GLRaV), enabling vineyard managers to map infection hotspots and target management interventions (Zhao et al., 2023). Expanding further, IoT-based monitoring has been tested in soybean systems in the United States for Soybean mosaic virus (SMV), where cloud-based analytics predict regional spread patterns. Similarly, IoT-assisted systems for banana plantations in Asia-Pacific have shown potential for managing Banana bunchy top virus (BBTV), which remains a major barrier to sustainable banana production. These examples highlight how IoT platforms not only support early detection but also provide predictive insights into the epidemiology of viral diseases across diverse agroecosystems.

9. CASE STUDIES AND APPLICATIONS

9.1. Detection of TMV, CMV, TYLCV, and Other Plant Viruses

Case studies on the deployment of biosensors for plant virus detection have demonstrated both the feasibility and advantages of nanotechnology-assisted platforms in real agricultural settings. One of the earliest and most studied examples is the detection of Tobacco mosaic virus (TMV), a model virus that has been targeted with electrochemical DNA sensors, fluorescence-based aptamer biosensors, and graphene-coated wearable sensors for early disease recognition (Jablonski et al., 2021; Takemura et al., 2024; Choi & Yoon, 2023). Similarly, biosensors for Cucumber mosaic virus (CMV) have been validated in cucumber fields using gold nanoparticle-labeled paper-based assays that deliver rapid visual detection (Yu et al., 2018; Patel et al., 2022). For Tomato yellow leaf curl virus (TYLCV), surface plasmon resonance (SPR)-based sensors and lateral flow devices have been tested in tomato-growing regions, achieving reliable detection even at low virus titers (Razmi et al., 2019; Takemura et al., 2024). Other notable viruses that have been successfully targeted include Potato virus Y (PVY) using amperometric and immunosensor platforms (Perdikaris et al., 2011), Rice tungro bacilliform virus (RTBV) using smartphone-integrated fluorescence assays and isothermal methods adapted to portable readouts (Samacoits et al., 2021), and Grapevine leafroll-associated virus (GLRaV) using IoT-enabled and remote sensing biosensing platforms (Mahlein, 2016; Pande, 2023). These case studies collectively demonstrate how biosensor technology has moved from proof-of-concept to targeted applications in multiple crop systems (Choi & Yoon, 2023; Patel et al., 2022).

9.2. In-Field Trials and Validation Data

The ultimate test for biosensor platforms lies in their performance under field conditions, where environmental variability, mixed infections, and resource limitations present significant challenges. Several devices have now undergone validation outside the laboratory. For instance, smartphone-based diagnostics for Cassava mosaic virus and Groundnut rosette virus (GRV) were field-tested across smallholder farms in East Africa using mobile-readout platforms and symptom-recognition apps that delivered high concordance with laboratory PCR benchmarks (Mrisho et al., 2020; Samacoits et al., 2021; Legg et al., 2011). Similarly, gold nanoparticle-based lateral flow assays (LFAs) for Banana bunchy top virus (BBTV) have been piloted in Southeast Asia, providing rapid on-site results in under 20 minutes and enabling immediate farmer response (Peng et al., 2012; Kuan et al., 2022). In cereals, IoT-enabled remote sensing platforms have been applied in wheat fields for early detection of Wheat streak mosaic virus (WSMV), identifying infections prior to visible symptom expression using hyperspectral and thermal indices (Mahlein, 2016). Validation studies in vineyards for Grapevine leafroll-associated virus (GLRaV) have confirmed that IoT-integrated biosensors and mapping systems can reliably identify disease hotspots and guide precision removal of infected vines (Bendel et al., 2020; MacDonald et al., 2016). Such trials confirm that portable biosensors not only match traditional assays in reliability but often surpass them in speed, accessibility, and field usability (Choi & Yoon, 2023; Patel et al., 2022).

9.3. Integration into Precision Farming Systems

Beyond diagnostics, biosensors are increasingly being integrated into precision agriculture frameworks that combine real-time data collection, predictive modelling, and automated responses. In maize production systems in Africa, IoT-linked biosensors for Maize streak virus (MSV) have been combined with drone imaging and cloud analytics to enable regional-scale surveillance of epidemics (Mahlein, 2016; Zhu et al., 2024). In sugarcane plantations, field-deployable sensors for Sugarcane mosaic virus (SCMV) have been linked with cloud-based analytics, allowing farm managers to track viral spread in near real-time and optimize application schedules (Shivashakarappa et al., 2022). Similarly, in soybean cultivation in the United States, IoT-enabled detection of Soybean mosaic virus (SMV) has been integrated with decision-support systems that predict seasonal outbreak patterns, reducing unnecessary interventions. These examples illustrate how biosensors are evolving from stand-alone diagnostic tools to central components of precision farming ecosystems, enabling proactive rather than reactive management of plant viral diseases (Choi & Yoon, 2023; Mahlein, 2016).

10. REGULATORY, BIOSAFETY, AND COMMERCIALIZATION ASPECTS

10.1. Risk Assessment and Regulatory Frameworks

The regulation of biosensors for plant virus detection is an evolving area, particularly because these devices often rely on emerging technologies such as nanomaterials, recombinant biorecognition elements, and digital connectivity. Risk assessment frameworks must consider both biosafety and environmental sustainability. Nanomaterials such as gold nanoparticles (AuNPs), carbon nanotubes (CNTs), and graphene derivatives, while highly effective in enhancing sensitivity, can persist in soils and interact with beneficial microorganisms, raising concerns over long-term ecological impacts (Sekhon, 2014; Rajput et al., 2020). For example, CNT-based biosensors for Cucumber mosaic virus (CMV) have shown excellent detection efficiency, but uncertainties about their environmental fate limit widespread regulatory approval (Narware et al., 2025). Similarly, biosensors incorporating CRISPR-Cas12 or Cas13 components for detecting RNA viruses like Potato virus Y (PVY) or Tomato spotted wilt virus (TSWV) require careful governance, since they involve engineered proteins derived from microbial systems (Liu & Dai, 2023; Robertson et al., 2022).

International organizations such as the Food and Agriculture Organization (FAO) and World Health Organization (WHO) have stressed the importance of establishing clear nanotechnology guidelines for agriculture (Kiran et al., 2024; Schlundt, 2014). The Cartagena Protocol on Biosafety also provides a framework for regulating transboundary movement of living modified organisms, which indirectly applies to biosensors that use engineered genetic elements (Convention on Biological Diversity, 2000). Regional differences persist: the European Union enforces stricter risk evaluations of nanomaterials, while the United States follows a more innovation-friendly regulatory pathway, potentially creating trade inconsistencies for virus-tested plant materials (Singh et al., 2021). Harmonizing these frameworks will be key to ensuring safe and equitable adoption of biosensors globally.

10.2. Market Potential and Startup Innovations

The commercialization of biosensors for plant viruses is rapidly gaining momentum as viral diseases continue to cause devastating crop losses worldwide. For instance, Cassava mosaic virus (ACMV, EACMV, ICMV) and Cassava brown streak virus (CBSV) together account for up to 50% yield losses in Africa, motivating investments into field-deployable diagnostics (Ndunguru et al., 2015). Similarly, Banana bunchy top virus (BBTV) threatens banana production in Asia and Africa, where startups and government agencies are collaborating to produce portable lateral flow assays (Peng et al., 2012). Global market forecasts estimate the agricultural biosensor sector will exceed USD 5–7 billion by 2030, with plant virus detection tools representing a growing niche.

Several companies and startups are pioneering this space. Agdia Inc. (USA) commercialized ELISA kits and lateral flow assays for viruses such as TMV, PVY, and Grapevine fanleaf virus (GFLV), which are widely used in certification programs, Plantix (Germany) integrates AI-based image recognition with diagnostic platforms for tomato viruses like TYLCV and Tomato mosaic virus (ToMV), supporting smallholder farmers in India and Africa (Verma et al., 2019). In China, quantum dot–based biosensors are being adapted for Rice stripe virus (RSV) detection to support paddy cultivation (Mousavi et al., 2022). Meanwhile, Indian startups supported by government initiatives are exploring IoT-enabled diagnostics for Sugarcane mosaic virus (SCMV) and BBTV (Khan et al., 2021). These examples illustrate that the commercialization of biosensors is no longer theoretical but is increasingly integrated into practical agricultural systems.

10.3. Challenges in Scaling and Farmer Adoption

Despite their potential, scaling biosensor technologies to widespread use among farmers remains difficult. Cost is a major barrier while simple paper-based assays for CMV or BBTV can cost less than USD 2 per test, advanced CRISPR-based systems or IoT-integrated sensors can exceed USD 50 per device, which is prohibitive for smallholder farmers in sub-Saharan Africa or South Asia (Robertson et al., 2022). Technical literacy also plays a role: farmers are often unfamiliar with interpreting biosensor readouts, necessitating strong extension services and farmer training programs (Mahlein, 2016). Infrastructure is another limiting factor; for example, IoT-based sensors for detecting Wheat streak mosaic virus (WSMV) or GLRaV in vineyards require reliable internet connectivity, which is not always available in rural areas (Zhang et al., 2019; Bendel et al., 2020).

Regulatory uncertainty can further delay adoption. Startups often struggle to navigate approval processes when devices incorporate nanomaterials, as guidelines vary between countries. This creates bottlenecks in commercial rollout and can discourage investment. Finally, socio-cultural acceptance matters: many farmers trust visual symptom diagnosis over novel biosensors, unless successful community-level demonstrations show clear benefits. Bridging this gap requires participatory approaches, subsidies for initial adoption, and integration with existing disease management frameworks. With coordinated efforts among governments, private companies, and international organizations, biosensors for plant virus detection could transition from niche innovations to mainstream tools within global precision agriculture.

11. FUTURE PROSPECTS AND RESEARCH DIRECTIONS

11.1. Multiplexed Detection Platforms

Multiplexed detection—the ability to detect many virus targets simultaneously from a single sample—is arguably the single most important near-term advance for plant-virus diagnostics. Multiplexing reduces time, cost, and sample volume, and is essential in crops where mixed infections (e.g., cassava infected with cassava mosaic begomoviruses plus cassava brown streak viruses; maize lethal necrosis involving MCMV + potyviruses) dictate disease outcome.

Optical approaches already exploit the narrow, size-tunable emission spectra of quantum dots for multiplex fluorescence readouts and have been used in laboratory panels to discriminate multiple tomato and grapevine viruses simultaneously (Zhao et al., 2019; Wu et al., 2015). Microarray and bead-based platforms allow high-plex nucleic-acid detection and have been adapted to plant virology for simultaneous screening of dozens of viruses in certification labs (Boonham et al., 2014). More recently, sequencing-based metaviromic workflows (shotgun RNA-Seq or targeted amplicon panels) provide exhaustive multiplex capability and can identify known and unknown viruses in a single run, though they traditionally require laboratory infrastructure (Roossinck, 2017).

On the molecular front, CRISPR-based diagnostics are being engineered for multiplexing by using orthogonal CRISPR effectors or barcoded reporters, allowing simultaneous detection of several RNA viruses such as TMV, TEV, and PVX in a single test (Marqués et al., 2022; Tanny et al., 2023; Zhan et al., 2024). Electrochemical sensor arrays interdigitated electrodes each functionalized with a different probe also permit parallel readouts with compact electronics, and prototypes have demonstrated multi-target detection in small-grain and horticultural crops (Cesewski & Johnson, 2020).

The key research directions for multiplexing are (1) robust probe design that avoids cross-reactivity in the field, (2) inexpensive readout hardware (smartphone optics, portable potentiostats, or low-cost imager chips), and (3) simplified sample-prep or direct-sap testing so multiplex tests are truly field-deployable rather than lab-bound (Boonham et al., 2014; Roossinck, 2017).

11.2. Self-Powered and Energy-Autonomous Sensors

Field deployment at scale requires energy autonomy. Self-powered biosensors eliminate or greatly reduce reliance on batteries and external power, a critical advantage for remote smallholder systems. Two main families of self-powering approaches are maturing: (i) mechanical-energy harvesters such as triboelectric and piezoelectric nanogenerators that convert wind, mechanical agitation, or plant motion into electrical power (Wang, 2017), and (ii) bioenergy harvesters including enzymatic biofuel cells and microbial fuel cells that oxidize plant exudates or sugars to produce current (Kjeang et al., 2009; Zhang et al., 2019).

Triboelectric nanogenerators (TENGs) have been demonstrated to drive low-power sensors and wireless transmitters, and recent studies have integrated TENG modules with agricultural monitoring sensors; analogous integrations with biosensors for virus detection are now feasible (Wang, 2017). Enzymatic biofuel cells using glucose oxidase or laccase can power portable electrochemical strips by harvesting chemical energy from plant sap, and pilot work has already shown proof-of-concept powering of lateral-flow readers and LED indicators (Kjeang et al., 2009).

11.3. Biodegradable, Sustainable Sensor Materials

Sustainability and biosafety concerns about nanomaterial persistence are driving the development of biodegradable sensor substrates and transient electronics. Paper-based and cellulose-derived platforms are already common for disposable lateral-flow tests and are increasingly functionalized with biodegradable conductive inks (carbon-based or polymer–metal composites) to create eco-friendly electrochemical strips (Alaswad et al., 2022).

Silk fibroin, cellulose nanofibrils, and polylactic acid (PLA) have been demonstrated as substrates for transient electronics and biodegradable RFID tags; when combined with biodegradable conductive traces (graphite, carbon nanotube inks with rapid environmental breakdown modifiers), they can minimize ecological footprint once sensors are discarded in fields (Gao et al., 2019). The pressing application here is single-use, wide-distribution screening: for example, farmers could deploy inexpensive, biodegradable test strips for TYLCV or BBTV during planting seasons, and these strips would degrade harmlessly if left in soil. Research is focusing on improving the stability-window (shelf life) of biodegradable sensors so they remain robust during expected use lifetimes but degrade thereafter, and on ensuring that sensing performance is not sacrificed for biodegradability (Gao et al., 2019; Alaswad et al., 2022).

11.4. Smart Integration: AI, Predictive Analytics, and Pathogen-Smart Cropping

Biosensors alone are valuable; when integrated into decision-support systems they become transformative. The next-generation vision is pathogen-smart agriculture, where in-field biosensor data, vector monitoring, remote sensing, and weather/soil data feed into AI models that predict outbreak risk and recommend actions. Pilot projects combining IoT biosensor alerts with microclimate models have shown improved early-warning lead times compared with any single modality (Zhang et al., 2025).

On a larger scale, networked biosensor arrays across landscapes can feed epidemiological models that account for vector movement and seasonal dynamics, supporting regional quarantine or seed certification programs (Jones, 2021). From the breeder's perspective, rapid, multiplexed field testing accelerates screening of resistant germplasm against multiple viruses, shortening breeding cycles. Research directions here include standardized data schemas, robust cloud platforms, and explainable AI that yields actionable recommendations rather than black-box outputs (Chowdhury et al., 2021; Jones, 2021).

11.5. Role in Climate-Resilient and Pathogen-Smart Agriculture

Climate change is altering host, vector, and virus distributions, increasing the risk of novel disease emergence in new geographies (Jones, 2016). Biosensors will play a critical role in climate-resilient agriculture by enabling rapid detection of invading viruses and by providing adaptive surveillance that tracks phenological changes in cropping systems.

Early detection allows managers to deploy climate-smart responses, such as adjusting planting dates to avoid peak vector pressure, deploying resistant cultivars in hotspot zones, or targeting integrated pest management measures more precisely. Real-time virus surveillance also contributes to seed-system resilience by ensuring certified virus-free seed movement.

11.6. Challenges and Future Outlook

Plant viruses pose a persistent threat to agricultural sustainability due to their genetic complexity, virulence, and adaptability. Conventional detection techniques ranging from visual inspection to molecular assays are often limited by their labor intensive nature, need for laboratory infrastructure, and inability to provide rapid, field-based diagnostics. Although serological and nucleic acid-based methods have improved detection accuracy, they remain insufficient for large-scale, on-site monitoring.

Despite significant advancements, several technical and socio economic barriers hinder the large scale deployment of nanobiosensors in plant virus detection. Achieving laboratory level sensitivity and specificity in cost-effective, portable, and multiplexed formats remains a major engineering challenge. Fabrication at the nanoscale requires precise control of material properties, making mass production, reproducibility, and cost optimization difficult. The integration of nanomaterials, microfluidics, and transducer systems further complicates manufacturing standardization and scalability.

Sample preparation from plant tissues also poses a limitation, simplifying or eliminating this step through more robust and self-contained assay chemistries is critical for field applicability. Moreover, long term field validation across diverse environments is necessary to ensure consistent performance and reliability, especially against rapidly evolving viral quasispecies. (Roossinck, 2017).

From a regulatory perspective, the deployment of nanobiosensors faces uncertainties regarding the approval of novel nanomaterials and CRISPR-based platforms. Clear guidelines are needed to evaluate environmental safety, material biocompatibility, and potential ecological impact. Without harmonized standards, commercialization and international adoption will remain constrained.

Socio economically, large scale implementation depends on affordability, accessibility, and integration into agricultural extension services. Effective adoption requires that detection data translate into actionable management decisions rather than isolated diagnostic outputs. Addressing these challenges will require transdisciplinary collaboration integrating materials science, molecular biology, nanofabrication engineering, and regulatory policy (Rajani et al., 2022).

Future research should focus on scalable fabrication techniques, modular sensor architectures, and regulatory frameworks tailored to agricultural nanodiagnostics. With continued innovation, nanobiosensors hold the potential to enable real-time, field-deployable, and economically viable virus detection systems, transforming plant health management and strengthening global food security.

12. CONCLUSION

The early and precise detection of plant viruses remains a cornerstone of sustainable agriculture and global food security. Plant viruses, including Tobacco mosaic virus (TMV), Cucumber mosaic virus (CMV), Potato virus Y (PVY), Tomato yellow leaf curl virus (TYLCV), Tomato brown rugose fruit virus (ToBRFV), and Banana bunchy top virus (BBTV), pose significant threats to crop productivity and quality worldwide. Traditional detection methods, such as ELISA, RT-PCR, and lateral flow assays, have long provided essential tools for virus surveillance and management. ELISA enables high-throughput screening for known viruses, while molecular methods like RT-PCR offer improved sensitivity and specificity. However, these approaches are limited by low detection capability during early infections, inability to detect unknown or highly divergent viral strains, dependency on laboratory infrastructure, and complications arising from mixed infections (Hull, 2013; Mumford et al., 2006). The emergence of rapidly evolving pathogens, exemplified by ToBRFV and diverse begomoviruses such as TYLCV, underscores the urgent need for advanced diagnostic platforms that can provide rapid, sensitive, and multiplexed detection in real-world field conditions.

In this context, nano-biosensors have emerged as transformative tools in plant virus diagnostics. By integrating specific bioreceptors with nanomaterials and versatile transducers, these devices achieve ultra-sensitive detection, rapid response, and, in many cases, multiplexed monitoring. Electrochemical nano-biosensors, enhanced with gold nanoparticles or carbon nanotubes, have enabled detection of viral proteins at femtomolar levels, demonstrating superior sensitivity over traditional ELISA and lateral flow assays for TMV, PVY, and CMV (Parkhe & Tiwari, 2024; Zhang et al., 2019). Optical platforms, including quantum dot-based fluorescence sensors and surface plasmon resonance (SPR) systems, facilitate simultaneous detection of multiple viruses, such as TYLCV, CMV, and TMV, in tomato, cucumber, and pepper crops (Zhang & Johnson, 2006; Wu et al., 2021). Piezoelectric and quartz crystal microbalance (QCM) sensors provide label-free, real-time monitoring of viral binding events, exemplified by TMV detection (Wu et al., 2019). Furthermore, magnetic nanoparticle-based and CRISPR-Cas integrated platforms enable virus concentration and sequence-specific recognition, allowing portable and field-deployable assays for viruses such as BBTV, PVY, TYLCV, and ToBRFV (Mahas et al., 2019; Wani et al., 2024).

The integration of nanomaterials enhances sensitivity and performance by increasing surface area for bioreceptor immobilization, amplifying signals, and enabling rapid electron transfer or optical response. Graphene field-effect transistors (GFETs) functionalized with virus-specific probes, for example, have demonstrated label-free detection of CMV and TMV in plant sap, even at low viral titres (Alam, 2024). Quantum dots and gold nanoparticles allow multiplexed optical detection, while paper-based and smartphone-integrated platforms facilitate real-time, on-site monitoring with minimal technical expertise. This convergence of nanotechnology and digital tools positions nano-biosensors as a critical component of precision agriculture, allowing timely interventions to limit virus spread and mitigate crop losses.

Despite these advancements, challenges remain. The high genetic variability of plant viruses necessitates frequent updates of bioreceptors and detection primers. Low virus titres during early infections, mixed infections, and sample complexity can complicate detection. Additionally, cost, scalability, regulatory approval, and farmer adoption present barriers to widespread implementation, particularly in resource-limited regions. Addressing these challenges will require continued innovation in bioreceptor design, nanomaterial functionalization, and integration with portable, low-cost devices, alongside regulatory guidance and extension efforts to facilitate adoption. Looking ahead, the development of multiplexed, self-powered, and biodegradable nano-biosensors promises to revolutionize plant virus diagnostics. Advances in CRISPR-based detection, nanomaterial engineering, and Internet of Things (IoT) integration can enable automated, real-time surveillance networks capable of predicting and mitigating virus outbreaks. Such platforms will be particularly valuable for climate-resilient agriculture, enabling proactive management of emerging pathogens while safeguarding global food systems.

In summary, the transition from conventional diagnostic methods to nano-biosensor-based platforms represents a paradigm shift in plant virus management. Nano-biosensors provide unprecedented sensitivity, rapid detection, field deployability, and multiplexing capabilities for economically important viruses, including TMV, CMV, PVY, TYLCV, ToBRFV, and BBTV. By enabling early intervention, reducing crop losses, and supporting precision agriculture, these next-generation diagnostics have the potential to transform plant disease management. Continued research, coupled with strategic implementation and stakeholder engagement, will be essential to fully harness the potential of nano-biosensors in ensuring global agricultural sustainability and food security.

REFERENCES

Adams, A. N., & Clark, M. F. (1977). Characteristics of the microplate method of enzyme-linked immunosorbent assay for the detection of plant viruses. *The Journal of General Virology*, *34*(3), 475–483. DOI: 10.1099/0022-1317-34-3-475

Alabi, O. J., Al Rwahnih, M., Karthikeyan, G., Poojari, S., Fuchs, M., Rowhani, A., & Naidu, R. A. (2011). Grapevine leafroll-associated virus 1 occurs as genetically diverse populations. *Phytopathology*, *101*(12), 1446–1456. DOI: 10.1094/PHYTO-04-11-0114

Alam, M. A. (2024). Graphene based electrochemical biosensors for the detection of cardiac biomarkers. *Biosensors & Bioelectronics: X*, *20*, 100515. DOI: 10.1016/j.biosx.2024.100515

Alaswad, S. O., Mahmoud, A. S., & Arunachalam, P. (2022). Recent advances in biodegradable polymers and their biological applications: A brief review. *Polymers*, *14*(22), 4924. DOI: 10.3390/polym14224924

Ardekani, L. S., & Thulstrup, P. W. (2022). Gold nanoparticle-mediated lateral flow assays for detection of host antibodies and covid-19 proteins. *Nanomaterials (Basel, Switzerland)*, *12*(9), 1456. DOI: 10.3390/nano12091456

Aydin, S., Emre, E., Ugur, K., Aydin, M. A., Sahin, İ., Cinar, V., & Akbulut, T. (2025). An overview of ELISA: A review and update on best laboratory practices for quantifying peptides and proteins in biological fluids. *The Journal of International Medical Research*, *53*(2), 03000605251315913. DOI: 10.1177/03000605251315913

Baldodiya, G. M., Saikia, R., Paswan, R. R., & Borah, B. K. (2022). Sustainable agriculture and viral diseases of plants: An overview. In Singh, H. B., & Vaishnav, A. (Eds.), *New and Future Developments in Microbial Biotechnology and Bioengineering* (pp. 419–434). Elsevier., DOI: 10.1016/B978-0-323-85577-8.00011-1

Bar-Joseph, M., & Lee, R. F. (1989). Citrus tristeza virus. *AAB Descriptions of plant viruses, 353*(7).

Bendel, N., Backhaus, A., Kicherer, A., Köckerling, J., Maixner, M., Jarausch, B., Biancu, S., Klück, H.-C., Seiffert, U., Voegele, R. T., & Töpfer, R. (2020). Detection of two different grapevine yellows in vitis vinifera using hyperspectral imaging. *Remote Sensing*, *12*(24), 4151. DOI: 10.3390/rs12244151

Besati, M., Safarnejad, M. R., Aliahmadi, A., Farzaneh, M., Ruiz, R., Montagud-Martínez, R., Rodrigo, G., & Rafati, H. (2025). Detection of tomato brown rugose fruit virus through CRISPR-Cas12a and CRISPR-Cas9 systems. *Scientific Reports*, *15*(1), 25638. DOI: 10.1038/s41598-025-11825-x

Bonyadi, F., Kavruk, M., Ucak, S., Cetin, B., Bayramoglu, G., Dursun, A. D., Arica, Y., & Ozalp, V. C. (2024). Real-time biosensing bacteria and virus with quartz crystal microbalance: Recent advances, opportunities, and challenges. *Critical Reviews in Analytical Chemistry*, *54*(8), 2888–2899. DOI: 10.1080/10408347.2023.2211164

Boonham, N., King, E., & Mumford, R. (2014). Methods in plant virus diagnostics: From ELISA to next-generation sequencing. *Virus Research*, *186*, 20–31.

Buja, I., Sabella, E., Monteduro, A. G., Rizzato, S., Bellis, L. D., Elicio, V., Formica, L., Luvisi, A., & Maruccio, G. (2022). Detection of ampelovirus and nepovirus by lab-on-a-chip: A promising alternative to elisa test for large scale health screening of grapevine. *Biosensors (Basel)*, *12*(3), 147. DOI: 10.3390/bios12030147

Candresse, T., Lanneau, M., Revers, F., Kofalvi, S., & Macquaire, G. (2000). Pcr-based techniques for the detection of plant viruses and viroids. *Acta Horticulturae*, (530), 61–67. DOI: 10.17660/ActaHortic.2000.530.5

Carpenter, A. C., Paulsen, I. T., & Williams, T. C. (2018). Blueprints for biosensors: Design, limitations, and applications. *Genes*, *9*(8), 375. DOI: 10.3390/genes9080375

Carrillo-Lopez, L. M., Villanueva-Verduzco, C., Villanueva-Sánchez, E., Fajardo-Franco, M. L., Aguilar-Tlatelpa, M., Ventura-Aguilar, R. I., & Soto-Hernández, R. M. (2024). Nanomaterials for plant disease diagnosis and treatment: A review. *Plants*, *13*(18), 2634.

Cesewski, E., & Johnson, B. N. (2020). Electrochemical biosensors for pathogen detection. *Biosensors & Bioelectronics*, *159*, 112214. DOI: 10.1016/j.bios.2020.112214

Chaturvedi, A., Tripathi, D., & Ranjan, R. (2025). Nano-enabled biosensors in early detection of plant diseases. *Frontiers in Nanotechnology*, *7*, 1545792. DOI: 10.3389/fnano.2025.1545792

Chen, Z., Mao, S., Zhang, W., Fan, X., Wu, W., Liu, C., Zhao, K., & Lu, R. (2021). Rapid visual detection method for barley yellow mosaic virus using reverse transcription loop-mediated isothermal amplification(Rt-lamp). *Plant Disease*, *105*(9), 2658–2663. DOI: 10.1094/PDIS-06-20-1216-RE

Choi, H. K., & Yoon, J. (2023). Nanotechnology-assisted biosensors for the detection of viral nucleic acids: An overview. *Biosensors (Basel)*, *13*(2), 208. DOI: 10.3390/bios13020208

Chowdhury, M. E. H., Rahman, T., Khandakar, A., Ibtehaz, N., Khan, A. U., Khan, M. S., Al-Emadi, N., Reaz, M. B. I., Islam, M. T., & Ali, S. H. M. (2021). Tomato leaf diseases detection using deep learning technique. In *Technology in Agriculture*. IntechOpen., DOI: 10.5772/intechopen.97319

Czosnek, H., & Laterrot, H. (1997). A worldwide survey of tomato yellow leaf curl viruses. *Archives of Virology*, *142*(7), 1391–1406. DOI: 10.1007/s007050050168

Dinant, G. J., Knottnerus, J. A., & Van Wersch, J. W. (1991). Discriminating ability of the erythrocyte sedimentation rate: A prospective study in general practice. *The British Journal of General Practice*, *41*(350), 365.

Fang, Y., & Ramasamy, R. (2015). Current and prospective methods for plant disease detection. *Biosensors (Basel)*, *5*(3), 537–561. DOI: 10.3390/bios5030537

Fauquet, C., & Fargette, D. (1990). African cassava mosaic virus: Etiology, epidemiology and control. *Plant Disease*, *74*(6), 404–411.

French, R., & Stenger, D. C. (2003). Evolution of Wheat streak mosaic virus: Dynamics of population growth within plants may explain limited variation. *Annual Review of Phytopathology*, *41*(1), 199–214.

Gao, G., Yu, J., Yang, X., Pang, Y., Zhao, J., Pan, C., Sun, Q., & Wang, Z. L. (2019). Triboiontronic Transistor of MoS2. *Advanced Materials*, *31*(7), 1806905. DOI: 10.1002/adma.201806905

Georgeous, J., AlSawaftah, N., Abuwatfa, W. H., & Husseini, G. A. (2024). Review of gold nanoparticles: Synthesis, properties, shapes, cellular uptake, targeting, release mechanisms and applications in drug delivery and therapy. *Pharmaceutics*, *16*(10), 1332. DOI: 10.3390/pharmaceutics16101332

Gong, X.-Y., Wang, Z.-H., Bashir, M., Tang, T., Gan, X., & Yang, W.-C. (2025). Recent application of CRISPR/Cas in plant disease detection. *Trends in Analytical Chemistry*, *189*, 118251. DOI: 10.1016/j.trac.2025.118251

Güller, H. K., & Paylan, İ. C. (2024). Simultaneous detection of grapevine viruses via multiplex reverse transcription polymerase chain reaction: An academic approach. *Applied Fruit Science*, *66*(3), 1027–1035. DOI: 10.1007/s10341-024-01076-x

Hajimorad, M. R., Domier, L. L., Tolin, S. A., Whitham, S. A., & Saghai Maroof, M. A. (2018). Soybean mosaic virus: A successful potyvirus with a wide distribution but restricted natural host range. *Molecular Plant Pathology*, *19*(7), 1563–1579.

Halabi, M. H., Oladokun, J. O., & Nath, P. D. (2022). Rapid detection of Potato leafroll virus and Potato virus Y by reverse transcription loop-mediated isothermal amplification method in north-east India. *Journal of Virological Methods*, *300*, 114363. DOI: 10.1016/j.jviromet.2021.114363

Harun-Ur-Rashid, M., Foyez, T., Jahan, I., Pal, K., & Imran, A. B. (2022). Rapid diagnosis of COVID-19 via nano-biosensor-implemented biomedical utilization: A systematic review. *RSC Advances*, *12*(15), 9445–9465.

Hibino, H. (1996). Biology and epidemiology of rice viruses. *Annual Review of Phytopathology*, *34*(1), 249–274.

Horser, C. L., Harding, R. M., & Dale, J. L. (2001). Banana bunchy top nanovirus DNA-1 encodes the 'master'replication initiation protein. *The Journal of General Virology*, *82*(2), 459–464.

Hull, R. (2013). *Plant virology*. Academic press.

Jablonski, M., Poghossian, A., Severins, R., Keusgen, M., Wege, C., & Schöning, M. (2021). Capacitive field-effect biosensor studying adsorption of tobacco mosaic virus particles. *Micromachines*, *12*(1), 57. DOI: 10.3390/mi12010057

Jarocka, U., Wąsowicz, M., Radecka, H., Malinowski, T., Michalczuk, L., & Radecki, J. (2011). Impedimetric immunosensor for detection of plum pox virus in plant extracts. *Electroanalysis*, *23*(9), 2197–2204. DOI: 10.1002/elan.201100152

Johansen, S. R. (1994). The role of the constant and linear terms in cointegration analysis of nonstationary variables. *Econometric Reviews*, *13*(2), 205–229.

Jones, R. A. C. (2016). Future scenarios for plant virus pathogens as climate change progresses. In *Advances in Virus Research* (Vol. 95, pp. 87–147). Elsevier., DOI: 10.1016/bs.aivir.2016.02.004

Jones, R. A. C. (2021). Global plant virus disease pandemics and epidemics. *Plants*, *10*(2), 233. DOI: 10.3390/plants10020233

Karasev, A. V., & Gray, S. M. (2013). Continuous and emerging challenges of Potato virus Y in potato. *Annual Review of Phytopathology*, *51*, 571–586. DOI: 10.1146/annurev-phyto-082712-102332

Katz, E., & Willner, I. (2004). Integrated nanoparticle–biomolecule hybrid systems: Synthesis, properties, and applications. *Angewandte Chemie International Edition*, *43*(45), 6042–6108. DOI: 10.1002/anie.200400651

Khan, A., Malik, S., Bilal, M., Ali, N., Ni, L., Gao, X., & Hong, K. (2021). 12—Polymer-coated magnetic nanoparticles. In Kanwar, S., Kumar, A., Nguyen, T. A., Sharma, S., & Slimani, Y. (Eds.), *Biopolymeric Nanomaterials* (pp. 275–292). Elsevier., DOI: 10.1016/B978-0-12-824364-0.00019-8

Kim, C., Kang, M. S., Raja, I. S., Joung, Y. K., & Han, D. W. (2024). Advancements in nanobiosensor technologies for in-vitro diagnostics to point of care testing. *Heliyon*, *10*(22).

Kiran, V., Harini, K., Thirumalai, A., Girigoswami, K., & Girigoswami, A. (2024). Nanotechnology's role in ensuring food safety and security. *Biocatalysis and Agricultural Biotechnology*, *58*, 103220. DOI: 10.1016/j.bcab.2024.103220

Kjeang, E., Djilali, N., & Sinton, D. (2009). Microfluidic fuel cells: A review. *Journal of Power Sources*, *186*(2), 353–369. DOI: 10.1016/j.jpowsour.2008.10.011

Kuan, C.-P., Tsai, C.-H., Tseng, C.-S., & Yang, T.-C. (2022). Development of a bead-based assay for detection of three banana-infecting viruses. *PeerJ*, *10*, e13409. DOI: 10.7717/peerj.13409

Kulabhusan, P. K., Tripathi, A., & Kant, K. (2022). Gold nanoparticles and plant pathogens: An overview and prospective for biosensing in forestry. *Sensors (Basel)*, *22*(3), 1259. DOI: 10.3390/s22031259

Kumar, H., Kumari, N., & Sharma, R. (2020). Nanocomposites (Conducting polymer and nanoparticles) based electrochemical biosensor for the detection of environment pollutant: Its issues and challenges. *Environmental Impact Assessment Review*, *85*, 106438. DOI: 10.1016/j.eiar.2020.106438

Lecoq, H., & Desbiez, C. (2012). Viruses of cucurbit crops in the Mediterranean region: An ever-changing picture. [). Academic Press.]. *Advances in Virus Research*, *84*, 67–126.

Legg, J. P., Jeremiah, S. C., Obiero, H. M., Maruthi, M. N., Ndyetabula, I., Okao-Okuja, G., Bouwmeester, H., Bigirimana, S., Tata-Hangy, W., Gashaka, G., Mkamilo, G., Alicai, T., & Lava Kumar, P. (2011). Comparing the regional epidemiology of the cassava mosaic and cassava brown streak virus pandemics in Africa. *Virus Research*, *159*(2), 161–170. DOI: 10.1016/j.virusres.2011.04.018

Lequin, R. M. (2005). Enzyme immunoassay (EIA)/enzyme-linked immunosorbent assay (ELISA). *Clinical Chemistry*, *51*(12), 2415–2418.

Li, C., Zhang, X., Li, J., Fang, T., & Dong, X. (2021). The challenges of modern computing and new opportunities for optics. *PhotoniX*, *2*(1), 20. DOI: 10.1186/s43074-021-00042-0

Liu, C. C., & Dai, Y. (2023). Application of crispr cas systems for biosensing. *Biosensors (Basel)*, *13*(7), 672. DOI: 10.3390/bios13070672

Liu, K., Lin, B., & Lan, X. (2013). Aptamers: A promising tool for cancer imaging, diagnosis, and therapy. *Journal of Cellular Biochemistry*, *114*(2), 250–255. DOI: 10.1002/jcb.24373

MacDonald, S. L., Staid, M., Staid, M., & Cooper, M. L. (2016). Remote hyperspectral imaging of grapevine leafroll-associated virus 3 in cabernet sauvignon vineyards. *Computers and Electronics in Agriculture*, *130*, 109–117. DOI: 10.1016/j.compag.2016.10.003

Madadelahi, M., Romero-Soto, F. O., Kumar, R., Tlaxcala, U. B., & Madou, M. J. (2025). Electrochemical sensors: Types, applications, and the novel impacts of vibration and fluid flow for microfluidic integration. *Biosensors & Bioelectronics*, *272*, 117099. DOI: 10.1016/j.bios.2024.117099

Mahas, A., Aman, R., & Mahfouz, M. (2019). CRISPR-Cas13d mediates robust RNA virus interference in plants. *Genome Biology*, *20*(1), 263. DOI: 10.1186/s13059-019-1881-2

Mahlein, A.-K. (2016). Plant disease detection by imaging sensors – parallels and specific demands for precision agriculture and plant phenotyping. *Plant Disease*, *100*(2), 241–251. DOI: 10.1094/PDIS-03-15-0340-FE

Majumder, S., & Johari, S. (2018). Development of a gold-nano particle based novel dot immunobinding assay for rapid and sensitive detection of Banana bunchy top virus. *Journal of Virological Methods*, *255*, 23–28. DOI: 10.1016/j.jviromet.2018.01.015

Malik, S., Singh, J., Goyat, R., Saharan, Y., Chaudhry, V., Umar, A., Ibrahim, A. A., Akbar, S., Ameen, S., & Baskoutas, S. (2023). Nanomaterials-based biosensor and their applications: A review. *Heliyon*, *9*(9), e19929. DOI: 10.1016/j.heliyon.2023.e19929

Marqués, M.-C., Sánchez-Vicente, J., Ruiz, R., Montagud-Martínez, R., Márquez-Costa, R., Gómez, G., Carbonell, A., Daròs, J.-A., & Rodrigo, G. (2022). Diagnostics of infections produced by the plant viruses tmv, tev, and pvx with crispr-cas12 and crispr-cas13. *ACS Synthetic Biology*, *11*(7), 2384–2393. DOI: 10.1021/acssynbio.2c00090

Martelli, G., Boracchi, P., Ardoino, I., Lozza, L., Bohm, S., Vetrella, G., & Agresti, R. (2012). Axillary dissection versus no axillary dissection in older patients with T1N0 breast cancer: 15-year results of a randomized controlled trial.

Medintz, I. L., Uyeda, H. T., Goldman, E. R., & Mattoussi, H. (2005). Quantum dot bioconjugates for imaging, labelling and sensing. *Nature Materials*, *4*(6), 435–446. DOI: 10.1038/nmat1390

Mehetre, G. T., Leo, V. V., Singh, G., Sorokan, A., Maksimov, I., Yadav, M. K., Upadhyaya, K., Hashem, A., Alsaleh, A. N., Dawoud, T. M., Almaary, K. S., & Singh, B. P. (2021). Current developments and challenges in plant viral diagnostics: A systematic review. *Viruses*, *13*(3), 412. DOI: 10.3390/v13030412

Miller and, W. A., & Rasochová, L. (1997). Barley yellow dwarf viruses. *Annual Review of Phytopathology*, *35*(1), 167–190.

Mishra, R., Shukla, A., & Singh, P. (2021). Quantum dot-based fluorescent biosensor for Rice tungro bacilliform virus detection. *Frontiers in Plant Science*, *12*, 665432. DOI: 10.3389/fpls.2021.665432

Miyazaki, C. M., Shimizu, F. M., & Ferreira, M. (2017). 6—Surface plasmon resonance (Spr) for sensors and biosensors. In Da Róz, A. L., Ferreira, M., de Lima Leite, F., & Oliveira, O. N. (Eds.), *Nanocharacterization Techniques* (pp. 183–200). William Andrew Publishing., DOI: 10.1016/B978-0-323-49778-7.00006-0

Moreno, P., Ambrós, S., Albiach-Martí, M. R., Guerri, J., & Peña, L. (2008). citrus tristeza virus: A pathogen that changed the course of the citrus industry. *Molecular Plant Pathology*, *9*(2), 251–268. DOI: 10.1111/j.1364-3703.2007.00455.x

Mousavi, S. M., Hashemi, S. A., Yari Kalashgrani, M., Omidifar, N., Lai, C. W., Vijayakameswara Rao, N., Gholami, A., & Chiang, W.-H. (2022). The pivotal role of quantum dots-based biomarkers integrated with ultra-sensitive probes for multiplex detection of human viral infections. *Pharmaceuticals*, *15*(7), 880. DOI: 10.3390/ph15070880

Mphuthi, N., Sikhwivhilu, L., & Ray, S. S. (2022). Functionalization of 2d mos2 nanosheets with various metal and metal oxide nanostructures: Their properties and application in electrochemical sensors. *Biosensors (Basel)*, *12*(6), 386. DOI: 10.3390/bios12060386

Mrisho, L. M., Mbilinyi, N. A., Ndalahwa, M., Ramcharan, A. M., Kehs, A. K., McCloskey, P. C., Murithi, H., Hughes, D. P., & Legg, J. P. (2020). Accuracy of a smartphone-based object detection model, plantvillage nuru, in identifying the foliar symptoms of the viral diseases of cassava–cmd and cbsd. *Frontiers in Plant Science*, *11*. Advance online publication. DOI: 10.3389/fpls.2020.590889

Mumford, R., Boonham, N., Tomlinson, J., & Barker, I. (2006). Advances in molecular phytodiagnostics – new solutions for old problems. *European Journal of Plant Pathology*, *116*(1), 1–19. DOI: 10.1007/s10658-006-9037-0

Mutegoa, E., & Sahini, M. G. (2025). Piezoelectric biosensors for healthcare applications. In Hasnain, M. S., Nayak, A. K., & Aminabhavi, T. M. (Eds.), *Fundamentals of Biosensors in Healthcare* (pp. 167–190). Academic Press., DOI: 10.1016/B978-0-443-21658-9.02001-5

Naidu, A. S., Bidlack, W. R., & Clemens, R. A. (1999). Probiotic spectra of lactic acid bacteria (LAB). *Critical Reviews in Food Science and Nutrition*, *39*(1), 13–126.

Narware, J., Chakma, J., Singh, S. P., Prasad, D. R., Meher, J., Singh, P., Bhargava, P., & Sawant, S. B., Pitambara, Singh, J. P., Manzar, N., & Kashyap, A. S. (2025). Nanomaterial-based biosensors: A new frontier in plant pathogen detection and plant disease management. *Frontiers in Bioengineering and Biotechnology*, *13*, 1570318. DOI: 10.3389/fbioe.2025.1570318

Ndunguru, J., Sseruwagi, P., Tairo, F., Stomeo, F., Maina, S., Djinkeng, A., Kehoe, M., & Boykin, L. M. (2015). Analyses of twelve new whole genome sequences of cassava brown streak viruses and ugandan cassava brown streak viruses from east africa: Diversity, supercomputing and evidence for further speciation. *PLoS One*, *10*(10), e0139321. DOI: 10.1371/journal.pone.0139321

Olmedo-Velarde, A., Wilson, J. R., Stallone, M., DeBlasio, S. L., Chappie, J. S., & Heck, M. (2023). Potato leafroll virus molecular interactions with plants and aphids: Gaining a new tactical advantage on an old foe. *Physiological and Molecular Plant Pathology*, *125*, 102015. DOI: 10.1016/j.pmpp.2023.102015

Pande, C. B., & Moharir, K. N. (2023). Application of hyperspectral remote sensing role in precision farming and sustainable agriculture under climate change: A review. In Pande, C. B., Moharir, K. N., Singh, S. K., Pham, Q. B., & Elbeltagi, A. (Eds.), *Climate Change Impacts on Natural Resources, Ecosystems and Agricultural Systems* (pp. 503–520). Springer International Publishing., DOI: 10.1007/978-3-031-19059-9_21

Pandey, P., Parmar, D., Parmar, R. G., & Borah, M. (2025). Soybean mosaic: Symptomatology, epidemiology, and management. In K. P. Singh, N. K. Singh, & A. T (Eds.), Soybean Production Technology: Crop Pests and Diseases (pp. 287–298). Springer Nature. https://doi.org/DOI: 10.1007/978-981-97-8104-1_16

Pappu, H. R., Jones, R. C., & Jain, R. K. (2009). Global status of tospovirus epidemics in diverse cropping systems: Successes achieved and challenges ahead. *Virus Research*, *141*(2), 219–236. DOI: 10.1016/j.virusres.2009.01.009

Park, H., & Jackman, J. A. (2025). Nanoplasmonic biosensors for detecting viruses and combating viral infections. *Npj Biosensing*, *2*(1), 1–16. DOI: 10.1038/s44328-025-00043-0

Parkhe, V. S., & Tiwari, A. P. (2024). Gold nanoparticles-based biosensors: Pioneering solutions for bacterial and viral pathogen detection—a comprehensive review. *World Journal of Microbiology & Biotechnology*, *40*(9), 269. DOI: 10.1007/s11274-024-04072-1

Patel, R., Mitra, B., Vinchurkar, M., Adami, A., Patkar, R., Giacomozzi, F., Lorenzelli, L., & Baghini, M. S. (2022). A review of recent advances in plant-pathogen detection systems. *Heliyon*, *8*(12), e11855. DOI: 10.1016/j.heliyon.2022.e11855

Peng, J., Zhang, J., Xia, Z., Li, Y., Huang, J., & Fan, Z. (2012). Rapid and sensitive detection of Banana bunchy top virus by loop-mediated isothermal amplification. *Journal of Virological Methods*, *185*(2), 254–258. DOI: 10.1016/j.jviromet.2012.06.026

Perdikaris, A., Vassilakos, N., Yiakoumettis, I., Kektsidou, O., & Kintzios, S. (2011). Development of a portable, high throughput biosensor system for rapid plant virus detection. *Journal of Virological Methods*, *177*(1), 94–99. DOI: 10.1016/j.jviromet.2011.06.024

Perestam, A. T., Fujisaki, K. K., Nava, O., & Hellberg, R. S. (2017). Comparison of real-time PCR and ELISA-based methods for the detection of beef and pork in processed meat products. *Food Control*, *71*, 346–352.

Pramanik, D., Shelake, R. M., Park, J., Kim, M. J., Hwang, I., Park, Y., & Kim, J.-Y. (2021). Crispr/cas9-mediated generation of pathogen-resistant tomato against tomato yellow leaf curl virus and powdery mildew. *International Journal of Molecular Sciences*, *22*(4), 1878. DOI: 10.3390/ijms22041878

Qazi, J. (2016). Banana bunchy top virus and the bunchy top disease. *Journal of General Plant Pathology*, *82*(1), 2–11. DOI: 10.1007/s10327-015-0642-7

Qin, R., Zhang, Y., Xu, H., & Nie, P. (2024). Gold nanoparticle-modified electrodes for electrochemical sensing of ammonia nitrogen in water. *ACS Applied Nano Materials*, *7*(1), 577–593. DOI: 10.1021/acsanm.3c04757

Rahman, M. A., Pal, R. K., Islam, N., Freeman, R., Berthiaume, F., Mazzeo, A., & Ashraf, A. (2023). A facile graphene conductive polymer paper based biosensor for dopamine, tnf-α, and il-6 detection. *Sensors (Basel)*, *23*(19), 8115. DOI: 10.3390/s23198115

Rajani, Mishra, P., Kumari, S., Saini, P., & Meena, R. K. (2022). Role of nanotechnology in management of plant viral diseases. *Materials Today: Proceedings*, *69*, 1–10. DOI: 10.1016/j.matpr.2022.06.355

Rajeshkumar, R., Moorthy, P. S., Raveendran, M., Karthikeyan, G., Gomathi, V., Djanaguiraman, M., Rajkishore, S. K., & Abinaya, K. (2025). Mitigating plant viruses through nanoparticles: Mechanisms, applications and future perspectives. *Plant Nano Biology*, *12*, 100155. DOI: 10.1016/j.plana.2025.100155

Rajput, V., Minkina, T., Mazarji, M., Shende, S., Sushkova, S., Mandzhieva, S., Burachevskaya, M., Chaplygin, V., Singh, A., & Jatav, H. (2020). Accumulation of nanoparticles in the soil-plant systems and their effects on human health. *Annals of Agricultural Science*, *65*(2), 137–143. DOI: 10.1016/j.aoas.2020.08.001

Ramadan, S., Lobo, R., Zhang, Y., Xu, L., Shaforost, O., Kwong Hong Tsang, D., Feng, J., Yin, T., Qiao, M., Rajeshirke, A., Jiao, L. R., Petrov, P. K., Dunlop, I. E., Titirici, M.-M., & Klein, N. (2021). Carbon-dot-enhanced graphene field-effect transistors for ultrasensitive detection of exosomes. *ACS Applied Materials & Interfaces*, *13*(7), 7854–7864. DOI: 10.1021/acsami.0c18293

Ramesh, M., Janani, R., Deepa, C., & Rajeshkumar, L. (2022). Nanotechnology-enabled biosensors: A review of fundamentals, design principles, materials, and applications. *Biosensors (Basel)*, *13*(1), 40. DOI: 10.3390/bios13010040

Ravi, K., Vadez, V., Isobe, S., Mir, R. R., Guo, Y., Nigam, S. N., & Varshney, R. K. (2011). Identification of several small main-effect QTLs and a large number of epistatic QTLs for drought tolerance related traits in groundnut (Arachis hypogaea L.). *Theoretical and Applied Genetics*, *122*(6), 1119–1132.

Razmi, A., Golestanipour, A., Nikkhah, M., Bagheri, A., Shamsbakhsh, M., & Malekzadeh-Shafaroudi, S. (2019). Localized surface plasmon resonance biosensing of tomato yellow leaf curl virus. *Journal of Virological Methods*, *267*, 1–7. DOI: 10.1016/j.jviromet.2019.02.004

Redinbaugh, M. G., & Stewart, L. R. (2018). Maize lethal necrosis: An emerging, synergistic viral disease. *Annual Review of Virology*, *5*(1), 301–322.

Regulation—Echa. (n.d.). Retrieved 17 September 2025, from https://euon.echa.europa.eu/url

Robertson, G., Burger, J., & Campa, M. (2022). CRISPR/Cas-based tools for the targeted control of plant viruses. *Molecular Plant Pathology*, *23*(11), 1701–1718. DOI: 10.1111/mpp.13252

Roossinck, M. J. (2001). Cucumber mosaic virus, a model for RNA virus evolution. *Molecular Plant Pathology*, *2*(2), 59–63. DOI: 10.1046/j.1364-3703.2001.00058.x

Roossinck, M. J. (2017). Deep sequencing for discovery and evolutionary analysis of plant viruses. *Virus Research*, *239*, 82–86. DOI: 10.1016/j.virusres.2016.11.019

S, A. N., R, H., S, R., Gaur, S., & R, S. (2023). A smart agriculturing IoT system for banana plants disease detection through inbuilt compressed sensing devices. Multimedia Tools and Applications, 82(29), 45519–45539. https://doi.org/DOI: 10.1007/s11042-023-15442-6

Saffa, M. D., Samura, A. E., Bah, M. A., Eni, A. O., Tibiri, E. B., Sagnon, A., Tiendrébéogo, F., Pita, J. S., Norman, P. E., & Johnson, R. A. B. (2025). Detection of cassava mosaic disease and assessment of selected agronomic traits of cassava(Manihot esculenta). *Horticulturae, 11*(6), 618. DOI: 10.3390/horticulturae11060618

Saifuddin, S. A., Rashid, R., Nor Azmi, N. J., & Mohamad, S. (2024). Colorimetric strategies applicable for loop-mediated isothermal amplification. *Journal of Microbiological Methods, 223*, 106981. DOI: 10.1016/j.mimet.2024.106981

Samacoits, A., Nimsamer, P., Mayuramart, O., Chantaravisoot, N., Sitthi-amorn, P., Nakhakes, C., Luangkamchorn, L., Tongcham, P., Zahm, U., Suphanpayak, S., Padungwattanachoke, N., Leelarthaphin, N., Huayhongthong, H., Pisitkun, T., Payungporn, S., & Hannanta-anan, P. (2021). Machine learning-driven and smartphone-based fluorescence detection for crispr diagnostic of sars-cov-2. *ACS Omega, 6*(4), 2727–2733. DOI: 10.1021/acsomega.0c04929

Sattar, M. N., Kvarnheden, A., Saeed, M., & Briddon, R. W. (2013). Cotton leaf curl disease–an emerging threat to cotton production worldwide. *The Journal of General Virology, 94*(4), 695–710.

Schachinger, F., Chang, H., Scheiblbrandner, S., & Ludwig, R. (2021). Amperometric biosensors based on direct electron transfer enzymes. *Molecules (Basel, Switzerland), 26*(15), 4525. DOI: 10.3390/molecules26154525

Schlundt, J. (2014). Institutions involved in food safety: World health organization(Who). Encyclopedia of Food Safety, 359–364. https://doi.org/DOI: 10.1016/B978-0-12-378612-8.00008-1

Scholthof, K.-B. G. (2004). Tobacco mosaic virus: A model system for plant biology. *Annual Review of Phytopathology, 42*, 13–34. DOI: 10.1146/annurev.phyto.42.040803.140322

Seal, S., & Muller, E. (2007). Molecular analysis of a full-length sequence of a new yam badnavirus from Dioscorea sansibarensis. *Archives of Virology, 152*(4), 819–825.

Sekhon, B. (2014). Nanotechnology in agri-food production: An overview. *Nanotechnology, Science and Applications, 31*. Advance online publication. DOI: 10.2147/NSA.S39406

Selvarajan, R., Kanichelvam, P. S., Balasubramanian, V., & Sethurama Subramanian, S. (2020). A rapid and sensitive lateral flow immunoassay (Lfia) test for the on-site detection of banana bract mosaic virus in banana plants. *Journal of Virological Methods, 284*, 113929. DOI: 10.1016/j.jviromet.2020.113929

Shi, Z., Li, Y., Wu, X., Zhang, K., Gu, J., Sun, W., Li, C. M., & Guo, C. X. (2023). Graphdiyne chelated AuNPs for ultrasensitive electrochemical detection of tyrosine. *Chemical Communications, 59*(91), 13647–13650. DOI: 10.1039/D3CC04148D

Shivashakarappa, K., Reddy, V., Tupakula, V. K., Farnian, A., Vuppula, A., & Gunnaiah, R. (2022). Nanotechnology for the detection of plant pathogens. *Plant Nano Biology, 2*, 100018. DOI: 10.1016/j.plana.2022.100018

Siddiqua, A., Kabir, M. A., Ferdous, T., Ali, I. B., & Weston, L. A. (2022). Evaluating plant disease detection mobile applications: Quality and limitations. *Agronomy (Basel), 12*(8), 1869. DOI: 10.3390/agronomy12081869

Singh, S., Sangwan, S., Sharma, P., Devi, P., & Moond, M. (2021). Nanotechnology for sustainable agriculture: An emerging perspective. *Journal of Nanoscience and Nanotechnology*, *21*(6), 3453–3465. DOI: 10.1166/jnn.2021.19012

Sklâdal, P. (2024). Piezoelectric biosensors: Shedding light on principles and applications. *Microchimica Acta*, *191*(4), 184. DOI: 10.1007/s00604-024-06257-9

Star, A., Tu, E., Niemann, J., Gabriel, J.-C. P., Joiner, C. S., & Valcke, C. (2006). Label-free detection of DNA hybridization using carbon nanotube network field-effect transistors. *Proceedings of the National Academy of Sciences of the United States of America*, *103*(4), 921–926. DOI: 10.1073/pnas.0504146103

Strausbaugh, C. A., Rearick, E., & Camp, S. (2008). Influence of curly top and Poncho Beta on storability of sugarbeet. *Journal of Sugar Beet Research*, *45*(1/2), 31.

Takemura, Y., Endo, H., Hibi, T., Nakano, Y., Seishima, R., Takeuchi, M., & Kitagawa, Y. (2024). Impact of the COVID-19 pandemic on short-term outcomes after pancreaticoduodenectomy for pancreatic cancer: A retrospective study from the Japanese National Clinical Database, 2018–2021. *Annals of Gastroenterological Surgery*, *8*(5), 877–887.

Tanny, T., Sallam, M., Soda, N., Nguyen, N.-T., Alam, M., & Shiddiky, M. J. A. (2023). Crispr/cas-based diagnostics in agricultural applications. *Journal of Agricultural and Food Chemistry*, *71*(31), 11765–11788. DOI: 10.1021/acs.jafc.3c00913

Towards a global guidance framework for the responsible use of life sciences: Summary report of consultations on the principles, gaps and challenges of biorisk management, May 2022. (n.d.). Retrieved September 18, 2025, from https://www.who.int/publications/i/item/WHO-SCI-RFH-2022.01

Uehara-Ichiki, T., Shiba, T., Matsukura, K., Ueno, T., Hirae, M., & Sasaya, T. (2013). Detection and diagnosis of rice-infecting viruses. *Frontiers in Microbiology*, *4*, 289. DOI: 10.3389/fmicb.2013.00289

Unit, B. (2025, September 4). The cartagena protocol on biosafety. The Biosafety Clearing-House (BCH). http://bch.cbd.int/protocol

Verma, S., Chug, A., Singh, A. P., Sharma, S., & Rajvanshi, P. (2019). Deep learning-based mobile application for plant disease diagnosis: A proof of concept with a case study on tomato plant. In *Applications of image processing and soft computing systems in agriculture* (pp. 242–271). IGI global.

Villamor, D. E. V., Ho, T., Al Rwahnih, M., Martin, R. R., & Tzanetakis, I. E. (2019). High throughput sequencing for plant virus detection and discovery. *Phytopathology*, *109*(5), 716–725. DOI: 10.1094/PHYTO-07-18-0257-RVW

Virk, V., Deepak, H., Taneja, K., Srivastava, R., & Giri, S. (2024). Amelioration in nanobiosensors for the control of plant diseases: Current status and future challenges. *Frontiers in Nanotechnology*, *6*, 1310165. DOI: 10.3389/fnano.2024.1310165

Wang, H., Ma, L., Jin, Z., Cui, Z., Yang, H., & Miao, M. (2023). Highly sensitive fluorescence detection of tobacco mosaic virus RNA based on polysaccharide and ARGET ATRP double signal amplification. *Talanta*, *257*, 124360. DOI: 10.1016/j.talanta.2023.124360

Wang, L., Liu, Z., Xia, X., Yang, C., Huang, J., & Wan, S. (2017). Colorimetric detection of Cucumber green mottle mosaic virus using unmodified gold nanoparticles as colorimetric probes. *Journal of Virological Methods*, *243*, 113–119. DOI: 10.1016/j.jviromet.2017.01.010

Wang, Q., Li, Q., Yang, X., Wang, K., Du, S., Zhang, H., & Nie, Y. (2016). Graphene oxide–gold nanoparticles hybrids-based surface plasmon resonance for sensitive detection of microRNA. *Biosensors & Bioelectronics*, *77*, 1001–1007. DOI: 10.1016/j.bios.2015.10.091

Wang, Y. M., Ostendorf, B., Gautam, D., Habili, N., & Pagay, V. (2022). Plant viral disease detection: From molecular diagnosis to optical sensing technology—a multidisciplinary review. *Remote Sensing*, *14*(7), 1542. DOI: 10.3390/rs14071542

Wang, Z. L. (2017). Catch wave power in floating nets. *Nature*, *542*(7640), 159–160. DOI: 10.1038/542159a

Wani, A. K., Akhtar, N., & Mir, T. ul G., Chopra, C., Singh, R., Hong, J. C., & Kadam, U. S. (2024). CRISPR/Cas12a-based biosensors for environmental monitoring and diagnostics. *Environmental Technology & Innovation*, *34*, 103625. DOI: 10.1016/j.eti.2024.103625

Ward, E., Foster, S. J., Fraaije, B. A., & Mccartney, H. A. (2004). Plant pathogen diagnostics: Immunological and nucleic acid-based approaches. *Annals of Applied Biology*, *145*(1), 1–16. DOI: 10.1111/j.1744-7348.2004.tb00354.x

Wei, Taiyun, and Yi Li. "Rice Reoviruses in Insect Vectors." Annual Review of Phytopathology, vol. 54, Aug. 2016, pp. 99–120. PubMed, https://doi.org/.DOI: 10.1146/annurev-phyto-080615-095900

Wong, Y. J., Luo, Y., Loh, X. J., & Chen, X. (2025). Nondestructive sensing of plant-borne chemicals: Biomarkers, agrochemicals, and pollutants. *Analysis & Sensing*, *2500037*. Advance online publication. DOI: 10.1002/anse.202500037

Wu, F., Yuan, H., Zhou, C., Mao, M., Liu, Q., Shen, H., Cen, Y., Qin, Z., Ma, L., & Song Li, L. (2016). Multiplexed detection of influenza A virus subtype H5 and H9 via quantum dot-based immunoassay. *Biosensors & Bioelectronics*, *77*, 464–470. DOI: 10.1016/j.bios.2015.10.002

Wu, Q., Ding, S.-W., Zhang, Y., & Zhu, S. (2015). Identification of viruses and viroids by next-generation sequencing and homology-dependent and homology-independent algorithms. *Annual Review of Phytopathology*, *53*(1), 425–444. DOI: 10.1146/annurev-phyto-080614-120030

Wu, S., Guo, Y., Wang, W., Zhou, J., & Zhang, Q. (2019). Label-free biosensing using a microring resonator integrated with poly-(Dimethylsiloxane) microfluidic channels. *The Review of Scientific Instruments*, *90*(3), 035004. DOI: 10.1063/1.5074134

Wu, W., Yu, X., Wu, J., Wu, T., Fan, Y., Chen, W., Zhao, M., Wu, H., Li, X., & Ding, S. (2021). Surface plasmon resonance imaging-based biosensor for multiplex and ultrasensitive detection of NSCLC-associated exosomal miRNAs using DNA programmed heterostructure of Au-on-Ag. *Biosensors & Bioelectronics*, *175*, 112835. DOI: 10.1016/j.bios.2020.112835

Xi, Q., Zhou, D.-M., Kan, Y.-Y., Ge, J., Wu, Z.-K., Yu, R.-Q., & Jiang, J.-H. (2014). Highly sensitive and selective strategy for microrna detection based on ws2 nanosheet mediated fluorescence quenching and duplex-specific nuclease signal amplification. *Analytical Chemistry*, *86*(3), 1361–1365. DOI: 10.1021/ac403944c

Xu, Q., Xiao, F., & Xu, H. (2023). Fluorescent detection of emerging virus based on nanoparticles: From synthesis to application. *Trends in Analytical Chemistry*, *161*, 116999. DOI: 10.1016/j.trac.2023.116999

Yadav, N., & Khurana, S. M. P. (2016). Plant virus detection and diagnosis: Progress and challenges. In Shukla, P. (Ed.), *Frontier Discoveries and Innovations in Interdisciplinary Microbiology* (pp. 97–132). Springer India., DOI: 10.1007/978-81-322-2610-9_7

Yu, X., Wei, L., Chen, H., Niu, X., Dou, Y., Yang, J., Wang, Z., Tang, Y., & Diao, Y. (2018). Development of colloidal gold-based immunochromatographic assay for rapid detection of goose parvovirus. *Frontiers in Microbiology*, *9*. Advance online publication. DOI: 10.3389/fmicb.2018.00953

Yuwen, T., Shu, D., Zou, H., Yang, X., Wang, S., Zhang, S., Liu, Q., Wang, X., Wang, G., Zhang, Y., & Zang, G. (2023). Carbon nanotubes: A powerful bridge for conductivity and flexibility in electrochemical glucose sensors. *Journal of Nanobiotechnology*, *21*(1), 320. DOI: 10.1186/s12951-023-02088-7

Zafar, S., D'Emic, C., Jagtiani, A., Kratschmer, E., Miao, X., Zhu, Y., Mo, R., Sosa, N., Hamann, H., Shahidi, G., & Riel, H. (2018). Silicon nanowire field effect transistor sensors with minimal sensor-to-sensor variations and enhanced sensing characteristics. *ACS Nano*, *12*(7), 6577–6587. DOI: 10.1021/acsnano.8b01339

Zhan, X., Zhang, F., Li, N., Xu, K., Wang, X., Gao, S., Yin, Y., Yuan, W., Chen, W., Ren, Z., Yao, M., & Wang, F. (2024). Crispr/cas: An emerging toolbox for engineering virus resistance in plants. *Plants*, *13*(23), 3313. DOI: 10.3390/plants13233313

Zhang, C.-Y., & Johnson, L. W. (2006). Quantum dot-based fluorescence resonance energy transfer with improved FRET efficiency in capillary flows. *Analytical Chemistry*, *78*(15), 5532–5537. DOI: 10.1021/ac0605389

Zhang, J., Huang, Y., Pu, R., Gonzalez-Moreno, P., Yuan, L., Wu, K., & Huang, W. (2019). Monitoring plant diseases and pests through remote sensing technology: A review. *Computers and Electronics in Agriculture*, *165*, 104943. DOI: 10.1016/j.compag.2019.104943

Zhang, P., Wang, Y., Yan, B., Wang, X., Zhang, Z., Wang, S., & Yang, J. (2025). Integration of hyperspectral imaging and deep learning for discrimination of fumigated lilies and prediction of quality indicator contents. *Foods*, *14*(5), 825. DOI: 10.3390/foods14050825

Zhang, X., Pint, C. L., Lee, M. H., Schubert, B. E., Jamshidi, A., Takei, K., Ko, H., Gillies, A., Bardhan, R., Urban, J. J., Wu, M., Fearing, R., & Javey, A. (2011). Optically- and thermally-responsive programmable materials based on carbon nanotube-hydrogel polymer composites. *Nano Letters*, *11*(8), 3239–3244. DOI: 10.1021/nl201503e

Zhao, C., Sun, F., Li, X., Lan, Y., Du, L., Zhou, T., & Zhou, Y. (2019). Reverse transcription-recombinase polymerase amplification combined with lateral flow strip for detection of rice black-streaked dwarf virus in plants. *Journal of Virological Methods*, *263*, 96–100. DOI: 10.1016/j.jviromet.2018.11.001

Zhao, G., Li, C., Wang, X., Liu, G., & Thuy, N. T. D. (2021). A reusable electrochemical aptasensor for the sensitive detection of pb (Ii) with an electrodeposited aunp-modified electrode based on the formation of a target-induced g-quadruplex. *International Journal of Electrochemical Science*, *16*(1), 150956. DOI: 10.20964/2021.01.09

Zhao, Z., Tian, Y., Xu, C., Xing, Y., Yang, L., Qian, G., Hua, X., Gong, W., Hu, B., & Wang, L. (2023). A monoclonal antibody-based immunochromatographic test strip and its application in the rapid detection of cucumber green mottle mosaic virus. *Biosensors (Basel)*, *13*(2), 199. DOI: 10.3390/bios13020199

Zhou, G., Xu, D., Xu, D., & Zhang, M. (2013). Southern rice black-streaked dwarf virus: A white-backed planthopper-transmitted fijivirus threatening rice production in Asia. *Frontiers in Microbiology*, *4*, 270. DOI: 10.3389/fmicb.2013.00270

Zhu, H., Lin, C., Liu, G., Wang, D., Qin, S., Li, A., Xu, J.-L., & He, Y. (2024). Intelligent agriculture: Deep learning in UAV-based remote sensing imagery for crop diseases and pests detection. *Frontiers in Plant Science*, *15*. Advance online publication. DOI: 10.3389/fpls.2024.1435016

ABBREVIATIONS

AuNPs: Gold nanoparticles
BBTV: Banana bunchy top virus
BSV: Banana streak virus
BYDV: Barley yellow dwarf virus
CaMV: Cauliflower mosaic virus
Cas: CRISPR-associated proteins (Cas12, Cas13, etc.)
CBSD: Cassava brown streak disease
CBSV: Cassava brown streak virus
CLCuV: Cotton leaf curl virus
CMD: Cassava mosaic disease
CMV: Cucumber mosaic virus
CNTs: Carbon nanotubes
CRISPR: Clustered regularly interspaced short palindromic repeats
CTV: Citrus tristeza virus
ELISA: Enzyme-linked immunosorbent assay
FAO: Food and Agriculture Organization
GFETs: Graphene field-effect transistors
GFLV: Grapevine fanleaf virus
GLRaV: Grapevine leafroll-associated virus
GPGV: Grapevine Pinot gris virus
GRBV: Grapevine red blotch virus
GRV: Groundnut rosette virus
GVT: Grapevine virus T
ICTV: International Committee on Taxonomy of Viruses
IoT: Internet of Things

LAMP: Loop-mediated isothermal amplification
LED: Light-emitting diode
LFA: Lateral flow assay
MCMV: Maize chlorotic mottle virus
MLN: Maize lethal necrosis
MSV: Maize streak virus
NGS: Next-generation sequencing
PCR: Polymerase chain reaction
PEG: Polyethylene glycol
PMMoV: Pepper mild mottle virus
PPV: Plum pox virus
PRSV: Papaya ringspot virus
PVX: Potato virus X
PVY: Potato virus Y
PYSV: Peanut yellow spot virus
QCM: Quartz crystal microbalance
QDs: Quantum dots
qPCR: Quantitative polymerase chain reaction
RPA: Recombinase polymerase amplification
RTBV: Rice tungro bacilliform virus
RT-PCR: Reverse transcription polymerase chain reaction
RTSV: Rice tungro spherical virus
scFvs: Single-chain variable fragments
SCMV: Sugarcane mosaic virus
SELEX: Systematic Evolution of Ligands by Exponential Enrichment
SMV: Soybean mosaic virus
SPR: Surface plasmon resonance
TENGs: Triboelectric nanogenerators
TMV: Tobacco mosaic virus
ToBRFV: Tomato brown rugose fruit virus
ToMMV: Tomato mottle mosaic virus
ToMV: Tomato mosaic virus
TSWV: Tomato spotted wilt virus
TYLCV: Tomato yellow leaf curl virus
WHO: World Health Organization
WSMV: Wheat streak mosaic virus
YMV: Yam mosaic virus

Chapter 9
Nanoparticle-Based Immunoassays and Rapid Field Diagnostic Tools

Batool Bashir
https://orcid.org/0009-0000-6915-3302
Chinese Academy of Agricultural Sciences, China

Anees Ali
Necmettin Erbakan Üniversitesi, Turkey

Rashid Abbas Khan
Prince of Songkla University, Thailand

Muhammad Iftikhar
University of Education, Pakistan

Vildan Altun
Necmettin Erbakan University, Turkey

ABSTRACT

Nanoparticle-based immunoassays and rapid diagnostic platforms are revolutionizing diagnostics by overcoming low sensitivity, long turnaround times, and dependence on centralized laboratories. Functionalized nanomaterials—gold nanoparticles, quantum dots, magnetic particles, carbon structures, and polymers—enable strong signal amplification, lower detection limits, and portable, field-ready use. Their integration into point-of-care devices, lateral flow assays, microfluidics, smartphone-based tools, and biosensors has enhanced detection of pathogens, cancer biomarkers, foodborne contaminants, and toxins with high precision. Applications extend beyond medicine to agriculture, veterinary health, food safety, and environmental monitoring. Remaining hurdles include reproducibility, scalable production, regulatory approval, and field stability. Emerging solutions such as multiplexed assays, CRISPR–nanoparticle hybrids, and AI-driven interpretation mark a paradigm shift toward accessible, precise, and sustainable global diagnostics.

DOI: 10.4018/979-8-3373-4913-8.ch009

9. Introduction

The health of the world is still threatened by the growing load of both infectious and non-infectious diseases (Naghavi et al., 2017; Vos et al., 2020). Infectious disease like flu, TB, HIV/acquired immune deficiency syndrome and most recent COVID-19 has proven the vulnerability of world's population and inevitability for rapid screening (Khojasteh-Kaffash et al., 2025). On the other hand, non-infectious diseases like cancer, heart disease, diabetics and so on carry not only high prevalence with them but also, they are deadly for health economy across the world. The twoness of the giant charmed arms attacking simultaneously in infection and chronic disease there offers responses that are new, fast, under control (Khojasteh-Kaffash et al., 2025; Prabhu & van Wagoner, 2023). Early and accurate diagnosis is essential for disease management, patient treatment, and for public health response (Nanone et al., 2024). However, the conventional diagnostic approaches are time-consuming, laboratory-based and require trained personnel, and as such they are not applicable for a variety of other settings such as at decentralized sites or in resource-limited regions (Sin et al., 2014). Delaying in fast/accurate diagnosis results in prolonging time to make decision on treatment, increasing the transmission of the disease and economic expense of patients (Sreeramareddy et al., 2009; Storla et al., 2008). Consequently, there is an escalating demand for rapid, sensitive, and specific, portable, and field-deployable diagnostic devices in the current medicine (Ambagala et al., 2017).

Nanotechnology has proved to be innovative in diagnosis and has the potential to provide solutions for limitations of existing assays (Dessale et al., 2022). Nanoparticles exhibit distinctive physicochemical properties such as high surface area, tunable optical properties and high catalytic activities which can be exploited to increase signal transduction, sensitivity and detection limit for immunoassays (Anas, Khattak, et al., 2025; Rivera-Gil et al., 2013). Acost-effective, and ultrasensitive devices for laboratories and on-site applications. These advancements are enabling new opportunities for point-of-care testing applications that may revolutionize clinical practice and disease surveillance (Thwala et al., 2023; Zhang et al., 2014). The purpose of this chapter is to summarize the current development in the field of Nanoparticle based immunoassays and their ease-of-use in rapid point-of-care test devices. It will discuss the principles, types of nanoparticle platforms and their applications in medical, agricultural and environmental fields. The current limitations, challenges, and prospects of nanoparticles-enabled diagnostics will also be discussed critically in the chapter. In summarizing the interdisciplinary achievements for the application of nanotechnology in immunoassay development, this chapter aims to emphasize their potential impact it will have on global healthcare and diagnostic monitoring.

9.1 Fundamentals of nanoparticles- Based immunoassays

Immunoassays are analytical methods that use the specificity of antigen–antibody binding to detect and quantify biomolecules, microorganisms, or other analytes. . The use of nanotechnology in immunoassay platforms has significantly enhanced the performance of such platforms, offered a level of sensitivity and multiplexed that were previously unattainable. Herein, we highlight the advancement of nanoparticles in immunoassay development by (1) description and principles of immunoassays; (2) application of NPs to amplify signals; and (3) various types of NPs designed, developed, utilized, compared eventually with conventional methods (W. Li et al., 2025; Xia et al., 2024).

9.1.1 Principles of Immunoassays

Immunoassays are based on the specific interaction between antibodies and their antigens. Such binding may, in turn, be detected by any of several labelling and signal amplification techniques including colorimetric, fluorescent, electrochemical or chemiluminescent signals (Dinis-Oliveira, 2014). Despite their drawbacks, classical immunoassays (e.g. ELISA, RIA) continue to be used extensively because of their robustness possessing strong dose recognition. Nevertheless, these approaches usually involve long periods of incubations, washing and advanced laboratory machinery. Utilization of nanomaterials in the immunoassays is a potential approach which offers improved sensitivity for detection at lower assay period at a point of care. Nano-traveled immuno-assays also follow the same principle of antigen-antibody, but it is using nanoparticles that signal is also generated or transformed (Gao et al., 2020; Hou et al., 2023).

9.1.2 Role of Nanoparticles in Signal Transduction

Nanoparticles represent one of the most desirable signal enhancers, which are based on the inherent properties of nanoparticles. They also possess a large surface-volume ratio with potential for the attachment of more than one biomolecule (e.g., antibody, aptamer and enzyme) to enhance target binding (Bisht et al., 2024). These optical and magnetic nanomaterials display atypical plasmonic and photoluminescent behaviors on the properties of the optical transducer, which have potential for amplified detection signal as well as a fast enrichment accessing to analytes thereby making these materials auspicious in advanced biosensor development (Chen et al., 2024). Gold nanoparticles (AuNPs) have intense colorimetric factors arising from localized surface plasmon resonance (LSPR) and can therefore be easily detected by the naked eye without the need of any sophisticated equipment. While Quantum dots are multicolor fluorescence which possess bright, tunable emission and photo-stability properties that has made them a new container for multiplexed detection (Figure 1). These NP tagged immunoassays have detection limits several orders of magnitude lower than have been established for an assay in which the NPs are replaced by mass alone; in many cases as low as femtomolar or picomoles (Singh et al., 2021; Zhang, Hou et al., 2025).

Figure 1. Key characteristics of nanoparticles used in immunoassays. Abbreviations: NP = nanoparticle; (AuNP) gold nanoparticle; (QDs) quantum dots; (MNP) magnetic nanoparticle; (CDs) carbon dots; (LSPR) localized surface plasmon resonance. These properties high surface-to-volume ratio, tunable optical/magnetic/electronic behaviours, and amenability to surface functionalization underpin signal enhancement, multiplexing and sample enrichment in modern biosensors

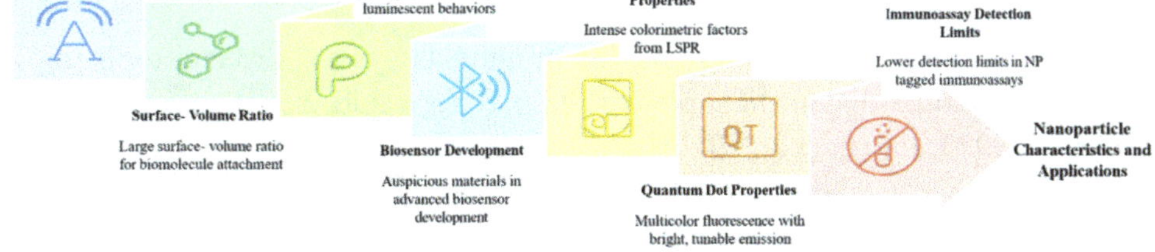

9.1.3 Types of Nanoparticles Used

Silver nanoparticles (AgNPs): They possess a higher surface plasmon absorption compared to AuNP, and can, with proper consideration, be used for ultrasensitive optical detection but have only limited chemical stability (Ferrari, 2023; Khan et al., 2025).

Magnetic nanoparticles (MNPs): Magnetic nanoparticles aids in the isolation, pre concentration and signal amplification of targets, and are used successfully for the magnetic resonance and electrochemical detection(Wang et al., 2024).

Quantum dots (QDs): Semiconductor nanocrystals with shape-dependent electronic structure and multicolor electroluminescence (Anas, Khan, et al., 2025; Bai et al., 2023).

Carbon based nanomaterials: (CNTs, graphene, graphene oxide): these possessed great electrical conductivity and specific surface area which were appropriate for construction of electrochemical immunoassays and sensors(Fahmy et al., 2022).

Synthetic Polymers: Biocompatible and custom synthesized nm-300 nm carrier method of immobilizing with bio molecules, to slow controlled release or targeting in immuno diagnostics(Yergök et al., 2024).

Hybrid Nanomaterials: Hybrid CDs with metallic, magnetic and polymeric composites, imparting enhanced sensitivity, specificity, and stability that supports immunoassays(Kidanemariam & Cho, 2025).

9.1.4 Advantages of Conventional Assays

Nanoparticles based immunoassays offer server distinct advantages compared to traditional diagnostic methods:

Enhanced sensitivity and specificity: due to signal amplification and higher loading capacity of recognition molecules(Kurdekar & Kamisetti, 2024).

Rapid detection: Reduced incubation and processing times enables near real time diagnostics(Ahmadsaidulu et al., 2023).

Multiplexing capability: Quantum dots and other nanomaterials allow simultaneous detection of multiple analytes(Z. Li et al., 2025).

Portability and field applicability: Integration into point of care devices, such as lateral flow strips and microfluidic chips, make them suitable for testing onsite(Sena-Torralba et al., 2022).

Cost effectiveness: Many nanoparticle-based assays minimize the need for complex instrumentation, lowering the diagnostic costs(Farka et al., 2024).

Scalability and versatility: application across medical, environmental and agricultural and food safety domains.

These advantages position nanoparticles enabled immunoassays as powerful tools to overcome the shortcomings of conventional diagnostics, paving the way for rapid, accurate and accessible disease detection.(Hou et al., 2023; Jabran et al., 2024)

9.2. Nanoparticles platforms for immunoassays

Nanoparticles present the opportunity for new generations of immunoassays due to their diverse physicochemical properties, biocompatibility and possible surface functionalization. Obviously, each type of nanoparticle has its specific contribution to the sensitivity, selectivity and multiplicity of diagnostics. The next sub-sections briefly describe the most utilized nanoparticle types in immunoassays and discuss their characteristics alongside their utilities(Xiao et al., 2023).

9.2.1. Gold Nanoparticles (AuNPs)

Their localized surface plasmon resonance (LSPR) provides drastic color responses, allowing them to be observable by the naked eye, rendering them ideal for lateral flow assays (LFA), as those that are widely used in COVID-19. AuNPs are also amenable for alternative detection modes, such as surface-enhanced Raman scattering (SERS) and electrochemical transduction. Due to their stability, low toxicity and well-defined conjugation chemistry, they represent a gold standard in the field of nanoparticle-based immunoassays(Kant et al., 2024; Kim et al., 2024).

9.2.2 Quantum dots (QDs)

Quantum dots (QDs), as a species of semiconductor nanocrystals, possess size-dependent fluorescence, high brightness, and exceptional photostability. Unlike traditional organic fluorophores, QDs are resistant to photobleaching and are well-suited for long-term, sensitive detection(Alsuraifi et al., 2025). Their capability of emitting several independent fluorescence signals allows them to be used as resources for multiplexed immunoassays in which several biomarkers can be simultaneously detected in one assay. This characteristic is particularly valuable in cancer diagnosis, panels of infectious disease tests, as well as environmental assays. However, concerns regarding the toward heavy metal members (such as cadmium) cytotoxicity remain a barrier to their clinical application(Le & Kim, 2023).

9.2.3 Magnetic nanoparticles (MNPs)

Magnetic nanoparticles (NPs) made of iron oxides (Fe_3O_4 or γ-Fe_2O_3) are known to possess peculiar magnetic properties. They are responsive to external magnetic fields and therefore are potential candidates for achieving efficient target capturing, separation, and enrichment(Diouf et al., 2025). After functionalization with antibodies or aptamers, MNPs can specifically recognize the target analytes and easily be separated from complex biological samples. They are commonly used for immunoassay applications in combination with optical or electrochemical detection methods, making them sensitive, fast and automated diagnostic tools. MNPs are also ideal for miniaturized point-of-care instruments because they are suitable for microfluidic and lab-on-a-chip systems(Pour et al., 2023).

9.2.4 Carbon nanotubes and Graphene-Based Nanomaterials

Among them, carbon-based nanomaterials such as carbon nanotubes (CNTs) and their derivatives, as well as graphene-related materials (GRMs), show excellent electrical conductivity, enormous specific surface area, and fine mechanic properties. With their one-dimensional (CNTs) or two-dimensional (graphene) structure, biomolecules can be densely anchored onto their surface and effectively interact with the electrode for electron transfer in electrochemical immunosensing(Singh et al., 2022). The presence of graphene oxide (GO) and reduced graphene oxide (rGO), as they are modified with additional functional groups, further enhances the complementarity of the assay. These nanomaterials are especially appealing candidates for the construction of low-cost, portable biosensors with high sensitivity for detecting biomolecules at ultralow concentrations(X. Li et al., 2025).

9.2.4 Polymer based nanoparticles

Polymeric nanoparticles, composed of either natural (e.g., chitosan, polylactic-co-glycolic acid (PLGA)) or synthetic (e.g., polyethylene glycol (PEG)) polymers, can represent a biocompatible and tunable support for immunoassays (Zhang, Zhang et al., 2025). They can also be used to contain signaling/intrinsic enzymes or fluorescent dyes, and in such a way, become versatile carriers for amplification of the signals. Surface decorated polymeric nanoparticles may also be modified to improve stability, minimize non-specific binding and extend shelf life. The flexibility and utility of these devices is highly applicable to point of care diagnostics where both cost and scalability are important factors (Abdelkawi et al., 2023).

9.2.5 Hybrid Nanomaterials

Hybrid nanomaterials had two or more nanoparticles with different properties, and were able to achieve integrated immunoassay performance. These hybrids were fabricated in the form of Au–plasmonic nanoparticles (Au–PNPs) to have optical sensitivity of AuNPs and magnetic separation power propagated by MNPs, or quantum dots/particles embedded with polymers to minimize release of fluorescent signals and achieve higher biocompatibility(Tai et al., 2022). These hybrid adsorbents might be useful for multimodal detection, sensitivity amplification and stability control in the next generation immunodiagnostic procedure. The flexibility of hybrid systems provides for the potential development of application-specific system from diagnostic to environmental measurements(Godja & Munteanu, 2024; Liu et al., 2025).

Table 1. Types of nanoparticles used in immunoassays and their key features

Nanoparticle Type	Unique Properties	Diagnostic Applications	Advantages Over Conventional Systems	References
Gold nanoparticles (AuNPs)	Strong surface plasmon resonance, easy functionalization	Lateral flow assays, POCT, cancer biomarker detection	High sensitivity, visible readout, stability	(Huang et al., 2017)
Quantum dots (QDs)	Size-tunable fluorescence, photostability	Multiplexed assays, imaging-based diagnostics	High brightness, multiplexing capability	(Yao et al., 2014)
Magnetic nanoparticles	Superparamagnetism, easy manipulation by magnets	Separation/enrichment, biosensors, immunoassays	Rapid sample preparation, low background noise	(J. Li et al., 2025; Zhong et al., 2023)
Carbon nanotubes & grapheme	High conductivity, large surface area	Electrochemical biosensors, pathogen detection	Fast electron transfer, miniaturization	(Fahmy et al., 2022)

continued on following page

Table 1. Continued

Nanoparticle Type	Unique Properties	Diagnostic Applications	Advantages Over Conventional Systems	References
Polymeric nanoparticles	Biodegradable, customizable chemistry	Drug–diagnostic hybrids, biosensors	Biocompatibility, tunable release properties	(Colson & Grinstaff, 2012; Geszke-Moritz & Moritz, 2024)
Hybrid nanomaterials	Combination of properties from multiple materials	Multifunctional assays, advanced biosensors	Enhanced sensitivity, multifunctionality	(Soto & Orozco, 2022)

9.3. Rapid field diagnostics Tools: Integration with nanoparticles

The current trend to develop fast diagnostic tools is increasing with the spread of point-of-care and low-resource disease detection. Nanoparticles are an essential part of this development due to their contribution towards increased sensitivity, portability and open-field detection capability. Nanomaterial-based point-of-care sensors offer health care in remote locations, environmental surveillance, and safe supply of foods regardless of centralized laboratories support. The following chapter presents key NP-driven platforms that are revolutionizing POC diagnostics(Sanyal et al., 2024). Nanoparticle-enabled formats for point-of-care (POC) diagnostics. Nanoparticles (NPs) are integrated into a small set of practical POC formats that together cover most field needs: (1) Lateral flow assays (LFAs) and dip-sticks, where AuNPs (colorimetric), QDs and fluorescent labels (including carbon dots) provide visual or reader-assisted signals for rapid antigen/antibody detection; (2) magnetically assisted assays, where magnetic nanoparticles (MNPs) enable rapid capture/enrichment from complex samples prior to optical or electrochemical readout—this markedly lowers matrix noise and improves LODs; (3) microfluidic lab-on-a-chip platforms, which combine on-chip sample handling with NP capture and multiplexed detection for automated, sensitive assays; and (4) smartphone-integrated readers and portable biosensors, which digitize colorimetric or fluorescence signals for quantitative interpretation and real-time data transmission. Hybrid devices that combine these formats (e.g., MNP preconcentration + LFA readout, or fluorescent CDs in paper sensors read by a smartphone) are increasingly common and deliver the best trade-off between sensitivity, speed and field robustness. The literature shows that AuNP-based LFAs remain the simplest and most widely deployed field format, whereas MNPs and fluorescent NPs (QDs, CDs) are preferred where enrichment or quantitative fluorescence readout is required (Table 2).

9.3.1. Point-of-Care Testing (POCT)

Point of care testing (POCT) is “diagnostic testing at or near the point of patient care. where Test results are produced at the point of care (or in a collection center) and within enough time for actions to be taken with respect to patient management(Song et al., 2021). In particular, the incorporation of nanoparticles can significantly enhance the sensitivity and performance of a POCT device, which can enhance properties including but not limited to (lower limits of detection), signal drifts and/or assay time. For instance, gold nanoparticles are commonly included in the POCT kits for infectious diseases. The latter being further optimized by finely adjusting the conditions of magnetic nanoparticle material to enable efficient capture analytes in complex biological fluid samples. The portable device of POCT system can provide the immunoassays based on nanoparticles with high sensitivity, low cost and user-friendly package, which is shockingly useful in resource-poor areas or exacerbation for emergency(Wang et al., 2025).

9.3.2. Lateral Flow Assays (LFAs) and Dipstick Tests

Lateral flow assays (LFAs) are the most typical rapid diagnostic format used globally, mainly due to its simplicity, low cost and easy translation. Enhanced performance of their visual read-out and sensitivity was achieved when gold, in particular, gold nanoparticles were used to promote the reaction(Wang et al., 2022). LFAs function through capillary flow of the sample across a nitrocellulose membrane containing immobilized specific antibodies and visible reaction with the target antigen. At present, quantum dots and up conversion nanoparticles have been applied in constructing quantitative LFAs for multiplex detection with enhanced sensitivity in recent years. Dipstick tests, which are much less sophisticated than the LFA but that have also shown enhanced performance with the use of nanoparticles, stand as fast and robust field portable methods for clinical and environmental diagnostics(Dey et al., 2023; Lu & Chan, 2024).

9.3.3. Lab-on-a-Chip and Microfluidic Devices

Lab-on-a-chip and microfluidic devices have scaled down complex laboratory procedures onto small compact devices that are portable and are able to conduct several analytic functions at once(Mehta et al., 2022). The application of nanoparticles in these devices has led to the improved performance of assays due to accurate analyte capture, effective liquid handling and sensitive detection. For example, magnetic NPs support on-chip target enrichment, while quantum dots (QDs) and graphene-like nano-materials offer strong signal transduction. These nanoparticle-based microfluidic assays specifically are extremely beneficial for portable, readily deployable high-throughput, automated and multiplexed diagnostics, and are particularly well suited for deployment as point-of-care assays or in low-resource settings such as field or clinical laboratories, where infrastructure and highly trained personnel may be scarce(Forouzanfar, 2021; Gupta et al., 2025).

9.3.4. Smartphone-Integrated Diagnostics

The proliferation of smartphones has created opportunities for decentralized diagnostics. Detection assays using nanoparticles can provide real-time, inexpensive diagnostic read-out with the aid of smartphone cameras and computation. Visual interpretation of results can also be reduced by using smartphone apps for quantitative analysis of the colourimetric(Younis et al., 2025). LFA-based on gold nanoparticles, for example, making it less subjective. Furthermore, smart phones can be coupled with a microfluidic chip, electrochemical sensor, or fluorescence assay to build a portable lab based on digital technology. This connectivity supports not only individualized medicine but also the surveillance of pandemics through the real-time collection and transmission of data to central databases(Baker et al., 2025).

Table 2. Comparison of NP-based immunoassay platforms for field and point-of-care (POC) applications

NP type	Typical assay formats (field/POC)	Representative LOD / detection limits (benchmark studies)	Typical reported sensitivity / specificity (range)	Notes
Gold nanoparticles (AuNPs)	Lateral flow assays (LFAs), colorimetric dipsticks, SERS-enhanced strips, smartphone-read LFAs	Typical LFA LODs: ~0.001–1 ng·mL^{-1} for protein targets; some optimized AuNP LFAs report < 0.001 ng·mL^{-1} (ultrasensitive examples); antigen LFAs often ≈ ng·mL^{-1} to tens of ng·mL^{-1} depending on target and antibody.	Visual-read LFAs typically ≈ 70–95% sensitivity; instrument-assisted AuNP LFAs > 90% sensitivity and > 95% specificity.	AuNPs provide strong LSPR color signal and simple visual readouts; sensitivity improves with optical readers, signal amplification, or alternative readouts (SERS, fluorescence).
Magnetic nanoparticles (MNPs / magnetic beads)	Magnetic capture + enrichment prior to optical / electrochemical readout; magneto-LFA; microfluidic capture with electrochemical or magnetoresistive detection	Reported LODs range from sub-µg·L^{-1} to pg·mL^{-1} depending on readout; e.g., magnetic immunoassay LOD ≈ 0.068 µg·L^{-1} (instrument) and ≈ 0.5 µg·L^{-1} (visual); advanced magneto-readouts reach pg·mL^{-1}.	Instrumented MNP assays typically > 90% sensitivity and > 95% specificity; field-visual formats depend on enrichment efficiency and operator (PMC).	MNPs excel at sample pre-concentration and low background in complex matrices; best performance when magnetic capture is paired with a sensitive detector (magnetometer or electrochemical transducer).
Carbon dots (CDs; fluorescent carbon nanodots)	Fluorescence immunoassays (paper-based turn-on sensors), fluorescence LFAs, microplate / fluorometric assays, paper sensors	LODs vary by target; examples: virus detection paper sensors ≈ 10^1 TCID$_{50}$·mL^{-1} (LSDV example); protein biomarkers in low ng·mL^{-1} to sub-ng·mL^{-1} range in optimized assays.	Fluorescence-read CD assays show > 90% specificity and sensitivities comparable to quantum dot systems in laboratory tests; field fluorescence depends on reader	

9.3.5. Portable Biosensors for Field Applications

Near-real-time detection of pathogens, toxins, and biomarkers in the field is becoming more important and thus portable biosensors that use nanoparticle (NP)-based recognition and transduction elements have potential for such applications(Hemdan et al., 2024). These sensors take advantage of a variety of signal detection mode (e.g., EC, optical and/or piezoelectric) that are magnified by nanomaterials such as AuNPs, QDs or carbon nanomaterials. Portable biosensors have become increasingly common in veterinary diagnostics, agricultural surveillance, environmental measurement and biothreat detection. Their durability and potential for ultrafast ultrasensitive analysis make them indispensable tools for providing diagnostic solutions outside traditional laboratory settings (Luckose et al., 2023).

9.3.6 Biosafety and Waste Disposal

Field-deployable nanoparticle-based diagnostic kits must follow strict biosafety and waste management measures to protect users and the environment. Used test strips, swabs, and sample containers may contain biological material and should be treated as biohazardous waste. All materials that come into contact with biological fluids must be collected in sealed biohazard bags and either autoclaved or incinerated at high temperature before disposal. Residual nanoparticles such as gold, magnetic, or carbon-based particles can persist in the environment; therefore, disposal through regulated medical-waste channels is essential. Field operators should receive basic biosafety training and use gloves, masks, and protective barriers during testing. Kits designed for remote or agricultural applications should include small waste-collection pouches with clear labeling and instructions for safe handling and return to designated facilities. Where centralized disposal is not possible, validated chemical or thermal inactivation methods should be followed to minimize risk. Implementing these simple yet critical measures ensures operator safety, prevents environmental contamination, and supports sustainable use of nanomaterials in field diagnostics. Proper biosafety practices also enhance community confidence in the use of portable nanoparticle-based assays in public health and agricultural monitoring programs.

9.4. Applications of Nanoparticles- base diagnostics

Nanoparticle-based immunoassays and diagnostics have exhibited great potential in numerous applications due to their high sensitivity, fast detection and small size. Due to their flexibility, they play an essential role in health care, agriculture, and environment monitoring and food safety (Figure 2). This section focuses on key application areas where nanotechnology-enabled diagnostics are making significant inroads (Puttasiddaiah et al., 2025).

Figure 2. Representative nanotechnology applications across sectors (health, agriculture, food safety, environment). (POCT) point-of-care testing; (LFA) lateral flow assay; (LOD) limit of detection. The figure illustrates common NP-platform pairings (AuNP-LFA for rapid antigen tests; MNP + electrochemical readout for enriched detection)

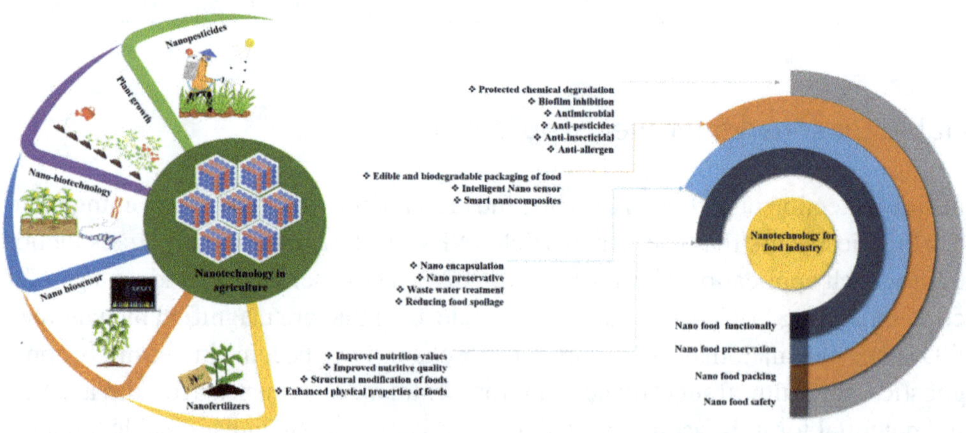

9.4.1. Infectious Diseases (Viral, Bacterial, Parasitic)

Early, accurate detection of the infectious agent is essential for effective patient management and outbreak management. Compared to traditional methods, nanoparticle-based assays can greatly improve sensitivity and decrease detection time(Thwala et al., 2023).

Viral infections: Gold nanoparticles have found extensive use over a decade in detecting viral diseases including but not limited to influenza, HIV, hepatitis B/C, Zika virus, and SARS-CoV-2. The quantum dot-based assays also enable highly accurate direct assessment of co-infections in a multiplex detection system(Mittal & Ali, 2022).

Bacteria: Antibody-coated magnetic nanoparticles are used to selectively capture and pre-concentrate bacterial pathogens (such as E. coli and Salmonella) from food and clinical samples. Rapid electrochemical readouts of multidrug-resistant bacterial strains using graphene-based sensors(Hulme, 2022; Risalvato et al., 2025).

Parasite infectivity: Darly diagnosis based on low parasitemia-level detection of malaria parasites have been enhanced using nanoparticle supported dipstick tests for treatment intervention, at endemic regions. Together, these systems offer cost-effective field-deployable rapid remedies, which are particularly critical in resource poor environments and times of epidemic emergencies (El Saftawy et al., 2024).

9.4.2. Agricultural and Veterinary Diagnostics

It is important to detect plant and animal diseases earlier to minimize economic casualties and to protect food safety.

Plant health: Immunoassays using AuNPs have also been designed for the in situ and in-field detection of viral infections in agricultural crops like maize, rice, and wheat. Carbon nanomaterial based portable biosensors are being used in the field for detection of fungal and bacterial phytopathogens from agricultural crops too(Chaturvedi et al., 2025).

These means help control the spread of disease and facilitate proper herd management. These in-field diagnostics technologies enable the monitoring of time depend on diagnosis, of great importance to sustainable farming and animal health in a cost-effective manner.

9.4.3 Environment Monitoring

Nanoparticle based biosensor has been applied to sensing pathogens, toxins and environmental pollutants in the environment due to its improved sensitivity and rapid response time (Mallick et al., 2025). Waterborne pathogen immunoassays have been greatly advanced by the use of gold as well as magnetic nanoparticles with the human element also playing a role in testing major water resources parasites including Cryptosporidium and Giardia, and practical methods to treat drinking water quality (Khodaparast et al., 2024). Similarly, carbon nanotube and graphene-based sensors were developed for pesticide, heavy metal and organic pollutant detection at soil or water resources, offering a convenient and rapid protocol compared with the traditional laboratory assays. Discovered Packets We also discovered nanoparticle immunosensors for on-line continuous systemic monitoring of air quality, which may identify allergens/antigens microbial pathogenic agents, and other airborne threats. Together they represent significant contributions to protection of the environment and public health, and rapid in situ discovery of any threat (Han et al., 2025).

9.4.4. Food Safety and Quality Control

Foodborne pathogens and pollutants present serious burdens to human health and world trade. Nanoparticle immunoassays have been developed to rapidly detect such threats with high sensitivity (Chen et al., 2021; Rodriguez-Quijada et al., 2020).

Detection of pathogens: AuNP-based LFAs and QD-assisted immunoassays enable the sensitive detection of Salmonella, Listeria monocytogenes and E. coli in food.
Chemical contamination: Nano-enabled biosensors have been used to measure pesticide residues, veterinary drug residues and heavy metals with high accuracy (Shivakumar, 2024).
Spoilage indicators: CNTs in smart packaging products can sense gases and metabolites related to spoilage of packaged foods to monitor the quality of the food in real-time. These advances help to protect food safety, regulatory compliance, and consumer trust through effective quality control (Chelliah et al., 2022; Thiruvengadam et al., 2024).

9.4.5. Cancer and Non-Communicable Diseases

In addition to infectious and environmental hazards, nanoparticle-based diagnostics are being applied to non-communicable diseases such as cancer at a remarkable pace.

Cancer biomarkers: AuNPs and QDs have been exploited for the detection of cancer biomarkers of prostate-specific antigen (PSA), carcinoembryonic antigen (CEA), cancer antigen 125 (CA-125) at ultralow levels (Numan et al., 2022; Türkcan & Akgöl, 2023).
Cardiovascular diseases: Electrochemical biosensors based on graphene (Dudala et al., 2022) and CNTs can be employed for the rapid detection of cardiac biomarkers, troponin I, for the early diagnosis of myocardial infarction (Zhang, Zhu et al., 2025).
Diabetes: The nanoparticle glucose biosensor delivers accurate and real-time glucose analysis through portable readers (Liu et al., 2022).

These strategies can help with personalized medicine and early disease intervention, so as to improve the survival and health condition of the patients.

Table 3. Applications of nanoparticle-based diagnostics across different sectors

Sector	Target Analytes/Pathogens	Diagnostic Platform Examples	Impact/Outcome	References
Human health (infectious diseases)	Viruses (HIV, SARS-CoV-2), bacteria (*E. coli*, *M. tuberculosis*)	AuNP-based LFAs, CRISPR–nanoparticle hybrids	Early detection, rapid outbreak control	(Dubey et al., 2022; Kinyua et al., 2025)
Cancer and NCDs	Tumor biomarkers (PSA, HER2, CEA)	QD-based multiplex assays, graphene biosensors	Precision oncology, improved patient monitoring	(Pandey et al., 2025)
Agriculture and veterinary	Plant viruses, zoonotic pathogens	Portable biosensors, POCT devices	Crop protection, animal health surveillance	(Manessis et al., 2022; Wilson, 2018)
Food safety	Salmonella, Listeria, aflatoxins, pesticides	AuNP dipstick assays, electrochemical sensors	Preventing foodborne illness, quality control	(Curulli, 2021)
Environmental monitoring	Heavy metals, pesticides, waterborne pathogens	Graphene-based electrochemical devices, magnetic nanoparticle assays	Pollution detection, real-time environmental assessment	(Zahra et al., 2023)

9.5. Challenges, Limitations, and Future Prospects

However, there are still challenges to the practical implementation of nanotechnology-related investigations and assay formats in short term, point-of-care diagnostic applications. Sensitivity and specificity remained the bottleneck to innovation as limitations in complex biological or environmental sample matrix effects challenged assay robustness, false positives/negatives. Therefore, it is a formidable challenge to achieve good stability across cross-environment testing. In addition, scalability and the cost of nanoparticle synthesis, functionalization, and device integration are challenging issues. While laboratory prototypes are often well performing, implementation into a mass-produced low-cost diagnostic kit demands straightforward but robust manufacturing methods and a high level of control of the quality. Another obstacle is related to the regulatory limitations and the absence of established norms for nanoparticle-mediated diagnostics. If the validation methodologies are not robust and there is no consensus on safety evaluations that is also the case for the approval and commercialization of these technologies. Meanwhile, discrepancy of laboratory and field performance introduces extra challenges such as environmental differences, operator manipulation and preservation situations might reduce reliability in field test. Recent advances in multiplexed immunoassays Quantum dots as well as Dendrimer peptide and Hybrid nanomaterial-based immunoassays for simultaneous detection of multiple markers on a single platform have been developed and may increase the efficiency of diagnosis. The application of AI and machine learning (ML) can allow for automation of data interpretation (to decrease human error) and the provision of predictive knowledge to aid treatment of the patients, and control of the outbreak. Meanwhile, the novel frontier NPs connected to CRISPR in hybrids have also been recently developed, for significantly improved sensitivity and specificity in nucleic acid detection communicating a highly low concentration of trace pathogen. Their application in the context of personalized or precision medicine is also expanding, based on patient specific biomarker panels that direct individual treatment responses. Efforts are being made to refine this value proposition, manufacturability and regulatory activities for successful transition to market and broad/lg scale translational success. As work for the NP platforms improvement and its integration to digital health strategies proceed, is expected that their potential for address next generations of diagnostic in a variety of scenarios (e.g.: Clinic, Food, Environment) will change lab efforts into solutions in the real world.

9.6. Conclusion

Nanoparticle-mediated immunoassays and rapid diagnostic devices are one of the most potential technologies that are revolutionizing the future for the diagnosis, monitoring and management of disease in different areas. Due to their unique optical, magnetic and electrical properties due to size control down to the molecular level nanomaterial-based platforms afford a much higher sensitivity, specificity, portability as well as cost compared with traditional diagnostics. Their possible application is not restricted to the clinical improvement of human and animal conditions, but pertains to farming and veterinary use, environmental security and food quality control, in general their broad societal benefit. The immobilization of nanoparticles on point-of-care testing, lateral flow assay and microfluidic device, smartphone-based platform and portable biosensors has further increased the possibility decentralized diagnostics in field deployable application for both emergencies as well as regular cases. However, there are several impeding bottlenecks, such as generalization efficiency on real hands. We can only overcome these obstacles through improved links between the research and clinical communities, with clinicians, engineers, industry and government. The authors conclude brightly with future diagnostics in multiplexed assays, AI-enabled data interpretation and systems for CRISPR–NP hybrids and precision medicine activities. As these technologies evolve, the path for commercialization and deployment is becoming increasingly defined. The NP-based immunoassay and RFT are expected to revolutionize worldwide healthcare and bio surveillance. Taken from lab to tap and next door they offer enormous potential for transforming diagnostics, building public health capacity, ensuring food and environmental safety, and laying a new cornerstone for sustainable development around the world. Advanced progress, not only in robust health systems but also opportunities for early diagnosis of diseases, personalized medicine and the confidence that disease preparedness is global in action in coming decades.

REFERENCES

Abdelkawi, A., Slim, A., Zinoune, Z., & Pathak, Y. (2023). Surface modification of metallic nanoparticles for targeting drugs. *Coatings*, *13*(9), 1660.

Ahmadsaidulu, S., Banik, O., Kumar, P., Kumar, S., & Banoth, E. (2023). Microfluidic point-of-care diagnostics for multi-disease detection using optical techniques: A review. *IEEE Transactions on Nanobioscience*, *23*(1), 140–147. PMID: 37399163

Ambagala, A., Fisher, M., Goolia, M., Nfon, C., Furukawa-Stoffer, T., Ortega Polo, R., & Lung, O. (2017, October). Field-Deployable Reverse Transcription-Insulated Isothermal PCR (RT-iiPCR) Assay for Rapid and Sensitive Detection of Foot-and-Mouth Disease Virus. *Transboundary and Emerging Diseases*, *64*(5), 1610–1623. DOI: 10.1111/tbed.12554

Alsuraifi, A., Mouzan, M. M., Mahmoud, A., Aqeel, Z., Al-Essa, U. I., Mohammed, N. A. R., & Ayad, A. (2025). Quantum Dots: Catalysts for a New Era of Precision Medicine and Biomedical Innovation.

Ambagala, A., Fisher, M., Goolia, M., Nfon, C., Furukawa-Stoffer, T., Ortega Polo, R., & Lung, O. (2017). Field-deployable reverse transcription-insulated isothermal PCR (RT-ii PCR) assay for rapid and sensitive detection of foot-and-mouth disease virus. *Transboundary and Emerging Diseases*, *64*(5), 1610–1623. PMID: 27589902

Anas, M., Khan, A., Falak, A., Shinwari, Z., Mohamed, H., Hkiri, K., Maaza, M. Sh. F., & Quraishi, U. (2025). Harnessing Biomedical Applications of Ascorbic Acid, Berberine, and Glutathione-Capped Quantum Dots Through Multifunctional Physico-chemical, Biochemical, and Anti-microbial Assays. *BioNanoScience*, *15*. Advance online publication. DOI: 10.1007/s12668-025-01864-z

Anas, M., Khattak, W. A., Fahad, S., Alrawiq, N., Alrawiq, H. S., Abdelsalam, N. R., Jaremko, M., & Quraishi, U. M. (2025). Mitigating heavy metal pollution in agriculture: A multi-omics and nanotechnology approach to safeguard global wheat production. *Journal of Hazardous Materials Advances*, *17*, 100584. https://doi.org/https://doi.org/10.1016/j.hazadv.2024.100584

Bai, B., Zhang, C., Dou, Y., Kong, L., Wang, L., Wang, S., Li, J., Zhou, Y., Liu, L., & Liu, B. (2023). Atomically flat semiconductor nanoplatelets for light-emitting applications. *Chemical Society Reviews*, *52*(1), 318–360. PMID: 36533300

Baker, D. V., Bernal-Escalante, J., Traaseth, C., Wang, Y., Tran, M. V., Keenan, S., & Algar, W. R. (2025). Smartphones as a platform for molecular analysis: Concepts, methods, devices and future potential. *Lab on a Chip*. PMID: 39918205

Bisht, T., Adhikari, A., Patil, S., & Dhoundiyal, S. (2024). Bioconjugation techniques for enhancing stability and targeting efficiency of protein and peptide therapeutics. *Current Protein & Peptide Science*, *25*(3), 226–243. PMID: 37921168

Chaturvedi, A., Tripathi, D., & Ranjan, R. (2025). Nano-enabled biosensors in early detection of plant diseases. *Frontiers in Nanotechnology*, *7*, 1545792.

Chelliah, R., Khan, I., Wei, S., Madar, I. H., Sultan, G., Daliri, E. B.-M., Swamidoss, C., & Oh, D. H. (2022). Intelligent packaging systems: food quality and intelligent medicine box based on nano-sensors. In *Smart Nanomaterials in Biomedical Applications* (pp. 555–587). Springer.

Chen, C.-H., Liang, H.-H., Wang, C.-C., Yang, Y.-T., Lin, Y.-H., & Chen, Y.-L. (2024). Unlocking early detection of Alzheimer's disease: The emerging role of nanomaterial-based optical sensors. *Yao Wu Shi Pin Fen Xi, 32*(3), 296. PMID: 39636776

Chen, K., Ma, B., Li, J., Chen, E., Xu, Y., Yu, X., Sun, C., & Zhang, M. (2021). A rapid and sensitive europium nanoparticle-based lateral flow immunoassay combined with recombinase polymerase amplification for simultaneous detection of three food-borne pathogens. *International Journal of Environmental Research and Public Health, 18*(9), 4574. PMID: 33925871

Colson, Y. L., & Grinstaff, M. W. (2012). Biologically responsive polymeric nanoparticles for drug delivery. *Advanced Materials, 24*(28), 3878–3886. DOI: 10.1002/adma.201200420 PMID: 22988558

Curulli, A. (2021). Electrochemical biosensors in food safety: Challenges and perspectives. *Molecules (Basel, Switzerland), 26*(10), 2940. PMID: 34063344

Dessale, M., Mengistu, G., & Mengist, H. M. (2022). Nanotechnology: A promising approach for cancer diagnosis, therapeutics and theragnosis. *International Journal of Nanomedicine, 17*, 3735. PMID: 36051353

Dey, M. K., Iftesum, M., Devireddy, R., & Gartia, M. R. (2023). New technologies and reagents in lateral flow assay (LFA) designs for enhancing accuracy and sensitivity. *Analytical Methods, 15*(35), 4351–4376. PMID: 37615701

Dinis-Oliveira, R. J. (2014). Heterogeneous and Homogeneous Immunoassays for Drug Analysis. *Bioanalysis, 6*(21), 2877–2896. DOI: 10.4155/bio.14.208 PMID: 25486233

Diouf, M., Kouyate, S., & Diakhoumpa, A. (2025). Inorganic nanostructures as carriers for doxorubicin in targeted cancer therapy: Advances in nanoformulation strategies. *Composite Interfaces*, •••, 1–32.

Dubey, A. K., Kumar Gupta, V., Kujawska, M., Orive, G., Kim, N.-Y., Li, C.-z., Kumar Mishra, Y., & Kaushik, A. (2022). Exploring nano-enabled CRISPR-Cas-powered strategies for efficient diagnostics and treatment of infectious diseases. *Journal of Nanostructure in Chemistry, 12*(5), 833–864. PMID: 35194511

Dudala, S., Dubey, S. K., Javed, A., Ganguly, A., & Goel, S. (2022). Electromicrofluidic device with integrated PDMS microchannel and laser-induced graphene electrodes for electrochemical detection of cardiac biomarker in a point-of-care platform. *Journal of Micromechanics and Microengineering, 32*(10), 104001.

El Saftawy, E., Farag, M. F., Gebreil, H. H., Abdelfatah, M., Aboulhoda, B. E., Alghamdi, M., Albadawi, E. A., & Abd Elkhalek, M. A. (2024). Malaria: Biochemical, physiological, diagnostic, and therapeutic updates. *PeerJ, 12*, e17084. PMID: 38529311

Fahmy, H. M., Abu Serea, E. S., Salah-Eldin, R. E., Al-Hafiry, S. A., Ali, M. K., Shalan, A. E., & Lanceros-Mendez, S. (2022). Recent progress in graphene-and related carbon-nanomaterial-based electrochemical biosensors for early disease detection. *ACS Biomaterials Science & Engineering*, *8*(3), 964–1000. PMID: 35229605

Farka, Z., Brandmeier, J. C., Mickert, M. J., Pastucha, M., Lacina, K., Skládal, P., Soukka, T., & Gorris, H. H. (2024). Nanoparticle-based bioaffinity assays: From the research laboratory to the market. *Advanced Materials*, *36*(3), 2307653. PMID: 38039956

Ferrari, E. (2023). Gold nanoparticle-based plasmonic biosensors. *Biosensors (Basel)*, *13*(3), 411. PMID: 36979623

Forouzanfar, S. (2021). Investigation of Carbon-based Label-Free Electrochemical Aptasensors for Point-of-Care Cancer Biomarker Detection.

Gao, L., Yang, Q., Wu, P., & Li, F. (2020). Recent advances in nanomaterial-enhanced enzyme-linked immunosorbent assays. *Analyst*, *145*(12), 4069–4078. PMID: 32391529

Geszke-Moritz, M., & Moritz, M. (2024). Biodegradable Polymeric Nanoparticle-Based Drug Delivery Systems: Comprehensive Overview, Perspectives and Challenges. *Polymers*, *16*(17), 2536. https://www.mdpi.com/2073-4360/16/17/2536 PMID: 39274168

Godja, N.-C., & Munteanu, F.-D. (2024). Hybrid nanomaterials: A brief overview of versatile solutions for sensor technology in healthcare and environmental applications. *Biosensors (Basel)*, *14*(2), 67. PMID: 38391986

Gupta, S., Mishra, V., Aljabali, A. A., Albutti, A., Kanday, R., El-Tanani, M., & Mishra, Y. (2025). Breaking barriers in cancer diagnosis: Unveiling the 4Ms of biosensors. *RSC Advances*, *15*(10), 8019–8052. PMID: 40098694

Han, Y., Tian, Y., Li, Q., Yao, T., Yao, J., Zhang, Z., & Wu, L. (2025). Advances in detection technologies for pesticide residues and heavy metals in rice: A comprehensive review of spectroscopy, chromatography, and biosensors. *Foods*, *14*(6), 1070. PMID: 40232082

Hemdan, M., Ali, M. A., Doghish, A. S., Mageed, S. S. A., Elazab, I. M., Khalil, M. M., Mabrouk, M., Das, D. B., & Amin, A. S. (2024). Innovations in biosensor technologies for healthcare diagnostics and therapeutic drug monitoring: Applications, recent progress, and future research challenges. *Sensors (Basel)*, *24*(16), 5143. PMID: 39204840

Hou, F., Sun, S., Abdullah, S. W., Tang, Y., Li, X., & Guo, H. (2023). The application of nanoparticles in point-of-care testing (POCT) immunoassays. *Analytical Methods*, *15*(18), 2154–2180. PMID: 37114768

Huang, X., O'Connor, R., & Kwizera, E. A. (2017). Gold nanoparticle based platforms for circulating cancer marker detection. *Nanotheranostics*, *1*(1), 80. PMID: 28217434

Hulme, J. (2022). Application of nanomaterials in the prevention, detection, and treatment of methicillin-resistant Staphylococcus aureus (MRSA). *Pharmaceutics*, *14*(4), 805. PMID: 35456638

Jabran, M., Ali, M. A., Muzammil, S., Zahoor, A., Ali, F., Hussain, S., Muhae-Ud-Din, G., Ijaz, M., & Gao, L. (2024). Exploring the potential of nanomaterials (NMs) as diagnostic tools and disease resistance for crop pathogens. *Chemical and Biological Technologies in Agriculture*, *11*(1), 75.

Kant, K., Beeram, R., Cao, Y., Dos Santos, P. S., González-Cabaleiro, L., García-Lojo, D., Guo, H., Joung, Y., Kothadiya, S., & Lafuente, M. (2024). Plasmonic nanoparticle sensors: Current progress, challenges, and future prospects. *Nanoscale Horizons*, *9*(12), 2085–2166. PMID: 39240539

Khan, A., Anas, M., Qadri, T. A., Algopishi, U. B., Fayad, E., Binjawhar, D. N., Khan, R. A., Maaza, M., & Shinwari, Z. K. (2025). Eco-friendly synthesis of copper oxide-silver bimetallic nanoparticles using Withania coagulans: Characterization and biomedical potential through antioxidant, antibacterial, and catalytic activities. *Inorganic Chemistry Communications*, *115131*, •••. https://doi.org/https://doi.org/10.1016/j.inoche.2025.115131

Khodaparast, M., Sharley, D., Marshall, S., & Beddoe, T. (2024). Advances in point-of-care and molecular techniques to detect waterborne pathogens. *NPJ Clean Water*, *7*(1), 74.

Khojasteh-Kaffash, S., Habibzadeh, A., Moghaddam, S., Afra, F., Samieefar, N., & Fateh, A. (2025). Tuberculosis Trends in the Post-COVID-19 Era: Is It Going to be a Global Concern? *Health Science Reports*, *8*(5), e70792. PMID: 40406648

Kidanemariam, A., & Cho, S. (2025). Recent Advances in Metal–Organic Framework-Based Nanozymes for Intelligent Microbial Biosensing: A Comprehensive Review of Biomedical and Environmental Applications. *Biosensors (Basel)*, *15*(7), 437. PMID: 40710087

Kim, M. J., Haizan, I., Ahn, M. J., Park, D.-H., & Choi, J.-H. (2024). Recent advances in lateral flow assays for viral protein detection with nanomaterial-based optical sensors. *Biosensors (Basel)*, *14*(4), 197. PMID: 38667190

Kinyua, D. M., Memeu, D. M., Mugo Mwenda, C. N., Ventura, B. D., & Velotta, R. (2025). Advancements and Applications of Lateral Flow Assays (LFAs): A Comprehensive Review. *Sensors (Basel)*, *25*(17), 5414. PMID: 40942842

Kurdekar, A., & Kamisetti, V. (2024). Advancing Diagnostics: Evaluating the Potential of Nanoparticle-Based Immunoassays as PCR Alternatives for Disease Confirmation. *Current Analytical Chemistry*, *20*(5), 287–294.

Le, N., & Kim, K. (2023). Current advances in the biomedical applications of quantum dots: Promises and challenges. *International Journal of Molecular Sciences*, *24*(16), 12682. PMID: 37628860

Li, J., Duan, L., Chen, Q., Huang, S., Li, W., Zuo, H., Li, S., Han, W., Fu, W., & Hou, S. (2025). Roles of nanomagnetic beads on biosensing: From fabrication to application. *Interdisciplinary Medicine*, •••, 20250030.

Li, W., Xu, Z., He, Q., Pan, J., Zhang, Y., El-Sheikh, E.-S. A., Hammock, B. D., & Li, D. (2025). Nanobody-Based Immunoassays for the Detection of Food Hazards—A Review. *Biosensors (Basel)*, *15*(3), 183. https://www.mdpi.com/2079-6374/15/3/183 PMID: 40136980

Li, W., Xu, Z., He, Q., Pan, J., Zhang, Y., El-Sheikh, E.-S. A., Hammock, B. D., & Li, D. (2025). Nanobody-Based Immunoassays for the Detection of Food Hazards—A Review. *Biosensors (Basel), 15*(3), 183. https://www.mdpi.com/2079-6374/15/3/183 PMID: 40136980

Li, X., Fu, L., Chen, F., Lv, Y., Zhang, R., Zhao, S., & Karimi-Maleh, H. (2025). Cyclodextrin-based architectures for electrochemical sensing: From molecular recognition to functional hybrids. *Analytical Methods, 17*(21), 4300–4320. PMID: 40392560

Li, Z., Guo, M., & Zhong, W. (2025). Multiplex Detection of Biomarkers Empowered by Nanomaterials. *Precision Chemistry*. PMID: 40574998

Liu, S., Shen, Z., Deng, L., & Liu, G. (2022). Smartphone assisted portable biochip for non-invasive simultaneous monitoring of glucose and insulin towards precise diagnosis of prediabetes/diabetes. *Biosensors & Bioelectronics, 209*, 114251. https://doi.org/https://doi.org/10.1016/j.bios.2022.114251 PMID: 35405503

Liu, Z., Huo, Y., Hu, Z., Xu, Y., Hao, B., Yao, X., Li, K., & Liu, S. (2025). Nanomaterials-enabled biosensing platforms for microcystin-LR detection: a review of analytical advancements: Z. Liu et al. Analytical and Bioanalytical Chemistry, 1-26.

Lu, Z.-Y., & Chan, Y.-H. (2024). The importance of antibody orientation for enhancing sensitivity and selectivity in lateral flow immunoassays. *Sensors & Diagnostics*.

Luckose, F., Bangera, M., Shetty, N. S., & Satyaprasad, A. U. (2023). Recent trends in nano biosensors for food testing. In *Engineering Aspects of Food Quality and Safety* (pp. 383–409). Springer.

Mallick, T., Joshi, R. K., & Mishra, R. (2025). Nanoparticle-Based Biosensors for Pathogen Detection: Current Updates and Future Prospects. *BioNanoScience, 15*(3), 523.

Manessis, G., Gelasakis, A. I., & Bossis, I. (2022). Point-of-care diagnostics for farm animal diseases: From biosensors to integrated lab-on-chip devices. *Biosensors (Basel), 12*(7), 455. PMID: 35884258

Mehta, M., More, B., Tamakuwala, T., & Shah, G. (2022). Laboratory-on-a-Chip: A Multitasking Device. Miniaturized Analytical Devices: Materials and Technology, 91-103.

Mittal, D., & Ali, S. A. (2022). Use of nanomaterials for diagnosis and treatment: The advancement of next-generation antiviral therapy. *Microbial Drug Resistance (Larchmont, N.Y.), 28*(6), 670–697. PMID: 35696335

Naghavi, M., Abajobir, A. A., Abbafati, C., Abbas, K. M., Abd-Allah, F., Abera, S. F., Aboyans, V., Adetokunboh, O., Afshin, A., Agrawal, A., Ahmadi, A., Ahmed, M. B., Aichour, A. N., Aichour, M. T. E., Aichour, I., Aiyar, S., Alahdab, F., Al-Aly, Z., Alam, K., & Murray, C. J. L. (2017). Global, regional, and national age-sex specific mortality for 264 causes of death, 1980–2016: A systematic analysis for the Global Burden of Disease Study 2016. *Lancet, 390*(10100), 1151–1210. DOI: 10.1016/S0140-6736(17)32152-9 PMID: 28919116

Nanone, G. T., da Cruz Júnior, M. S., Bianchi, M. M., Dias, C. C., Macêdo, L. F., Simões, N. S., Passaglia, A. F. C., dos Santos, G. T., & Leite, K. B. G. (2024). Manejo da Doença Renal Crônica: Importância do diagnóstico precoce e monitoramento contínuo. *LUMEN ET VIRTUS, 15*(39), 3184–3191.

Numan, A., Singh, S., Zhan, Y., Li, L., Khalid, M., Rilla, K., Ranjan, S., & Cinti, S. (2022). Advanced nanoengineered—Customized point-of-care tools for prostate-specific antigen. *Microchimica Acta, 189*(1), 27. PMID: 34905090

Pandey, S. N., Afzal, M., Goyal, A., Priya, G. P., Mohanty, B., Goyal, K., Rana, M., & Imran, M. (2025). Quantum dot biosensors for Glioblastoma: Merging nanotechnology with precision oncology. *Inorganic Chemistry Communications*, •••, 114671.

Pour, S. R. S., Calabria, D., Emamiamin, A., Lazzarini, E., Pace, A., Guardigli, M., Zangheri, M., & Mirasoli, M. (2023). Electrochemical vs. optical biosensors for point-of-care applications: A critical Review. *Chemosensors (Basel, Switzerland), 11*(10), 546.

Prabhu, S., & van Wagoner, N. (2023). Human immunodeficiency virus infection and acquired immunodeficiency syndrome (HIV/AIDS): An overview. Sexually transmissible. *Oral Diseases*, •••, 51–71.

Puttasiddaiah, R., Basavegowda, N., Lakshmanagowda, N. K., Raghavendra, V. B., Sagar, N., Sridhar, K., Dikkala, P. K., Bhaswant, M., Baek, K.-H., & Sharma, M. (2025). Emerging nanoparticle-based diagnostics and therapeutics for cancer: Innovations and challenges. *Pharmaceutics, 17*(1), 70. PMID: 39861718

Risalvato, J., Sewid, A. H., Eda, S., Gerhold, R. W., & Wu, J. J. (2025). Strategic Detection of Escherichia coli in the Poultry Industry: Food Safety Challenges, One Health Approaches, and Advances in Biosensor Technologies. *Biosensors (Basel), 15*(7), 419. PMID: 40710069

Rivera-Gil, P., Jimenez De Aberasturi, D., Wulf, V., Pelaz, B., Del Pino, P., Zhao, Y., De La Fuente, J. M., Ruiz De Larramendi, I., Rojo, T., & Liang, X.-J. (2013). The challenge to relate the physicochemical properties of colloidal nanoparticles to their cytotoxicity. *Accounts of Chemical Research, 46*(3), 743–749. PMID: 22786674

Rodriguez-Quijada, C., Lyons, C., Santamaria, C., Quinn, S., Tlusty, M. F., Shiaris, M., & Hamad-Schifferli, K. (2020). Optimization of paper-based nanoparticle immunoassays for direct detection of the bacterial pathogen V. parahaemolyticus in oyster hemolymph. *Analytical Methods, 12*(23), 3056–3063. PMID: 32930166

Sanyal, A., Mitra, P., Dey, T., Dutta, D., Saha, K., Pandey, A., & Pattnaik, R. (2024). Development in Biosensor-Based Diagnostics for Bacterial Diseases: Opportunities and Challenges. Functional Smart Nanomaterials and Their Theranostics Approaches, 197-239.

Sena-Torralba, A., Álvarez-Diduk, R., Parolo, C., Piper, A., & Merkoçi, A. (2022). Toward next generation lateral flow assays: Integration of nanomaterials. *Chemical Reviews, 122*(18), 14881–14910. PMID: 36067039

Shivakumar, N. (2024). Recent advances in biological nanodevices and biosensors: Insights into applications and technological innovations. *Malaysian NANO-An International Journal, 4*(1), 86–101.

Sin, M. L., Mach, K. E., Wong, P. K., & Liao, J. C. (2014). Advances and challenges in biosensor-based diagnosis of infectious diseases. *Expert Review of Molecular Diagnostics, 14*(2), 225–244. PMID: 24524681

Singh, A., Ahmed, A., Sharma, A., & Arya, S. (2022). Graphene and its derivatives: Synthesis and application in the electrochemical detection of analytes in sweat. *Biosensors (Basel)*, *12*(10), 910. PMID: 36291046

Singh, H., Bamrah, A., Bhardwaj, S. K., Deep, A., Khatri, M., Brown, R. J., Bhardwaj, N., & Kim, K.-H. (2021). Recent advances in the application of noble metal nanoparticles in colorimetric sensors for lead ions. *Environmental Science. Nano*, *8*(4), 863–889.

Song, Q., Sun, X., Dai, Z., Gao, Y., Gong, X., Zhou, B., Wu, J., & Wen, W. (2021). Point-of-care testing detection methods for COVID-19. *Lab on a Chip*, *21*(9), 1634–1660. PMID: 33705507

Soto, D., & Orozco, J. (2022). Hybrid Nanobioengineered Nanomaterial-Based Electrochemical Biosensors. *Molecules (Basel, Switzerland)*, *27*(12), 3841. https://www.mdpi.com/1420-3049/27/12/3841 PMID: 35744967

Sreeramareddy, C. T., Panduru, K. V., Menten, J., & Van den Ende, J. (2009). Time delays in diagnosis of pulmonary tuberculosis: A systematic review of literature. *BMC Infectious Diseases*, *9*(1), 91. DOI: 10.1186/1471-2334-9-91 PMID: 19519917

Storla, D. G., Yimer, S., & Bjune, G. A. (2008). A systematic review of delay in the diagnosis and treatment of tuberculosis. *BMC Public Health*, *8*(1), 15. DOI: 10.1186/1471-2458-8-15 PMID: 18194573

Tai, J., Fan, S., Ding, S., & Ren, L. (2022). Gold nanoparticles based optical biosensors for cancer biomarker proteins: A review of the current practices. *Frontiers in Bioengineering and Biotechnology*, *10*, 877193. PMID: 35557858

Thiruvengadam, M., Kim, J.-T., Kim, W.-R., Kim, J.-Y., Jung, B.-S., Choi, H.-J., Chi, H.-Y., Govindasamy, R., & Kim, S.-H. (2024). Safeguarding public health: Advanced detection of food adulteration using nanoparticle-based sensors. *Critical Reviews in Analytical Chemistry*, •••, 1–21. PMID: 39269682

Thwala, L. N., Ndlovu, S. C., Mpofu, K. T., Lugongolo, M. Y., & Mthunzi-Kufa, P. (2023). Nanotechnology-based diagnostics for diseases prevalent in developing countries: Current advances in point-of-care tests. *Nanomaterials (Basel, Switzerland)*, *13*(7), 1247. PMID: 37049340

Türkcan, C., & Akgöl, S. (2023). Polymeric nanoparticle-based electrochemical sensor for the detection of CA 125. *Turkish Journal of Chemistry*, *47*(1), 137–147. PMID: 37720868

Vos, T., Lim, S. S., Abbafati, C., Abbas, K. M., Abbasi, M., Abbasifard, M., Abbasi-Kangevari, M., Abbastabar, H., Abd-Allah, F., Abdelalim, A., Abdollahi, M., Abdollahpour, I., Abolhassani, H., Aboyans, V., Abrams, E. M., Abreu, L. G., Abrigo, M. R. M., Abu-Raddad, L. J., Abushouk, A. I., & Murray, C. J. L. (2020). Global burden of 369 diseases and injuries in 204 countries and territories, 1990–2019: A systematic analysis for the Global Burden of Disease Study 2019. *Lancet*, *396*(10258), 1204–1222. DOI: 10.1016/S0140-6736(20)30925-9 PMID: 33069326

Vos, T., Lim, S. S., Abbafati, C., Abbas, K. M., Abbasi, M., Abbasifard, M., Abbasi-Kangevari, M., Abbastabar, H., Abd-Allah, F., Abdelalim, A., Abdollahi, M., Abdollahpour, I., Abolhassani, H., Aboyans, V., Abrams, E. M., Abreu, L. G., Abrigo, M. R. M., Abu-Raddad, L. J., Abushouk, A. I., & Murray, C. J. L. (2020). Global burden of 369 diseases and injuries in 204 countries and territories, 1990–2019: A systematic analysis for the Global Burden of Disease Study 2019. *Lancet, 396*(10258), 1204–1222. DOI: 10.1016/S0140-6736(20)30925-9 PMID: 33069326

Wang, M., Jin, L., Hang-Mei Leung, P., Wang-Ngai Chow, F., Zhao, X., Chen, H., Pan, W., Liu, H., & Li, S. (2024). Advancements in magnetic nanoparticle-based biosensors for point-of-care testing. *Frontiers in Bioengineering and Biotechnology, 12*, 1393789. PMID: 38725992

Wang, S., Guan, X., & Sun, S. (2025). Microfluidic biosensors: Enabling advanced disease detection. *Sensors (Basel), 25*(6), 1936. PMID: 40293099

Wang, Z., Zhao, J., Xu, X., Guo, L., Xu, L., Sun, M., Hu, S., Kuang, H., Xu, C., & Li, A. (2022). An overview for the nanoparticles-based quantitative lateral flow assay. *Small Methods, 6*(1), 2101143. PMID: 35041285

Wilson, A. D. (2018). Applications of electronic-nose technologies for noninvasive early detection of plant, animal and human diseases. *Chemosensors (Basel, Switzerland), 6*(4), 45.

Xia, N., Li, Y., He, C., & Deng, D. (2024). Nanolabels Prepared by the Entrapment or Self-Assembly of Signaling Molecules for Colorimetric and Fluorescent Immunoassays. *Biosensors (Basel), 14*(12). Advance online publication. DOI: 10.3390/bios14120597 PMID: 39727862

Xiao, Y., Zhang, Z., Yin, S., & Ma, X. (2023). Nanoplasmonic biosensors for precision medicine. *Frontiers in Chemistry, 11*, 1209744. PMID: 37483272

Yao, J., Yang, M., & Duan, Y. (2014). Chemistry, biology, and medicine of fluorescent nanomaterials and related systems: New insights into biosensing, bioimaging, genomics, diagnostics, and therapy. *Chemical Reviews, 114*(12), 6130–6178. PMID: 24779710

Yergök, R. A., Derman, S., Alakbarli, J., & Eskhıta, A. S. (2024). Brain-Targeted Nano-Drug Delivery for the Treatment of Parkinson's Disease. *Current Research in Health Sciences, 1*(2), 77–92.

Younis, M. I., Sallam, Y. I., Mahmoud, K. F., Ruan, Z., Tlay, R. H., & Abedelmaksoud, T. G. (2025). Smartphones as Catalysts for Synergistic Nutrition: A New Era in Bioactive Detection, Personalization, and Food System Intelligence. *Food Science & Nutrition, 13*(9), e70880. PMID: 40905019

Zahra, Q. u. A., Fang, X., Luo, Z., Ullah, S., Fatima, S., Batool, S., Qiu, B., & Shahzad, F. (2023). Graphene based nanohybrid aptasensors in environmental monitoring: Concepts, design and future outlook. *Critical Reviews in Analytical Chemistry, 53*(7), 1433–1454. PMID: 35085047

Zhang, J., Zhang, J., Li, H., Zhang, H., & Meng, H. (2025). Research progress on biodegradable polymer-based drug delivery systems for the treatment of knee osteoarthritis. *Frontiers in Bioengineering and Biotechnology, 13*, 1561708. PMID: 40276032

Zhang, N., Zhu, Y., Li, F., Wang, Y., Fu, Z., Jia, M., Zhan, X., & Zhang, W. (2025). Research progress on nanomaterial-empowered electrochemical biosensors for the detection of cardiac troponin I. *RSC Advances*, *15*(32), 26473. PMID: 40703066

Zhang, P., Hou, H., Xu, S., Wen, Y., Zhang, Y., & Xing, F. (2025). Localized surface plasmon resonance sensing based on monometallic gold nanoparticles: From material preparation to detection of bioanalytes. *Analytical Methods*. PMID: 39693100

Zhang, S., Cai, J., & Wu, L. (2014). Application of nanomaterials in in-vitro diagnostics. Zhongguo yi Liao qi xie za zhi=. *Chinese Journal of Medical Instrumentation*, *38*(2), 118–121. PMID: 24941776

Zhong, Z., He, J., Li, G., & Xia, L. (2023). Recent advances in magnetic nanoparticles-assisted microfluidic bioanalysis. *Chemosensors (Basel, Switzerland)*, *11*(3), 173.

Chapter 10
Nanoformulations for Targeted Antiviral Delivery in Plants

Fahadul Alam
Department of Biological Sciences, Texas Tech University, Lubbock, USA

Rahmat Ali
http://orcid.org/0009-0007-8777-1148
Department of Biological Sciences, Texas Tech University, Lubbock, USA

Tahlil Nafi
Department of Biotechnology, Bangladesh Agricultural University, Bangladesh

Sajid Ali
Department of Chemistry, Bacha Khan University, Charsadda, Pakistan

Shumaila Ijaz
https://orcid.org/0009-0003-9614-9802
School of Biomedical Engineering, Shenzhen University, China

ABSTRACT

Plant viruses cause crop losses and over half of emerging plant diseases are viral, yet effective cures are scarce. Nanoformulations offer a targeted strategy to deliver antivirals or boost plant defenses—an emerging field called nanophytovirology. Nanoparticles can inactivate virions, enhance innate immunity, or carry genetic and chemical agents into cells. Classes include metal/metal-oxide NPs (Ag, Au, Cu, ZnO), carbon materials, silica/nano-clays, polymeric and lipid carriers (chitosan, dendrimers, liposomes), and biogenic NPs. This chapter reviews virus classes and impacts; core mechanisms (virion inactivation, RNA interference, immune priming) and delivery routes (foliar spray, root uptake, vascular injection, seed/tissue application). We integrate laboratory and field results, highlight environmental and regulatory issues (toxicity, safety, scalability), and map future directions—NP-mediated CRISPR delivery, smart carriers, and sustainable nano-agriculture. The goal is to turn lab advances into practical tools that strengthen virus surveillance and plant-health security.

DOI: 10.4018/979-8-3373-4913-8.ch010

1. INTRODUCTION

Plant viruses are major, often hidden, causes of agricultural loss worldwide. Plant viral diseases are thought to cost between $30 and $50 billion in lost yields each year, this is about half of all plant disease cases around the world. (Hilaire et al., 2022). For example, tomato spotted wilt virus (TSWV) alone costs more than $1 billion in annual losses across many hosts (Koundal et al., 2011). Entire families of viruses inflict multi-billion-dollar damage: the whitefly-transmitted geminiviruses (e.g. African cassava mosaic virus, *Tomato yellow leaf curl virus*) have caused devastating epidemics in tropical staples like cassava, cotton, and tomato (Varma & Singh, 2024). In Asia and the Middle East, *tomato yellow leaf curl begomovirus* has equally devastated tomato and vegetable harvests (Varma & Singh, 2024). The burden is significant even in temperate regions: in Europe, *Luteoviridae* viruses in cereals (e.g., Barley yellow dwarf viruses) result in annual losses amounting to tens of millions of dollars (Hilaire et al., 2022), while a recent study approximates that *Potato virus Y* (PVY) incurs costs of approximately €187 million per year to the European Union (Dupuis et al., 2024). These losses disseminate throughout food systems, jeopardizing agricultural revenues and regional food security.

Most plant viruses move quickly from one plant to another by vectors, which are usually insects (including aphids, whiteflies, thrips, and leafhoppers) or nematodes. They can also spread by infected seeds and sap. There is no chemical remedy for a plant that has been infected (Nishina et al., 1989). Farmers need to use preventive measures, such as breeding virus-resistant crop types, keeping vector populations in check, cleaning tools and fields, and getting rid of diseased plants to lower the amount of inoculum (Nishina et al., 1989). These were the only techniques that worked for decades, and they are still very important. Improvements in diagnostics have made it easier to find viruses, for example, next-generation sequencing can swiftly find novel or dormant viruses (Maina & Jones, 2023), but modern agriculture faces accelerating viral threats. When crops are naive and there isn't enough containment, this can cause huge outbreaks (Maina & Jones, 2023). Climate change changes disease-carrying insect ranges and lengthens growing seasons, making viral outbreaks difficult to predict and control. (Maina & Jones, 2023).

Since viruses are responsible for about half of new plant disease outbreaks (Jones, 2021), it is crucial to find new, multidisciplinary ways to work with existing methods. Nanotechnology-enabled plant virus management is a potential new area of research. Researchers are investigating nanomaterials for the early detection of viruses, targeted antiviral administration, and the induction of immunity in plants (Farooq et al., 2021). For example, virus-specific components like double-stranded RNAs or antiviral chemicals can be put into nanocarriers and sent directly to affected tissues. This is called "nanophytovirology." (Farooq et al., 2021). These innovative strategies, still largely in experimental stages, have been found to lower virus loads and protect plants in controlled studies. As they keep getting better, nano-based antivirals and vaccines could be useful tools for fighting plant viruses when traditional breeding or pesticides don't work. Also, intensive vegetable production still relies heavily on synthetic pesticides, and farmers report substantial concerns about health and environmental risks associated with their misuse (Alam et al., 2020). As pharmaceutical remedies are still not accessible, prevention and emerging technologies, such as nanotechnology, will be very important for keeping crops safe from the "hidden" pandemic of plant virus diseases (Farooq et al., 2021).

2. PLANT VIRUS DIVERSITY AND CLASSIFICATION

Plant viruses are incredibly diverse, yet most of the time they are grouped by the type of genome they have (Baltimore classification). Around 75% of the plant viruses we know of have RNA genomes (Kheyrodin, 2017). Most of these viruses are single-stranded positive-sense RNA (+ssRNA) viruses, which act directly as mRNA in host cells, whereas about 10% of plant viruses are negative-sense ssRNA (−ssRNA) viruses[REMOVED HYPERLINK FIELD] (Kheyrodin, 2017). Viruses such as *Rhabdoviridae* and *Tospoviridae* (the latter currently classed as *Orthotospovirus*, including TSWV) necessitate the transcription of their RNA genome into positive-sense RNA prior to protein synthesis. Rice ragged stunt virus and Rice tungro spherical virus are members of the *Reoviridae* family. They can cause big losses in rice and other crops. Plant *pararetroviruses*, such as *Caulimoviridae* (e.g., Cauliflower mosaic virus) and *Badnaviridae*, possess double-stranded DNA genomes yet replicated through an RNA intermediate via reverse transcription.

Finally, about 17% of plant viruses are single-stranded DNA (ssDNA) viruses (Kheyrodin, 2017). The main plant viruses in this group are geminiviruses (family *Geminiviridae*) and the related *Nanoviridae*; *geminiviruses* have short, circular ssDNA genomes and are spread by insects such as whiteflies and leafhoppers. Some of the most harmful crop viruses in the tropics are the African cassava mosaic virus, the Maize streak virus, and the *Tomato yellow leaf curl virus* (Varma & Singh, 2024). It's important to note that true double-stranded DNA (dsDNA) viruses are almost completely missing from higher plants. They only infect lesser plants, such some algae (LeClair & Boundless, 2024). This is very different from animal virology, where a quarter of animal viruses are dsDNA viruses (Kheyrodin, 2017). Mosaic patterns of light and dark green on leaves, mottling and chlorosis (yellowing), leaf deformation or curling, general stunting of the plant, and sometimes necrotic lesions or streaks on leaves and stems are all common indicators of a viral infection (Sánchez-Tovar et al., 2025).

A weak viral strain in a crop variety that can handle it might just cause minor cosmetic harm. On the other hand, a virulent virus in a crop variety that can't handle it can destroy the whole crop. Mixed infections, in which a plant has two or more distinct viruses, generally make the disease worse than if it were just one virus. One virus can work with another to raise its titer or lower the host's defences, which can make symptoms worse and lower the yield (Sánchez-Tovar et al., 2025). Such interactions are common; a recent review noted that mixed viral infections in the field frequently result in more severe symptoms and higher yield losses, and importantly no chemical control strategies are effective once plants are virus-infected (Sánchez-Tovar et al., 2025). Traditional methods have helped stop many outbreaks, but new viral strains are always appearing and spreading to new areas, so the pressure is always on (Maina & Jones, 2023). Thus, emerging plant virus control methods like transgenic resistance (coat protein-mediated resistance and RNA interference transgenes), RNA silencing sprays, and nanotechnologies are of great interest.

3. NANOTECHNOLOGY APPROACHES AND NANO FORMULATION STRATEGIES

3.1 Nanoparticle Synthesis: Synthetic and Biogenic Approaches

Recent advances in nanotechnology have produced a variety of nanoparticles (NPs) for plant virus management, with both synthetic and biogenic synthesis routes. Conventional chemical synthesis methods (e.g. precipitation, thermal decomposition) can yield metallic and metal-oxide NPs like silver (Ag), gold (Au), copper oxide (CuO), zinc oxide (ZnO), titanium dioxide (TiO_2), iron oxide (Fe_3O_4), silica (SiO_2), and ceria (CeO_2). However, these chemical methods often involve toxic reagents, prompting interest in "green" or biogenic synthesis (Notions et al., 2015). Biogenic approaches use biological agents such as plant extracts or microorganisms to reduce metal salts into NPs, avoiding harsh chemicals and imparting eco-friendly coatings on NP surfaces (Assel et al., 2024). For example, biosynthesized silver nanoparticles using *Bacillus* bacteria have been demonstrated to possess antiviral activity (Elbeshehy et al., 2015). Likewise, plant extracts from *Haloxylon salicornicum* have been employed to produce CuO nanoparticles rich in surface phytochemicals (Assel et al., 2024). These biogenic NPs often carry capping biomolecules that can enhance stability and may synergistically contribute to antiviral effects. By tailoring synthesis methods, NPs can be optimized for high reactivity against viruses while being safer for plants and the environment (Nations et al. 2015, Assel et al. 2024). This convergence of biotechnology and nanotechnology (bio-nanotechnology) is opening sustainable avenues for nanomaterials in crop protection.

Biogenic or "green" synthesis techniques can minimize input costs and make nanoformulations easier to produce on a large scale since they use plant extracts or microbial metabolites instead of harsh reducing chemicals (Aseel et al., 2024). Furthermore, the biomolecule-rich coatings produced during green synthesis may alleviate environmental and food safety issues, offering a natural connection to the economic, safety, and regulatory constraints outlined in Section 7.

3.2 Nanocarriers for Antiviral Delivery

A wide spectrum of nanocarrier types has been explored for plant virus control, each with unique properties. The choice of nanoparticle depends on the mode of action needed (e.g. direct virus neutralization vs. gene silencing delivery) and on practical considerations like plant uptake and safety. Below we survey major classes of nanomaterials being developed as antiviral agents or delivery vehicles in plants.

Table 1. Comparative summary of nanocarrier categories for plant antiviral delivery

Nanoparticle Type	Examples	Antiviral Mechanisms	Pros	Cons/Safety
Metallic	Ag, CuO	Direct virucidal (inactivates virions); induces plant immune defenses	High potency, broad-spectrum action	Potential phytotoxicity at high doses; possible environmental accumulation
Polymeric	Chitosan NPs, liposomes	Delivers antiviral payloads (e.g. dsRNA for RNAi); can boost plant immunity (chitosan)	Biocompatible & biodegradable; protects and slowly releases cargo	Stability issues (aggregation, short shelf-life); lower intrinsic antiviral activity (requires active payload)
Carbon-based	CNTs, carbon quantum dots (CQDs)	Carriers for genes/RNA (RNAi delivery); bind viruses & prime defenses	High stability; good tissue penetration; easily functionalized	Unknown long-term toxicity; potential environmental persistence
Hybrid	Chitosan-coated AgNP, BioClay (dsRNA-clay), etc.	Multi-modal (combines virucidal, delivery, immune activation)	Synergistic effects; improved efficacy/stability (e.g. sustained release) and reduced toxicity	Complex fabrication and higher cost; unclear regulatory classification

3.2.1 Metallic Nanoparticles. Metal-based nanoparticles are the most widely studied in plant virus management. Silver nanoparticles (AgNPs) in particular have broad-spectrum antimicrobial properties and have shown potent antiviral effects in plants. AgNPs can bind to virion surface proteins and genomic RNA, physically destabilizing the virus particle and preventing infection. This direct virucidal action also has an epidemiological benefit as treated plants become less infectious to insect vectors. In the BYMV study, aphids that fed on AgNPs-treated plants acquired far fewer viral particles, effectively reducing secondary spread (Ei Gamal et al., 2022). Beyond silver, copper nanoparticles (CuNPs) and their oxides have emerged as strong antiviral agents. Copper has long been used in agriculture as a broad biocide, and at the nanoscale it can directly interact with and disrupt viruses. A recent work used biosynthesized cupric oxide nanoparticles (CuO NPs) made from orange peel extract against tobacco mosaic virus. At 100 mg/L, these CuO NPs bound to TMV virions and ruptured their structure, rendering them non-infectious (Liu et al., 2024). Inside the plant, CuO NPs were observed to form complexes with TMV coat proteins, likely due to electrostatic attraction between positively charged CuO and the negatively charged virus (Vargas-Hernandez et al., 2020). Such findings show that metal NPs can offer a dual mode of action, direct virion destruction and host defense induction.

Gold nanoparticles (AuNPs) by contrast, are generally chemically inert and have been used more as nanocarriers or sensing agents than as direct antivirals. AuNPs can be functionalized to carry molecules that target viruses or to improve NP uptake. Alkubaisi & Aref (2017) reported that dispersed AuNPs damaged the virus-like particles of *Barley yellow dwarf virus* (BYDV, an ssRNA luteovirus), effectively eliminating infectivity in vitro. In addition to elemental metals, metal oxide nanoparticles are being explored for their antiviral and plant-protective properties. Zinc oxide nanoparticles (ZnO NPs) and silica (SiO_2) NPs are generally regarded as safe and have unique surface chemistry that can generate reactive oxygen species (ROS) under light or interact with biological macromolecules.

3.2.2 Polymeric Nanocarriers. Biodegradable polymer nanoparticles are another cornerstone of nano-delivery in plant systems. Polymers offer a tunable matrix to encapsulate and protect payloads like nucleic acids or chemical antivirals. A leading example is chitosan nanoparticles, derived from chitosan (a natural polysaccharide obtained from chitin). Chitosan is inherently biocompatible and is known to elicit plant defense on its own. Nanoscale chitosan formulations can bind nucleic acids due to their cationic amino groups and can facilitate uptake into plant cells. Researchers have leveraged this to deliver virus-targeted RNA molecules: positively charged chitosan NPs carrying virus-specific double-stranded RNA (dsRNA) have achieved efficient gene silencing of viruses inside plants. For example, chitosan NPs loaded with dsRNA against *Potato virus Y* (PVY) were sprayed on tobacco leaves, resulting in significant knockdown of the viral genes and protection of the plants. Because chitosan can slow the degradation of dsRNA and help it cross the plant cuticle, the antiviral effect persisted for weeks. Xu et al. (2023) directly compared chitosan-RNA NPs to other carriers and found chitosan-based nanoparticles induced strong RNA interference and systemic resistance against PVY, especially when delivered via roots or spray. Notably, Chitosan's innate defense-triggering ability as it is recognized by plant pattern-recognition receptors as a microbe-associated molecule means that Chitosan NPs can double up as both a delivery vehicle and an immune booster.

Beyond chitosan, synthetic polymers like PLGA (poly-lactic-co-glycolic acid), PEG (polyethylene glycol), and dendritic polymers have been formulated into nano-capsules for plant use. These polymers are valued in medicine for their slow-release properties and low toxicity, and they can be adapted for agrochemical delivery. For example, antiviral compounds such as ribavirin (a nucleoside analog) or plant-derived antivirals (e.g. quercetin) can be encapsulated in PLGA nanoparticles to improve their stability under field conditions.

An exciting recent development is the creation of hybrid nanocomposites that blend a biopolymer with a known defense signal. A prime example is a chitosan–salicylic acid nanocomposite (Chitosan/SA NC) designed to gradually release salicylic acid (SA), a plant hormone that triggers systemic acquired resistance. In greenhouse trials, a chitosan–SA nanoparticle (90 ppm, foliar applied) provided superior protection to cucumber plants against CMV, compared to equivalent treatment with free SA (El-Ganainy et al., 2025). The Chitosan/SA nanocomposite applied 24 h before inoculation was the most effective, significantly reducing disease severity and viral load. Free SA also protected plants to some degree, but the nanocomposite's slow release and better leaf adhesion made it more potent. These results underscore that polymeric nanocarriers can amplify the effects of conventional antivirals or resistance inducers by improving their delivery. A comparative study on garlic found that biogenic AgNPs and nanosilica both mitigated Leek yellow stripe virus infection, with combinations of the two giving the best outcomes in reducing symptom incidence (Abdelkader et al., 2025). Together, these findings highlight the versatility of polymers, whether by carrying genetic material for RNA Interference (RNAi), serving as a depot for chemical antivirals, or inherently eliciting plant defenses, polymeric nanoparticles are a cornerstone of nano-enabled virus management.

3.3.3 Carbon-Based Nanomaterials. Carbon-based nanomaterials bring a completely different set of properties to the table. These include carbon nanotubes (CNTs), carbon quantum dots (CQDs etc. Carbon nanomaterials are generally lightweight, highly stable, and can be made very pure (containing only carbon and hydrogen/oxygen functional groups). They often have high aspect ratios or extremely small size, which aids in penetrating plant tissues. For example, single-walled carbon nanotubes (SWCNTs) can easily traverse the plant cell wall (which has a pore size of 5–20 nm) due to their needle-like shape and ~1 nm diameter. Kwak et al. (2019) demonstrated that functionalized SWCNTs can deliver genes into plant cells without external biolistic or chemical aid: CNTs were used to carry plasmid DNA and even CRISPR/Cas9 components into *N. benthamiana* leaves, achieving stable expression of foreign genes. This same approach can be adapted for antivirals – e.g. delivering a DNA encoding an inhibitory RNA or directly delivering small RNAs that silence viral genes. The ability of CNTs to ferry cargo into chloroplasts and other organelles is particularly intriguing for targeting viruses that hide in those compartments. Aside from delivery applications, CNTs themselves may affect virus infections. Multi-walled CNTs have been reported to bind viral particles and also to induce subtle stress responses in plants that prime immunity. Another study found that CNT treatment suppressed TMV systemic infection in tobacco by enhancing the plant's photosynthetic efficiency and activating defense pathway (Farooq et al., 2024).

Carbon quantum dots (CQDs), small (~10 nm) fluorescent carbon nanoparticles, are emerging as effective antiviral agents for plant viruses. They can be easily functionalized and used to deliver dsRNA for virus gene silencing. A study by Xu et al. (2023) showed that amine-functionalized CQDs provided over 21 days of protection against *Potato virus Y* (PVY). Farooq et al. (2024) used CQDs alone to treat plants infected with Cotton leaf curl Multan virus (CLCuMuV), significantly reducing viral titer, alleviating leaf curl symptoms, and improving plant health. Mechanistically, CQDs triggered immune responses, upregulating genes related to pattern-triggered immunity (PTI) and effector-triggered immunity (ETI), enhancing plant defense. The CQDs may also interfere with the virus replication cycle by binding to viral components or altering host cellular processes as the study found evidence of CQD-induced alternative splicing events in the host that could hinder the virus.

3.3.4 Hybrid and Composite Nanostructures. Hybrid nanomaterials combine two or more components to achieve synergistic properties beyond what a single material could offer. In plant virus management, the motivation for hybrids is often to get the benefits of different nanoparticle types, for instance the strong antiviral potency of a metal NP with the biocompatibility of a polymer, or the stability of an inorganic core with the specificity of a biomolecule coating. One example is polymer-coated metal nanoparticles, such as chitosan-capped AgNPs. Here, a silver core provides the virucidal action, while the chitosan shell improves dispersibility in water and can stimulate plant immunity. Studies have shown that chitosan-coated AgNPs are less phytotoxic and more effective in reducing virus symptoms than non-coated AgNPs, presumably because the chitosan moderates the release of $Ag+$ ions and also amplifies defense signaling in the plant (El-Ganainy et al., 2023). A study by Niño-Sánchez et al. (2022) explored the use of BioClay, a formulation combining layered double hydroxide (LDH) nanosheets with double-stranded RNA (dsRNA), for plant protection. The study found that BioClay extended the window of protection against viruses for up to 20 days and improved protection against whitefly eggs and nymphs. Another emerging area is metal–organic frameworks (MOFs), porous crystalline nanoparticles made of metal nodes and organic linkers. MOFs have extremely high surface area and tunable pore sizes; they can be loaded with antiviral compounds or even enzymes (Mishra et al., 2023).

4. MECHANISMS OF ANTIVIRAL ACTION

Nanoparticles can combat plant viruses by several complementary mechanisms, often simultaneously. Understanding these modes of action is key to optimizing their use and combining them effectively.

4.1 Direct Virion Inactivation

Nanoparticles can bind to and denature virus particles directly upon contact. This "virus-static" or virucidal action is most pronounced with metal-based NPs. As discussed, AgNPs have a strong affinity for viral surface proteins and genomic RNA due to their charged surface and catalytic properties. They can attach to coat proteins, cause protein unfolding or cross-linking, and even promote local generation of reactive oxygen species that oxidize viral components (Cai et al., 2019). In bean plants infected with BYMV, AgNP foliar sprays physically inactivated virions to the point that virus replication was halted and no systemic spread occurred. Similarly, copper oxide NPs exhibited direct virucidal activity against TMV, as TEM imaging confirmed disrupted rod-shaped particles after treatment (Liu et al., 2024). Even gold NPs can, under certain conditions, irreversibly attach to virus particles such as AuNPs coated with citrate have been shown to bind TMV and cause clumping. The net effect of such interactions is that the virus can no longer recognize or enter host cells (Alkubaisi et al., 2017). An additional benefit of inactivating virions on plants is reduced vector transmission: if a whitefly or aphid probes an NP-treated leaf, the virus it uptakes might already be neutralized.

4.2 RNA Interference (RNAi) Delivery

RNA interference is a natural antiviral defense in plants where double-stranded RNA is processed into small interfering RNAs that guide the degradation of viral RNAs. Spraying naked dsRNA on plants can trigger RNAi against viruses, but naked dsRNA is quickly broken down by UV light and nucleases, and it struggles to enter cells. Nanotechnology solves this by protecting dsRNA and helping it traverse the plant surface. Several types of NPs like silica, chitosan, layered clays, carbon dots have been used to chaperone antiviral dsRNA into plants. Xu et al. (2023) systematically compared three: a. silica nanopowder with amine groups; b. quaternized chitosan nanoparticles, and c) carbon quantum dots. All three could successfully deliver dsRNA targeting *Potato virus Y* and achieved at least 14 days of protection in treated tobacco plants, whereas naked dsRNA gave only a few days of effect. The nanoparticles likely adhere to the leaf surface and slowly release dsRNA, which is then protected from degradation. Once inside plant cells, the dsRNA is processed by the plant's Dicer enzymes, which break it down into siRNAs. These siRNAs guide the RNA-induced silencing complex (RISC) to complementary viral RNA sequences, where they cleave and silence the virus's replication. This method is highly specific, and with the correct dsRNA sequence, it can target any virus, even new emerging strains.

4.3 Immune Priming and Elicitation

Many nanoparticles have the ability to enhance the plant's innate immune responses, a phenomenon known as immune priming. This process prepares the plant to react more effectively to pathogen attacks by triggering an initial defense response, often without causing full-scale damage. Some nanoparticles induce mild oxidative or osmotic stress, which the plant interprets as a signal of pathogen invasion, leading to the activation of defense pathways. For example, silica (SiO_2) nanoparticles have been shown to activate systemic acquired resistance (SAR) in *Arabidopsis* by triggering salicylic acid (SA) signaling. In a study by El-Shetehy et al. (2021), SiO_2 NPs were taken up by the plant through stomatal openings and localized in intercellular spaces. Instead of causing toxicity, the nanoparticles activated a defense cascade, including increased SA levels, expression of pathogenesis-related (PR) genes, and fortification of cell walls. These treated plants became more resistant to subsequent pathogen challenges, highlighting the potential of nanoparticles as immune priming agents. Metal nanoparticles like AgNPs and ZnO have immune-priming effects. Low nanoparticle concentrations trigger pathogenesis-related proteins and boost crop antioxidant defenses. For instance, AgNP treatment of BYMV-infected bean plants inactivated the virus and enhanced PR-1 gene and defense enzyme expression, improving the plant's virus resistance (Ei Gamal et al., 2022). This dual role of nanoparticles; both as virucidal agents and immune priming agents, is particularly valuable when dealing with plant viruses for which no specific chemical treatment exists.

4.4 Carrier-Mediated Delivery of Antiviral Agents

Nanocarriers can also enhance the delivery of traditional antivirals or bioactive substances that otherwise have limited efficacy in plants. This mechanism enhances the formulation of antiviral agents by improving their stability, solubility, and targeted delivery. A notable example is the quercetin nanoliposome field trial, which demonstrated that encapsulating quercetin, a naturally occurring antiviral compound, in lipid nanoparticles significantly increased its effectiveness against Tobacco Mosaic Virus (TMV). Quercetin, which is poorly water-soluble and prone to rapid degradation, was stabilized by the nanoliposome formulation, leading to enhanced virus control in the field. Other plant-derived antivirals, such as neem or ginger extracts, could also benefit from nanoparticle encapsulation, improving their stability and targeted release. Additionally, nanoparticles can be used to deliver synthetic antiviral drugs, such as salicylic acid derivatives or nucleoside analogs, in a controlled manner. This slow-release approach ensures that antiviral compounds remain active over extended periods, providing prolonged protection against plant viruses (Farooq et al., 2024).

In future applications, nanoparticles could even deliver CRISPR/Cas systems to target and edit viral genomes directly within plant cells. This would allow precise antiviral gene editing without transgenically modifying the plant's genome, making it a non-GMO solution to combating plant viruses.

Figure 1. Action of nanoparticles on plants

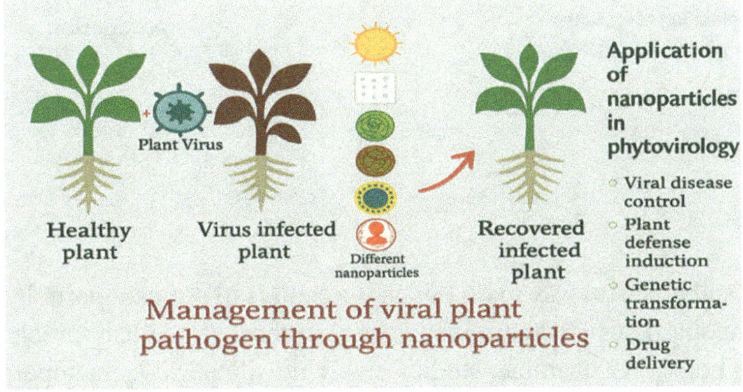

5. DELIVERY METHODS AND UPTAKE PATHWAYS

An important obstacle when employing nanotechnology to control plant viruses is getting nanoparticles (NPs) and their antiviral payloads into the right plant tissues quickly and effectively. Plants have obstacles like the waxy cuticle and cell wall that make delivery significantly different from how animals do it. Researchers have been looking into many ways to deliver NPs to plants, each with its advantages and limitations:

Figure 2. Mechanisms by which nanoformulations combat plant viruses and their delivery routes. Foliar sprays, soil/root applications, and seed or trunk treatments introduce nanoparticles into plant tissues, where they (i) directly inactivate virions (especially metallic NPs), (ii) deliver antiviral dsRNA to trigger RNA interference, (iii) prime innate immunity (e.g., SiO_2, CQDs, chitosan-based NPs), and (iv) act as carriers for conventional antiviral compounds, collectively reducing virus load and enhancing systemic protection

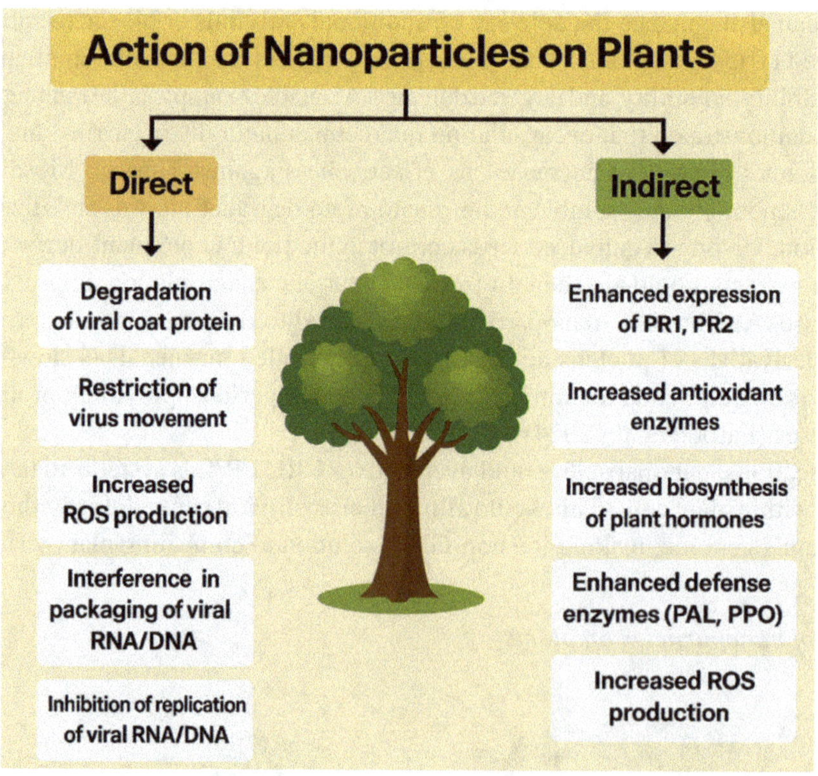

5.1 Foliar Sprays

The most popular and practical way to do this in the field is to put nanoparticle compositions on the surfaces of leaves. Nanoparticles <100 nm in size can get through the leaf cuticle and cell wall, especially through natural holes. For example, studies involving silica (SiO_2) nanoparticles in *Arabidopsis thaliana* demonstrated that the nanoparticles penetrate leaves solely through open stomatal pores, thereafter residing in the intercellular spaces of the spongy mesophyll (El-Shetehy et al., 2021). Stomata (the microscopic openings on leaves for gas exchange) are key entry points. They are important entry points because NPs can disperse or be carried in when they are open. Researchers discovered that in high humidity conditions (when stomata are more open), foliar absorption of nanoparticles (NPs) is increased. Furthermore, incorporating surfactants or spreader-stickers into the spray can enhance NP leaf coverage and penetration. Once within the leaf, nanoparticles usually stay in the apoplast, which is the region outside of cells. Small nanoparticles (tens of nm) have been seen sticking to cell walls in mesophyll tissue (El-Shetehy et al., 2021), and some can be taken up by cells by endocytosis. Foliar spraying is appealing due to its ease and scalability, as it can be performed with standard spray equipment. In fact, foliar NP treatments have been shown to inhibit viruses in both greenhouse and field experiments. One important study showed that spraying leaves with dsRNA-loaded clay nanosheets (named "BioClay") protected them from viruses for more than 20 days in the environment (Mitter et al., 2017). The clay nanoparticles slowly produced virus dsRNA, which the plant took up and started antiviral RNA interference. This protected even new growth that came up after the spray. In trials, spraying carbon quantum dots with viral siRNAs has also protected tobacco and cowpea from viruses. One study found that carbon dot-dsRNA complexes applied to leaves protected unsprayed upper leaves for 21 days after application (Xu et al., 2023). Generally, however, several inorganic NPs (such metal particles) don't migrate very far from where they were applied and mostly work in that area. So, foliar sprays might work best for localised antiviral action on the leaves that were treated, or if the NPs are made to provide a mobile signal (like a silencing signal) that the plant can move. Nonetheless, foliar distribution has proven effective against various viruses (CMV, TMV, PVY, among others) in experimental contexts (Mitter et al., 2017; Xu et al., 2023), indicating its potential for field application.

5.2 Soil/Root Application

Delivering nanoparticles to plants via the root system is another strategy, often used for systemic effects. Roots of plants can take up nanoparticles from soil or hydroponic solution, especially if the nanoparticles are very small or contain coatings that target roots. NPs can go up through the xylem to sections of the plant that are above ground once they are inside the roots (Wang et al., 2023). For example, studies have shown that watering or root-dipping plants in a suspension of NPs can result in uptake through root hairs or at sites of lateral root emergence. In one notable study, researchers utilized viral dsRNA conjugated to different nanoparticles (chitosan, silica, and carbon quantum dots) on *Nicotiana* plants through root soaking, resulting in pronounced antiviral efficacy against Potato (Xu et al., 2023). They noticed that root delivery of a chitosan–dsRNA complex yielded superior therapeutic outcomes compared to foliar spray of the identical formulation (Xu et al., 2023). The nanoparticles that were put to the roots spread throughout the plant and silenced virus genes. Root uptake can thereby facilitate a more homogeneous distribution of the therapeutic agent, extending to newly formed shoots and leaves that develop post-treatment. But how well nanoparticles go to the roots depends on their qualities and the soil conditions. For instance, positively charged nanoparticles may stick to the negatively charged parts of soil and root surfaces (Wang et al., 2023), whereas neutrally charged or well-coated nanoparticles can get in more easily. It's also important to consider that using nanomaterials in a field could mean using a lot of them and cause problems with leaching or persistence. In practice, root treatment may be more effective in controlled circumstances or for high-value crops (e.g., immersing seedling roots or utilizing drip irrigation in greenhouses) rather than broadcasting in an open field. The successful application of root-delivered NP–dsRNA in the PVY study indicates that this method can efficiently transport antivirals into phloem-connected tissues (since PVY is phloem-limited) and eradicate systemic infections (Xu et al., 2023).

5.3 Trunk or Vascular Injection

For large plants like fruit trees or vines, putting nanoparticle solutions directly into the stem or trunk is similar to how arborists give trees vaccines or antibiotics. The goal is to force the NPs into the plant's vascular system (xylem/phloem) so that they quickly spread across the plant's circulation. This method has generally been used to give trees nutrients or pesticides, but it works for viruses too. A recent study on citrus trees demonstrated that trunk injection of a polymeric nanoparticle formulation could disseminate silver nanoparticles throughout the tree and into the phloem tissue (Su et al., 2020). In practice, trunk injection means making a tiny hole and using a pressurized injector or drip bag to put the solution in. There hasn't been a large-scale field experiment of antiviral NP injection in trees yet, but it might be used in high-value circumstances. For example, injecting an antiviral NP cocktail into grapevines or orchard trees that are afflicted with viruses could inhibit systemic infections like grapevine leafroll or plum pox. This could be very helpful in quarantines or as a treatment when other options aren't available. The benefits are that the material is supplied within (no spray spread, and environmental exposure is kept to a minimum) and directly to the conductive tissues. However, it requires labor and wounding the plant, and nanoparticles must be formulated to avoid clogging the vasculature. If these challenges are met, vascular injections could become a viable delivery route for perennial crops.

5.4 Seed or Propagule Treatment

Another way to protect plants before they are planted in the field is to treat them first. Nanoparticles can be used to coat or soak seeds, and vegetative propagules like cuttings, tubers, or budwood can be dipped in NP solutions. The goal is to either kill any surface or transmissible viruses on the planting material or to provide the young plant an antiviral drug that will protect it during its early growth. One way to do this is to coat seeds with dsRNA-loaded NPs. This way, when the seed germinates, the seedling picks up the dsRNA and is already immune to a virus. Chitosan nanoparticles, which can stick to seed coverings and break down naturally, could work well in this function by delivering dsRNA or even CRISPR components to seedlings that are just starting to grow. In the same way, adding antiviral NPs to the media for tissue-culture propagation of virus-free plantlets should help guard against reinfection after transplantation. Although research in this field is nascent, the notion of "vaccinating" seeds is attractive for viruses that are challenging to manage. Whether nanoparticles are used to treat seeds, they will need to be carefully tested to see whether they have any influence on the seeds' viability or vigour, and they will also need to be approved by the government if they are used in food.

5.5 Laboratory Methods

Plant scientists also use experimental delivery methods like vacuum infiltration, syringe infiltration or biolistic particle bombardment. These approaches are very useful for proof-of-concept investigations. For example, vacuum infiltration was used to prove that carbon nanotubes could transfer genetic material into leaf cells, and syringe agroinfiltration is often used to introduce viruses or VLPs for studies of temporary expression. These methods have shown how nanoparticles function in plants, but they are not useful for large-scale farming. For instance, injecting quantum dots into cells with a syringe helped scientists figure out how they migrate. It also showed that most quantum dot NPs stay in the place where they were injected unless they are smaller than about 5 nm, in which case some can transfer from cell to cell or across the vascular system. Such findings inform the development of enhanced field delivery techniques, such as creating nanoparticles capable of naturally penetrating stomata or root tips, given the impracticality of manually infiltrating every plant in a field.

In general, the way nanoparticles are taken up by plants varies a lot depending on their size, surface chemistry, and charge. Smaller NPs (<50 nm) are more likely to get through cell barriers. They may go via pores in cell walls or be actively taken up by endocytosis if they reach the plasma membrane (Wang et al., 2023). On the other hand, neutral or anionic NPs may be able to move more easily through the cell wall matrix. A study utilizing synthetic polymer nanoparticles with varying charges revealed that neutral and negatively charged nanoparticles were absorbed by *Arabidopsis* root cells more efficiently than positively charged ones, which were retained at the root surface. (Wang et al., 2023). This indicates that too adhesive nanoparticles will not be effectively internalized. Very small NPs (<10 nm) or those with biological coatings can sometimes get through plasmodesmata (intercellular channels), or at least their movement is less restricted. On the contrary, larger NPs are stuck in extracellular spaces and may build up at vein termination or other "dead ends." Numerous foliar spray studies have observed that nanoparticles frequently clump at the leaf surface or within the initial cell layers; hence, attaining deep tissue penetration may necessitate the optimization of nanoparticle formation, such as employing dispersants, size regulation, or biodegradable carriers that release smaller entities.

Figure 3. Schematic representation of delivery methods and uptake pathways

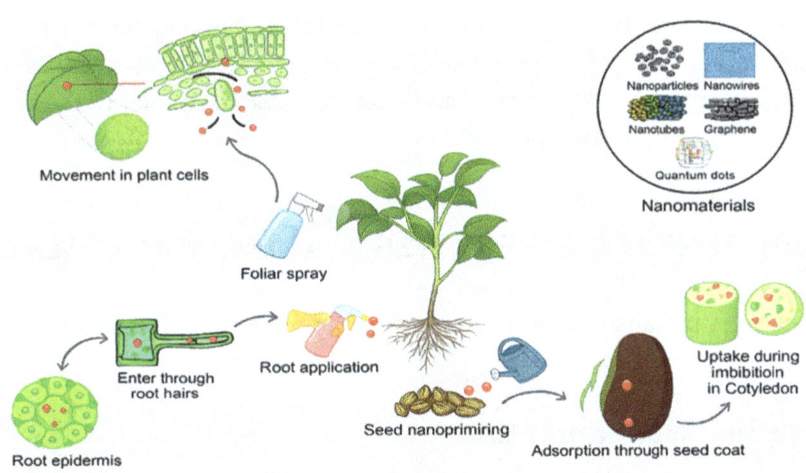

6. LABORATORY AND FIELD STUDIES OF NANODRIVEN ANTIVIRAL CONTROL

A significant portion of nanophytovirology research thus far has been performed in regulated laboratories or greenhouse environments. Laboratory investigations have shown that this is possible in many systems. For example, *Hilaire et al., 2022* demonstrated in greenhouse-cultivated tobacco that various nanoparticles containing PVY-dsRNA provided prolonged viral resistance beyond two weeks. In vitro cell culture or detached-leaf experiments have demonstrated that silver and copper nanoparticles inhibit many viruses (TMV, CMV, *Cucumber mosaic virus*, etc.) through direct viral inactivation. Numerous research utilise model plants (*Nicotiana benthamiana, Arabidopsis*) and model viruses (TMV, GFP-tagged TMV, etc.) to evaluate the efficiency of nanomaterials. These investigations frequently employ leaf infiltration or hoover application, attaining nearly total viral suppression with optimized NP formulations (El Gamal et al., 2022).

Field trials and greenhouse trials are fewer but emerging. A notable example is the study by Wang *et al.*, 2022 investigated quercetin-loaded nanoliposomes against TMV in *N. benthamiana*. In a field trial, nano-quercetin demonstrated approximately 38% more control efficacy compared to free quercetin or traditional pesticide formulations (Wang et al., 2022). This is one of the first times that nanoparticle formulations have been shown to work better than regular therapies in the field. This success indicates nanoformulations can translate from controlled environments to real-world farming, though such examples are still rare. Notably, Mitter et al. (2017) demonstrated that a single application of BioClay (LDH–dsRNA) conferred over 20 days of protection against viral infection in greenhouse plants subjected to simulated outdoor conditions, providing semi-field evidence that nano-enabled RNA interference can maintain efficacy in a more realistic setting.

To date, most published studies on antiviral nanotechnology only get as far as the greenhouse level because of problems with regulations and scaling up, highlighting the need for more translational research. The trend is speeding up, though. We anticipate to see more field trials in the coming few years as interest and investment grow. It's important to note that the global context is active: groups in Asia, Europe, North America, Africa, and Latin America are all looking into nano-antivirals for their main crops. The slow progress toward field deployment is not only due to limited trials, but also due to regulatory and scalability hurdles, as discussed next.

7. CHALLENGES: TOXICITY, SAFETY, REGULATION, AND SCALABILITY

Despite the potential, the transition of nano-antiviral strategies into practice is impeded by several challenges:

7.1 Phytotoxicity and Non-target Effects

Many nanoparticles can be harmful to plants at large quantities. If not given the right amount, metal NPs (Ag, Cu, Zn) might stop plants from growing or mess up photosynthesis. For instance, SiO_2 NPs were safe at effective antiviral dosages, whereas soluble silicon at high concentrations produced stress (El-Shetehy et al., 2021). Carbon and polymeric nanoparticles are usually less harmful, but the way they are made (surfactants, solvents) can make them more dangerous. Sublethal toxicity can cause oxidative stress or change the genes in plants. Nanoparticles that are released into the environment may also have an effect on soil bacteria, earthworms, and aquatic life. We need more thorough toxicological investigations; for now, there isn't much data on agro-NPs. Regulatory agencies will need environmental safety studies, which are similar to pesticide registration. This might delay the process of getting people to use the product.

7.2 Human and Food Safety

Any agrochemical that uses nanoparticles must be tested to make sure it is safe for people to eat or feed. It is necessary to test the toxicity of residues of NPs on food parts. In certain instances, the nanoparticles themselves (e.g., silver) have antibacterial properties and may influence human microbiota upon ingestion; conversely, in other circumstances, transporters (such as lipids or chitosan) are innocuous. Because plant viruses can't infect people, people's worries about nanomaterials have mostly been about not knowing enough. Nanoformulated antivirals require evaluation not only for their effectiveness in plants but also for the concentration and behavior of nanoparticle residues on consumable tissues and any possible long-term impacts on consumers. For regulatory approval and public acceptability, clear disclosure of risks and benefits and thorough safety testing will be necessary.

7.3 Regulatory and Policy Hurdles

At this time, there aren't many rules for nanoparticles in farming. In many places, a nanoformulated pesticide or biocontrol would have to be registered in the same way as regular pesticides. However, there are some questions about what they are (for example, is a chitosan NP a pesticide or a fertiliser?). Nanomaterials may have different laws since they have new characteristics. There are some rules from the EU and the US EPA for nanoparticles in pesticides, but criteria that can be enforced (such labelling and testing for nano-specific toxicity) are still being worked out. This lack of clear rules can make people less likely to invest and start businesses. A nano-RNA biopesticide, for instance, could be seen as a GMO or not a GMO depending on where you live. Overcoming these hurdles requires dialogue with regulators and clear demonstration of safety.

7.4 Production and Scalability

Manufacturing nanoparticles at agricultural scale (tons per year) at low cost is nontrivial. Chemical synthesis usually requires costly reagents and methods that use a lot of energy. Green synthesis is easier to scale up (for example, using plant waste extracts), but it still needs processes to clean it up. Polymeric nanoparticles and liposomes can be made in lab-scale batches but scaling to tons (for large fields) requires adaptation of pharmaceutical processes. The shelf life and stability of NP formulations are also problems. Some NPs clump together or settle down over weeks, which means further formulation work is needed. Also, the cost-benefit must make sense for farmers: new antivirals must not cost more than the value of the crop. Because of this, existing nano-antiviral treatments will probably show up first in crops that are worth a lot of money or where there aren't many other options and then move to markets for staple crops.

Nanotechnology has a lot of useful tools, but safety and sustainability must always come first. Some nanomaterials, such SiO_2, lipids, and chitosan, are naturally low in toxicity and can break down in the environment. For instance, El-Shetehy *et al.2021* concluded that SiO_2 NPs represent a "inexpensive, highly efficient, safe and sustainable alternative" for disease control.

8. FUTURE PROSPECTS

Looking ahead, several trends are expected to shape the future of nanotechnology in agriculture. Nanoparticles, such as carbon nanotubes and lipid nanoparticles, are being explored to deliver CRISPR/Cas components for viral resistance in plants. These "non-GMO" gene-editing methods, including Cas9 or Cas13 for RNA viruses, could create virus-resistant plants or target viral genomes in infected cells, without altering the plant's DNA. Researchers are already investigating nanoparticles to deliver base editors or ribonucleoproteins directly to plants. Advances in nanofabrication are leading to the development of smart or targeted carriers. Stimuli-responsive nanoparticles, which release their cargo upon encountering virus-induced pH changes or enzymes, are emerging as powerful tools. Nano-biosensors integrated with antiviral release mechanisms could detect viral RNA and trigger the release of antiviral agents precisely at the infection site.

The focus is shifting toward green and circular approaches in nanotechnology, using natural or waste-derived materials for nanoparticle synthesis. The emphasis will be on ensuring biodegradability, minimizing environmental footprints, and developing biodegradable nanocarriers. This aligns with the broader goal of integrating nanotechnology into agroecological practices, combining it with biopesticides and breeding strategies for sustainable pest management. In conclusion, nanotechnology, especially in the form of nanoformulations, presents new possibilities for combating plant viral diseases, enhancing immunity, and enabling gene-editing approaches. However, achieving widespread adoption will require continued research, interdisciplinary collaboration, and careful management to ensure safety and effectiveness.

COMPETING INTERESTS

The authors declare that they have no competing interests.

AUTHORS' CONTRIBUTIONS

All authors made contributions to the development of this book chapter. FA, RA, and TN jointly carried out the literature review, data collection, and analytical work. SA and SI worked in reviewing and formatting the chapter. Each author participated in drafting and critically revising the text and approved the final version for publication.

Table 2. Abbreviations

A. thaliana	*Arabidopsis thaliana*
ACMV	African cassava mosaic virus
AgNPs	Silver nanoparticles
AMV	Alfalfa mosaic virus
AuNPs	Gold nanoparticles
BYDV	Barley yellow dwarf virus

continued on following page

Table 2. Continued

A. thaliana	*Arabidopsis thaliana*
BYMV	Bean yellow mosaic virus
Cas9	CRISPR-associated protein 9 endonuclease
CMV	*Cucumber mosaic virus*
CNTs	Carbon nanotubes
CQDs	Carbon quantum dots
CRISPR	Clustered regularly interspaced short palindromic repeats
CTV	*Citrus tristeza virus*
CuO	Copper(II) oxide (often as CuO NPs)
dsDNA	Double-stranded DNA
dsRNA	Double-stranded RNA
GMO	Genetically modified organism
LDH	Layered double hydroxide
MOFs	Metal–organic frameworks
NPs	Nanoparticles
PEG	Poly(ethylene glycol)
PLGA	Poly(lactic-co-glycolic acid)
RNAi	RNA interference
ssDNA	Single-stranded DNA
TMV	*Tobacco mosaic virus*
TSWV	Tomato spotted wilt virus
TYLCV	*Tomato yellow leaf curl virus*
VLP	Virus-like particle
ZnO	Zinc oxide

REFERENCES

Abdelkader, H. S., Kheder, A. A., Amin, H. A., & Shafie, R. M. (2025). A comparative study of the antiviral effects of biogenic silver nanoparticles and nanosilica ($nSiO_2$) against leek yellow stripe virus on *Allium sativum* L. *European Journal of Plant Pathology*, *171*(3), 509–530.

Alam, F., Saha, N. R., Islam, M. S., Ahmed, M. S., & Haque, M. S. (2020). Perception on environmental concern of pesticide use in relation to farmers' knowledge. *Journal of Environmental Science & Natural Resources*, *13*(1–2), 94–99.

Alkubaisi, N. A., & Aref, N. M. (2017). Dispersed gold nanoparticles potentially ruin gold barley yellow dwarf virus and eliminate virus infectivity hazards. *Applied Nanoscience*, *7*(1), 31–40.

Aseel, D. G., Rabie, M., El-Far, A., & Abdelkhalek, A. (2024). Antiviral properties and molecular docking studies of eco-friendly biosynthesized copper oxide nanoparticles against alfalfa mosaic virus. *BMC Plant Biology*, *24*(1), 1089. PMID: 39551727

Cai, L., Liu, C., Fan, G., Liu, C., & Sun, X. (2019). Preventing viral disease by ZnO NPs through directly deactivating TMV and activating plant immunity in *Nicotiana benthamiana. Environmental Science. Nano*, *6*(12), 3653–3669.

Dupuis, B., Nkuriyingoma, P., & Ballmer, T. (2024). Economic impact of *potato virus Y* (PVY) in Europe. *Potato Research*, *67*(1), 55–72.

El Gamal, A. Y., Tohamy, M. R., Abou-Zaid, M. I., Atia, M. M., El Sayed, T., & Farroh, K. Y. (2022). Silver nanoparticles as a viricidal agent to inhibit plant-infecting viruses and disrupt their acquisition and transmission by their aphid vector. *Archives of Virology*, *167*(1), 85–97. PMID: 34738153

El-Ganainy, S. M., Shafie, R. M., Soliman, A. M., Mazyad, A. A., Sattar, M. N., Kesba, H. H., & Farroh, K. Y. (2025). Differential effects of chitosan–salicylic acid nanocomposite and salicylic acid on *cucumber mosaic virus* in cucumber. *Polymers*, *17*(16), 2195. PMID: 40871142

El-Shetehy, M., Moradi, A., Maceroni, M., Reinhardt, D., Petri-Fink, A., Rothen-Rutishauser, B., Mauch, F., & Schwab, F. (2021). Silica nanoparticles enhance disease resistance in Arabidopsis plants. *Nature Nanotechnology*, *16*(3), 344–353. https://www.nature.com/articles/s41565-020-00812-0.pdf PMID: 33318639

Elbeshehy, E. K., Elazzazy, A. M., & Aggelis, G. (2015). Silver nanoparticles synthesis mediated by new isolates of *Bacillus* spp., nanoparticle characterization and their activity against bean yellow mosaic virus and human pathogens. *Frontiers in Microbiology*, *6*, 453. PMID: 26029190

Farooq, T., Adeel, M., He, Z., Umar, M., Shakoor, N., da Silva, W., Elmer, W., White, J. C., & Rui, Y. (2021). Nanotechnology and plant viruses: An emerging disease management approach for resistant pathogens. *ACS Nano*, *15*(4), 6030–6037. PMID: 33761237

Farooq, T., Hussain, M. D., Wang, Y., Kamran, A., Umar, M., Tang, Y., & She, X. (2024). Enhanced antiviral defense against begomoviral infection in *Nicotiana benthamiana* through strategic utilization of fluorescent carbon quantum dots to activate plant immunity. *Journal of Nanobiotechnology*, *22*(1), 707. PMID: 39543670

Hilaire, J., Tindale, S., Jones, G., Pingarron-Cardenas, G., Bačnik, K., Ojo, M., & Frewer, L. J. (2022). Risk perception associated with an emerging agri-food risk in Europe: Plant viruses in agriculture. *Agriculture & Food Security*, *11*(1), 21. PMID: 35310134

Jones, R. A. (2021). Global plant virus disease pandemics and epidemics. *Plants*, *10*(2), 233. https://pmc.ncbi.nlm.nih.gov/articles/PMC7911862/ PMID: 33504044

Kheyrodin, H. (2017). Importance of Plant Viruses and Tobacco Mosaic Virus. *International Journal of Research in Engineering, IT and Social Sciences (IJREISS)*, *7*(11), 80-88. https://indusedu.org/pdfs/IJREISS/IJREISS_1433_57083.pdf

Koundal, V., Mitter, N., Williams, S., & Pappu, H. (2011). Characterization of small RNAs derived from Tomato spotted wilt virus infection by deep sequencing. *Phytopathology*, *102*(7), 65.

Kwak, S. Y., Lew, T. T. S., Sweeney, C. J., Koman, V. B., Wong, M. H., Bohmert-Tatarev, K., & Strano, M. S. (2019). Chloroplast-selective gene delivery and expression *in planta* using chitosan-complexed single-walled carbon nanotube carriers. *Nature Nanotechnology*, *14*(5), 447–455. PMID: 30804482

LeClair, R. J., & Boundless. (2024). 15.10.1: Plant DNA Viruses. *Medicine LibreTexts*. https://med.libretexts.org/Courses/Virginia_Tech_Carilion_School_of_Medicine/Block_V_Preparation_Materials/15%3A_Acellular_Pathogens/15.10%3A_DNA_Viruses_in_Eukaryotes/15.10.01%3A_Plant_DNA_Viruses

Liu, S., Tian, W., Liu, Z., Wei, X., Yuan, K., Du, W., & Cai, L. (2024). Biosynthesis of cupric oxide nanoparticles: Its antiviral activities against TMV by directly destroying virion and inducing plant resistance. *Phytopathology Research*, *6*(1), 30.

Maina, S., & Jones, R. A. (2023). Enhancing biosecurity against virus disease threats to Australian grain crops: Current situation and future prospects. *Frontiers in Horticulture*, *2*, 1263604.

Mishra, N. O., Quon, A. S., Nguyen, A., Papazyan, E. K., Hao, Y., & Liu, Y. (2023). Constructing physiological defense systems against infectious disease with metal–organic frameworks: A review. *ACS Applied Bio Materials*, *6*(8), 3052–3065. PMID: 37560923

Mitter, N., Worrall, E. A., Robinson, K. E., Li, P., Jain, R. G., Taochy, C., Fletcher, S. J., Carroll, B. J., Lu, G., & Xu, Z. P. (2017). Clay nanosheets for topical delivery of RNAi for sustained protection against plant viruses. *Nature Plants*, *3*, 16207. PMID: 28067898

Nations, S., Long, M., Wages, M., Maul, J. D., Theodorakis, C. W., & Cobb, G. P. (2015). Subchronic and chronic developmental effects of copper oxide (CuO) nanoparticles on *Xenopus laevis*. *Chemosphere*, *135*, 166–174. PMID: 25950410

Niño-Sánchez, J., Sambasivam, P. T., Sawyer, A., Hamby, R., Chen, A., Czislowski, E., & Jin, H. (2022). BioClay™ prolongs RNA interference-mediated crop protection against *Botrytis cinerea*. *Journal of Integrative Plant Biology*, *64*(11), 2187–2198. PMID: 36040241

Nishina, M., Nishijima, W., Zee, F., Chia, C., Mau, R., & Evans, D. (1989). *Papaya ringspot virus* (PRV): *A serious disease of papaya* (Commodity Fact Sheet PA-4(A)). University of Hawaii, College of Tropical Agriculture and Human Resources.

Sánchez-Tovar, M. R., Rivera-Bustamante, R. F., Saavedra-Trejo, D. L., Guevara-González, R. G., & Torres-Pacheco, I. (2025). Mixed plant viral infections: Complementation, interference and their effects, a review. *Agronomy (Basel)*, *15*(3), 620.

Su, Y., Ashworth, V. E., Geitner, N. K., Wiesner, M. R., Ginnan, N., Rolshausen, P., Roper, C., & Jassby, D. (2020). Delivery, fate, and mobility of silver nanoparticles in citrus trees. *ACS Nano*, *14*(3), 2966–2981. PMID: 32141736

Vargas-Hernández, M., Macias-Bobadilla, I., Guevara-González, R. G., Rico-García, E., Ocampo-Velázquez, R. V., Ávila-Juárez, L., & Torres-Pacheco, I. (2020). Nanoparticles as potential antivirals in agriculture. *Agriculture*, *10*(10), 444.

Varma, A., & Singh, M. K. (2024). The role of satellites in the evolution of begomoviruses. *Viruses*, *16*(6), 970. https://pmc.ncbi.nlm.nih.gov/articles/PMC11209209/ PMID: 38932261

Wang, J., Hao, K., Yu, F., Shen, L., Wang, F., Yang, J., & Su, C. (2022). Field application of nanoliposomes delivered quercetin by inhibiting specific hsp70 gene expression against plant virus disease. *Journal of Nanobiotechnology*, *20*(1), 16. https://jnanobiotechnology.biomedcentral.com/counter/pdf/10.1186/s12951-021-01223-6.pdf PMID: 34983536

Wang, X., Xie, H., Wang, P., & Yin, H. (2023). Nanoparticles in plants: Uptake, transport and physiological activity in leaf and root. *Materials (Basel)*, *16*(8), 3097. https://pmc.ncbi.nlm.nih.gov/articles/PMC10146108/ PMID: 37109933

Xu, X., Yu, T., Zhang, D., Song, H., Huang, K., Wang, Y., Shen, L., Li, Y., Wang, F., & Zhang, S. (2023). Evaluation of the anti-viral efficacy of three different dsRNA nanoparticles against potato virus Y using various delivery methods. *Ecotoxicology and Environmental Safety*, *255*, 114775. PMID: 36933482

Chapter 11
Nanocarriers for RNAi-Mediated Plant Virus Suppression

Sonia Navvuru
https://orcid.org/0009-0005-5762-6025
University of Alberta, Edmonton, Canada

ABSTRACT

Plant diseases caused by viruses pose a significant threat to food production, resulting in an annual loss of approximately $ 30 billion USD globally. RNA interference (RNAi) and nanocarrier-based delivery systems have emerged as transformative technologies for the sustainable suppression of plant viruses. This chapter centers on recent advances in nano-carrier-mediated RNAi for plant viral disease control, detailing how nanocarriers protect fragile RNA molecules and enable efficient delivery past plant cellular barriers. Chapter provide a comprehensive overview of the mechanisms, innovations, and practical applications of these delivery systems within the interdisciplinary fields of plant molecular biology, virology, nanotechnology, and crop protection. Emphasizing RNAi as a precision-based, environmentally friendly, and non-GMO strategy and how nanocarriers have addressed major RNAi delivery challenges in natural crop settings.

1. INTRODUCTION

Plant viruses have caused significant economic losses worldwide, declining both yield and quality of major field and horticultural crops such as tomato, papaya, cassava, rice, wheat, etc. With rapid climate changes and a burgeoning human population, virus diseases of crops are becoming harder to eradicate or control. Further, positive intentions to expand cultivation resulted in introducing crops to newer regions, which exposed the crops to damaging virus diseases that they had never faced previously. Viruses account for 47% of the reoccurring plant disease epidemics worldwide, causing $220 billion USD of economic losses (Anderson et al. 2004, Jones et al. 2019, Food and Agriculture Organization of the United Nations, 2025). Some of the major recurring virus disease epidemics include Maize lethal necrosis disease, rice tungro, wheat streak mosaic, Tobacco mosaic virus, Cucumber mosaic disease, Potato virus Y, Cassava mosaic disease, and Tomato yellow leaf curl disease.

The major strategies to control viral disease so far have been cultural practices, chemical control against the insect vectors, crop genetic resistance, and biotechnological approaches. Conventional breeding

DOI: 10.4018/979-8-3373-4913-8.ch011

approaches depend on the natural genetic resistance present in the crop, which is then used to develop cultivars carrying the original resistance genes (R genes). However, effective antiviral strategies remain limited, particularly for non-transgenic systems in crops where genome editing or breeding approaches are impractical and slow. Furthermore, Genetic engineering is relatively modern and includes tools such as RNAi, genome editing (CRISPR/Cas), and cross protection. In this chapter, we focus exclusively on the present and future potential of RNAi-mediated virus suppression.

RNAi is an emerging precision-based approach for viral disease management. RNAi is a gene silencing mechanism relies on a strongly conserved sequence-based double-stranded RNA (dsRNA). Exogenous application of virus-derived dsRNA or siRNA molecules is effective in silencing target viral genes, limiting virus replication and spread. RNAi has two main approaches: regulation of cellular gene expression carried by miRNA and inhibition of incoming viruses by siRNA (Vance et al. 2001, Akbar et al. 2022). RNAi-mediated virus suppression can silence viral genetic material in the plant or target essential genes of the vector. This affects the survivability of the virus carriers (Jones-Rhoades and Bartel, 2004). Despite the rising significance of RNAi technology, there are several hindrances to its implementation, such as instability of RNA in open environments, poor cellular uptake across plant cell walls, and inefficient systemic spread of applied RNA molecules.

Nano- carriers have emerged as a crucial innovation to overcome the hurdles in using RNAi for virus suppression (Mujtaba et al., 2021). Nano- carriers encapsulate RNA, protecting it from degradation, enhance its penetration and uptake into plant cells, and control its release in a time-dependent manner (Ray et al. 2022, Wang et al. 2023). We will be discussing the RNAi carrying efficiency, structure, and action mechanism of Nano- carriers, such as clay nano sheets, liposomes, carbon nanotubes, and silica. The synergistic approach of nano- carrier - RNAi mediated virus silencing is a sustainable solution for plant virus management.

However, successful utilization of these integrated innovations need to pass the Environmental and Biosafety Considerations, including Nano toxicology (Degradation and persistence of nanomaterials in the environment) and regulatory considerations. This needs their advocacy at the policy, regulation, commercialization, and farmer adoption level. There is huge unexplored possibilities and research potential for scalable production and cost efficient integration of RNAi - Nano- carrier-based implementations into IPM for Crop Protection Programs. This chapter directly addresses this global concern by evaluating nano- carrier- RNAi technologies as a next-generation alternative to chemical-based control measures. By merging plant molecular virology with nanotechnology, nano- carrier-mediated RNAi offers a scalable, biodegradable, and policy-compliant approach to crop protection.

2. RNA INTERFERENCE IN PLANT ANTIVIRAL DEFENSE

"RNA interference" or "RNAi" is a eukaryotic RNA-mediated gene silencing mechanism to defend against a wide range of pests and pathogens. The RNAi mechanism is based on double-stranded RNA (dsRNA) or small interfering RNA (siRNA) to downregulate a specific gene expression. To carry out the silencing, plants synthesize dsRNA by copying the complementary genes of the pathogen. Further, the dsRNA associates with a group of proteins that target the target gene expression in the pathogen via complementary base pairing and degradation of the mRNA, ultimately leading to the suppression of the protein required for infection. Dicer protein associates with DRB (double-stranded binding) protein and interacts with dsRNA or hpRNA to break it into 20-24 nucleotide-sized fragments called siRNA. In addition to siRNAs, microRNAs (miRNAs) constitute another important class of small RNA (sRNAs). Unlike siRNAs, miRNAs are produced from single-stranded RNA molecules that fold into characteristic hairpin loop structures before being processed (Padilla-Roji et al. 2023). Dicer (Ribonuclease III) has a double- stranded RNA (dsRNA) binding domain on its N terminus and two RNase III motifs and helicase domain on its C terminus (Padilla-Roji et al. 2023). Hence, during the siRNA fragmentation, the dicer attaches it to the RNA- induced silencing complex (RISC) and following a endo-nucleolytic protein, argonaute (AGO) joins in, producing a multi-component RNA- protein complex that carries out the RNA cleavage and gene silencing (Puyam et al. 2017).

2.1 RNAi Approaches for Crop Protection

There are three major RNA silencing technologies in crop biotechnology, including Virus-Induced Gene Silencing (VIGS), Host-Induced Gene Silencing (HIGS), and Spray-Induced Gene Silencing (SIGS). VIGS is a RNA- mediated reverse genetics technology that uses recombinant viruses to initiate in planta RNAi machinery, which in turn suppresses specific target genes in the host via post- transcriptional gene silencing. HIGS is a plant genetic engineering approach that engineers plants to synthesize dsRNA with affinity for the target genes of pests or pathogens, triggering RNAi and providing plant protection. SIGS is a non- transgenic approach, where small RNA (sRNA) or dsRNA are applied as a spray on the plant surfaces.

VIGS is popularly used in plants to study rapid, transient gene knockdown. It also helps to characterize or study gene functions for defense and to induce high-throughput heritable epigenetic modifications (Zulfiqar et al. 2023). The major drawback of VIGS is its transient effect, making it a temporary solution for plant protection. Additionally, it applies to only a few plants and has biosafety issues for field application. HIGS technology is based on stable genetic modification and engineering of the plant itself. The plant produces dsRNA targeting the gene of interest in the invading pest or pathogen. This technology needs genetic modification and stable transformation, which can be time consuming and have regulatory concerns. In SIGS, dsRNA or siRNA is directly sprayed or applied onto the plant surface. Plants take up the RNA, resulting in targeted gene silencing for crop protection without genetically modifying the crop. However, SIGS is not cost-effective. It requires consistent reapplication due to environmental degradation and limited uptake efficiency of the dsRNA applied (Zhang et al. 2020, Ray et al. 2022). Therefore, an efficient method of dsRNA delivery into plants is essential for reducing losses caused by viral diseases. Table 1 displays the summarized comparison between VIGS, HIGS, and SIGS.

Table 1. Comparison between VIGS, HIGS, and SIGS RNA silencing technologies in crop biotechnology

Features	VIGS (Virus-Induced Gene Silencing)	HIGS (Host-Induced Gene Silencing)	SIGS (Spray-Induced Gene Silencing)
Mode of Action	Virus genome is engineered to carry recombinant sequence of target gene, which triggers RNAi defense in the plant.	Plant is genetically modified to express a dsRNA of a critical gene of a virus or vector. When the pest ingests the dsRNA during feeding, RNAi gets initiated to silence the target gene expression.	dsRNA are synthesized externally and then sprayed or applied directly to the plant. The dsRNA initiates the RNAi in plant or the pest vector.
Genetic Modification in the host plant	No, Virus is a transient vector and it does not permanently change the plant	Yes, stable genetic transformation of the plant to produce dsRNA	No, Topical application of the dsRNA. Not a transgenic approach
Target Organism	Plant, silences the gene in the host plant itself	Pest/ Vector/ pathogen that feeds on the transgenic plant	dsRNA is absorbed into the plant or vector to initiate the RNAi
Field Applicability	Generally used for functional genomics studies in Laboratory. However, typically not used for field applications due to biosafety concerns regarding recombinant viruses.	Suitable for long-term pest and pathogen resistance in commercial agriculture. Requires public and regulatory acceptance of Genetically modified organisms (GMOs).	It is a fast, targeted alternative to chemical pesticides. Highly applicable for field use due to its non-transgenic nature and environmental safety profile.
Efficiency/ speed	Fast. Powerful tool for functional studies	Slow. Development of transgenic plants takes time.	Fast. dsRNA can be directly sprayed on the plant for quick results
Duration of Effect	Transient, with the silencing effect lasting for a few weeks to months before the plant recovers	Stable heritable across generations. Long lasting	Transient, as the sprayed dsRNA is subject to environmental degradation. The duration depends on stability and delivery method.
Advantages	No stable genetic transformation required. Hence, useful for studying genes with lethal mutant phenotypes.	Permanent and heritable resistance	Non-transgenic, environmentally friendly and biodegradable. Has high specificity with minimal off-target effects
Disadvantages	Host-range limitations of the viral vector. Transient effect, not a permanent solution for protection and has biosafety concerns.	Takes significant time to develop transgenic lines. Needs genetic modification and stable transformation. Raises public and regulatory concerns about GMOs	Degradation of dsRNA is a common issue and requires consistent reapplication, which adds to the cost. Efficacy can be limited by dsRNA uptake efficiency. Has some environmental concerns.

3. CHALLENGES IN RNAI DELIVERY TO PLANTS

The plant virus genome is primarily single-stranded RNA, but some viruses also contain double-stranded DNA or RNA. RNAi is a promising tool for plant antiviral defense engineering. Topical application of artificial microRNA and trans-active siRNA molecules on plants is a potential strategy to develop GMO-free virus-resistant crops. Unfortunately, the dsRNA, when sprayed directly on crops, is prone to degradation and may cause environmental concerns. dsRNA is highly unstable and degrades rapidly in soil or water with a half-life of around 30-72 hours (Zhang et al. 2020, Ray et al. 2022). In addition to the field environments, enzymatic damage, and UV damage, the dsRNA also faces hindrance from the plant structural barriers, including the cuticle and cell wall barriers. However, even if the dsRNA passes the cell wall barrier, the plant biosystem produces systemic barriers and nucleases, which degrade the dsRNA/siRNA, reducing their efficiency and stability. In order to overcome these physical and systemic barriers, higher doses of exogenous RNA need to be applied for effective silencing, which further adds to the cost and environmental risks. Approaches like sprays, soaking, agroinfiltration, and viral vectors have been used; however, each has constraints in efficiency, biosafety, and applicability. Hence, the introduction of nano- carriers emerged as an innovative approach for a stable, non-toxic, and scalable approach for delivering RNAi into the plant (Mujtaba et al. 2021; Avila-Quezada et al. 2022; Kolge et al. 2023). The following Figure 1 depicts the challenges posed with using RNAi in virus suppression and how nano- carriers can overcome these constraints.

Figure 1. Overcoming challenges in RNAi delivery by nano-carriers

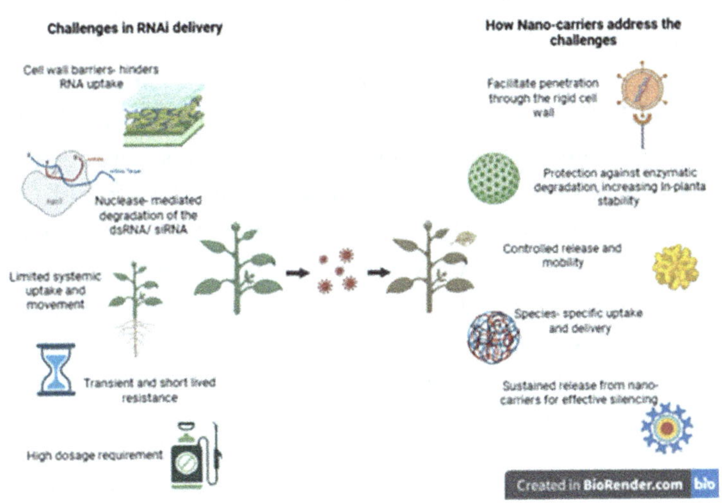

4. INTRODUCTION TO NANO- CARRIERS

Nano- carriers have recently emerged as vehicles of efficient RNAi delivery by overcoming conventional methods (e.g., sprays, injections). Each Nano- carrier, carbon-based nano- carrier, chitosan, cationic dendrimers, layered double hydroxides, mesoporous silica nanomaterials, has exclusive advantages enabling their application in improving ds RNA delivery interference in plant systems. Nano- carriers like carbon dots, clay nanosheets, and liposomes increase dsRNA permeability through the cell wall and membrane. Chitosan Nanoparticles (NPs) have high adhesive and biocompatibility properties that facilitate and promote dsRNA uptake by the plant surface and further into its system. Similarly, Cationic dendrimers carry a positive charge on their surface, which provides strong protection of dsRNA against enzymatic degradation and improves their uptake and stability in planta (Mat Jalaluddin et al. 2023).

Nano- carriers of layered double hydroxides and carbon-based NPs enable delivery of dsRNA with stability and control, enhancing their targeted delivery with high precision and low concentration (Hernández-Soto and Chacón-Cerdas 2021, Ray et al. 2022). Mesoporous silica nanomaterials have a large protected surface are to load and carry dsRNA for a long period with stability to regulate gene expression (Mat Jalaluddin et al. 2023). Further, engineered Nano- carriers can also be tailored species-specific to develop non-transgenic and field-friendly cultivars. The following sections discuss the different nano- carriers used for dsRNA delivery for potential Virus suppression in plants.

4.1 Clay Nano Sheets/ Layered Double Hydoxide (LDH)

Bioclay structure typically consists of positively charged brucite-like layers of trivalent and divalent metal cations (M^{2+} and M^{3+}) (i.e., Aluminum and Magnesium), which are combined with exchangeable anions (A^{m-}) and water molecules. These that are organized in an octahedral coordination pattern (general structural formula: $[(M^{2+})_{1-x} (M^{3+})_x (OH)_2]^{x+}(A^{m-})_{x/m} \cdot nH_2O]$ (Kuang et al. 2010; Bao et al., 2016). The presence of both positive surface and exchangeable negative anion interlayer of hydration allows easy loading and unloading of negatively charged dsRNA or dsDNA molecules. Bioclays carrying pepper mild mottle virus (PMMoV) and cucumber mosaic virus (CMV) have been effective in delivering intact dsRNA persistently in *N. tabacum* and cowpea (Mitter et al. 2017). Thus, bioclays naturally breakdown into its constituent ions and dsRNA stably and facilitate slow and sustained dsRNA delivery (Yan et al. 2020a). Worrall et al. (2019) found superior potential of Bioclay - potyvirus bean common mosaic virus (BCMV) over both aphid mediated and mechanical inoculation. The use of LDH for dsRNA delivery improved the percentage of titer reduction (protection duration extension compared to naked dsRNA), increasing the protection duration from 1 week to 3 weeks (Niño-Sánchez et al. 2022).

Distinguishing features of bioclays include anion exchange capacity, hydro-thermal stability, UV-protective, biodegradable, eco-friendliness, and prominent biocompatibility (Mitter et al. 2017). Bioclays carrying ds RNA when topically sprayed, successfully stick to the plant surfaces, including the leaves, petioles, pollen, and roots, and stably delivers dsRNA into the plant (Mitter et al. 2017, Yong et al. 2021, Mosa et al. 2021). In comparison to naked dsRNA, bioclays embedded with dsRNA delay degradation and improve effectiveness and persistent silencing effects after considerable spray time. However, there are constraints to their application, including the complex integration process of the dsRNA in bioclay. Further, the cost and complexity involved in its structure deter its practical application, adoption, and scalability. Moreover, lots of awareness, marketing, and regulatory considerations need to be undertaken to use it as a viable solution in virus suppression in crops.

4.2 Carbon Nanotubes (CNTs)

Carbon nanotubes (CNTs) are one-carbon allotropes having a distinguishable ultra-light weight, large surface and hollow tubular structure (Chakrabarti et al. 2015). CNTs have a unique multifunctional surface chemistry and large surface- to- volume ratio, enabling high cargo loading capacity (Mehra et al. 2014). CNTs based their number of lamellar graphite layers can be either be single-walled (SW-CNTs) or multi walled (MW- CNTs) (Marchesan et al. 2014, Demirer et al. 2020). SW-CNTs are the smallest carbon-based NPs with ~1 nm in size, 0.8-1.2 nm diameter, and 500- 1000 nm diameter (Demirer et al. 2020).

SW- CNTs with their small size and structural stiffness have proven to efficiently pass intact plant cell walls and into various cell organelle membranes (Giraldo et al. 2014, Kwak et al. 2019, Demirer et al. 2020). Edwards et al. (2020) proved easy and quick penetration of CNTs- dsRNA in combination with a dendrimer to deliver dsRNA. Demirer et al. (2020) developed a platform to separately load SW-CNTs with either sense or antisense siRNA, which then after entering the plant cell, thermodynamically hybridize to further initiate the gene-silencing mechanism. A concentration of 200ng dsRNA per plant application via CNT microinjection or foliar application has shown to be optimum (Edwards et al., 2020). RNA protected and delivered via SW-CNTs have a lifetime of 24 hrs without degradation in comparison to the six hrs of free siRNA. In comparison to cationic NPs, SW-NTs are biocompatible and have a lower cellular toxicity or stress induction, enabling their increased scalability (Demirer et al. 2020). Further, the same study found percentage titer reduction of 74- 81% against Tobacco Mosaic Virus (TMV).

4.3 Liposomes and Lipid NPs

Nanoliposomes are amphiphilic lipid-based vesicles with a bilayer membrane structure. Nanoliposomes encapsulate hydrophilic dsRNA molecules in their aqueous core. Due to their high similarity to cell membrane, Nanoliposomes are reliable nucleic acid transporters. Nanoliposomes' features, such as size, lipid composition, and surface charge, can be tailored according to our desired dsRNA delivery requirements (Sharma et al. 2023, Arjunan et al. 2024). Hao et al. (2024) employed Nanoliposomes to encapsulate (ZmFd3)-dsRNA into corn plants, which successfully suppressed damage caused by maize chlorotic mottle virus (MCMV) and/or sugarcane mosaic virus (SCMV). Similarly, Avital et al. (2021) modified and developed a foliar spray compatible PEI (lmPEI)-based long dsRNA delivery nanoplatform to silience grapevine leafroll-associated virus 3 (GLRaV-3) genes; *RdRp* and *CP* in grapevines. These studies prove the Nanoliposomes as promising dsRNA carriers with biocompatibility, prolonged stability, and robust modifiability.

4.4 Chitosan NPs (CNPs)

Chitosan NPs are biodegradable natural polysaccharides. The main structure of chitosan includes chitin, which is composed of random repeating units of copolymers (β1→4) 2-amino-2-deoxy-D-glucopyranose (GlcN) and (β1→4)-2- 2-acetamido-2-deoxy-D-glucopyranose (GlcNAc) (Petrônio et al., 2022). Their organic structure and biodegradable nature make them highly biocompatible for different functions (Negm et al. 2020). Chitason is amiable for easy to engineer and is very versatile. For instance, adding mild acid protonates the amino group $-NH_2$ to form NH_3^+, this positive charge facilitates the electrostatic interaction with the negative charge of the phosphate group of the nuclei acid backbone, enabling both efficient encapsulation in nanostructure and gene delivery (Kashyap et al. 2015).

The dsRNA-chitosan complex is 69 ± 12 nm in size and has a characteristic spherical or ellipsoidal shape. dsRNA-chitosan has low toxicity, stability up to 48h, and resistance against ph, temperature, and deterioration (gut hydrolysis and nuclease). dsRNA-chitosan complex has been used to suppress G-protein-coupled receptor kinase gene in piercing sucking insect, Aphid lucorum, which resulted in reduced weight, development time, and reduced mortality (Qiao et al. 2023). Further, ds-RNA -chitosan with a sequence homologous to a partial sequence in phytopathogen tomato mosaic virus (ToMV) has been effectively studied for Nano formulation and efficiency (Petrônio et al., 2022). Throughout dsRNA-chitosan application in diverse pests and plants, no phytotoxicity has been reported. However, research needs to be carried out for dsRNA-chitosan synthesis, formulations, and reducing the cost, which will further promote widespread adoption.

4.5 Dendrimers

Dendrimers are highly branched man- made polymers with 100 nm size inner cores and twisted side chains on the periphery. Dendrimers have a large positive charge on their surface and quick dissolvability in water, making them suitable for dsRNA delivery into biological systems, including plants. Bioactive components such as dsRNA can be carried on the surface or with pores of dendrimers via interactions between the surface amines and nucleic acid phosphates (Dufès et al., 2005, Voit & Lederer, 2009).

Properties of the Dentrimers' surface groups can be changed easily by adding hydrophilic co-polymers or targeted ligands. For instance, positive charge from the primary amines on Polyamidoamine (PAMAM) dendrimer surface can hold dsRNA electrostatically (Cheng et al. 2007, Lu et al. 2019). PAMAM dendrimer formulation with fluorescent perylenediimide (PDI) cores and four amino acid-functionalized arms on their structural periphery was sprayed to deliver dsRNA into soybean to control Soybean Mosaic Virus and Alfalfa Mosaic Virus vector, A. glycines (Zheng et al. 2019). Cationic dendrimer- dsRNA/ detergent formulation was effective in penetrating into the aphid system, resulting in 95.4% knockout effect and 80.5% population reduction (Zheng et al. 2019). Zhang et al. (2022) used a similar formulation for RNAi of key genes vestigial (vg) and Ultrabithorax (Ubx)) for wing development in the aphid vector, *Myzus persicae*.

4.6 Mesoporous Silica NPs (MSNs)

Mesoporous Silica NPs (MSNs) have a porous structure, which provides a huge surface area. Additionally, it has a distinct solid framework and excellent thermostability, enabling tunable pore size and versatile functional group binding. MSNs can be engineered for efficient dsRNA loading, site-specific targeting, and enhanced dispensability. Amino-mesoporous silica NPs with a positive surface brought about by 3-aminopropyltriethoxysilane ligands enables dsRNA adsorption through electrostatic interaction. Additionally, MSMs have silanol groups (-Si-OH) on their exposed surface, which facilitates biocompatibility and wide functionality for biological applications. Amine-functionalized mesoporous silica NPs was used as a vehicle to deliver dsRNA targeting the AC2 gene into plant to control Tomato leaf curl New Delhi virus infection (Sangwan et al. 2023). Furthermore, MSNs are highly biocompatible and have various phyto- biotechnology applications. Table 2 summarizes key features of Nano- carriers reported in RNAi virus suppression.

Table 2. Key features and structure of potential nano-carriers

Features	LDH	Chitosan	Gold nanocluster	SWNTs	Minicell	Nanoliposomes	Star polycation	DNA nanostructure	Carbon dots	CPPs
Sustained silencing period	20-60 days	5 days	-	7 days	12 days	5–10 days	-	7 days	20 days	36-72 hours
RNA: NC ratio/ Concentration	1:3	1:200	1:1	100nMsiRNA NC (2mg/L)	500ng/ ml	1:5–1:20	1:1	-	50:40	-
Cytotoxicity	Low-moderate	low	Low-moderate	Not detected	Not detected	Low-moderate	-	Not detected	-	-
Cost of synthesis	low	low	low	low	Very low	Moderate to high	low	low	low	low
Figure	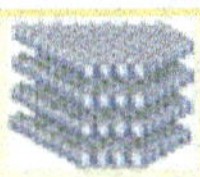		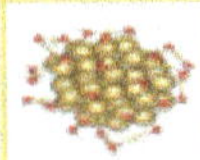			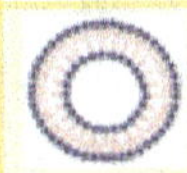	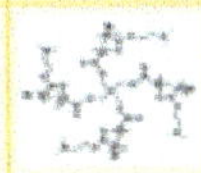		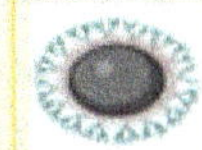	

4.7 Uptake and Systemic Translocation Mechanisms of dsRNA- Nano- Carrier Complexes in Plants

The effective use of dsRNA- nano- carriers for virus suppression fundamentally depends on the dsRNA in the plant. In order for the RNAi to work, dsRNA should cross physical barriers and immune biochemistry, achieving cellular uptake and systemic distribution before it gets degraded and ineffective. On entering the plant system, the complex has to translocate via the vascular system. The cellular uptake is mainly through passive penetration or diffusion, endocytosis, thiol-mediated uptake, and pi-pi stacking and non-covalent binding.

Primary hurdle in nano- carrier mediated dsRNA delivery is to pass the complex multi-layered plant cell wall. Plant cell walls have 5-20 nm pores, acting as main gateways through which the nano-carrier has to pass, underlining the importance of its optimum size, charge, and surface chemistry. Smaller nanoparticles or their dsRNA cargo can enter inside the cell through the cell wall pores through passive penetration (Liu et al. 2009, Chang et al. 2013). In the endocytosis mechanism, the dsRNA-nano-carrier complexes of less than 1 μm size are engulfed by the inward folding of the plasma membrane, forming vesicles (Ma et al. 2018). Endocytosis and its combined action with transport proteins on the outer epidermis have been reported to be effective for carbon nanotubes (Grillo et al. 2021). Nano-carriers containing guanidium use thiol-mediated pathways to enter cells (Mou et al. 2022). Further, non-covalent bonds, or Pi-Pi stacking, between dsRNA and nano-carriers like CNTs can mediate cell wall penetration and entry into the cytosol.

5. RESOURCES AND APPLICATION FRAMEWORK FOR NANO- CARRIER- MEDIATED RNAI CONSTRUCTS

5.1 Databases for Nano- Carriers

Nano- carrier databases are crucial resources for studying and researching the different Nanomaterials that can be used as vehicles for dsRNA delivery to induce virus suppression. They are curated platforms for data on nanomaterial structure, biological interaction, and potential applications. These resources act as a publically available nanoinformatics tool for safety assessment, data computation, and model designing for improved nano- carrier development. Some of the major knowledge resources for studying RNAi and its application include: GenomeRNAi, PubChem BioAssay, FlyRNAi.org, DRSC/ TRiP Functional Genomics Resources, WormBase, E-RNAi, RNAi Atlas, siRecord, VIRsiRNAdb, and RNAi Codex. The following Table 3 displays the different nano- carrier databases. These resources are essential for understanding and applying nano- carriers for RNAi delivery and virus suppression.

Table 3. Resources/ databases for nanomaterial research and use

Database	Focus	Information	References
Nanomaterial-Biological Interactions (NBI) Knowledgebase	Nanomaterial-biological interactions	Analysis tools for nanomaterial properties and their biological effects. Promotes design and risk assessment	Liu et al. 2013, Ji et al. 2021
caNanoLab	Biomedical and cancer nanotechnology	Composition, function, physicochemical properties, and experimental protocols for nanomaterials	Morris et al. 2015, Ke et al. 2025
eNanoMapper	Safety assessment and computational toxicology of engineered nanomaterials	Toxicological, physicochemical and structural data	Mat Jalaluddin et al. 2023
Nanomaterial Registry (NR)	General nanomaterial info	Curated, web-based data on properties and biological/environmental effects. Compliance ratings, MIAN standards.	Ostraat et al. 2013, Pavlicek et al. 2021
PubVINAS	Nanomaterial simulation and modeling	Annotated nanostructures, property, bioactivity data; PDB files.	Yan et al. 2020b, Ji et al. 2021
Nano-HUB	Computational Nano informatics and education	Simulation tools, educational resources	Haley et al., 2009, Kim et al. 2011

5.2 Framework and Workflow for Nano-Carrier-Mediated RNAi in Plant Viral Disease Management

To efficiently employ nano- carrier- RNAi-mediated Virus suppression, a structured workflow that combines bioinformatics, molecular design, and nanotechnology and delivery optimization is necessary. The initial execution discussion begins with target choice, a conserved indispensable viral gene that facilitates broad- spectrum and durable silencing effects. Ideal gene targets for RNAi could be involved in basic biological functions, including replication, movement, and coat protein synthesis. Further, to prevent any off- target silencing of the host plant genes or other beneficial organisms, a thorough in- silico off- target screening of the host genome is conducted using bioinformatics to ensure both specificity and biosafety. Following the assurance of specificity, dsRNA is designed with optimal dsRNA length and end structure (overhangs vs. blunt ends) to ensure uptake, stability, and silencing efficiency. The next step would be to effectively deliver these dsRNA into the host system or virus vector. Nano- carriers (such as carbon nanotubes, layered double hydroxide, or chitosan NPs) are versatile, and their cellular uptake and systemic movement depend on characteristics such as particle size, surface charge, biocompatibility, and release kinetics. The next phase involves the dosage optimization for balanced efficiency while maintaining minimal phytotoxicity through experimentation and modeling to effectively balance cost and robust RNAi mediated virus suppression. However, before their approval, the designed formulation needs to undergo biological assay and validation. Here, the anti- viral efficiency of the formulation is evaluated through molecular, biological and phenotypic analysis. This iterative strategy will ensure a rational, data driven understanding and assurance on the nano- carrier-enabled RNAi formulation for sustainable management of plant viral diseases. This comprehensive nano- carrier- RNAi mediated virus suppression framework is illustrated in Figure 2. as a workflow.

Figure 2. Workflow for nano-carrier -RNAi mediated virus suppression strategy

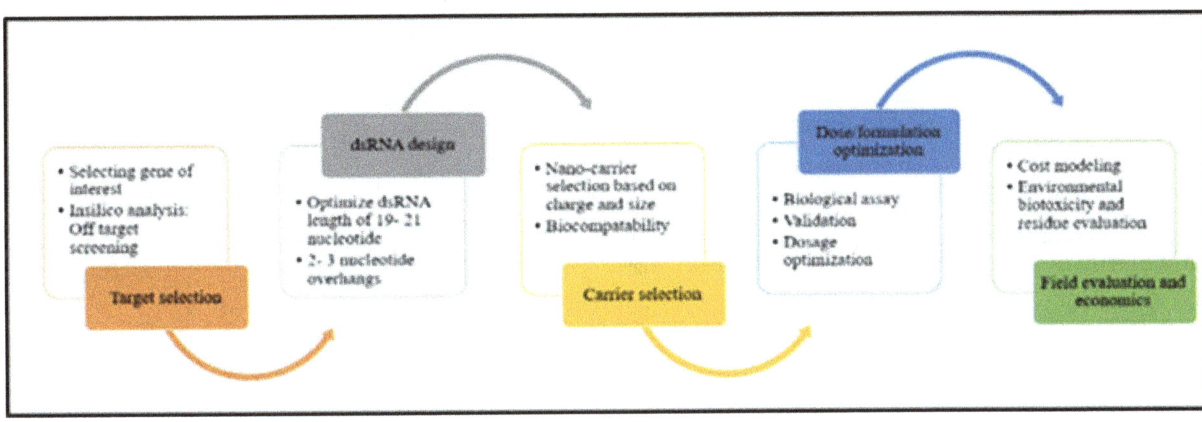

6. CASE STUDIES OF RNAI MEDIATED VIRAL DISEASE SUPPRESSION VIA NANO- CARRIERS.

RNAi-mediated virus suppression utilizing nanomaterials as vehicles have been successful in different crop pathosystems. Tabien et al. (2021) used Layered double hydroxide (LDH) nano- carriers to introduce dsRNA as a topical spray to target Tomato Spotted Wilt Virus (TSWV) genes (such as the nucleocapsid N gene). In a greenhouse experiment, LDH carrying dsRNA to suppress 2b silencing suppressor gene ensured robust suppression of local and systemic infection of Cucumber Mosaic Virus (CMV) in Cucurbits (Ntui et al. 2014). Table 4 summarizes other Nano- carriers- RNAi mediated Virus suppression successes.

Table 4. Nano- carriers- RNAi mediated virus suppression in model and crop plants

Nano- carrier	Mode of application	Result	Reference
MgAl Layered Double Hydroxide (LDH)	Topical application in Arabidopsis thaliana or spray atomizer on *Vigna unguiculata* and *Nicotiana tabacum*	Silences PMMoV (Pepper mild mottle virus) replicase gene and CMV (Cucumber Mosaic Virus) target gene	Mitter et al. 2017
	Pollen drenching of *S. lycopersicum*	CMV (Cucumber Mosaic Virus) target gene	Yong et al. 2021
	Topical spray on detached leaves and/ or seedlings	Bean common mosaic virus (Nib, CP) in *N. bentahmiana*	Worrall et al. 2019
Chitosan Nanoparticle		Silences coat protein gene of Tomato mosaic virus	Petrônio et al. 2022
Carbon Dots	Spray	464 bp coat protein (cp) gene of cucumber green mild mottle virus	Delgado-Martín et al. 2022
Mesoporous silica NPs	Rub inoculation and roots soaking	*AC2* (dsAC2) gene of Tomato leaf curl New Delhi virus (ToLCNDV) in *N. benthamiana*	Sangwan et al. 2023
Star Polycations (SPc)	Topical spray on seedlings, root soaking	cowpea aphid-borne mosaic virus (CABMV). *Aphis glycines* (*TREH, ATPD, ATPE* and *CHS1*)	Li et al. 2019; Yan et al. 2020a
	Topical leaf spray	*Myzus persicae* (green peach aphid) (ATPD, ATPG) in oilseed rape	Ma et al. 2022
CD-Branched Polyethylenimine (CD- bPEI)	Petiole immersion and topical leaf spray	Silences coat protein genes and RNA polymerase of Grapevine leaf roll associated virus-3 (GLRaV-3)	Avital et al. 2021

7. ENVIRONMENTAL AND BIOSAFETY ASPECTS OF USING RNAI - NANO- CARRIER SYSTEMS

There are several aspects of environment and biosafety considerations that need to be addressed before RNAi use. Prominent concern in agricultural applications is biotoxicity as it enters the food chain and can affect a wide range of organisms. Another major worry is the persistence and degradation of nano- carriers. Main concerns associated with RNAi silencing are the impact of off-target silencing on both non-target and target organisms (Romeis and Widmer, 2020; Mat Jalaluddin et al. 2023). In order to predict the off-targets beforehand, RNAi targets need to be evaluated using bioinformatics tools. These tools help in finding sequence similarity between the target and off- target transcripts. It is crucial for accurately designing the RNAi construct and experiment protocol for minimal off- targets. Further, dsRNA- nano- carriers when sprayed on the plants, humans, and non-target organism can also come in contact in either or both of the potential risks; dsRNA and Nano- carriers. The dsRNA- Nano-carrier can enter the non- target biological system via absorption on skin, inhalation, and ingestion (Baum et al. 2007). However, the innate immune system in humans can detect and degrade the foreign dsRNA

(Jensen et al. 2013, Chen et al. 2018, Fletcher et al. 2020). Similarly, Food Standards Australia New Zealand (FSANZ) also reported no scientifically proven incidences of genetically modified foods causing risks higher than the naturally occurring dsRNA in common foods in the supply chain (FSANZ, 2013).

With respect to the use of Nano- material use in biological systems, two major factors to be considered before confidently using nano- carriers. First is the size of the nano- carriers that enter the biosystem, and second, their post- consumption localization (Mu et al. 2010, Bensoussan et al. 2018, Agathokleous et al. 2020, Yan et al. 2023). NPs to deliver dsRNA into plants for targeted virus suppression can also lead to some unanticipated results. However, extensive studies need to be carried out for phytotoxicity and impact on beneficial microbes and insects. Moreover, Biosafety regulations and policy need to be organized for changing the public perception. Furthermore, efforts are needed to direct and improve the commercialization, regulatory, and economic outlook of RNAi and nanomaterial use.

8. REGULATORY AND STEWARDSHIP CONSIDERATIONS

Under the scope of regulation, dsRNA- nano- carrier-mediated RNAi formulations are usually categorized as bio-based plant protection products or bio pesticides and are increasingly interesting to major jurisdictions, such as the Pest Management Regulatory Agency (Canada), the Environmental Protection Agency (USA), and the European Food Safety Authority (EU). Regulatory evaluation is concerned with the environmental performance of dsRNA molecules, including features such as soil and water mobility and degradation rate, as well as the safety of the carrier nano-materials (e.g., LDH, chitosan, silica NPs). The residual expectations of dsRNA are low. dsRNA confer lower risk due to their rapid degradation and minimal persistence, and bioaccumulation. However, maximum residue levels can be fixed for food crops and consistent risk assessment to reiterate low environment mobility and rapid breakdown. Non-target testing, a central factor for regulatory reviews, can also be assessed through in-silico genome- wide screening for off-target effects. Additionally, laboratory assays, environmental fate assessment, and field studies tracking can be carried out for dsRNA and nano- carrier toxicity in key biological systems and broader ecosystems (e.g., primary stratum in the food chain, humans, beneficial insects, pollinators, soil biota). These assessments will not only initiate responsible development and deployments leading to improve regulatory bodies' approval but also will convince public acceptance. Furthermore, effective stewardship will create a transparent framework between the commercialization regulatory bodies and the public. Therefore, effective stewardship will ensure long-term support and ecosystem health through viral disease suppression in crops through RNAi and nano- carriers.

9. CONCLUSION

Viral disease mitigation in crops using Nano- carriers for RNAi-Mediated Plant Virus Suppression is a groundbreaking approach in plant virus management. Nano- carriers can overcome cell wall and tissue barriers in plants. They mediate transformation by transporting exogenous molecules across the plasma membrane. Recently, RNAi and nanomaterials have received more attention. Building on these advancements, this chapter explores the diverse applications of nano- carriers in greater depth. Throughout this chapter, the diverse spectrum of nano- carriers (layered double hydroxides, chitosan NPs, carbon nanomaterials, liposomes) and their potential as capable vehicles for RNAi to suppress virus load in crops has been detailed. The progress in RNAi technology, including detailed dissection of each method, nano- carrier, their stability, bioavailability, delivery, dosage and efficiency has been discussed. RNAi- nano- carrier mediated virus suppression Case studies and quantitative matrix have reinforced their consistency in effect, minimal dosage, lower cost and rapid environment degradation. Attention to regulatory and stewardship considerations emphasize the potential of RNAi - Nano- carrier in suppressing the viral diseases; through careful evaluation of the environment and non- target effects. Further, these regulatory cautionary will facilitate meeting national and international standards for bio-based plant protection products.

In future, research should focus on reducing dsRNA synthesis and dsRNA-nano-carrier formulation costs. Ready-to-use RNAi- nano sprays need development to improve marketability. The potential of combining nano- carrier-RNAi with biocontrol agents and pheromone traps should be explored. Additionally, targeting multiple viral genes could enhance resistance durability and acceptability. To overcome regulatory hurdles, detailed classification and field-testing are needed. This will help increase farmer acceptance and adoption. Integrating digital farming, sensor-based systems, and precision technologies is also required for climate-smart agriculture.

ABBREVIATIONS

AGO2 - Argonaute 2
CNPs - Chitosan Nanoparticles
CDs - Carbon Dots
CPPs - Cell-Penetrating Peptides
CRISPR - Clustered Regularly Interspaced Short Palindromic Repeats
CRISPR-Cas - CRISPR-associated Protein System
DNA - Deoxyribonucleic Acid
GMO - Genetically Modified Organism
HIGS - Host-Induced Gene Silencing
hrs - hours
miRNA – MicroRNA
μm – micro meter (10^{-6} m)
MSNs - Mesoporous Silica Nanoparticles
NP - Nanoparticle
nm - nano meter (10^{-9} m)
RNA - Ribonucleic Acid
SIGS - Spray-Induced Gene Silencing
sRNA - small RNA
siRNA – small interfering RNA
SWNTs - Single-Walled Carbon Nanotubes
ToMV - Tomato Mosaic Virus
VIGS - Virus-Induced Gene Silencing

REFERENCES

Agathokleous, E., Feng, Z., Iavicoli, I., & Calabrese, E. J. (2020). Nano-pesticides: A great challenge for biodiversity? The need for a broader perspective. *Nano Today*, *30*, 100808.

Akbar, S., Wei, Y., & Zhang, M. Q. (2022). RNA interference: Promising approach to combat plant viruses. *International Journal of Molecular Sciences*, *23*(10), 5312. PMID: 35628126

Anderson, P. K., Cunningham, A. A., Patel, N. G., Morales, F. J., Epstein, P. R., & Daszak, P. (2004). Emerging infectious diseases of plants: Pathogen pollution, climate change and agrotechnology drivers. *Trends in Ecology & Evolution*, *19*(10), 535–544. PMID: 16701319

Arjunan, N., Thiruvengadam, V., & Sushil, S. N. (2024). Nanoparticle mediated dsRNA delivery for precision insect pest control: A comprehensive review. *Molecular Biology Reports*, *51*, 355. PMID: 38400844

Avila-Quezada, G. D., Golinska, P., & Rai, M. (2022). Engineered nanomaterials in plant diseases: Can we combat phytopathogens? *Applied Microbiology and Biotechnology*, *106*(1), 117–129. PMID: 34913996

Avital, A., Muzika, N. S., Persky, Z., Bar, G., Michaeli, Y., Fridman, Y., & Schroeder, A. (2021). Foliar delivery of siRNA particles for treating viral infections in agricultural grapevines. *Advanced Functional Materials*, *31*(44), 2101003. PMID: 34744552

Bao, W., Wang, J., Wang, Q., O'Hare, D., & Wan, Y. (2016). Layered double hydroxide nanotransporter for molecule delivery to intact plant cells. *Scientific Reports*, *6*, 26738. PMID: 27221055

Baum, J. A., Bogaert, T., Clinton, W., Heck, G. R., Feldmann, P., Ilagan, O., & Roberts, J. (2007). Control of coleopteran insect pests through RNA interference. *Nature Biotechnology*, *25*(11), 1322–1326. PMID: 17982443

Bensoussan, N., Zhurov, V., Yamakawa, S., O'Neil, C. H., Suzuki, T., Grbić, M., & Grbić, V. (2018). The digestive system of the two-spotted spider mite, Tetranychus urticae Koch, in the context of the mite-plant interaction. *Frontiers in Plant Science*, *9*, 1206. PMID: 30271412

Chakrabarti, M., Kiseleva, R., Vertegel, A., & Ray, S. K. (2015). Carbon nanomaterials for drug delivery and cancer therapy. *Journal of Nanoscience and Nanotechnology*, *15*, 5501–5511. PMID: 26369109

Chang, F. P., Kuang, L. Y., Huang, C. A., Jane, W. N., Hung, Y., Hsing, Y. I. C., & Mou, C. Y. (2013). A simple plant gene delivery system using mesoporous silica nanoparticles as carriers. *Journal of Materials Chemistry. B*, *1*(39), 5279–5287. PMID: 32263331

Chen, X., Mangala, L. S., Rodriguez-Aguayo, C., Kong, X., Lopez-Berestein, G., & Sood, A. K. (2018). RNA interference-based therapy and its delivery systems. *Cancer and Metastasis Reviews*, *37*(1), 107–124. PMID: 29243000

Cheng, Y., Qu, H., Ma, M., Xu, Z., Xu, P., Fang, Y., & Xu, T. (2007). Polyamidoamine (PAMAM) dendrimers as biocompatible carriers of quinolone antimicrobials: An in vitro study. *European Journal of Medicinal Chemistry*, *42*, 1032–1038. PMID: 17336426

Delgado-Martín, J., Delgado-Olidén, A., & Velasco, L. (2022). Carbon dots boost dsRNA delivery in plants and increase local and systemic siRNA production. *International Journal of Molecular Sciences, 23*(10), 5338. PMID: 35628147

Demirer, G. S., Zhang, H., Goh, N. S., Pinals, R. L., Chang, R., & Landry, M. P. (2020). Carbon nanocarriers deliver siRNA to intact plant cells for efficient gene knockdown. *Science Advances, 6*, eaaz0495. PMID: 32637592

Dufès, C., Uchegbu, I. F., & Schätzlein, A. G. (2005). Dendrimers in gene delivery. *Advanced Drug Delivery Reviews, 57*, 2177–2202. PMID: 16310284

Edwards, C. H., Christie, C. R., Masotti, A., Celluzzi, A., Caporali, A., & Campbell, E. M. (2020). Dendrimer-coated carbon nanotubes deliver dsRNA and increase the efficacy of gene knockdown in the red flour beetle Tribolium castaneum. *Scientific Reports, 10*(1), 12422. PMID: 32709999

Fletcher, S. J., Reeves, P. T., Hoang, B. T., & Mitter, N. (2020). A perspective on RNAi-based biopesticides. *Frontiers in Plant Science, 11*, 51. PMID: 32117388

Food and Agriculture Organization of the United Nations. (2025). OECD-FAO agricultural outlook 2025-2034: Full report. OECD Publishing. https://www.oecd.org/en/publications/2025/07/oecd-fao-agricultural-outlook-2025-2034_3eb15914/full-report.html

FSANZ. (2013). *Response to Heinemann et al. on the regulation of GM crops and foods developed using gene silencing*. Food Standards Australia New Zealand.

Giraldo, J. P., Landry, M. P., Faltermeier, S. M., McNicholas, T. P., Iverson, N. M., Boghossian, A. A., Reuel, N. F., Hilmer, A. J., Sen, F., & Brew, J. A.. (2014). Plant nanobionics approach to augment photosynthesis and biochemical sensing. *Nature Materials, 13*, 400–408. PMID: 24633343

Grillo, R., Mattos, B. D., Antunes, D. R., Forini, M. M., Monikh, F. A., & Rojas, O. J. (2021). Foliage adhesion and interactions with particulate delivery systems for plant nanobionics and intelligent agriculture. *Nano Today, 37*, 101078.

Haley, B. P., Klimeck, G., Luisier, M., Vasileska, D., Paul, A., Shivarajapura, S., & Beaudoin, D. L. (2009). Computational nanoelectronics research and education at nanoHUB. org. *Journal of Computational Electronics, 8*(2), 124–131.

Hao, K., Gao, X., Yang, M., Wang, Z., An, M., Liu, H., Xia, Z., & Wu, Y. (2024). A nanomaterial for the delivery of dsRNA as a strategy to alleviate viral infections in maize. *Chemical Engineering Journal, 488*, 150923.

Hernández-Soto, A., & Chacón-Cerdas, R. (2021). RNAi crop protection advances. *International Journal of Molecular Sciences, 22*(22), 12148. PMID: 34830030

Jensen, P. D., Zhang, Y., Wiggins, B. E., Petrick, J. S., Zhu, J., Kerstetter, R. A., & Ivashuta, S. I. (2013). Computational sequence analysis of predicted long dsRNA transcriptomes of major crops reveals sequence complementarity with human genes. *GM Crops and Food: Biotechnology in Agriculture and the Food Chain, 4*(2), 90–97. PMID: 23787988

Ji, Z., Guo, W., Sakkiah, S., Liu, J., Patterson, T. A., & Hong, H. (2021). Nanomaterial databases: Data sources for promoting design and risk assessment of nanomaterials. *Nanomaterials (Basel, Switzerland)*, *11*(6), 1599. PMID: 34207026

Jones, R. A., & Naidu, R. A. (2019). Global dimensions of plant virus diseases: Current status and future perspectives. *Annual Review of Virology*, *6*(1), 387–409. PMID: 31283443

Jones-Rhoades, M. W., & Bartel, D. P. (2004). Computational identification of plant microRNAs and their targets, including a stress-induced miRNA. *Molecular Cell*, *14*(6), 787–799. PMID: 15200956

Kashyap, P. L., Xiang, X., & Heiden, P. (2015). Chitosan nanoparticle based delivery systems for sustainable agriculture. *International Journal of Biological Macromolecules*, *77*, 36–51. PMID: 25748851

Ke, W., He, R., Jensen, M. A., & Dobrovolskaia, M. A. (2025). Transforming Cancer Nanotechnology Data Analysis and User Experience. Part I: Current Challenges and Solutions Provided by caNanoLab. *Wiley Interdisciplinary Reviews. Nanomedicine and Nanobiotechnology*, *17*(4), e70030. PMID: 40826521

Kim, M., Rabbi, M., & Marszalek, P. E. (2011). Probing the Strength of a Bio-Nano Hub by Single-Molecule Force Spectroscopy. *Biophysical Journal*, *100*(3), 485a.

Kolge, H., Kadam, K., & Ghormade, V. (2023). Chitosan nanocarriers mediated dsRNA delivery in gene silencing for Helicoverpa armigera biocontrol. *Pesticide Biochemistry and Physiology*, *189*, 105292. PMID: 36549819

Kuang, Y., Zhao, L., Zhang, S., Zhang, F., Dong, M., & Xu, S. (2010). Morphologies, preparations and applications of layered double hydroxide micro-/nanostructures. *Materials (Basel)*, *3*, 5220–5235. PMID: 28883378

Kwak, S. Y., Lew, T. T. S., Sweeney, C. J., Koman, V. B., Wong, M. H., Bohmert-Tatarev, K., Snell, K. D., Seo, J. S., Chua, N. H., & Strano, M. S. (2019). Chloroplast-selective gene delivery and expression in planta using chitosan-complexed single-walled carbon nanotube carriers. *Nature Nanotechnology*, *14*, 447–455. PMID: 30804482

Li, J., Qian, J., Xu, Y., Yan, S., Shen, J., & Yin, M. (2019). A facile-synthesized star polycation constructed as a highly efficient gene vector in pest management. *ACS Sustainable Chemistry & Engineering*, *7*, 6316–6322.

Liu, Q., Chen, B., Wang, Q., Shi, X., Xiao, Z., Lin, J., & Fang, X. (2009). Carbon nanotubes as molecular transporters for walled plant cells. *Nano Letters*, *9*(3), 1007–1010. PMID: 19191500

Liu, X., Tang, K., Harper, S., Harper, B., Steevens, J. A., & Xu, R. (2013). Predictive modeling of nanomaterial exposure effects in biological systems. International journal of nanomedicine, 8(sup1), 31-43.

Lu, C., Li, Z., Chang, L., Dong, Z., Guo, P., Shen, G., Xia, Q., & Zhao, P. (2019). Efficient delivery of dsRNA and DNA in cultured silkworm cells for gene function analysis using PAMAM dendrimers system. *Insects*, *11*, 12. PMID: 31877645

Ma, X., & Yan, J. (2018). Plant uptake and accumulation of engineered metallic nanoparticles from lab to field conditions. *Current Opinion in Environmental Science & Health*, *6*, 16–20.

Ma, Z., Zhang, Y., Li, M., Chao, Z., Du, X., & Yan, S.. (2022). A first greenhouse application of bacteria-expressed and nano- carrier-delivered RNA pesticide for Myzus persicae control. *Journal of Pest Science*.

Marchesan, S., Kostarelos, K., Bianco, A., & Prato, M. (2014). The winding road for carbon nanotubes in nanomedicine. *Materials Today*, *18*, 12–19.

Mat Jalaluddin, N. S., Asem, M., Harikrishna, J. A., & Ahmad Fuaad, A. A. H. (2023). Recent progress on nanocarriers for topical-mediated RNAi strategies for crop protection—A review. *Molecules (Basel, Switzerland)*, *28*(6), 2700. PMID: 36985671

Mehra, N. K., Mishra, V., & Jain, N. K. (2014). A review of ligand tethered surface engineered carbon nanotubes. *Biomaterials*, *35*, 1267–1283. PMID: 24210872

Mitter, N., Worrall, E. A., Robinson, K. E., Li, P., Jain, R. G., Taochy, C., & Xu, Z. P. (2017). Clay nanosheets for topical delivery of RNAi for sustained protection against plant viruses. *Nature Plants*, *3*(2), 1–10. PMID: 28067898

Morris, S. A., Gaheen, S., Lijowski, M., Heiskanen, M., & Klemm, J. (2015). Experiences in supporting the structured collection of cancer nanotechnology data using caNanoLab. *Beilstein Journal of Nanotechnology*, *6*(1), 1580–1593. PMID: 26425409

Mosa, M. A., & Youssef, K. (2021). Topical delivery of host induced RNAi silencing by layered double hydroxide nanosheets: An efficient tool to decipher pathogenicity gene function of Fusarium crown and root rot in tomato. *Physiological and Molecular Plant Pathology*, *115*, 101684.

Mou, Q., Xue, X., Ma, Y., Banik, M., Garcia, V., Guo, W., & Lu, Y. (2022). Efficient delivery of a DNA aptamer-based biosensor into plant cells for glucose sensing through thiol-mediated uptake. *Science Advances*, *8*(26), eabo0902. PMID: 35767607

Mu, Q., Yang, L., Davis, J. C., Vankayala, R., Hwang, K. C., Zhao, J., & Yan, B. (2010). Biocompatibility of polymer grafted core/shell iron/carbon nanoparticles. *Biomaterials*, *31*(19), 5083–5090. PMID: 20378165

Mujtaba, M., Wang, D., Carvalho, L. B., Oliveira, J. L., Espirito Santo Pereira, A. D., Sharif, R., & Fraceto, L. F. (2021). Nano- carrier-mediated delivery of miRNA, RNAi, and CRISPR-Cas for plant protection: Current trends and future directions. *ACS Agricultural Science & Technology*, *1*(5), 417–435.

Negm, N. A., Hefni, H. H. H., Abd-Elaal, A. A. A., Badr, E. A., & Abou Kana, M. T. H. (2020). Advancement on modification of chitosan biopolymer and its potential applications. *International Journal of Biological Macromolecules*, *152*, 681–702. PMID: 32084486

Niño-Sánchez, J., Sambasivam, P. T., Sawyer, A., Hamby, R., Chen, A., Czislowski, E., & Jin, H. (2022). BioClay™ prolongs RNA interference-mediated crop protection against Botrytis cinerea. *Journal of Integrative Plant Biology*, *64*(11), 2187–2198. PMID: 36040241

Ntui, V. O., Kong, K., Azadi, P., Khan, R. S., Chin, D. P., Igawa, T., & Nakamura, I. (2014). RNAi-mediated resistance to Cucumber mosaic virus (CMV) in genetically engineered tomato. *American Journal of Plant Sciences*, *5*(5), 554–572.

Ostraat, M. L., Mills, K. C., Guzan, K. A., & Murry, D. (2013). The Nanomaterial Registry: facilitating the sharing and analysis of data in the diverse nanomaterial community. International journal of nanomedicine, 8(sup1), 7-13.

Padilla-Roji, I., Ruiz-Jimenez, L., & Bakhat, N.. (2023). RNAi Technology: A New Path for the Research and Management of Obligate Biotrophic Phytopathogenic Fungi. *International Journal of Molecular Sciences*, *24*, 9082. PMID: 37240427

Pavlicek, A., Part, F., Rose, G., Praetorius, A., Miernicki, M., Gazsó, A., & Huber-Humer, M. (2021). A European nano-registry as a reliable database for quantitative risk assessment of nanomaterials? A comparison of national approaches. *NanoImpact*, *21*, 100276. PMID: 35559789

Petrônio, M. S., Barros-Alexandrino, T. T., Lima, A. M., Assis, O. B., Nagata, A. K., Nakasu, E. Y., & Pilon, L. (2022). Physicochemical and toxicity investigation of chitosan-based dsRNA nanocarrier formation. *Biointerface Research in Applied Chemistry*, *12*, 5266–5279.

Puyam, A., Sharma, S., & Kashyap, P. L. (2017). RNA interference-a novel approach for plant disease management. *Journal of Applied and Natural Science*, *9*(3), 1612–1618.

Qiao, H., Zhao, J., Wang, X., Xiao, L., Zhu-Salzman, K., Lei, J., & Hao, D. (2023). An oral dsRNA delivery system based on chitosan induces G protein-coupled receptor kinase 2 gene silencing for Apolygus lucorum control. *Pesticide Biochemistry and Physiology*, *194*, 105481. PMID: 37532313

Ray, P., Sahu, D., Aminedi, R., & Chandran, D. (2022). Concepts and considerations for enhancing RNAi efficiency in phytopathogenic fungi for RNAi-based crop protection using nano- carrier-mediated dsRNA delivery systems. *Frontiers in Fungal Biology*, *3*, 977502. PMID: 37746174

Romeis, J., & Widmer, F. (2020). Assessing the risks of topically applied dsRNA-based products to non-target arthropods. *Frontiers in Plant Science*, *11*, 679. PMID: 32582240

Sangwan, A., Gupta, D., Singh, O. W., Roy, A., Mukherjee, S. K., Mandal, B., & Singh, N. (2023). Size variations of mesoporous silica nanoparticle control uptake efficiency and delivery of AC2-derived dsRNA for protection against tomato leaf curl New Delhi virus. *Plant Cell Reports*, *42*(10), 1571–1587. PMID: 37482559

Sharma, B., Attri, S., Syal, J., & Sharma, U. (2023). Liposomal NPs: A viable nanoscale drug carriers for the treatment of cancer. In Tyagi, R. K. (Ed.), *Liposomes*. IntechOpen.

Vance, V., & Vaucheret, H. (2001). RNA silencing in plants--defense and counterdefense. science, 292(5525), 2277-2280.

Voit, B. I., & Lederer, A. (2009). Hyperbranched and highly branched polymer architectures synthetic strategies and major characterization aspects. *Chemical Reviews*, *109*, 5924–5973. PMID: 19785454

Wang, Y., Yan, Q., Lan, C., Tang, T., Wang, K., Shen, J., & Niu, D. (2023). Nanoparticle carriers enhance RNA stability and uptake efficiency and prolong the protection against Rhizoctonia solani. *Phytopathology Research*, *5*(1), 2.

Worrall, E. A., Bravo-Cazar, A., Nilon, A. T., Fletcher, S. J., Robinson, K. E., & Carr, J. P.. (2019). Exogenous application of RNAi-inducing double-stranded RNA inhibits aphid-mediated transmission of a plant virus. *Front. Plant Sci.*, *10*, 265. PMID: 30930914

Yan, S., Qian, J., Cai, C., Ma, Z., Li, J., & Yin, M.. (2020a). Spray method application of transdermal dsRNA delivery system for efficient gene silencing and pest control on soybean aphid Aphis glycines. *Journal of Pest Science*, *93*, 449–459.

Yan, S., Yin, M. Z., & Shen, J. (2023). Nanoparticle-based nontransformative RNA insecticides for sustainable pest control: Mechanisms, current status and challenges. *Entomologia Generalis*, *43*, 21–30.

Yan, X., Sedykh, A., Wang, W., Yan, B., & Zhu, H. (2020b). Construction of a web-based nanomaterial database by big data curation and modeling friendly nanostructure annotations. *Nature Communications*, *11*(1), 2519. PMID: 32433469

Yong, J., Zhang, R., Bi, S., Li, P., Sun, L., Mitter, N., & Xu, Z. P. (2021). Sheet-like clay nanoparticles deliver RNA into developing pollen to efficiently silence a target gene. *Plant Physiology*, *187*(2), 886–899. PMID: 34608968

Zhang, H., Goh, N. S., Wang, J. W., Pinals, R. L., González-Grandío, E., Demirer, G. S., Butrus, S., Fakra, S. C., Del Rio Flores, A., & Zhai, R.. (2022). Nanoparticle cellular internalization is not required for RNA delivery to mature plant leaves. *Nature Nanotechnology*, *17*, 197–205. PMID: 34811553

Zheng, Y., Hu, Y., Yan, S., Zhou, H., Song, D., Yin, M., & Shen, J. (2019). A polymer/detergent formulation improves dsRNA penetration through the body wall and RNAi-induced mortality in the soybean aphid Aphis glycines. *Pest Management Science*, *75*, 1993–1999. PMID: 30610748

Zulfiqar, S., Farooq, M. A., Zhao, T., Wang, P., Tabusam, J., Wang, Y., & Gu, A. (2023). Virus-induced gene silencing (VIGS): A powerful tool for crop improvement and its advancement towards epigenetics. *International Journal of Molecular Sciences*, *24*(6), 5608. PMID: 36982682

KEY TERMS AND DEFINITIONS

Aphid: A small sap-sucking insect with soft bodies, often found in colonies on plants. Aphids can damage plants by feeding on sap and can transmit plant viruses.

Apoplast: The space between the plant cell wall and the cell membrane. Some delivery strategies aim to keep the dsRNA in the apoplast, where it can be taken up directly by pathogens like fungi.

Argonaute: 2 (AGO2): The core component of the RISC. It has an “endonucleolytic” or “slicer” function, meaning it cleaves the target viral mRNA, rendering it nonfunctional.

Biodegradability: The ability of the nano- carrier material to break down naturally within the body, reducing potential toxicity and accumulation over time.

Biopolymer NPs: Materials like chitosan can form NPs that encapsulate and protect dsRNA from degradation.

Carbon Dots: Carbon dots are small carbon-based NPs with fluorescent properties, used in imaging, sensing, and biomedical applications.

Carbon-based nanomaterials: This category includes carbon dots (CDs) and single-walled carbon nanotubes (SWNTs), which are also used to deliver nucleic acids into plant cells.

Chitosan: A linear polysaccharide derived from the shells of crustaceans. In plant protection, chitosan is used as a natural nano- carrier to form complexes with dsRNA. It protects the dsRNA and can also enhance a plant's natural defense responses.

CPPs (Cell-Penetrating Peptides): CPPs are short peptides that facilitate the delivery of therapeutic molecules across cellular membranes into cells for drug and gene delivery.

CRISPR: Clustered Regularly Interspaced Short Palindromic Repeats, a natural bacterial defense system that has been adapted as a gene-editing technology to precisely modify DNA sequences in living organisms.

CRISPR-Cas: A gene-editing system combining the CRISPR DNA sequences and Cas (CRISPR-associated) proteins, such as Cas9, which act as molecular scissors to cut DNA at specific locations guided by RNA molecules, enabling targeted genetic modifications.

Dendrimer: A synthetic, highly branched spherical polymer that can efficiently encapsulate siRNA.

Dicer: A cellular enzyme (an RNase III) that recognizes and cleaves long dsRNA, including vRI-dsRNA, into shorter, 21-23 nucleotide-long siRNA duplexes.

DNA Nanostructure: DNA nanostructures are precisely engineered molecular assemblies made from DNA strands that self-assemble into defined shapes for biomedical and nanotechnological applications.

dsRNA (double-stranded RNA): An RNA molecule with two complementary strands. In RNAi-based plant protection, dsRNA is the trigger for gene silencing. Delivered either through a transgene (HIGS) or external application (SIGS), the plant's cellular machinery processes it into smaller interfering RNAs (siRNAs).

dsRNA delivery: The process of introducing dsRNA into target cells, whether they are in plants, insects, or other organisms. Effective delivery is a major challenge in applied RNAi, especially in field settings, and often requires protection from environmental degradation.

dsRNA- nano- carrier complexes: Structures formed when dsRNA molecules are encapsulated within or bound to NPs. These complexes protect the dsRNA from premature degradation and facilitate its uptake into plant cells or target pathogens.

dsRNA nanomaterials: A class of materials, such as layered double hydroxides (nanoclays) or chitosan-based NPs, used to enhance the stability and delivery of dsRNA for plant protection. They improve environmental stability against factors like UV light and enzymes.

Endosomal escape: A crucial step in nano- carrier delivery. After a cell internalizes a nano- carrier via endocytosis, the nano- carrier must escape the endosome (a membrane-bound vesicle) into the cell's cytoplasm for the siRNA to be effective.

Environmental stability: Without the protection of a nano- carrier, naked dsRNA is quickly degraded in the field by UV light, nucleases, and rain. Nano- carrier encapsulation is crucial to ensure the dsRNA persists long enough to be effective.

Gold Nanocluster: Gold nanoclusters are ultra-small gold NPs composed of a few to a hundred atoms, exhibiting unique optical and electronic properties leveraged in bioimaging, sensing, and catalysis.

Guide strand and passenger strand: During RISC assembly, the siRNA duplex is unwound. The guide (or antisense) strand is incorporated into the RISC, while the passenger (or sense) strand is cleaved and degraded.

HIGS (Host-Induced Gene Silencing): The RNAi mechanism in which a host plant is genetically engineered to produce a specific double-stranded RNA (dsRNA) molecule. When a pathogen or pest feeds on the plant, it takes up the dsRNA, which then silences the pest or pathogen's genes.

Immunogenicity: The potential for the nano- carrier or siRNA to trigger an immune response, leading to inflammation or a reduction in therapeutic efficacy.

Layered double hydroxides (LDH): Also known as "nanoclay," these nanosheets are a highly effective nano- carrier for stabilizing and delivering dsRNA via foliar spray, offering prolonged protection.

Leafhopper: A small, agile insect that feeds on plant sap, known for its jumping ability. Leafhoppers can also be plant pests and vectors of plant diseases.

Lipid NPs (LNPs): The most widely used nano- carriers for siRNA delivery. They are a mixture of lipids that form a protective sphere around the RNA. The success of mRNA vaccines accelerated their development.

Lipoplex: A specific type of LNP formed from the combination of cationic lipids and nucleic acids via electrostatic interactions.

Minicell: A minicell is an anucleate, small cell-like structure produced by some bacteria, lacking chromosomal DNA, used as a vehicle in drug delivery and synthetic biology applications.

miRNA (microRNA): Small, non-coding RNA molecules that play a role in regulating gene expression, similar to siRNAs. However, miRNAs are derived from shorter stem-loop RNA products and primarily function by repressing translation rather than cleaving the target mRNA.

Nano- carrier: A nanoscale delivery vehicle (typically <100 nm) used to transport, protect, and release therapeutic cargo, such as siRNA.

Nanoliposomes: Nanoliposomes are nanoscale vesicles composed of lipid bilayers capable of encapsulating drugs or molecules, used for targeted drug delivery and controlled release.

Nanoparticle (NP): The general term for the various types of nano- carriers, including lipid, polymer, and inorganic NPs.

Non-transformative biopesticide: A biologically derived pesticide, such as dsRNA delivered externally via spraying, that does not involve genetically modifying the crop plant itself. This approach avoids public perception issues associated with genetically modified organisms (GMOs) and offers environmental benefits like higher target specificity and less chemical residue.

Off-target effects: A concern is the potential for the delivered dsRNA to accidentally silence host genes with similar sequences, causing unintended effects on the plant.

Plant disease: Any dynamic process that interferes with the normal functions of a plant over time. It can be caused by living organisms (e.g., fungi, bacteria, viruses) or nonliving factors (e.g., nutrient deficiencies).

Plant protection: A field of agricultural science focused on the prevention, control, and elimination of harmful organisms like pests and pathogens that can damage plants and cause yield loss. RNAi-based strategies are a modern tool for achieving plant protection.

Polymeric NPs (polyplexes): Carriers formed by a complex between cationic polymers and the negatively charged phosphate backbone of siRNA.

R Genes: Resistance genes in plants that encode proteins enabling the plant to recognize and defend against specific pathogens, often by triggering immune responses to block infection.

RNA interference (RNAi): A natural cellular process that regulates gene expression. It is triggered by double-stranded RNA (dsRNA) and results in the sequence-specific degradation of messenger RNA (mRNA).

RNA-induced silencing complex (RISC): An enzyme complex that incorporates one strand of the siRNA (the guide strand). It then uses this guide strand to find and bind to complementary viral mRNA.

siRNA (small interfering RNA): Short (20-24 base pairs) double-stranded RNA fragments produced by the cleavage of longer dsRNA. In RNAi, siRNA guides the RISC complex to target and destroy complementary messenger RNA (mRNA), silencing the gene.

Small interfering RNA (siRNA): A synthetic or naturally occurring double-stranded RNA molecule, typically 21-23 nucleotides long, that is designed to match and silence a specific target gene.

Spray-Induced Gene Silencing (SIGS): This is the method of application where dsRNA-nano-carrier formulations are sprayed directly onto plant leaves. The nano- carrier aids in dsRNA penetration through the waxy cuticle and cell wall into plant cells.

Star Polycation: Star polycation refers to a polymer with multiple positively charged arms extending from a central core, used in gene delivery and as carriers for biological molecules.

SWNTs (Single-Walled Carbon Nanotubes): SWNTs are cylindrical nanostructures made of a single layer of carbon atoms arranged in a hexagonal lattice, with extraordinary mechanical, electrical, and thermal properties used in nanotechnology and materials science.

Systemic barriers: Nano- carriers must navigate biological obstacles, such as blood circulation and the body's immune clearance mechanisms, to reach their target.

Targeting ligand: Molecules like antibodies, peptides, or aptamers attached to the surface of a nano-carrier to direct it to a specific cell type, tissue, or organ.

Vector (plant disease): In the context of plant disease, a vector is an organism, typically an insect, that transmits a pathogen (like a virus or bacterium) from an infected plant to a healthy one. Many plant viruses are spread by insect vectors such as aphids and whiteflies.

VIGS (Virus-Induced Gene Silencing): A reverse genetics technique that uses a modified plant virus to trigger a plant's natural RNA interference (RNAi) defense system, leading to the silencing of a specific host gene. VIGS is used to study gene function by observing the effects of knocking down a particular gene.

Virus suppression: Virus suppression refers to the process of reducing or inhibiting the replication, spread, or activity of a virus within a host organism (eg., plant or vector).

Virus: A tiny infectious agent consisting of genetic material (DNA or RNA) enclosed in a protein coat, which can only replicate inside living cells. Viruses cause diseases in plants, animals, and humans.

Chapter 12
Field Application Approaches:
Foliar Sprays, Soil Additives, and Seed Treatments

Zakir Ullah
https://orcid.org/0000-0002-1819-5753
University of Bradford, UK

Sidra Shafqat
https://orcid.org/0009-0009-2243-2051
University of Peshawar, Pakistan

Ghaida Mustafa
University of Bradford, UK

Sajid Ali
Bacha Khan University, Pakistan

ABSTRACT

Plant viral diseases represent a major global threat to food security, necessitating the development and implementation of innovative, effective, and sustainable management strategies. Nanoparticles (NPs) have become a significant candidate to plant viral disease management using several application methods. Among them, foliar spray is the most practical approach since it allows the direct interaction of NPs with the plant surfaces along with systematic translocation, Soil amendment strategies utilize the soil matrix as a NPs reservoir, and Nanopriming of seeds is not only a protective seedling way of management but also a more proactive method of seedling management. The present research outlines the potential of the future based on NPs-based strategies as a potential solution to the challenge of plant viral pathogens management. These treatments enforce plant defense and improve the absorption of nutrients. Future studies should focus on the customizable delivery mechanisms, and the traditional management strategies to define their efficacy and survival in crop protection.

DOI: 10.4018/979-8-3373-4913-8.ch012

1. INTRODUCTION

Agricultural crops are highly susceptible to a wide range of biotic and abiotic stresses (Bibi et al., 2025). Abiotic stresses include drought, temperature extremes, salinity, water logging, and heavy metal stress while biotic stresses comprise of bacterial, fungal, and viral infections, as well as insect pests. Abiotic stresses alone can cause losses of 20–50% of crop yields annually, with drought potentially reducing yields by over 50% in some regions (Rahman et al., 2022). Plant viruses are particularly significant among biotic stresses, accounting for over 47% of crop losses and thus substantially reducing agricultural productivity (Dutta et al., 2022). Viral infections cause over 50 percent of crops disease losses in developing countries, underscoring the fact that there is a dire need to develop effective management approaches (Rahman et al., 2022). Plant viruses are particularly difficult to manage due to their inevitable nature and regular transmission through insect vectors (Gul et al., 2025). Plants deal with the stress using intricate physiological and molecular responses, such as hormonal signalling (e.g., abscisic acid), metabolic modifications and the production of protective substances. Growth, photosynthesis, susceptibility to pathogens, and product quality are affected by these responses (Dutta et al., 2022). Different strategies have been discussed to alleviate such pressures; they include, crop breeding, biotechnology (genetic engineering and genetic editing), bio-stimulants, precision agriculture, and nanotechnology (Rahman et al., 2022).

Over the last few years, nanotechnology has become a viable solution towards enhancing the health of plants, especially in diagnosis, prevention and infection to the viral conditions (Mittal et al., 2020). Nanophytovirology in plant is broadly explained as disease diagnostics, therapeutic delivery, inactivation of plant defense systems, genetic transformation, and bio-stimulation (Rahman et al., 2022). Nanotechnology is the method of designing, preparing, and using materials and machines at the nanoscale, which is less than 100 nanometers (Ullah et al., 2023; Ullah et al., 2024). NPs have a high surface area-to-volume ratio that can volatilize their reactivity relative to bulk counterpart (Ullah et al., 2025). This characteristic allows active compounds with extended protection capabilities to be delivered more effectively and with greater specificity (Dutta et al., 2022).

The growing world population, coupled with rising food demand, also stresses the importance of efficient and sustainable food farming solutions (Ullah et al., 2025b). Interferences based on NPs can achieve this by enhancing the prospects of controlling the disease, cutting issued dependence on chemical pesticides, and growing more crops (Mittal et al., 2020). NPs act either as autonomous protectants, or as vectors of pesticides or RNA interference (RNAi) molecules, to provide high-specificity delivery of active functionalities (Ullah et al., 2024). NPs are also powerful bio-stimulants or biosensors to identify plant viruses at great accuracy (Awasthi et al., 2025). Metallic and non-metallic NPs (silver (Ag), copper (Cu), zinc oxide (ZnO), gold (Au), titanium oxide (TiO) and Iron oxide (FeO), have shown to possess antibacterial, antifungal, and antiviral properties and have been successful against pathogenic agents like potato virus Y (PVY), cucumber mosaic virus (CMV) and bean yellow mosaic virus (BYMV) (Ahmad et al., 2022).

NPs which are used in field applications are usually applied as a foliar spray, soil additive, or seed treatment, increasing NPs intake, boosting the delivery of the active compounds in these applications, and offering shield against viral diseases (Awasthi et al., 2025). Nano-encapsulation may also increase the effectiveness of pest management and minimize off-target, and achievement of sustainable agriculture (Mittal et al., 2020). Though there is increasing evidence on the potential of engineered nanomaterials in plant pathogen combating there is an evident lack of research that specifically looks at the antiviral effect of such materials (Ahmad et al., 2022). This research aims at giving a detailed account of the application of NPs in inculcation of resilience to plant viral infections. We focus on the antimicrobial capabilities of NPs, practical field applications (foliar sprays, soil amendments, and seed treatments), and advances in virus detection and control using NPs. Additionally, we discuss associated patents and innovations to inform future strategies for sustainable crop protection (Figure 1).

Figure 1. Nanoparticles enhance plant growth, improve stress resilience, offer antimicrobial defense, and advance sustainable agriculture

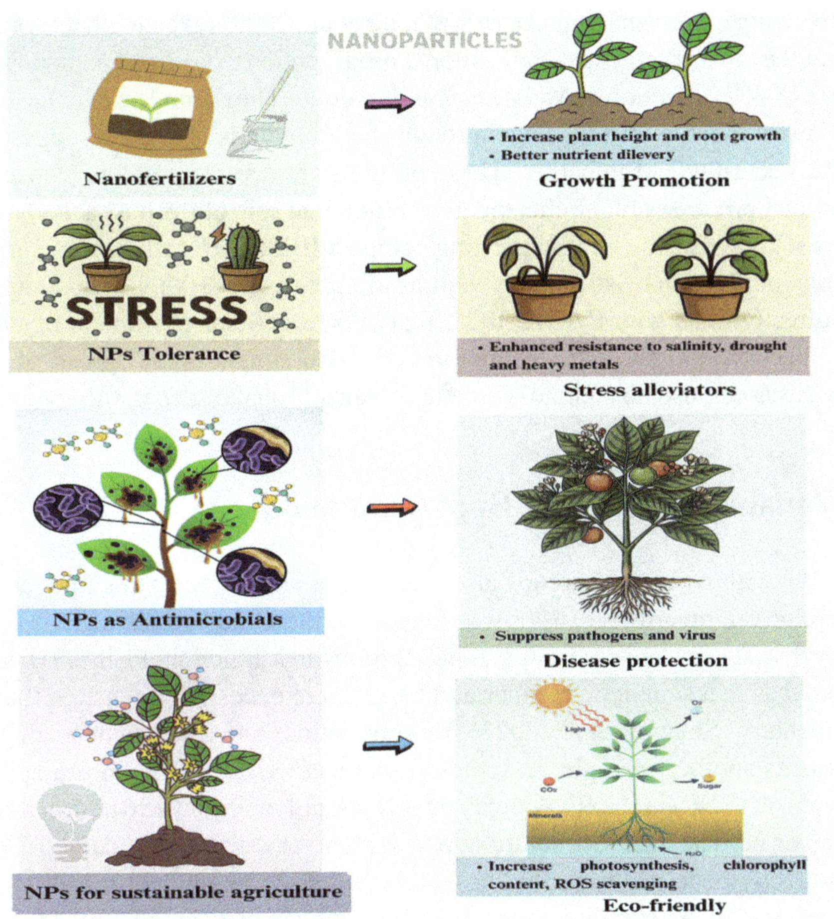

2 AGRICULTURAL PLANT VIRAL DISEASES

Plant pathogenic viruses in agricultural systems significantly reduce crop yields and pose a serious threat to global food security (Ahmad et al., 2022 and Ullah et al., 2025). The emergence of *viromics* has uncovered an immense diversity of viruses, many of which remain poorly understood with respect to their ecological roles and potential impacts on food security (Awasthi et al., 2025). Viruses are increasingly recognized not only for their direct ability to infect crops but also for their indirect effects on agricultural ecosystems (Ullah et al., 2025b). They have the potential to manipulate plant resistance to biotic and abiotic stresses and interact with bacterial species, weeds, pathogenic and symbiotic fungi, insect pollinators, and other species, either positively or negatively affecting crop productivity (Awasthi et al., 2025).

2.1. Viral Distribution, Diversity, and Pathogenesis

Recent research has emphasized the depth of the interaction of the viruses with the hosts in plants and other accompanying organisms. Indicatively, (Gong et al., 2023) revealed that protein lysine acetylation (Kac), one of the most crucial post-translational modifications, has a dramatic influence on tomato spotted wilt virus (TSWV) infection in *Nicotiana benthamiana*. They found an average of three thousand cis-pliesified lysines contented by a thousand proteins and hundreds of tissues suggested negatively or favorably after the occurrence of infection. These proteins are diverse proteins whose co-centrality in metabolic and cellular processes highlights the multifunctional role of the host acetylation in the pathogenesis of the virus. Likewise, the light-harvesting complex II-like LHC a/b-binding proteins have been found to play a role in infection by strawberry vein banding virus (SVBV). P1 illustrates that the SVBV passage protein directly links with FvLHC II-1L that aids in cellular movement of viruses and foster effectiveness of infection. FvLHC II-1L overexpression also enhances the spread of SVBV and does so to show that viruses can use host photosynthetic proteins to make successful colonization decisions (Xu et al., 2022).

2.2. Genetic Variability and Virus–Host Adaptation

Genetic variation within the viral groups also contributes significantly to pathogenicity. According to a WSSV research work on crustaceans, it was found that there are two distinct viral sub populations Type A and Type B within a stable 1:3 ratio, containing distinct functional genes (Hasan *et al.*, 2022). These variants acted as in a symbiotic manner and hence were necessary to sustain the infection hence the viral complementary (Shingote et al., 2022). Viral mutations can result in host adaptation in insect-associated viruses. As another example, Acyrthosiphon pisum virus (APV) in pea aphids had one nucleotide polymorphism (SNP) in the RNA-dependent RNA polymerase (RdRp) domain and it led to an increase in enzymatic activity and viral solution in host plants that were less adapted (Holt et al., 2024). This mutation indirectly benefited the aphid host by suppressing plant defenses, illustrating how viral evolution can improve host adaptability under stress conditions (Zhang et al., 2024).

2.3. Virus Transmission by Insect Vectors

Insect vectors play a central role in agricultural virus epidemiology. *Bemisia tabaci* is a notorious vector of chili leaf curl virus (ChiLCV), a major constraint on chili production in South Asia (Shingote et al., 2022). Transcriptome profiling of *B. tabaci* following ChiLCV acquisition identified 80 differentially expressed genes involved in signaling, metabolism, and cellular processes. Several candidate genes including replication factor A protein 1 and dynein heavy chain were implicated in facilitating viral circulation, providing potential molecular targets for vector control strategies (Nekkanti et al., 2022).

2.4. Mycoviruses as Biocontrol Agents

Mycoviruses are gaining attention as eco-friendly alternatives for managing fungal diseases in crops (Paria et al., 2025). *Alternaria alternata* apple pathotype, the causal agent of Alternaria leaf blotch, was found to harbor two mycoviruses: *Alternaria alternata chrysovirus 1* (AaCV1-QY2) and *A. alternata magoulivirus 1* (AaMV1). These viruses, particularly AaCV1-QY2, conferred hypovirulence by reducing fungal growth and mycotoxin production, suggesting their potential as natural biocontrol agents (Paria et al., 2025; Li et al., 2022).

2.5. Novel Plant Viruses and Mixed Infections

New viral species continue to be discovered in agricultural ecosystems (Li et al., 2022). Co-infections of tobamoviruses such as frangipani mosaic virus (FrMV) and plumeria mosaic virus (PluMV) were reported in *Plumeria rubra f. acutifolia* (Hamim et al., 2023). PluMV, with a genome structure like other tobamoviruses, could infect a broad host range and exhibited a distinct but related evolutionary profile compared to FrMV (Kumar et al., 2022). Such miscellaneous infections make difficult disease outcomes and demand precise diagnostic strategies (Hamim et al., 2023).

2.6. Diagnostics and Detection Technologies

Accurate diagnosis is very important for controlling plant viral diseases. Traditional and molecular methods require high-quality nucleic acid preservation, which is often challenging in field conditions (Muller et al.,, 2016). Recent research work has shown that virus-infected leaf tissues preserved in RNAlater remain suitable for RT-PCR, PCR, ELISA, Sanger sequencing, and high-throughput sequencing (Li et al., 2022). This approach enables reliable detection of viral pathogens even in remote or resource-limited agricultural sites, significantly strengthening disease surveillance and management efforts (Hamim et al., 2023). Agricultural plant viral diseases represent a dynamic and multifaceted challenge to global food security. Beyond direct crop infection, viruses effect ecological interactions, vector biology, and even host adaptation (Hu et al., 2020). Advances in viromics, functional genomics, and molecular diagnostics are steadily improving our understanding of viral diversity, pathogenesis, and transmission.

3 NANOTECHNOLOGY-BASED CROP PROTECTION

Crop protection through nanotechnology has revolutionized modern agriculture, overcoming the limitations of conventional pesticide formulations such as low aqueous solubility, short half-life, development of pest resistance, and environmental pollution due to excessive use of chemicals (Kumar et al., 2022). Nanoformulations of agrochemicals improve the delivery and bio-availability of active ingredients, thereby reducing dosage requirements and application rate, lowering production costs, and minimizing ecological footprints (Hu et al., 2020).

3.1 Advantages of NPs in Crop Protection

NPs offer unique physicochemical properties, including high surface-area-to-volume ratios, controllable sizes, and surface reactivity (Li et al., 2022). These attributes enable:

- Targeted delivery of pesticides, herbicides, fungicides, bactericides, and insecticides.
- Slow-release formulations, enhancing persistence and reducing non-target exposure.
- Nanofertilizers that improve nutrient uptake, decrease leaching losses, and synergize plant nutrition with pest control (Paria et al., 2025).

3.2 Types of Nanomaterials in Crop Protection

A wide range of nanocarriers and active nanomaterials are being developed for agricultural applications:

- **Polymeric NPs:** Chitosan, PLGA, and starch-based NPs encapsulate pesticides and fertilizers, providing stability and biodegradability (Li et al., 2022).
- **Lipid-Based Nanocarriers:** Liposomes and solid lipid nanoparticles enhance delivery of hydrophobic pesticides with high encapsulation efficiency (Li et al., 2022).
- **Metal and Metal Oxide NPs:** AgNPs, ZnONPs, and TiONPs possess inherent antimicrobial and insecticidal activities while serving as carriers and active agents (Hamim et al., 2023).
- **Silica Nanoparticles:** Mesoporous silica nanoparticles (MSNPs) allow controlled pesticide release through adjustable pore sizes and surface modifications (Kumar et al., 2022).
- **Nanoemulsions and Nanocapsules:** These systems increase pesticide solubility, stability, and penetration through plant cuticles and insect exoskeletons (Hamim et al., 2023).

3.3 Uptake and Mechanisms of Action

NPs enter plants via stomatal openings, root epidermis, and vascular tissues, enabling systemic protection. Foliar uptake is facilitated by nanoscale cuticular penetration, while root absorption involves endocytosis and apoplastic transport (Muller et al., 2016). Their enhanced adhesion reduces wash-off and improves field persistence. The interaction of NPs with pests and pathogens occurs through, Physical disruption of membranes, Induction of oxidative stress via reactive oxygen species (ROS), Suppression of metabolic functions critical for survival (Hu et al., 2020).

3.4 Green Synthesis and Environmental Safety

Conventional chemical synthesis of NPs may generate hazardous by-products. Hence, green synthesis using plants, algae, fungi, and bacteria is increasingly adopted to produce eco-friendly nanomaterials (Ullah et al., 2025b). These biologically derived NPs are often more biocompatible and less toxic. However, comprehensive toxicological studies are essential to assess potential risks to soil microbiota, pollinators, aquatic systems, and human health (Muller et al., 2016).

3.5 Integration with Precision Agriculture

Nanotechnology complements precision agriculture by enabling real-time crop monitoring and site-specific input management. Nanobiosensors—based on quantum dots, carbon nanotubes, and metallic nanoparticles—can detect pathogens, pesticide residues, and nutrient deficiencies at trace levels (Muller et al., 2016). Combined with drones for optimized spraying, these systems minimize chemical waste and environmental damage while enhancing efficiency (Hamim, et al., 2023).

3.6 Nano-Enabled Crop Protection

- **Fungicides:** Chitosan NPs improved delivery of fungicides against *Botrytis cinerea* and *Fusarium oxysporum* (Muller et al., 2016).
- **Insecticides:** Silica NPs coated with essential oils demonstrated enhanced insecticidal activity (Attia et al., 2023).
- **Antimicrobial Agents:** AgNPs and ZnONPs inhibited bacterial wilt and powdery mildew due to their intrinsic antimicrobial properties (Hu et al., 2020).
- **Herbicides:** Nanoclays and layered double hydroxides (LDHs) provided controlled herbicide release, reducing leaching and improving weed control (Kumar et al., 2022).

Nanotechnology offers an advanced platform for sustainable crop protection by combining targeted delivery, enhanced efficacy, and reduced environmental impact. With ongoing progress in green synthesis, smart biosensors, and precision farming integration, nanotechnology-based solutions hold great promise for transforming agricultural practices and ensuring global food security (Fathi *et al.*, 2025).

4 NANOPARTICLE BASED FOLIAR APPLICATIONS

Foliar NPs represent an emerging technology in agriculture with significant potential to enhance crop productivity and resilience (Kumar et al., 2022). Compared to root or soil-based applications, foliar sprays allow for more efficient delivery of NPs directly to plant tissues. However, the possibility of toxicity remains an important concern that must be carefully addressed (Hu et al., 2020).

4.1 Delivery Efficiency and Uptake Mechanisms

The physical properties of NPs play a key role in their delivery effectiveness. Positively charged particles (>15 mV) and smaller sizes (< 20 nm in cotton, < 11 nm in maize) have demonstrated superior penetration, reaching guard cells (100%), extracellular spaces (90.3%), and chloroplasts (55.8%). Surfactants that reduce surface tension to ~22 mN/m enable rapid absorption via stomatal and cuticular pathways within 10 minutes (Hu et al., 2020). Research indicates that NPs can directly enter stomata and be translocated through phloem between leaves and roots. Uptake rates vary depending on particle shape, application method, and plant tissue type (Wahane et al., 2020). While current first-generation agricultural NPs are primarily metal and biopolymer-based, future generations are expected to feature improved porosity and surface characteristics for more precise and efficient agrochemical delivery (Kumar et al., 2022).

4.2 Agricultural Benefits of Foliar NP

Foliar NP applications provide a direct route for transporting nutrients, pesticides, and bioactive compounds to plant leaves without soil interaction, thereby reducing nutrient immobilization and leaching. This approach offers several benefits like, Faster absorption and quicker plant responses, Increased nutrient-use efficiency with reduced input requirements, and Mitigation of abiotic stresses such as drought, salinity, and heavy metal toxicity (Hu et al., 2020). For instance, zinc oxide (ZnO) NPs enhance antioxidant activity during drought stress, while nano-selenium and nano-silica reduce cadmium accumulation in wheat grains (Wahane et al., 2020). Silver (Ag) and copper (Cu) nanoparticles also provide antimicrobial protection, improving pest and disease management through targeted delivery (Hu et al., 2020).

5. MECHANISMS OF ACTION: LEAF SURFACE PENETRATION AND SYSTEMIC TRANSLOCATION

NPs in agriculture possess the unique capability of penetrating leaf surfaces and undergoing systemic translocation, thereby enabling precise delivery of agrochemicals and antiviral agents (Singh et al., 2024). Foliar-applied NPs primarily enter through natural openings such as stomata, cuticular cracks, and hydrophilic pores within the cuticle. The efficiency of this uptake is strongly influenced by particle size, hydrophobicity, and surface charge (Hernandez *et al.*, 2021). For instance, positively charged nanoparticles smaller than 20 nm readily penetrate stomatal pores and guard cells, achieving near 100% internalization into epidermal and mesophyll tissues within minutes (Hong et al., 2021).

The use of surfactants further enhances foliar absorption by lowering surface tension, thereby improving nanoparticle adhesion and diffusion across the waxy cuticle. This allows uptake even when stomata are closed or access is limited to cuticular microchannels (Singh et al., 2024). Once inside the leaf, NPs exploit both apoplastic and symplastic transport pathways. Movement through the xylem supports upward (root-to-shoot) transport, while phloem translocation facilitates downward (shoot-to-root) delivery, ensuring NPs distribution to distant tissues such as meristems, key sites of viral replication (Hernandez *et al.*, 2021; Subhan *et al.*, 2022).

The systemic mobility of NPs depends largely on their chemical properties. Metal oxide NPs such as ZnO and CuO can release ions and regulate ROS, conferring antiviral activity. Meanwhile, polymeric and lipid-based nanocarriers are particularly suited for transporting nucleic acids, siRNAs, and chemical antivirals with high tissue specificity (Wahane et al., 2020).

Functional modifications, including surface ligands and peptides, further enhance targeted delivery, while controlled-release coatings improve nanoparticle stability under field conditions (Islam et al., 2025). These mechanisms collectively enable NPs to disperse bioactive compounds uniformly throughout the plant, provide sustained release, and trigger host defense pathways (Singh et al., 2024). Engineered NPs can inactivate viral particles, silence viral genes, or stimulate plant immunity, making them a promising next-generation agricultural platform. When integrated into foliar sprays, soil additives, or seed treatments, NPs-based systems have the potential to overcome current limitations in plant protection, nutrient delivery, and systemic disease management (Hong et al., 2021).

6. NANOPARTICLE ROOT-MEDIATED UPTAKE AND SOIL ADDITIVES

NP uptake through roots begins at the rhizosphere, where soil composition, pH, and root exudates play a critical role in determining NP stability and bioavailability (Ullah et al., 2024c). When used as soil additives, NPs may adsorb onto root surfaces or dissolve to release bioactive ions, enabling the targeted delivery of nutrients and antiviral compounds (Islam et al., 2025). Uptake efficiency largely depends on particle size: ultrasmall NPs can penetrate directly through root epidermal cell walls, whereas larger NPs typically enter via endocytosis or through natural openings such as lateral root emergence zones (Ullah et al., 2023c). Functionalization of NPs surfaces with amino, carboxyl, or polymeric groups reduces aggregation, thereby improving NP dispersion, mobility, and penetration from the apoplast to the symplast (Arunachalam et al., 2025).

Once internalized, NPs are transported within root tissues via apoplastic (cell wall) and symplastic (plasmodesmata) pathways until they reach the vascular tissues (Islam et al., 2025). Movement through the xylem enables upward translocation to aerial parts, while the phloem distributes NPs to developing tissues and storage organs (Hong et al., 2021). Studies on MONPs, such as ZnO, Fe_3O_4, and TiO_2, demonstrate that gradual dissolution releases essential micronutrients, enhances redox balance, and strengthens plant defenses, thereby reducing viral infection risks (Ullah et al., 2023a). Root uptake efficiency is further influenced by plant species, transpiration rates, and soil microbiota, as microbial activity can alter NPs dissolution, transformation, and subsequent plant responses (Arunachalam et al., 2025). Beyond nutrient delivery, nanoparticles introduced into soils act as carriers of antiviral and antimicrobial agents. For example, silica and chitosan NPs can induce systemic acquired resistance (SAR), while Ag and CuNPs exhibit direct antiviral activity by binding to viral particles or interfering with replication (Ullah et al., 2023c). In addition, nanofertilizers and nanopesticides offer improved nutrient-use efficiency with minimal leaching losses, making them more environmentally sustainable than conventional agrochemicals. This dual functionality—as both delivery vehicles and defense modulators—positions NPs-based soil additives as a powerful tool for precision agriculture, sustainable nutrient management, and improved plant immunity (Figure 2).

Figure 2. NPs enhance plant antiviral defense by regulating physiology, strengthening barriers, and modulating gene expression

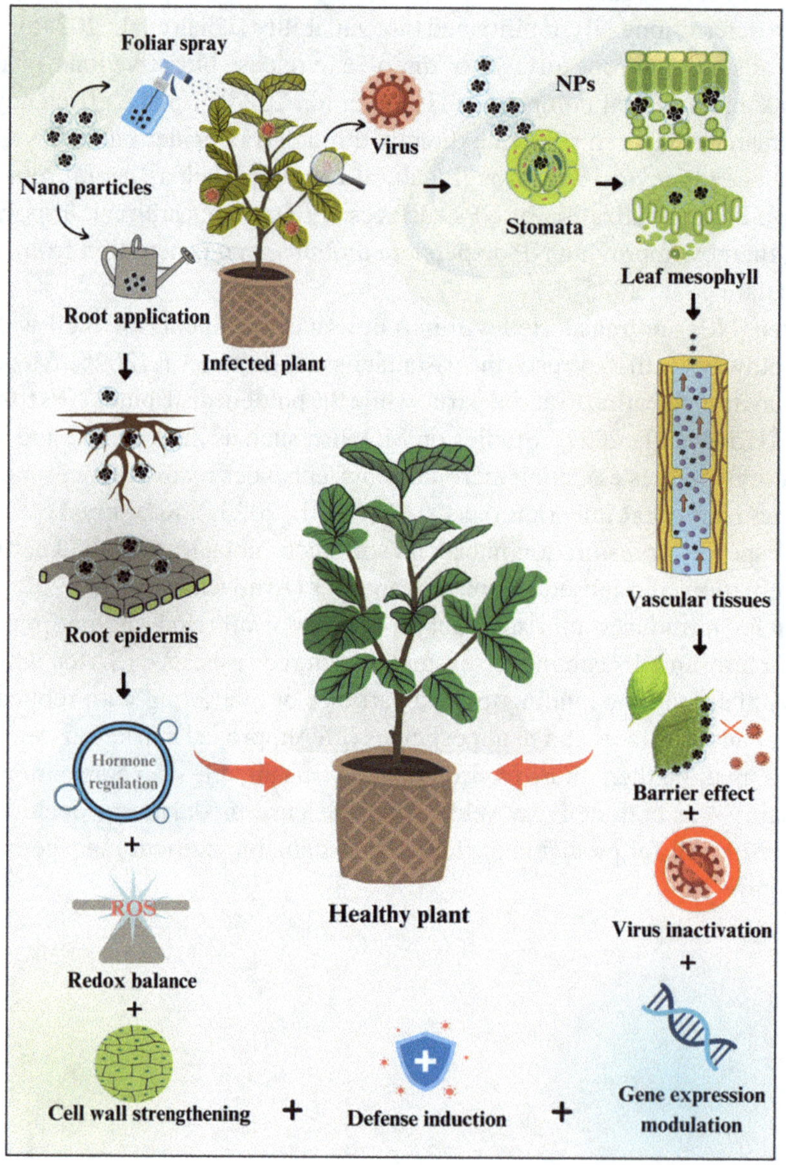

7. NANOPARTICLE AND SOIL PHYSICOCHEMICAL INTERACTION

When NPs are applied to plant surfaces or soils, they encounter a highly complex soil matrix composed of minerals, dissolved ions, and natural organic matter. These components collectively govern the fate, mobility, and bioavailability of nanoparticles (Arunachalam et al., 2025). Homo and heteroaggregation, adsorption, dissolution, redox reactions, and sulfidation are some of the critical mechanisms that affect engineered nanoparticles (ENPs) in soils; these are highly dependent on soil pH, ionic strength, and the existence of divalent cations (Islam et al., 2025). AgNPs could become Ag2S, a less soluble and less bioavailable derivative and copperoxyde and ironoxide NPs could release micronutrient ions via a ligand-mediated dissolution reaction, shifting their reactivity and antiviral activity (Jaramillo-López et al., 2024).

Rhizosphere has proven to have a potent effect on NP behavior because root exudates, as well as microbial metabolites could modify the state of aggregation, speciation, and redox conditions that ultimately influence uptake, systemic movement and antiviral efficacy. (Jaramillo-López et al., 2024). Additionally, soil constituents such as clay minerals, Fe/Al oxides, humic substances, and biochar form "environmental coronas" around NPs, providing charged surfaces that stabilize or immobilize them. This corona formation directly affects NP persistence, transport, and interactions with plants and soil microorganisms (Warghane et al., 2024). NPs applied to soil whether metallic, metal oxide, or polymeric—can exhibit both direct and indirect antiviral activity (Ullah et al., 2025b). Direct mechanisms include disrupting viral structures, interfering with replication, or reducing the activity of soil-borne vectors such as aphids and white flies, thereby lowering virus transmission rates (Warghane et al., 2024). Indirectly, NPs stimulate plant defense responses by activating antioxidant enzymes (e.g., superoxide dismutase, peroxidase), modulating phytohormones such as salicylic acid (SA), jasmonic acid (JA), and abscisic acid (ABA), and reinforcing physical barriers like silicon-enriched cell walls (Warghane et al., 2024; Asghar et al., 2024). In addition, layered double hydroxide (LDH) nanocarriers, such as BioClay, can deliver double-stranded RNA (dsRNA) for RNA interference (RNAi), enabling targeted and sustained antiviral action (Singh et al., 2024). Specific nanoparticles studied include ZnO for immune stimulation, SiO_2 for structural reinforcement, Fe_3O_4 for antioxidant activation, chitosan NPs for antiviral delivery and gene transfer, and nanoclays for controlled release of dsRNA (Žigrayová et al., 2023).

The antiviral efficacy of soil-applied NPs depends on properties (size, surface charge, and surface coatings) as well as soil parameters such as pH, redox potential, organic matter, and sulfide levels (Warghane et al., 2024). Additives like biochar and humic compounds can help immobilize NPs and control their transport, reducing the risks of leaching and toxicity (Ukwattage et al., 2025). Application methods include soil drenching, seed treatment, and co-formulation with nanocarriers to achieve localized and sustained effects. While laboratory and greenhouse studies have demonstrated promising results, large-scale field application remains constrained by challenges including potential ecotoxicological impacts on beneficial microorganisms, risks of bioaccumulation, inter-soil variability, and the absence of standardized protocols (Alkhaza'leh et al., 2025). Optimizing the antiviral benefits of NPs in soil requires careful consideration of NP chemistry, soil properties, and crop-specific factors to balance efficacy with environmental safety (Wang et al., 2024).

8. NANO PRIMING AND SEED TREATMENTS

Nano priming is an advanced seed treatment technology where NPs are applied to seeds to enhance germination, early seedling vigor, and tolerance against biotic and abiotic stresses (Alkhaza'leh etv al., 2025). Unlike conventional chemical priming, nano priming allows direct, controlled, and targeted delivery of nutrients, growth regulators, and antiviral agents to the seed surface or seed coat (Alkhaza'leh et al., 2025). Due to their large surface-area-to-volume ratio, NPs provide high reactivity, improved penetration, and uniform coating, leading to enhanced water uptake, enzyme activity, and metabolic priming during germination (Arunachalam et al., 2025). NPs seed treatments are used as carriers of antiviral molecules (dsRNA, plant extracts, metallic ions etc), modulators of phytohormones, and activators of antioxidant defense systems, while also preventing infection by seed or soil-borne viral pathogens at early developmental stages (Wang et al., 2024).

8.1 Mechanisms of Action

Nanopriming enhances plant performance and viral suppression through multiple interrelated mechanisms, including the activation of germination pathways where NPs penetrate the seed coat and stimulate key metabolic processes such as α-amylase activity, ROS signaling, and phytohormonal regulation involving abscisic acid, gibberellins, and salicylic acid, leading to faster and more resilient seedling growth (Ullah et al., 2023c). In addition to growth promotion, metallic NPs like Ag, Cu, and ZnO, as well as chitosan NPs, exert direct antiviral effects by binding to viral proteins or nucleic acids to block replication and restrict viral spread. RNA interference (RNAi)-based protection is achieved through nanocarriers such as layered double hydroxides (LDHs) and polymeric NPs, which enable controlled release of double-stranded RNA (dsRNA) to trigger RNAi against major plant viruses including Cucumber mosaic virus (CMV), Tobacco mosaic virus (TMV), and Tomato yellow leaf curl virus (TYLCV) (Hernandez *et al.*, 2021). Furthermore, nanopriming may modulate the seed-associated microbiome, promoting beneficial microbes that suppress viral vectors and soil-borne pathogens (Wang et al., 2024). Several types of NPs are used in seed treatments: metallic NPs (Ag, Cu) with strong antiviral and antibacterial properties that lower viral loads in crops such as tomato and lettuce (Elsharkawy, 2024); metal oxide NPs (ZnO, TiO_2) that enhance germination, seedling vigor, and antioxidant enzyme activity while providing moderate antiviral effects (Alkhaza'leh et al., 2025); polymeric NPs (starch and chitosan) serving as biodegradable carriers of antiviral molecules and growth enhancers, with chitosan also capable of inducing SAR and hybrid NPs (e.g., ZnO–chitosan or SiO_2–polymer composites) that integrate nutrient delivery with antiviral protection, thereby offering multifunctional benefits (Mousavi et al., 2024).

8.3 Agronomic and Stress-Related Benefits

Nano-priming offers multiple advantages for sustainable agriculture by improving germination and seedling vigor through enhanced water uptake, enzyme activation, and early growth, while also strengthening abiotic stress tolerance against drought, salinity, heavy metals, and oxidative stress to ensure stable crop emergence under challenging conditions (Nile et al., 2022). Moreover, it provides antiviral protection as metallic NPs and RNAi-based nanocarriers help restrict the spread of seed or soil-borne viruses, and it reduces chemical inputs by lowering dependence on conventional seed treatments and pesticides, thereby minimizing environmental impacts and promoting eco-friendly farming practices (Ehsan et al., 2024).

Despite its promise, nano priming faces several challenges. High NP concentrations may cause phytotoxicity, reduce seed viability, or alter germination dynamics (Ullah, et al., 2025d). Uptake and residual effects are strongly influenced by NP size, surface charge, and coatings, as well as soil and water chemistry (Ullah et al., 2025b). Standardization of protocols including NP type, concentration, soaking time, and drying conditions remains a major hurdle for reproducibility and large-scale implementation (Alkhaza'leh et al., 2025). Moreover, long-term environmental risks such as bioaccumulation and effects on seed-associated microbiota require further study (Asghar et al., 2024). Future directions include the integration of nanopriming with RNAi-based antivirals, nanonutrient delivery, and combinations of NPs with plant extracts (Warghane et al., 2024). Controlled-release systems using layered nanoclays and biodegradable polymers can extend protection during germination and early seedling growth. Advances in molecular imaging and tracing technologies are expected to clarify NP penetration, translocation, and mechanisms of action, accelerating the rational design of next-generation nano priming strategies for managing viral diseases in crops (Pal et al., 2022).

9. COMPARATIVE ANALYSIS OF NPS DELIVERY

Field application approaches for NP delivery in plants include soil additives, foliar sprays, and seed treatments (Islam et al., 2025). Each method has distinct uptake mechanisms, formulation considerations, advantages, and limitations (Wang et al., 2017). Optimizing these strategies based on crop type, pathogen load, and agronomic conditions can significantly improve the management of plant viral diseases (Alkhaza'leh et al., 2025).

9.1 Foliar Sprays Rapid Virus Suppression

Foliar based NP delivery enables uptake through apoplastic and symplastic pathways, allowing penetration via stomata and cuticle into systemic tissues, where metal NPs such as Ag, CuO, and ZnO can directly destabilize viral capsids while activating plant defense responses. Additionally, nanocarriers like mesoporous silica nanoparticles (MSNs) and lipid-based systems safeguard dsRNA therapeutics from UV and enzymatic degradation, facilitating RNAi-mediated antiviral activity (Warghane et al., 2024 and Arunachalam et al., 2025). Effective formulation requires UV-protective strategies such as core–shell nanostructures, antioxidants, or UV-absorbing coatings; the inclusion of adjuvants like non-ionic surfactants, humectants, and sticker-spreader agents; and optimization of NP size ($\leq$100 nm) with a mild positive zeta potential for enhanced adhesion and penetration (Warghane et al., 2024). This method offers the advantages of rapid viral symptom suppression and localized action with minimal soil contamination, though its limitations include short residual activity due to UV and RNase degradation as well as risks of environmental drift. Foliar NP applications are best suited for emergency interventions, foliar-feeding viral infections, and localized outbreaks (Arunachalam et al., 2025).

9.2. Soil Additives Microbiome-Directed Defense and Sustained Delivery

Soil-based NP delivery leverages root exudate interactions that facilitate NP entry into roots, followed by xylem-mediated transport to aerial tissues, where they act as slow-release depots for season-long antiviral protection while also modulating root microbe virus interactions to enhance immunity and nutrition (Singh et al., 2024). Formulation strategies include controlled release systems using mesoporous silica, polymeric coatings, or alginate matrices; surface chemistry modifications to optimize mobility, reduce leaching, and stabilize NPs in the rhizosphere; and biodegradable carriers such as PLA or chitosan to safeguard beneficial microbes. This approach offers advantages such as prolonged systemic coverage, efficient root-to-shoot dispersion, and reduced need for repeated foliar applications (Hu et al., 2020). However, challenges include potential non-target impacts on soil microbiota and variability in NP uptake due to heterogeneous soil conditions. Overall, soil-applied NP formulations are best suited for long-term management of soilborne and persistent viral infections (Islam et al., 2025).

9.3 Seed Treatments

Seed-based NP delivery offers a promising strategy for plant protection by enabling uptake during imbibition, where NPs penetrate seed coats, reach embryonic tissues, and prime systemic resistance (Singh et al., 2024). Coatings can be engineered to carry dsRNA, micronutrients, or elicitors, thereby providing prophylactic protection prior to infection. Advanced formulation strategies include triggered release systems responsive to moisture, pH, or enzymes; precision dosing to ensure uniform NP coverage while minimizing phytotoxicity; and bioadhesive materials that enhance coating adhesion and extend shelf life. This approach is economical, durable, and compatible with existing seed treatment practices, while also ensuring early and systemic resistance against pathogens (Nile et al., 2022). However, limitations such as fixed dose constraints, the need for rigorous germination and vigor testing, and regulatory approval requirements must be addressed. Overall, NP-based seed treatments are best suited for prophylactic use in virus-endemic regions and large-scale farming systems (Jaramillo-López et al., 2024).

9.4 Nanoparticle Techniques for Viral Disease Control

Nanotechnology offers innovative and sustainable strategies for combating viral diseases in plants by enabling targeted delivery, enhanced stability, and improved efficacy of antiviral agents. Several NP-based techniques have been explored, which act through diverse mechanisms such as capsid destabilization, ROS generation, interference with viral replication, and improved delivery of antiviral biomolecules (Muller et al., 2016 and Kumar et al., 2022). Some key NPs-Based Approaches:

- **Mesoporous Silica Nanoparticles (MSNs):** MSNs exhibit a high surface area and tunable pore size, making them excellent carriers for double-stranded RNA (dsRNA) and RNA interference (RNAi) therapeutics. Their stability and controlled release properties allow for precise delivery to infected plant tissues (Asghar et al., 2024).
- **Polymeric Nanoparticles:** Biodegradable polymers such as chitosan, PLGA, and PEG-based systems act as carriers for nucleic acids, peptides, and proteins. They not only protect payloads from degradation but also enhance cellular uptake, thereby increasing the effectiveness of antiviral strategies (Žigrayová et al., 2023).
- **Lipid-Based Nanocarriers (LNPs):** LNPs improve encapsulation efficiency and facilitate fusion with plant cell membranes. Their ability to enhance the uptake and bioavailability of RNA therapeutics, including siRNA and miRNA, makes them promising tools for plant viral disease management (Wang et al., 2024).
- **Metal and Metal-Oxide Nanoparticles:** Metallic NPs such as Ag, Cu, and ZnO have intrinsic antiviral activity through capsid destabilization, ROS production, and direct interaction with viral nucleic acids. These NPs can act synergistically with RNAi or other biologically active agents for enhanced protection (Warghane et al., 2024). The integration of metal-based antivirals, RNA therapeutics, and smart nanocarriers represents a paradigm shift in managing plant viral diseases (Awasthi et al., 2025). Foliar sprays offer immediate defense, soil applications deliver prolonged systemic protection, and seed treatments establish early-stage immunity (Arshad et al., 2025). Together, these approaches provide multilayered, precise, and environmentally sustainable protection, reducing dependence on chemical pesticides while enhancing crop productivity and resilience (Figure 3).

Figure 3. Methods of nanoparticle applications in plants including seed priming, foliar spray, root application, and trunk injection for growth, protection, and stress management

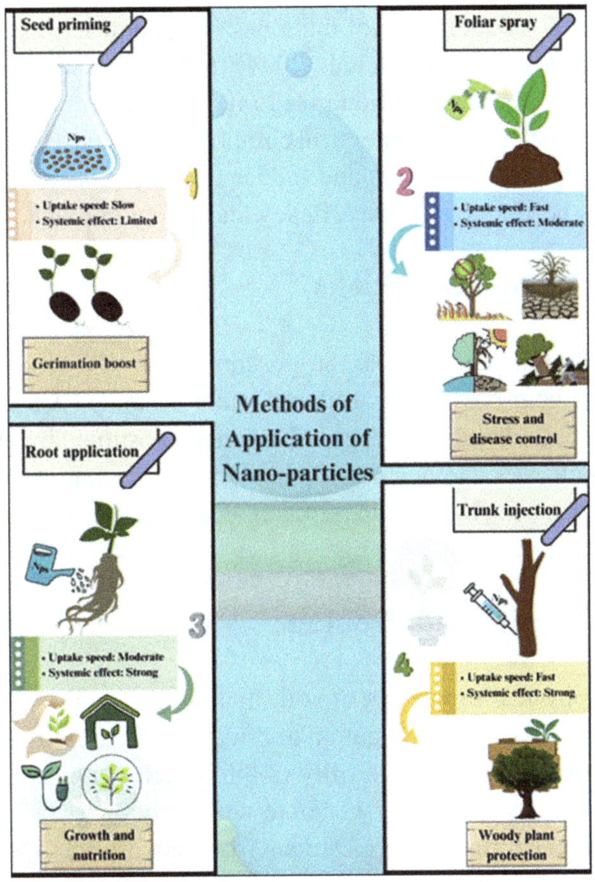

10. ENVIRONMENTAL EFFECTS AND REGULATORY MATTERS

10.1 Soil Microbial Health and Biodiversity

The introduction of NPs into soil ecosystems can significantly affect microbial communities. For instance, AgNPs have been shown to disrupt microbial populations, alter enzymatic activity, and reduce microbial diversity (Ullah et al., 2023d). Such interferences can impair vital soil processes, including nutrient cycling and organic matter decomposition. In contrast, carbon-based NPs have demonstrated positive effects by enhancing microbial activity and nutrient (N and P) availability, thereby improving soil fertility and promoting plant growth (Wei et al., 2024).

10.2 Phytotoxicity and Plant Health

NPs could increase the ability of the plants to resist the viral infections as well as triggering a phytotoxic effect in some cases. Over-saturation of metallic NPs may lead to root inhibitor, a reduction in chlorophyll, and oxidative stress (Warghane et al., 2024). In research work, it is found that the Oxides of Cerium (CeO2) NPs disrupt nitrogen fixation in legumes, which reduces growth. Bigger NP toxicity relies on the size, the surface charge, and the chemical composition because smaller NPs with high-surface area have greater chances to penetrate tissues and induce toxicity (Ukwattage et al., 2025). Therefore, NP features manipulation is needed to achieve efficacy and safety in plants.

10.3 Ecological Bioaccumulation Risks

Another significant environmental issue is the possibility of NPs bioaccumulating in food webs. Wastewater or runoff would help NPs access soil and water systems (Alkhaza'leh et al., 2025). They could then move to herbivores and even to higher levels when consumed by plants. Research shows that the gold and cadmium selenide NPs may accumulate in plants and predators, increasing human exposure by consuming contaminated foods (Ullah et al., 2023c). This trophic transfer emphasizes the need for long-term ecological monitoring.

10.4 Soil Structural Integrity and Erosion

NPs may also alter the physical properties of soil, including porosity and aggregation. While materials like nanoclays can improve soil aggregation and water retention, excessive NP use may acidify soils, reduce microbial diversity, and lower fertility (Alkhaza'leh et al., 2025). Moreover, interactions between NPs and existing contaminants may generate new, unknown compounds with unpredictable toxicity (Wang et al., 2017). Such risks complicate environmental hazard assessments and necessitate careful management of NP applications

11. REGULATORY AND ETHICAL CONSIDERATIONS

11.1 Lack of Standardized Regulations

The rapid progress of nanotechnology has outpaced the development of regulatory frameworks. Currently, there is no global consensus on nanomaterial nomenclature, characterization, or testing protocols (Ullah et al., 2023c). In the United States, the Environmental Protection Agency (EPA) regulates nanomaterial-based pesticides under the Federal Insecticide, Fungicide, and Rodenticide Act (FIFRA) (Subhan *et al.*, 2022). However, agricultural use of nanotechnology remains constrained by regulatory uncertainty and ethical concerns about environmental and human safety (Wang et al., 2024).

11.2 Ethical Concerns and Public Perception

The integration of NPs in agriculture raises ethical issues, particularly regarding health, environment, and transparency. Public acceptance is heavily influenced by perception, and without clear labeling and communication, consumers may reject food products containing nanomaterials (Hong et al., 2021). Excessive use of nanoparticle-based formulations can also contribute to the development of resistant pests, similar to the conventional pesticides, thus lessening the effectiveness in the long term (Isibor et al., 2024). These concerns need to be met with transparency and responsible usage.

11.3 Comprehensive Risk Assessment

In order to ensure safe agricultural application, the risk assessment studies need to be done very well. These are to assess the physicochemical characteristics of NPs, their environmental behavior and exposure pathways and the long-term environmental impact (Wang et al., 2024). The global acceptability of nanotechnology in agriculture needs harmonized regulation to achieve global acceptance of the tools at standardized testing platforms. Responsible deployment entails a moderated strategy comprising strict scientific investigation, regulatory standardization, and mass involvement (Bhattacharya et al., 2024). The interdisciplinary partnership between scientists, policymakers, and stakeholders will be important in creating a sustainable future of nanotechnology in agriculture (Isibor et al., 2024).

12 CHALLENGES

12.1 Environmental Effects of ENPs in Soil

The introduction of ENPs into the soil creates a complex interaction of the soil with its constituents which defines their mobility, bioavailability, and potential toxicity (Islam et al., 2025). Other factors that have significant effects on the behavior of ENP are their pH, the level of organic matter, the levels of ionic strength, and the presence of divalent cations (Ijaz et al., 2024). Indicatively, high ionic strength and divalent cations have the ability of compressing the electrical double layer of ENPs, increasing aggregation and reducing mobility. Conversely, the ENPs can also be coated with the natural organic matter (NOM) and become stable in suspension, which can be moved through the soil matrix. Transformation of ENPs in soil is also influenced by redox conditions and the presence of ligands. However, in the context of lowered conditions, metal based ENPs, e.g., AgNPs, can undergo sulfidation to create less soluble, less bioavailable compounds, e.g., silver sulfide (Ag2S), during which their toxicity and uptake in plants are altered (Wang et al., 2017). Similarly, copper oxide (CuO) nanoparticles are also reducible and oxidable and metal ions can be released, in turn potentially affecting the health of plants and soil microbial communities (Ullah et al., 2023c).

12.2 Impact on Soil Ecosystems

The introduction of ENPs into soil can result in far-reaching effects on soil biota, including microorganisms, invertebrates, and plants (Isibor et al., 2024). Investigations have established that ENPs could alter the ratio of microbiome, which affects the cycling of nutrients and soil health. To demonstrate this, the presence of ENPs may inhibit proliferation and activity of positive soil bacteria and fungi and establish low soil fertility and stunted plant growth (Sun et al., 2021). The plants are also able to take up ENPs; it might lead to bioaccumulation and transfer through the food chain. ENP also can be appear to soil invertebrates such as earthworms and nematodes (Ijaz et al., 2025). ENPs have been demonstrated to be toxic to these organisms, which impairs their survival, reproductive fauna and soil ecology through decomposition and aerating of soil. Disruption of these processes can lead to low soil productivity and health (Ijaz et al., 2024).

12.3 Regulatory Challenges and Considerations

The lack of standard nomenclature, characterization procedures, and testing procedures is a huge problem in regulation of ENPs in agriculture (Alizadeh et al., 2025). Such inconsistency makes risk assessment and regulatory decisions hard. As an example, ENP size, shape, surface charge, and coating may differ, which may result in varying environmental behaviors and toxicity profiles and requires standardized guidelines to evaluate and regulate them. Regulatory Frameworks Current regulatory policies are usually ineffective in dealing with the peculiarities and characteristics of ENPs (Isibor et al., 2024). Although various nations have taken the step of regulation, including making the safety test of nanomaterials a mandatory condition, not all countries have enacted these laws, and these laws may not fully address the possible risks of ENP use in farming. Harmonized international standards are required that take life cycle ENPs, production to disposal, into account to guarantee environmental and common health safeguards (Ijaz et al., 2025).

12.4 Ethical and Socioeconomic Considerations

The use of ENPs in agriculture is associated with ethical issues pertaining to equity, access and cultural concerns (Bhattacharya et al., 2024). Smaller and resource constrained farmers might find it hard to access and afford ENP-based technologies and this might lead to the further increase in inequalities in agricultural practices. In addition, some communities are suspicious of embracing nanotechnology due to lack of clarity about the effects of nanotechnology on the environment in the long run and the sufficiency of existing regulation (Ijaz et al., 2024).

12.5 Challenges in Implementing Nanotechnology in Agriculture

The possible environmental hazards of ENPs include their perseverance in soil and water systems, bioaccumulation in food chains and their toxicity to non-target organism (Alizadeh et al., 2025). It has also been proved, as an example, that ENPs may accumulate in aquatic organisms with adverse effects on biodiversity and services to the ecosystems (Bhattacharya et al., 2024). Moreover, the shift in the environment of ENPs also may lead to the formation of secondary nanoparticles with different properties and functions that complicate the evaluation of the risks and management strategies (Ijaz et al., 2024). Despite the extensive research, it remains to be filled in on the long-term fate of ENP and its effect on agricultural systems. The individual studies ought to be conducted in such a way that they examine interactions of ENPs with various soil components, their transformations and their impacts on soil health and performance of the ecosystem. This form of study is necessary in the development of evidence-based practices and regulatory frameworks to safely use ENPs in agriculture (Chidiamassamba et al., 2024).

12.6 Public Perception and Acceptance

Nanotechnology has been applied to agriculture, and this depends on the thinking of the people. This fact on the safety, efficacy and environmental effect of ENP may not be received well by consumers and farmers because of the issues of its safety, efficacy and environmental impacting effects (Chidiamassamba et al., 2024). It requires communication and public involvement and the stakeholder's involvement to build trust and make informed decisions on use of ENPs in agricultural activity. Introduction of engineered nanoparticles into agricultural practices is potentially in future to enhance yield of crops and make them more sustainable (Arshad et al., 2025). However, their effects on the environment, regulatory aspects and the assessment posed in their use makes them worth considering and administering. These problems should be addressed in a multidisciplinary manner that incorporates scientific research, regulation and involvement of the people in the process so that the application of nanotechnology has attained its objectives without causing harm to the environment and human health (Tawiah et al., 2024).

13 FUTURE RESEARCH DIRECTIONS

Future studies on the application of NPs to treat the parasome in plant viruses should proceed by the understanding of the mechanistic characteristics of interactions between NPs and plants. The mechanisms through which NPs penetrate leaves, roots, and seeds are not well understood even though the effectiveness thereof has been demonstrated. Knowledge of uptake of cells, its systemic translocation via vascular vessels and regulation of host defense mechanisms, such as reactive oxygen species (ROS), phytohormone signals and RNA interference will be vital to maximizing stable antiviral defense in crop. Careful study of the ecotoxicology and long-term environmental fate of NPs is to be made. Their chemical modification, i.e. the sulfidation of AgNPs, redox reactions in CuONPs and FeNPs, etc. can be used to influence their bioavailability and antiviral effects. Bioaccumulation in tissues of edible plants and potential trophic transfer by the food chains are critical to human and ecological safety. Also, an evaluation of impacts on soil microbial diversity, nutrient cycling and viral vectors populations will guide sustainable field use and reduce off-target toxicity. NPs formulations and delivery systems must be optimized to be practically deployable. Released antiviral agents can be provided and antiviral can be safeguarded by stimuli-responsive nanocarriers, including layered double hydroxides, chitosan, or polymeric NPs. Modifications to the surface to change charge, hydrophobicity, or steric stabilization can increase soil mobility, foliar retention and plant uptake. To enhance efficacy and NP load, crop- and virus-specific studies and synergistic methods involving NPs and standard antiviral agents or useful microbes are required.

Future study will be directed by advanced analytical and modeling tools. NP uptake and translocation can be visualized using imaging technologies such as TEM, confocal microscopy, and fluorescence tracking. NP prediction NP models that forecast NP transport, dissolution, and bioavailability in different soil and plant systems will be used to create the best strategies of applying NP in the field. Omics technologies, such as transcriptomics, proteomics, and metabolomics, can uncover how NPs alters plant immunity and viral suppression and serve as a basis of high-throughput screening and design. Lastly, regulatory, socioeconomic, and policy issues should be incorporated into research planning. Having standard procedures of risk assessment, safe dose and environmental monitoring will help in the approval of regulations. An assessment of cost-efficiency, scalability, and accessibility will guarantee its commercial adoption in agriculture, whereas the involvement of the population and the transfer of knowledge will contribute to solving the ethical issue and making nanotechnology-based viral management acceptable. In general, nanotechnology, soil science, regulatory frameworks, and plant virology will be multidisciplinary tools needed to transform nanoparticle strategies into sustainable, field-ready solutions to plant viral disease suppression.

14 CONCLUSION

The NPs-based strategies are an innovative concept to treat the plant viral diseases and have provided both specific delivery, improved bioavailability and multifunctional antiviral activities. NPs applications have shown remarkable efficacy in the suppression of viral replication, stimulation of systemic plant responses, and the regulation of rhizosphere microbiome. Direct antiviral activity is demonstrated by metal and MONPs (i.e., AgNPs, CuO and ZnO but controlled delivery of antiviral molecules (i.e., a dsRNA) or other antiviral agents to prolong their stability and efficacy in the field is facilitated by polymeric and layered NPs i.e., chitosan and LDH. Comparative analyses indicate that each of the delivery routes, foliar sprays, soil amendments, or seed treatments, has distinct advantages in the uptake and systemic transport and persistence, and crop and virus specific optimization is necessary. Environmental interactions and soil physicochemistry critically influence NPs fate, bioavailability, and antiviral potency. Transformation to sulfidation, redox reaction, and complexation to organic matter can decrease the activity, however, it also could decrease the off-target toxicity. Nanoparticles mobility and dissolution are also controlled by root exudates and microbial metabolites and determine antiviral activity and uptake by plants following oral delivery. Such findings suggest that prior to trying any large-scale application, one should consider integrated measuring rhizosphere measurements and soil to ensure that NPs chemistry is matched with soil firms during pre-crop phenology and ecological microbes during pre-crop ecology to assure results are consistent and reliable. Nanoparticle technologies show potential, however, to implement controllable settings to real-life agriculture is challenging. Severe problems include NPs phytotoxicity, bioaccumulation, long-term environmental impacts, price and regulatory restriction. The inclusion of nanotechnology, plant physiology, virology and environmental sciences in future studies are needed to ameliorate NPs formulations, understand mechanistic pathways better and establish safe and standard application procedures. This interdisciplinary effort will permit scaleable and field-prepared NPs intervention, a powerful, ecof-concentrated substitute on the old-fashioned chemical antivirals and a long-term system of defending crops against disease vectors.

REFERENCES

Ahmad, A., Muzammil, I., Munir, T., Naseer, M. A., Aqib, A. I., Ali, M. M., & Arshad, M. I. (2022). Metallic and Non-metallic Nanoparticles against Viruses. In *Nanomaterials in the Battle Against Pathogens and Disease Vectors* (pp. 205–228). CRC Press.

Alizadeh, M., Musazade, E., Qaderi, S., Qarachal, J. F., Siahpoush, S., Abbod, M., & Ghasemi, H. (2025). The environmental and anthropogenic impacts of nanoparticles on forest trees. *Journal of Nanoparticle Research*, *27*(6), 143.

Alkhaza'leh, H., Rahbeh, M., & Hamadneh, I., AL-Mashakbeh, H., & Albalawna, Z. (2025). Nanoparticles in soil reclamation: A review of their role in reducing soil compaction. *Air, Soil and Water Research*, *18*, 11786221241311725.

Arshad, M. A., Akbar, B. A., Shehzadi, N., Iqbal, N., Mushtaq, M. Z., Rouf, S., & Jawad, A. (2025). Nanoparticles in Plant Genetic Engineering: Innovative Tools and Future Prospects for Enhanced Crop Traits and Agricultural Sustainability. *Journal of Plant Sciences*, *13*(2), 38–58.

Arunachalam, A., Perraki, M., Knegt, B., Macel, M., Voigt, D., & Kamperman, M. (2025). Turning over a new leaf: Innovative pest control from a materials science perspective. *Chemical Society Reviews*. PMID: 40485430

Asghar, A., Ullah, Z., Ali, T., Iqbal, J., Abbasi, B. A., Ijaz, S., . . . Mahmood, T. (2024). Introduction to nc-RNA and Their Crucial Role in Biological Regulation Under Environmental Stress. In *ncRNAs: Mediated Regulation: The Unarguable Biological Function to Environmental Constrains* (pp. 1-46). Cham: Springer Nature Switzerland.

Attia, R. G., Khalil, M. M., Hussein, M. A., Fattah, H. M. A., Rizk, S. A., & Ma'moun, S. A. (2023). Cinnamon oil encapsulated with silica nanoparticles: Chemical characterization and evaluation of insecticidal activity against the rice moth, Corcyra cephalonica. *Neotropical Entomology*, *52*(3), 500–511. PMID: 36913124

Awasthi, S., Chalapathy, C. V., Khandelwal, L., & Rajesh, P. (2025). Nanocarrier-Enabled Agricultural Biotechnology and Crop Breeding: Advances and Prospects. *Nanocarriers in Plant Science and Agriculture*, 297-312.

Bhattacharya, B., Mondal, R., Mandal, A. K., & Biswas, D. (2024). Ethical Issues Surrounding the Utilization of Gold Nanoparticles in Plant and Soil Systems. In *Plant Response to Gold Nanoparticles: Plant Growth, Development, Production, and Protection* (pp. 341–349). Springer Nature Singapore.

Bibi, S., Ullah, R., Burni, T., Ullah, Z., Uddin, J., Nur-e-Alam, M., & Kazi, M. (2025). Morphophysiological attributes of AMF inoculated tomato (Lycopersicon esculentum Mill.) ameliorated by resorcinol, biochar and nanobiochar. *Discover Nano*, *20*(1), 1–24. PMID: 40721547

Chidiamassamba, S. B., Gomes, S. I., Amorim, M. J., & Scott-Fordsmand, J. J. (2024). Considering safe and sustainable by design alternatives–Environmental hazards of an agriculture nano-enabled pesticide to non-target species. *Chemosphere*, *367*, 143582. PMID: 39454771

Dutta, P., Kumari, A., Mahanta, M., Biswas, K. K., Dudkiewicz, A., Thakuria, D., & Mazumdar, N. (2022). Advances in nanotechnology as a potential alternative for plant viral disease management. *Frontiers in Microbiology*, *13*, 935193. PMID: 35847105

Ehsan, M., Raja, N. I., Mashwani, Z. U. R., Fatima, N., Abasi, F., Wattoo, F. H., & Iqbal, M. (2024). The Role of Green Synthesized Nanoparticles in Biotic Stress Resistance in Vegetables. In *Nanoparticles in Plant Biotic Stress Management* (pp. 383–402). Springer Nature Singapore.

Elsharkawy, M. M. (2024). Nanoparticles in Phytovirus Management. In *Nanoparticles in Plant Biotic Stress Management* (pp. 181–206). Springer Nature Singapore.

Fathi Qarachal, J., & Alizadeh, M. (2025). Harnessing precision agriculture and nanotechnology for sustainable farming: A review. *Nanotechnology for Environmental Engineering*, *10*(2), 37.

Gong, Y., Li, Y., Liu, D., Jiang, L., Liang, H., Wu, Y., & Yang, J. (2023). Analysis of lysine acetylation in tomato spot wilt virus infection in Nicotiana benthamiana. *Frontiers in Microbiology*, *14*, 1046163. PMID: 36819054

Gul, F., Ullah, Z., Iqbal, J., Abbasi, B. A., Ali, S., Kanwal, S., & Mahmood, T. (2025). Ecofriendly synthesis characterization and biological activities of Eruca sativa mediated silver oxide nanoparticles. *Scientific Reports*, *15*(1), 13466. PMID: 40251221

Hamim, I., Komatsu, K., Prager, S. M., & Wu, B. (2023). Viruses in agricultural systems: Interactions with plants, insect pollinators and fungi. *Frontiers in Microbiology*, *14*, 1170402. PMID: 37032904

Hasan, M. M. (2022). *Prevalence and Relative Virulence Properties of Circulatory Phylotypes of White Spot Syndrome Virus in Crustaceans* (Doctoral dissertation, © University of Dhaka).

Hernandez, E. S. (2021). *Layered double hydroxide (LDH)-mediated topical delivery of dsRNA for protection against Tomato yellow leaf curl virus (TYLCV) in Nicotiana benthamiana* (Doctoral dissertation).

Holt, J. R., Cavichiolli de Oliveira, N., Medina, R. F., Malacrinò, A., & Lindsey, A. R. (2024). Insect–microbe interactions and their influence on organisms and ecosystems. *Ecology and Evolution*, *14*(7), e11699. PMID: 39041011

Hong, J., Wang, C., Wagner, D. C., Gardea-Torresdey, J. L., He, F., & Rico, C. M. (2021). Foliar application of nanoparticles: Mechanisms of absorption, transfer, and multiple impacts. *Environmental Science. Nano*, *8*(5), 1196–1210.

Hu, P., An, J., Faulkner, M. M., Wu, H., Li, Z., Tian, X., & Giraldo, J. P. (2020). Nanoparticle charge and size control foliar delivery efficiency to plant cells and organelles. *ACS Nano*, *14*(7), 7970–7986. PMID: 32628442

Ijaz, S., Iqbal, J., Abbasi, B. A., Ullah, Z., Ijaz, N., Yaseen, T., & Mahmood, T. (2024). Regulatory and Ethical Concerns of Nanotechnology in Agriculture. In *Revolutionizing Agriculture: A Comprehensive Exploration of Agri-Nanotechnology* (pp. 395–427). Springer Nature Switzerland.

Iqbal, J., Ijaz, S., Ijaz, N., Abbasi, B. A., Yaseen, T., Ullah, Z., . . . Kazi, M. (2025). Microbial synthesis of metal nanoparticles for nanomedicinal and catalytic applications. In *Expanding Nanobiotechnology: Applications and Commercialization* (pp. 88-124). CRC Press.

Isibor, P. O., Kayode-Edwards, I. I., Oyewole, O. A., Agbontaen, D. O., Onwaeze, O. O., Paul, E., & Chimbekujwo, K. I. (2024). Risk assessment and management in nanotoxicology. In *Environmental Nanotoxicology: Combatting the minute contaminants* (pp. 267–293). Springer Nature Switzerland.

Islam, A. N., Ridwan, A., Neon, A. K., Rab, M. F., & Uddin, M. N. (2025). *Engineered Nanoparticles (ENPs) in Aquatic Environments and Soil-Plant Ecosystems: Transformation*. Toxicity, and Environmental Challenge.

Jaramillo-López, P. F., Romero, J. B., Sarabia, M., Fonteyne, S., Saldivia-Tejeda, A., Verhulst, N., & Larsen, J. (2024). Non-target effects of pesticide and microbial seed treatments in maize and barley on the resident soil microbiota under conservation agriculture. *European Journal of Soil Biology*, *122*, 103653.

Kumar, A., Solanki, V., Katiyar, A., & Mandal, B. (2022). Host biology and genomic properties of Plumeria mosaic virus, a tobamovirus discovered in a temple tree in India co-infecting with frangipani mosaic virus. *Frontiers in Microbiology*, *13*, 1030042. PMID: 36406428

Li, B., Cao, Y., Ji, Z., Zhang, J., Meng, X., Dai, P., & Wang, Y. (2022). Coinfection of two mycoviruses confers hypovirulence and reduces the production of mycotoxin alternariol in Alternaria alternata f. sp. mali. *Frontiers in Microbiology*, *13*, 910712. PMID: 35756001

Mittal, D., Kaur, G., Singh, P., Yadav, K., & Ali, S. A. (2020). Nanoparticle-based sustainable agriculture and food science: Recent advances and future outlook. *Frontiers in Nanotechnology*, *2*, 579954.

Mousavi, S. M., Hashemi, S. A., Yari Kalashgrani, M., Zare, I., Rahmanian, V., Chiang, W. H., & Mostafavi, E. (2024). Synthetic polymers as antibacterial and antiviral agents. In *Antibacterial and Antiviral Functional Materials* (Vol. 2, pp. 1–46). American Chemical Society.

Muller, R., Betsou, F., Barnes, M. G., Harding, K., Bonnet, J., & Kofanova, O. & A Review from the International Society for Biological and Environmental Repositories (ISBER) Biospecimen Science Working Group. (2016). Preservation of biospecimens at ambient temperature: Special focus on nucleic acids and opportunities for the biobanking community. *Biopreservation and Biobanking*, *14*(2), 89–98. PMID: 26886348

Nekkanti, A., Chakraborty, P., Ghosh, A., Iquebal, M. A., Jaiswal, S., & Baranwal, V. K. (2022). Transcriptomic changes of Bemisia tabaci Asia II 1 induced by chilli leaf curl virus trigger infection and circulation in its vector. *Frontiers in Microbiology*, *13*, 890807. PMID: 35572639

Nile, S. H., Thiruvengadam, M., Wang, Y., Samynathan, R., Shariati, M. A., Rebezov, M., & Kai, G. (2022). Nano-priming as emerging seed priming technology for sustainable agriculture—Recent developments and future perspectives. *Journal of Nanobiotechnology*, *20*(1), 254. PMID: 35659295

Pal, A., Kaur, P., Dwivedi, N., Rookes, J., Bohidar, H. B., Yang, W., & Manna, P. K. (2022). Clay-Nanocomposite based smart delivery systems: A promising tool for sustainable farming. *ACS Agricultural Science & Technology*, *3*(1), 3–16.

Paria, K., Pyne, S., Khamarai, R., & Chakraborty, S. K. (2025). Eco-sustainable agriculture practices through Mycoviruses. *Journal of Natural Pesticide Research*, 100129.

Rahman, M. U., Zulfiqar, S., Raza, M. A., Ahmad, N., & Zhang, B. (2022). Engineering abiotic stress tolerance in crop plants through CRISPR genome editing. *Cells, 11*(22), 3590. PMID: 36429019

Shingote, P. R., Wasule, D. L., Parma, V. S., Holkar, S. K., Karkute, S. G., Parlawar, N. D., & Senanayake, D. M. J. B. (2022). An overview of chili leaf curl disease: Molecular mechanisms, impact, challenges, and disease management strategies in Indian subcontinent. *Frontiers in Microbiology, 13*, 899512. PMID: 35847087

Singh, R., Choudhary, P., Kumar, S., & Daima, H. K. (2024). Mechanistic approaches for crosstalk between nanomaterials and plants: Plant immunomodulation, defense mechanisms, stress resilience, toxicity, and perspectives. *Environmental Science. Nano, 11*(6), 2324–2351.

Subhan, M. A., & Subhan, T. (2022). Safety and global regulations for application of nanomaterials. In *Nanomaterials recycling* (pp. 83–107). Elsevier.

Sun, W., Dou, F., Li, C., Ma, X., & Ma, L. Q. (2021). Impacts of metallic nanoparticles and transformed products on soil health. *Critical Reviews in Environmental Science and Technology, 51*(10), 973–1002.

Tawiah, B., Ofori, E. A., & George, S. C. (2024). Nanotechnology in societal development. In *Nanotechnology in societal development* (pp. 1–64). Springer Nature Singapore.

Ukwattage, N. L., & Zhiyong, Z. (2025). Impacts of Cerium Dioxide Nanoparticles on the Soil–Plant System and Their Potential Agricultural Applications. *Nanomaterials (Basel, Switzerland), 15*(12), 950. PMID: 40559314

Ullah, R., Ullah, Z., Iqbal, J., Chalgham, W., & Ahmad, A. (2023). Aspartic acid-based nano-copper induces resilience in Zea mays to applied lead stress via conserving photosynthetic pigments and triggering the antioxidant biosystem. *Sustainability, 15*(16), 12186.

Ullah, Z.. (2025). *0007, CABI Biotechnology Series*. CABI, Smart and Stress-resilient Agriculture Achieved by Engineered Nanomaterials and Nanobiotechnology., DOI: 10.1079/9781800627307.0007

Ullah, Z., Gul, F., Iqbal, J., Abbasi, B. A., Kanwal, S., Chalgham, W., & Mahmood, T. (2023). Biogenic synthesis of multifunctional silver oxide nanoparticles (Ag2ONPs) using Parieteria alsinaefolia delile aqueous extract and assessment of their diverse biological applications. *Microorganisms, 11*(4), 1069. PMID: 37110492

Ullah, Z., Iqbal, J., Abbasi, B. A., Akhtar, W., Kanwal, S., Ali, I., & Mahmood, T. (2023). Assessment of gus expression induced by anti-sense Os PPO gene promoter and antioxidant enzymatic assays in response to drought and heavy metal stress in transgenic Arabidopsis thaliana. *Sustainability, 15*(17), 12783.

Ullah, Z., Iqbal, J., Abbasi, B. A., Ali, I., Ijaz, S., Yaseen, T., & Mahmood, T. (2025). Global Food Security and Bioenergy Production. In *Forage Crops in the Bioenergy Revolution: From Fields to Fuel* (pp. 15–48). Springer Nature Singapore.

Ullah, Z., Iqbal, J., Abbasi, B. A., Gul, F., Ali, S., Kanwal, S., & Mahmood, T. (2025). Eco-friendly synthesis of iron oxide nanoparticles using parietaria alsinifolia extracts and evaluation of biological applications. *Applied Biochemistry and Biotechnology, 197*(4), 2328–2355. PMID: 39752127

Ullah, Z., Iqbal, J., Abbasi, B. A., Ijaz, S., Munir, M., Yaseen, T., ... & Mahmood, T. (2025). Applications of Nanoparticles in Biofortification of Crops: Amplifying Nutritional Quality. *Crop Biofortification: Biotechnological Approaches for Achieving Nutritional Security Under Changing Climate*, 321-350.

Ullah, Z., Iqbal, J., Gul, F., Abbasi, B. A., Kanwal, S., Elsadek, M. F., & Mahmood, T. (2024). Biogenic synthesis, characterization, and in vitro biological investigation of silver oxide nanoparticles (AgONPs) using Rhynchosia capitata. *Scientific Reports*, *14*(1), 10484. PMID: 38714767

Wahane, A., Waghmode, A., Kapphahn, A., Dhuri, K., Gupta, A., & Bahal, R. (2020). Role of lipid-based and polymer-based non-viral vectors in nucleic acid delivery for next-generation gene therapy. *Molecules (Basel, Switzerland)*, *25*(12), 2866. PMID: 32580326

Wang, F., Zhou, L., Mu, D., Zhang, H., Zhang, G., Huang, X., & Xiong, P. (2024). Current research on ecotoxicity of metal-based nanoparticles: From exposure pathways, ecotoxicological effects to toxicity mechanisms. *Frontiers in Public Health*, *12*, 1390099. PMID: 39076413

Wang, P., Lombi, E., Sun, S., Scheckel, K. G., Malysheva, A., McKenna, B. A., & Kopittke, P. M. (2017). Characterizing the uptake, accumulation and toxicity of silver sulfide nanoparticles in plants. *Environmental Science. Nano*, *4*(2), 448–460. PMID: 32802334

Warghane, A., Saini, R., Shri, M., Andankar, I., Ghosh, D. K., & Chopade, B. A. (2024). Application of nanoparticles for management of plant viral pathogen: Current status and future prospects. *Virology*, *592*, 109998. PMID: 38301447

Wei, X., Xie, B., Wan, C., Song, R., Zhong, W., Xin, S., & Song, K. (2024). Enhancing soil health and plant growth through microbial fertilizers: Mechanisms, benefits, and sustainable agricultural practices. *Agronomy (Basel)*, *14*(3), 609.

Xu, S., Zhang, X., Xu, K., Wang, Z., Zhou, X., Jiang, L., & Jiang, T. (2022). Strawberry vein banding virus movement protein P1 interacts with light-harvesting complex II Type 1 like of fragaria vesca to promote viral infection. *Frontiers in Microbiology*, *13*, 884044. PMID: 35722273

Zhang, L., Chen, C., Li, Y., Suo, C., Zhou, W., Liu, X., & Yang, X. (2024). Enhancing aphid resistance in horticultural crops: A breeding prospective. *Horticulture Research*, *11*(12), uhae275. PMID: 39712868

Žigrayová, D., Mikušová, V., & Mikuš, P. (2023). Advances in antiviral delivery systems and chitosan-based polymeric and nanoparticulate antivirals and antiviral carriers. *Viruses*, *15*(3), 647. PMID: 36992356

Chapter 13
Phytotoxicity and Environmental Impact of Nanoparticles in Agriculture

Erdogan Esref Hakki
https://orcid.org/0000-0001-7147-7875
Selcuk University, Turkey

Tamay Şeker
https://orcid.org/0000-0001-7606-7226
Central Laboratory, Molecular Biology and Biotechnology R&D Center, Middle East Technical University, Turkey

Ulku Baykal
Giresun University, Turkey

ABSTRACT

This chapter examines the dual role of nanotechnology in agriculture, highlighting its potential to enhance crop productivity, nutrient efficiency, and pest control, alongside risks of phytotoxicity and ecological disruption. It discusses how engineered nanoparticles—metal, metal oxide, and carbon-based—enable targeted delivery and controlled release of agrochemicals, while factors such as size, surface chemistry, and concentration influence toxicity. The chapter also explores environmental effects on soil microbes, nutrient cycling, and bioaccumulation, proposing safer approaches like green synthesis, biodegradable carriers, and functionalized nanoparticles to ensure sustainable and responsible agricultural use.

14.1 INTRODUCTION

Nanotechnology has emerged as a transformative platform across various scientific domains, including medicine, energy, environmental sciences, and, more recently, agriculture (Deshmukh et al., 2023). Nanoparticles (NPs) have been investigated in the agricultural sector and various applications have been reported, such as nutrient delivery, controlling pests and diseases, reducing stress, bio sensing and improving crop productivity (Kamal et al., 2025). Their distinctive physicochemical characteristics, like

DOI: 10.4018/979-8-3373-4913-8.ch013

large ratio of surface and volume, reactivity, tunable size, and controllable release solution, make them appealing substitutes for conventional agrochemicals, mostly synonymous with ineffectiveness, toxicity, and resilience to the environment. Nano-fertilizers may increase the resistance to nutrient uptake, minimize leaching losses and diminish the excessive exploitation of chemical fertilization, whereas nano-pesticides provide specific pest control during which there may be reduced residue issues (Anas, Khattak et al., 2025). Although they have these benefits, there is growing alarm about the non-target effects of nanoparticles on plants, soil life, non-target organisms, and the health of the entire agro ecosystem. In comparison to bulk materials, nanoparticles have new biological interactions because of their nanoscale size that enable them to penetrate cellular barriers, disrupt biochemical pathways, and stay within environmental compartments. This bivalence of use, so healthy on the one hand and so poisonous on the other, imperatively demands the revaluation of the use of GMFs in farming. Phytotoxicity is defined as the adverse or negative effect of chemical compounds, materials and environmental stress factors on the development, health and reproduction of plants (Goncharuk, E. A., & Zagoskina, N. V. 2023). The scope of phytotoxicity, when extended to nanoparticles, refers to the group structure, physiologic, biochemical and molecular changes that have been noted in plants when exposed to these engineered nanomaterials. Up to the point of plant death, the symptoms of phytotoxicity are an intellectual disability: seed germination, stunted seedling growth, chlorosis, oxidative injury, photosynthesis intervention, and blockage of nutrient uptake and, in the worst case, death (Geetha, 2019). Nanoparticles could affect plants in unexpected manners unlike the conventional agricultural chemicals. They are artists in terms of cell wall and membrane crossing, which allow them to localize in intracellular compartments (including chloroplasts, mitochondria, and nuclei) and there they are capable of disrupting electron transport chains, producing reactive oxygen species (ROS), and causing genotoxicity (Abbasi et al., 2025). The phytotoxicity is not constant in all the types of nanoparticles; phytotoxicity varies with particle size, composition, surface charge, concentration, length of exposure and plant species. The approach to phytotoxicity study is therefore a very complicated yet important endeavor in agricultural nanoscience. Farming experiments are naturally coupled with the surrounding environment, and any intervention, be it chemical or nanotechnological, has ripple effects on the soil, water, and the biodiversity (Rana et al., 2024). Nanoparticles deposited in the agricultural ecosystem might linger in the soils and cause a change in the microbial ecosystem, either filtered into groundwater, or accumulating in food chains. Zinc oxide nanoparticles (ZnONPs) could be beneficial to plant nutrition at lower dosages but detrimental at higher dosages, also impairing soil microbial and enzyme functioning (Liu et al., 2021). Likewise, silver nanoparticles (AgNPs), popular as used as anti-microbial agents, may harm useful soil bacteria and earthworms, which are proven to ensure soil fertility (Khan et al., 2023). Interactions of nanoparticles with organic matter, pH changes, and ionic strength and redox conditions influence the environmental fate and transport of nanoparticles. Nanoparticles within runoff have the potential to induce bioaccumulation in aquatic ecosystems within the algae, fish and high trophic tiers (Babaei et al., 2022). As farming is one of the significant interfaces between human action and the natural environment, research awareness of the ecological impact of nanoparticles use is important to guarantee ecological sustainability (Atanda et al., 2025).

14.1.1 Background of Nanotechnology in Agriculture

Nanotechnology can be described as the handling of matter at a nanoscale (1–100 nm) and its application has been an extremely quick, cutting-edge scientific industry that is now developing into a multidisciplinary stage with hypothesis transformations in agriculture (Das et al., 2025). Traditional agricultural systems are still overly dependent on chemical fertilizers, pesticides, and growth regulators which, although effective, have been criticized as inefficient in their efficiency in using nutrients, large-scale leaching, resistance to pests and environmental pollutants. Nanotechnology proposes new ways to counter down such challenges by using the physicochemical advantages of nanoparticles (NPs), which comprise their reduced dimensions, vast surface area, increased possibility of solubility, and reactivity (Szczyglewska et al., 2023). Nanotechnology has a wide range of uses in the agricultural field. Nano-fertilizers are also able to carry the nutrients to the present plants with efficiency due to either a slow or controlled release, cutting down the loss of nutrients and enhancing the efficiency of plants to take up nutrients. Nano-pesticides, in turn, have the potential to promote the effectiveness of pest control with the deployment of better adherence to the plant surface, controlled release of the active component and/or low toxicity of non-targets (Mishra & Kumar, 2023). In addition to crop protection and nutrition, nanoparticles have been explored as real-time soil health, plant stress and pathogen detector sensors. Metal and metal oxide nanoparticles, including zinc oxide (ZnO), titanium dioxide (TiO2) and silver (Ag) have exhibited a benefit in enhancing seed germination, photosynthesis, and resistance to biotic and abiotic stressors (Silva et al., 2022; Faisal et al., 2024). It is the properties which contribute to the attractiveness of nanoparticles which can become hazardous. They are capable of penetrating biological barriers and thus accumulating in the tissues of plants with an impact on biochemical processes and their response to the soil and water systems, unlike bulk materials. Therefore, whereas nanotechnology is seen as promising solutions to transform sustainable agriculture, there are uncertainties concerning the unintended effects, which touch on phytotoxicity through food safety, ecological balance, and food safety. Due to this dual approach, which is innovation vs. risk, the role of using nanoparticles in agriculture can be evaluated (Fu et al., 2020).

14.1.2 Relevance of Phytotoxicity Studies

Phytotoxicity investigation with respect to nanoparticles is critical in the balance between agricultural innovation and (ecological) safety. Phytotoxicity is negatively felt on the health of plants and is therefore defined by decline in germination, growth retardation, chlorosis, decreased photosynthetic rate, or oxidative injury (Geetha, 2019). As plants are the main consumers of agricultural inputs, they are the initial contact of the nanoparts put in the agricultural systems. It is thus important to understand the effect of nanoparticles on plant physiology in order to determine both their effectiveness and safety. Nanoparticles approach plants differently to the manner in which traditional agrochemicals approach plants (Fu et al., 2020). They can pass through cell walls and membranes due to their nanoscale size, built up in organelles like chloroplasts and mitochondria and alter metabolic pathways. Indicatively, overexpression of ZnO and Ag nanoparticles has been reported to result in reactive oxygen generation (ROS), oxidative stress, and DNA damage and cell death. Simultaneously, at this initial level of low doses, some nanoparticles can enhance growth or bind antioxidant defenses, and thus play a dual dose-dependent role which makes the risk evaluation difficult (Ashrafpour, S., & Ashrafpour, M. 2025). Phytotoxicity experiments are also suitable for predicting ecological. Nanoparticles in soil or water may remain, be Ba-transformed, or bioaccumulated, which could have bio modifications such as obstructing microorganisms in soil, insects which are beneficial, as well as other un-targeted organisms. Plants serve as the links between the environment and the next trophic levels. Therefore, comprehensive knowledge about nanoparticle-caused phytotoxicity conditions sheds some light on the processes on large scales in food chains and on food safety issues. In addition to this, there is beaming interest in nanotechnology producing agro-products in the world, but the regulatory mechanisms are not yet established (Kumari et al., 2023). In the absence of standardized phytotoxicity investigations, there would be no common basis upon which to determine safe concentration limits, procedures, as well as risk evaluation agreements. Therefore, it is both a scientific requirement and a regulatory and societal requirement to assess the relevancy of phytotoxicity in ensuring sustainable use of nanotechnology in agriculture (Mishra, et al., 2017).

14.1.3 Scope and Objectives

The study of phytotoxicity and environmental impact of nanoparticles in agriculture is wide and spans molecular, physiological, ecological, and regulatory levels. This chapter tries to dissect these facets in the best way to give consistent insights into the role that has been held by the laboratory experiment, description of greenhouse and field trials, in a bid to give a multidimensional picture of the transactions between nanoparticles and plants. Metal and metal oxide nanoparticles (ZnO, TiO_2, Fe_2O_3, and Ag) are of the greatest interest, as they are the most actively investigated and have high applications as an agronogative, but also are associated with toxicity. This chapter has threefold objectives. This is why, first, it attempts to explain the aspects by which nanoparticles engage plants, such as the route of uptake in plants, translocation between plants, and the localization. Through the knowledge of such mechanisms, it becomes feasible to distinguish between the harmful and the beneficial concentrations. Second, the chapter seeks to capture the range of phytotoxic effects encompassing, inhibition of germination and seedling development, photosynthetic inhibition, oxidative stress and damage to genes. The review indicates both a positive and negative effect, which stress the complexity of the interactions that nanoparticles have on plants. Third, the chapter does not limit itself to plant-level impacts but goes to discussing environmental impacts. It goes on to examine the effects of nanoparticles on soil ecosystem, microbial diversity, and water quality, and non-target organisms, thus placing phytotoxicity in the ecological context. Moreover, the discourse has combined insights into the risk evaluation techniques, regulatory schemes, and risk reduction mechanisms, like green synthesis and designing biodegradable nanoparticles. The gist of this chapter is finally, that the chapter gives the reader its balance point on what is properly known as an uncritical synthesis of what happens to be the currently existing knowledge concerning the subject of phytotoxicity and environmental consequences of nanoparticles in agriculture. In this way, it will be able to recognize research gaps, propose sustainable approaches to the application of nanoparticles and play a role in establishing the guidelines on the safe use of nanotechnology in current farming practice.

14.2 MECHANISMS OF NANOPARTICLE INTERACTION WITH PLANTS

Nanoparticles (NPs) introduced into agricultural systems interact with plants through complex processes that determine their uptake, distribution, and subsequent effects on plant physiology (Chen, 2018). These interactions are influenced by the physicochemical properties of the nanoparticles such as size, shape, charge, solubility, and surface modifications, as well as plant-specific traits including root architecture, leaf morphology, and metabolic activity (Sembada & Lenggoro, 2024). Understanding these mechanisms is fundamental for predicting both the beneficial and toxicological outcomes of nanoparticle exposure. Broadly, nanoparticle–plant interactions can be divided into three sequential phases: uptake (via roots or leaves), translocation and accumulation within plant tissues, and cellular/molecular interactions that govern physiological responses (Haq et al., 2024).

14.2.1 Uptake Pathways (Root Absorption, Foliar Entry)

Roots represent the primary gateway for nanoparticle entry into plants, as they are in constant contact with soil or nutrient solutions. Root absorption of nanoparticles is facilitated by apoplastic and symplastic pathways (Zaman et al., 2025). The apoplastic route allows nanoparticles to move through the cell wall and intercellular spaces until they reach the endodermis, where the Casparian strip may restrict further passage (Afzal et al., 2024). However, nanoparticles smaller than the pore size of cell walls (~5–20 nm) can traverse these barriers. The symplastic pathway involves entry into root epidermal cells via endocytosis or transporter-mediated processes, after which nanoparticles can move cell-to-cell through plasmodesmata (Sembada & Lenggoro, 2024). Factors such as root exudates, pH, ionic strength, and presence of organic ligands strongly influence nanoparticle dissolution and uptake efficiency (Yang et al., 2022). Foliar entry constitutes another important uptake pathway, particularly when nanoparticles are applied as sprays (Ding et al., 2023). Leaves provide access through stomata, trichomes, cuticular cracks, or hydrophilic channels in the cuticle. Stomata apertures (~10–25 μm) readily permit the penetration of nanoparticle suspensions, while hydrophilic pores within the cuticle can facilitate uptake of ultra-small particles (<5 nm). Surfactants and carriers often enhance foliar uptake by reducing surface tension and improving adhesion to leaf surfaces (Elahi et al., 2025). Once inside, nanoparticles may localize in the mesophyll and influence photosynthetic machinery. The efficiency of root versus foliar uptake is highly dependent on nanoparticle properties, plant species, and environmental conditions (Anas, Falak et al., 2025). For instance, metal oxide nanoparticles like ZnO and TiO_2 can be taken up efficiently through roots, whereas carbon-based nanomaterials may show higher foliar penetration (Hanif et al., 2024). These entry routes not only determine nanoparticle bioavailability but also their potential to induce phytotoxic or beneficial effects. (Table 1).

Table 1. Summary of nanoparticle (NP) types, concentrations, soil/plant systems, and reported effects. The table highlights how different NPs influence microbial activity, nutrient cycling, and soil fertility in agricultural contexts.

Sr No	NP Type	Plant / System	Uptake Routes	Key Findings / Relevance	References
1	Multiple (Ag, ZnO, TiO_2, CNTs)	Various model & crop plants	Root apoplast/ symplast; stomatal & cuticular foliar uptake	Review: size, charge, coating control uptake and transport patterns.	Wang X., et al. (2023). Nanoparticles in Plants: Uptake, Transport and Role of Size/Charge. Materials. https://www.mdpi.com/1996-1944/16/8/3097
2	Multiple	Various species	Apoplast & symplast root; stomatal-dominant foliar pathway	Explains crossing of cuticle and stomata as main foliar route; root apoplast/symplast detailed.	Wu H., et al. (2022). Nano-enabled agriculture: How do nanoparticles cross plant barriers? Current Opinion / PMC9700125 https://pmc.ncbi.nlm.nih.gov/articles/PMC9700125/
3	Metal oxides & carriers	Leaves of various crops	Stomata, cuticle, hydathodes; surfactant effects	Details surfactant and wax effects on foliar uptake; stomata size key determinant.	Hong, J., Wang, C., Wagner, D. C., Gardea-Torresdey, J. L., He, F., & Rico, C. M. (2021). Foliar application of nanoparticles: mechanisms of absorption, transfer, and multiple impacts. *Environmental Science: Nano*, *8*(5), 1196-1210.

continued on following page

Table 1. Continued

Sr No	NP Type	Plant / System	Uptake Routes	Key Findings / Relevance	References
4	Multiple (Au, Ag, TiO_2, ZnO)	Various crops	Comparison: foliar vs root translocation patterns	Shows different systemic patterns and phytophysiological impacts from foliar vs root exposures.	Djanaguiraman, M., Anbazhagan, V., Dhankher, O. P., & Prasad, P. V. (2024). Uptake, translocation, toxicity, and impact of nanoparticles on plant physiological processes. *Plants*, *13*(22), 3137.
5	TiO_2	Lettuce	Root and foliar uptake; stomata entry	Demonstrates dose- and exposure-route dependent outcomes in lettuce.	Hu, J., Wu, X., Wu, F., Chen, W., Zhang, X., White, J. C., ... & Wang, X. (2020). TiO 2 nanoparticle exposure on lettuce (Lactuca sativa L.): Dose-dependent deterioration of nutritional quality. *Environmental Science: Nano*, *7*(2), 501-513.
6	CeO_2	Hydroponic cucumber	Root uptake; xylem & phloem transport	Split-root shows CeO2 can translocate via xylem and phloem to shoots.	Ma, Y., He, X., Zhang, P., Zhang, Z., Ding, Y., Zhang, J., ... & Yang, K. (2017). Xylem and phloem based transport of CeO2 nanoparticles in hydroponic cucumber plants. Environmental Science & Technology, 51(9), 5215-5221.
7	CeO_2	Cucumber	Root uptake and biotransformation; translocation	Reports biotransformation of CeO2 and evidence for plant-mediated redox changes.	Zhang P., et al. (2012). Biotransformation of ceria nanoparticles in cucumber. ACS Nano https://pubs.acs.org/doi/10.1021/nn303543n
8	Au NPs	Barley	Root penetration; size-dependent root fate	Size-dependent penetration of AuNPs into root tissues and internal localization.	Milewska-Hendel A., et al. (2017). Fate of neutral-charged AuNPs in barley roots. Sci Rep. https://www.nature.com/articles/s41598-017-02965-w
9	AuNPs	Barley	Root cell wall & apoplastic modification	AuNPs induce cell-wall changes that could affect NP uptake and root function.	Milewska-Hendel A., Sala K., Gepfert W., Kurczyńska E. (2021). AuNPs modify barley root cell walls. Cells. https://pmc.ncbi.nlm.nih.gov/articles/PMC8393560/
10	Metal oxides	Leaves & plants (model)	Stomatal & cuticular pathways discussed	Conceptual and experimental overview of metal-oxide foliar entry routes.	Li, R., Zhang, R., Li, Y., Liu, C., Wang, P., Sun, H., & Wang, L. (2024). Foliar uptake and distribution of metallic oxide nanoparticles in maize (zea mays L.) leaf. Environmental Science & Technology, 58(38), 16994-17003.
11	Nanocarriers	Leaves (model species)	Leaf-side and stomatal pathways control delivery	Leaf-side (adaxial/abaxial) and stomatal aperture influence foliar delivery efficiency.	Kohay H., et al. (2024). Nanocarrier foliar uptake pathways affect delivery. Communications Biology. PMC11494269
12	ZnO	Rice	Stomatal foliar uptake and translocation	Foliar ZnO efficiently enters leaves and moves into developing tissues improving yield.	Wang, S., Fang, R., Yuan, X., Chen, J., Mi, K., Wang, R., ... & Zhang, H. (2023). Foliar spraying of ZnO nanoparticles enhanced the yield, quality, and zinc enrichment of rice grains. Foods, 12(19), 3677.
13	ZnO	Rice	Phyllosphere-assisted foliar uptake	Phyllosphere interactions enhance foliar ZnO retention and uptake, boosting efficacy.	Guo S., et al. (2024). ZnO NPs cooperate with phyllosphere to boost rice (PNAS 2024). https://www.pnas.org/doi/10.1073/pnas.2414822121

continued on following page

Table 1. Continued

Sr No	NP Type	Plant / System	Uptake Routes	Key Findings / Relevance	References
14	Multiple	Multiple crops	Review: root vs foliar uptake mechanisms	Useful synthesis for designing foliar nanofertilizers and understanding uptake limits.	Bhati, V. S., Singh, P. K., Parmar, Y. J., & Rokins, P. G. (2025). Foliar uptake and translocation of nanoparticles in the plants: a review. *Journal of Plant Nutrition*, 1-38.
15	MWCNTs	Cucumber seedlings	Root uptake; endocytosis of CNTs	13C labeling confirmed MWCNT root uptake and internal movement in cucumber.	Montes de Oca-Vásquez, G. (2021). Impacto de concentraciones ambientalmente relevantes de nanopartículas de plata en la comunidad microbiana del suelo.
16	CNTs	Various	Foliar and root CNT interfaces; penetration mechanisms	Summarizes CNT uptake routes and physical interactions with plant tissues.	Tiwari DK., et al. (2014). Interfacing CNTs with plants: uptake and effects. Appl Nanoscience. https://link.springer.com/article/10.1007/s13204-013-0236-7
17	Multiple	Multiple species	Apoplast/symplast; stomatal & cuticle	Provides broad mechanistic framework across NP classes for plant uptake.	Yang J., et al. (2017). Interactions between nanoparticles and plants: mechanisms. J Nanobiotechnology https://www.tandfonline.com/doi/full/10.1080/17429145.2017.1310944
18	Ag, CeO_2	Model systems	Root retention; limited shoot translocation	Tracing studies show strong root retention; translocation varies by NP chemistry.	Bhaduri, B., & Nath, J. (2021). Tracing of Ag-and CeO2 based engineered nanoparticles in cucumber plant system. *Journal of environmental chemical engineering*, *9*(4), 105778.
19	CuO	Wheat	Rhizosphere metabolites increase CuO dissolution and influence uptake	Shows rhizosphere metabolites increase CuO dissolution and influence uptake.	Ibrahim, A. S., Ali, G. A., Hassanein, A., Attia, A. M., & Marzouk, E. R. (2022). Toxicity and uptake of CuO nanoparticles: evaluation of an emerging nanofertilizer on wheat (Triticum aestivum L.) plant. *Sustainability*, *14*(9), 4914.
20	CuO	Wheat	Root adsorption; uptake into root tissues	CuO NPs strongly interact with root surfaces; uptake influenced by dissolution.	Huang, Y., Bai, X., Li, C., Kang, M. E., Weng, Y., & Gong, D. (2022). Modulation mechanism of phytotoxicity on Ipomoea aquatica Forssk. by surface coating-modified copper oxide nanoparticles and its health risk assessment. Environmental Pollution, 314, 120288.
21	AgNPs	Various crops	Stomatal, cuticular, hydathodes for foliar uptake	Review indicates foliar Ag often results in stem accumulation and systemic effects.	Tripathi, D. K., Tripathi, A., Shweta, Singh, S., Singh, Y., Vishwakarma, K., ... & Chauhan, D. K. (2017). Uptake, accumulation and toxicity of silver nanoparticle in autotrophic plants, and heterotrophic microbes: a concentric review. *Frontiers in microbiology*, *8*, 07.
22	AgNPs	Aquatic/wetland plants	Uptake & translocation in aquatic plants	AgNP behavior in aquatic plants shows uptake and potential trophic transfer.	Lopez-Herrera A., et al. (2024). AgNP uptake in aquatic/wetland plants. Environments. MDPI https://www.mdpi.com/2076-3298/11/12/297
23	ZnO & CuO	Wheat	Root vs foliar dominance; stomatal importance for foliar	Metabolomics links root uptake of ZnO/CuO to specific metabolic pathway changes.	Wu, P., Wang, Z., Adusei-Fosu, K., Wang, Y., Wang, H., & Li, X. (2024). Integrative chemical, physiological, and metabolomics analyses reveal nanospecific phytotoxicity of metal nanoparticles. *Journal of Environmental Management*, *354*, 120338.
24	TiO_2	Multiple species	Root uptake & metabolomic shifts after NP exposure	TiO2 shows hormetic uptake: low-dose beneficial, high-dose toxic effects.	Pandey, K., & Dasgupta, C. N. (2025). Role of nanobionics to improve the photosynthetic productivity in plants and algae: an emerging approach. *3 Biotech*, *15*(4), 1-21.

continued on following page

Table 1. Continued

Sr No	NP Type	Plant / System	Uptake Routes	Key Findings / Relevance	References
25	ZnO	Rice	Size/charge-dependent uptake; hormetic responses	Foliar ZnO spray can be agronomically effective via stomatal uptake.	Wang, R., Mi, K., Yuan, X., Chen, J., Pu, J., Shi, X., ... & Zhang, H. (2023). Zinc oxide nanoparticles foliar application effectively enhanced zinc and aroma content in rice (Oryza sativa L.) grains. *Rice, 16*(1), 36.
26	CeO_2	Cucumber	Foliar spray more effective than root for some responses	Foliar CeO2 sprays more effective than root for mitigating salt stress in cucumber.	Chen, L., Peng, Y., Zhu, L., Huang, Y., Bie, Z., & Wu, H. (2022). CeO2 nanoparticles improved cucumber salt tolerance is associated with its induced early stimulation on antioxidant system. *Chemosphere, 299*, 134474.
27	MWCNTs	Cucumber	Root absorption; apoplastic and endocytosis	MWCNTs alter root uptake of hydrophobic organics and can enter roots.	Shen, X., Li, S., Zhang, H., Chen, W., Yang, Y., Li, J., ... & Wang, X. (2018). Effect of multiwalled carbon nanotubes on uptake of pyrene by cucumber (Cucumis sativus L.): Mechanistic perspectives. *NanoImpact, 10*, 168-176.
28	Inorganic NPs	Leaves (review)	MWCNT effects on uptake of co-contaminants; root route	Review highlighting key NP properties (size, charge, coating) that control foliar uptake.	Avellan, A., Yun, J., Morais, B. P., Clement, E. T., Rodrigues, S. M., & Lowry, G. V. (2021). Critical review: role of inorganic nanoparticle properties on their foliar uptake and in planta translocation. *Environmental science & technology, 55*(20), 13417-13431.
29	ZnO	Pepper	NP physico-chemical drivers of foliar uptake	Zinc-phosphate coating increases foliar mobility and uptake compared to bare ZnO.	Rodrigues S., et al. (2024). Zinc phosphate-coated ZnO NPs: uptake & mobility in pepper. Environ Sci Technol. https://pmc.ncbi.nlm.nih.gov/articles/PMC11494269/
30	CeO_2	Hydroponic cucumber	Foliar and root; coating alters mobility	Split-root data corroborates xylem/phloem systemic transport of CeO_2.	Khan, S., & Musarrat, J. Bilal Ahmed, Asfa Rizvi, Khursheed Ali, Jintae Lee, Almas Zaidi, Mohammad.

14.2.2 Translocation and Accumulation in Plant Tissues

In agricultural methods, nanoparticles are frequently applied by spraying them onto the surface of leaves, where they settle and are absorbed by plants through the cuticle or stomata (Xueran Wang et al., 2023). The waxy cuticle of the leaf epidermis, composed mainly of wax, cutin, and pectin, acts as a protective layer to minimize water loss during plant growth and serves as the primary natural barrier against the entry of nanoparticles (X. Wang et al., 2023). However, the waxy *stratum corneum* contains two distinct types of channels: one hydrophilic and the other lipophilic, with diameters ranging from 0.6 nm to 4.8 nm (Singh, Sharma, et al., 2024). Hydrophilic channels allow the diffusion of hydrophilic nanoparticles smaller than 4.8 nm, while lipophilic channels enable the absorption of lipophilic nanoparticles through diffusion and infiltration. Recently, (Sidhu et al., 2024) demonstrated that carbon dots smaller than 2 nm could penetrate cotton leaves via the cuticular pathway using confocal fluorescence microscopy with high spatial and temporal resolution (Sidhu et al., 2024). Due to the limited size of the cuticle's pore channels, the absorption of nanoparticles through the epidermis is restricted. Some studies have reported that nanoparticles can accumulate in the leaf epidermis and vascular tissue after being applied to the leaf surface (Singh, Agrawal, et al., 2024). Additionally, numerous studies have observed that nanoparticles can be transported to other plant tissues (Khan et al., 2025). Researchers have suggested that nanoparticles might also be absorbed through an alternative route, known as the stomata pathway (X. Wang et al., 2023) (**Figure 1**).

Figure 1. Nanoparticle uptake and transport pathways in plants through root absorption, leaf penetration, and vascular system movement via apoplastic and symplastic routes

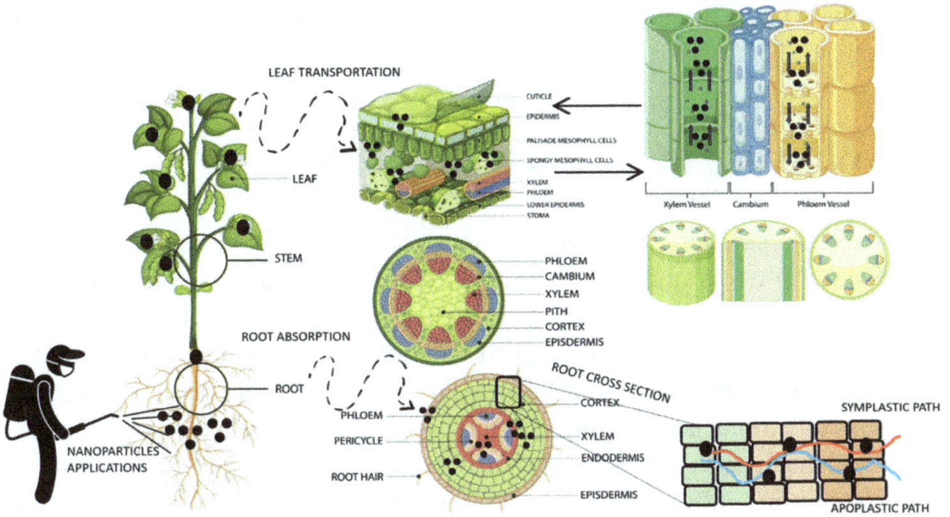

14.2.3 Cellular and Molecular Interactions

The biological effects of exposure are ascertained at the molecular level and at the cellular level of interaction of nanoparticles with plants (Khan, Adam et al., 2019). Upon internalization, nanoparticles may be targeted to organelles (e.g. chloroplasts, mitochondria, and nuclei), ultimately perturbing vital metabolic processes. Using TiO_2 nano-particles as an example, they can modulate electron flux in chloroplasts, thus changing photosynthetic production, and Ag nano-particles can manipulate processes in mitochondria by coupling to the enzymes and triggering the release of reactive oxygen species (ROS) (Moshirian Farahi et al., 2023). It also has mediated plant signaling through the nanoparticles. The most prevalent response is the generation of ROS, leading to oxidative stress, completely rupturing lipids, proteins and DNA. Much lower concentrations of ROS, however, can also serve as signal molecules, with superoxide dismutase (SOD), catalase (CAT), and peroxidases (POD) subsequently being activated. It is this duality of ROS that reveals how sensitive interactions are between the two that are profitable and, in other cases, poisonous (Bardaweel et al., 2018). Genetically, nanoparticles may induce breaks of DNA strands or chromosomal alterations or modifications within the genes (Ramos-Pan et al., 2024). Transcriptomic study indicates that nanoparticles can cause a dose of the nanoparticles to elicit stress-responsive genes, or secondary metabolism, and defense-related proteins. As an example, the ZnO nanoparticles have been shown to cause transcription factors related to the protective effect of oxidative stress, but in abnormally high concentrations, have been shown to causally induce genes for apoptosis through hydrosolvency (Akhtar et al., 2012; Wang et al., 2014). The cellular reactions are also made complex by the interaction between proteins and nanoparticles. It is possible that by coating surfaces of nanoparticles with films of protein, one can change patterns of cellular uptake, and also reactivity (Saptarshi et al., 2013). The nanoparticles, too, can chelate essential nutrition, modify ion homeostasis, and disrupt hormone signaling and modify plant growth and physiological development. Cellular and molecular responses, taken in totality, are the point of contact, wherein the exposure to nanoparticles has physiological effects (Rasheed et al., 2022). It is essential that, besides prediction of phytotoxicity, mechanisms rely on these interactions to design nanoparticles with reduced risks and maximum farming benefits (Saleem et al., 2025).

14.3 PHYTOTOXIC EFFECTS OF NANOPARTICLES

When engineered nanoparticles (NPs) interact with plants, various biological effects may be noticed that can be diverse in range and can be minute effects on growth moderate at low doses through extreme effects on morphology and physiology at high doses (YingYing et al., 2025). This kind of dualistic nature of nanoparticles; they are good, and at other times, they are very harmful, making the role of nanoparticles in the agricultural world an object of hope and, at the same time, few things to fear. Nanoparticles may also enhance planting, nutrient absorption or stress resistance due to their small size and reactivity, which permits them to augment the bioavailability of specific nutrients in the lowest exposure ranges (Zhang et al., 2024). However, when concentrations are excessive or exposure is excessive, the same properties will lead to toxicity and disrupt essential function of cells and impede normal development of plants. Phytotoxic effects of the nanoparticles also appear at different levels of biologic functions. Morphologically plants can exhibit a stunted development of roots and shoots, chlorosis or necrotic leaves (Budhani et al., 2019). Sometimes, physiologically, there are processes that are normal (photosynthesis, respiration, and transpiration), but are usually deficient. At the level of biochemistry, nanoparticles can interfere with antioxidant machines, enzyme activity and nutrient processing but, at the genetic and molecular one, damage in the DNA, and changes in gene transcription and in chromosomes are well-reported (Etesami et al., 2021). They are no longer consistently uniform in terms of extent and severity, instead ironically dependent on the innate properties of the nanoparticles, including their size, makeup, surface charge, solubility, and toxic metal ion liberation ability. In addition, plant factors such as plant species, level of growth, and plant genotype are a definite predictor of sensitivity (Fischer et al., 2014). Environmental factors that influence the bioavailability and toxicity of the nanoparticles also include the soil type, pH, organic matter and co-contaminants. This complexity necessitates the use of integrated study as nanoparticle-plant-environmental interactions would be incorporated when studying phytotoxicity. Outlined in the subsections that follow are some of the main phytotoxic impacts reported exposing plants when they are subjected to nanoparticles.

14.3.1 Germination and Seedling Growth Inhibition

Seed germination and early seedling are these critical points in the life cycle of plants, which are sensitive in the case of nanoparticle toxicity (Szőllősi et al., 2020). Various studies have demonstrated that nanoparticles do have the potential to stimulate or inhibit germination according to the type and concentration. Low concentrations of metal oxide nanoparticles of ZnO, CuO and TiO_2 frequently improve germination through actions like stimulating water uptake, stimulating metabolic activity, or supplying of micronutrients (Donia & Carbone, 2023; Rudi et al., 2025). When used in high concentrations, they retard germination, decrease the percentage of germination and inhibit the radicle and plumule extension. The physiological processes that potentially act to suppress germination involve restriction of seed acculturation of water by nanoparticle coats on seeds, inhibition of protein enzyme reactions required to mobilize starch, truncation of hormonal signaling such as gibberellins and abscisic acid (Fu et al., 2024). Duly, silver nanoparticles have been described to inhibit amylase activity, hence, reducing the amount of energy required to initiate seedling emergence (Khan et al., 2023). In addition, nanoparticles can cause oxidative stress of the embryonic tissues, further depriving cells of cell division and elongation. The seedling growth inhibition is particularly observed in studies where seedling root length is tested directly as the root directly touches the nanoparticles in the soil (Wohlmuth et al., 2022). Symptoms are observed in the shortening of the primary roots, decreased lateral root density and poor formation of root hair. The effects have an effect of decreasing the nutrient and water uptake, which diminish the seedling vigor and growth (Zulfiqar, 2021). In agriculture, nanoparticles may be used as growth stimulants at low concentrations, whereas serious possibilities of toxicity at high concentrations imply the significance of analyzing cortex concentration-dependent in high concentration scenarios.

14.3.2 Alterations in Photosynthesis and Chlorophyll Content

One of the major processes affected by exposure to nanoparticles in physiology is photosynthesis. Plants receiving nanoparticles usually exhibit significant alterations in chlorophyll content, photosynthetic efficiency and gas exchange parameters (Li et al., 2024). Optimal doses of some nanoparticles such as TiO_2 and CeO_2 have the potential to enhance the absorption of light, transport of electrons, and stability of chlorophyll, thus increasing photosynthetic rate (Preetha et al., 2023; Waani et al., 2021). Nonetheless, an over accumulation has mostly negative implications. Given that nanoparticles have the possibility to directly contact chloroplast structures, they are able to cause structural deformity, disorganizing thylakoids, and a decrease in stacks of grana. This affects the photorespiratory processes of photosynthesis. Indicatively, ZnO nanoparticles have been known to cause a reduction in the values of chlorophyll a and b, reduced Fv/Fm values (which is a measure of PSII efficiency) and an impairment of CO_2 assimilation at high concentrations (Wang et al., 2016). Such disruptions are found as chlorosis and necrosis of leaves. Nanoparticles cause oxidative stress indirectly, upper arrested photosynthetic processes (e.g. chlorophyll degradation) and damage photosynthetic proteins (Rubisco). Moreover, nanoparticles can cause a disturbance in stomata conductance, transpiration and influence the fixation of carbon and water use efficiency. Nanoparticles deposited by foliar are also able to block stomata, resulting in low CO2 uptake. Interestingly, the effects of nanoparticles on photosynthesis are found to be extremely species specific (Bhati et al.). Increase in protective pigments (carotenoids, anthocyanins) production, and stimulation of antioxidant enzymes, which alleviate chlorophyll breakdown, are some examples of tolerance mechanisms seen in certain plants. Nevertheless, a challenge to photosynthetic efficiency due to nanoparticles is a matter of serious phytotoxic concern that has direct consequences on crop yields and biomass (Figure 2).

Figure 2. Positive and negative impacts of nanoparticles on plants, highlighting growth promotion, nutrient uptake, and photosynthesis enhancement versus oxidative stress, enzyme imbalance, and yield reduction

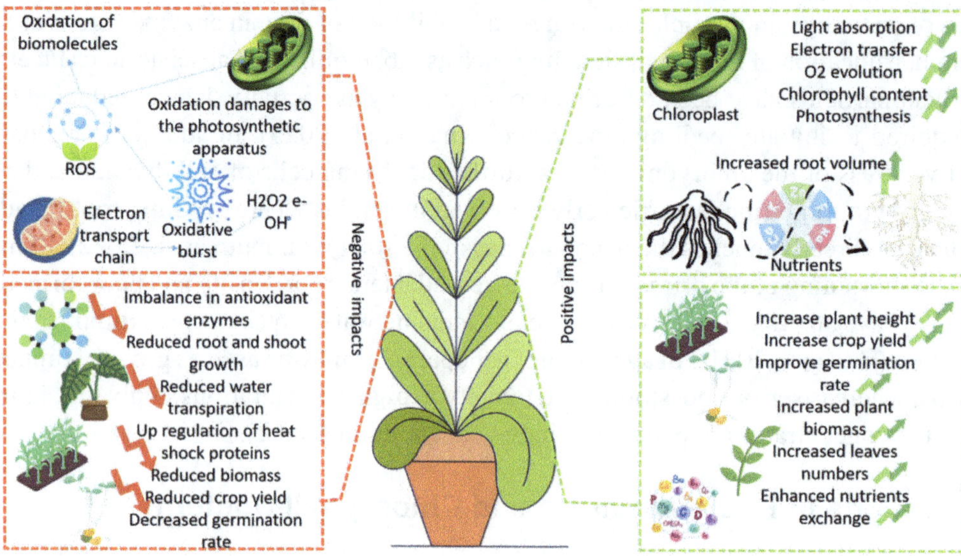

14.3.3 Oxidative Stress and Reactive Oxygen Species (ROS) Generation

Overproduction of reactive oxygen species (ROS) can be considered one of the most universal phytotoxic mechanisms of nanoparticles. Ag, ZnO, and CuO nanoparticles have redox potential and may be used to produce superoxide anions, hydroxyl radical, and hydrogen peroxide in plant tissues (Bhati et al.). Although, at low basal levels, ROS are involved in signaling, their uncontrolled build-up causes oxidative stress which harms cellular particles. ROS damages the membranes in lipids, leading to lipid peroxidation and disruption of the membrane integrity, expressed in elevated amounts of malondialdehyde (MDA). The oxidation of proteins and inactivations of the enzymes disrupt metabolism and negatively affect genetic stability, respectively. Symptoms that will be visible are yellowing of leaf, necrosis, and premature senescence. Enzymatic and non-enzymatic antioxidant defense systems are activated by the response of plants to ROS stress (Chinmayi et al., 2025). Most of the enzymes unregulated include superoxide dismutase (SOD), catalase (CAT) and peroxidases (POD) following exposure to nanoparticles. Naive antioxidants also play a role in neutralizing the ROS, though they are non-enzymatic in nature, i.e. ascorbate, glutathione, carotenoids. But at levels where production of ROS surpasses the ability to detoxify, oxidative damage will take preeminence, causing irreversible injury (Rudenko et al., 2023). The levels of ROS formation in relation with antioxidant defense can make nanoparticles either stress inducers or stimulants. An example is that a low dose of nanoparticles can prime plants by increasing their antioxidant capacity and subsequent tolerance to liability. This homeotic activity makes a determination of toxicity more difficult, which highlights the significance of dose-response designs when examining the effects of nanoparticles.

14.3.4 DNA Damage and Genotoxicity

Nanoparticles have gained growing concern with regard to genotoxicity in plants with concern about inheriting genetic changes. Nanoparticles may cause DNA damage due to direct contact between the genetic material and the nanoparticles or indirectly by oxidative stress, which is produced by ROS (Rim et al., 2013). The typical indicators of genotoxicity that are reported are chromosomal aberrations, formation of micronucleus, breakages of DNA strands, and abnormal expression profiles of the genes. Application-based research conducted to investigate the application of comet assays and root tip cell RAPD-PCR methods demonstrates that nickel nanoparticles (AgNPs), zinc nanoparticles (ZnONPs) and titanium nanoparticles (TiO_2NPs) can result in severe fragmentation in the DNA of root tip cells when exposed (Fais et al., 2024). Mitosis has been reported to bring about chromosome abnormality that comprises delaying chromosomes, bridges and stickiness, suggesting the disruption of spindle apparatus, and chromosomal segregation (Potapova & Gorbsky, 2017). Alternatively, nanoparticles may react with DNA on the molecular scale, binding to DNA or interrupting DNA replication and transcription, or even harming epigenetic processes such as DNA methylation. Transcriptomic data reveal an up regulation of an incretion response gene, and down regulation of a cell regulation gene, photosynthetic and developmental genes (Ali et al., 2022). Reliance on genotoxic, too, questions the ecological and food safety effects of nanoparticles in the long run, even when the mutations are inherited into the next generation. Despite the mechanism of the methods of plant DNA repair, chronic exposure might overwhelm the mechanism within the system that induces progressive genetic instability, despite the manner by which it oversees reacting to the SoA (Raina et al., 2021). These are not exclusive genotoxicity to the growth of plants, as they may affect crop genetic resources, as well as plant biodiversity and the sustainability of agriculture.

14.3.5 Effects on Nutrient Uptake and Metabolism

The effects of nanoparticles on nutrient intake and metabolism could be very serious and either result in additional sources of nutrients or disturbing ion homeostasis. ZnO, Fe_2O_3, Mn_3O_4, are metallic nanoparticles that could release essential micronutrients in ionic efflloresce, and improve the plant nutrition when the doses are low (Khanizadeh et al., 2024; Ravishankar et al., 2025). These are usually discovered to enhance nitrogen assimilation, synthesis of chlorophyll and the activity of enzymes. However, at very high concentrations, nanoparticles may exert a competitive blockage effect on uptake of vital nutrients in the body by either blocking transporters, or altering membrane permeability (Brandelli, 2020). A case in point is with AgNPs and CuO NPs, making the nutrients phosphorus and potassium be upset, leading to nutrient imbalance. Some nanoparticles entrap relevant ions within the rhizosphere that deteriorates their bioavailability further. The other impact that nanoparticles have is their ability to disrupt the metabolic pathways through cofactor-binding or attaching to enzymes. The affected tissues, which are soft tissues, can encompass, among others, impaired protein production, disturbed carbohydrate metabolism, and impaired hormone signaling. Such would be an exogenous addition of ZnONPs with resultant root architecture deformation and cytokinin signaling system disruption in the shoot of the plant (Zafar et al., 2016). Information Nanoparticle bi-phasic surprisingly, the interaction between nanoparticles with nutrients takes a bi-phasic geometry, stimulatory and then inhibitory above the optimum concentration. Such dualistic quality underscores the significance of nanoparticles as nano-fertilizer even when it is established that the application of nanoparticles is properly regulated. This is critical to understanding how to create nanoparticles with desired outcomes of enhancing efficiency of nutrient use without producing an unbalanced social agreement or toxicity.

14.4 FACTORS INFLUENCING PHYTOTOXICITY

14.4.1 Type and Composition of Nanoparticles (Metal, Metal Oxide, Carbon-Based)

The reactivity and toxicity of nanoparticles depend significantly on the chemical composition of nanoparticles. Nanoparticles made of metal (e.g., silver (AgNPs), copper (CuNPs), and gold (AuNPs) can leak ions that can disregard cellular redox homeostasis and the activity of enzymes (Ameh et al., 2022). Among the typically studied metal oxide nanoparticles are zinc oxide (ZnONPs) and titanium dioxide (TiO_2NPs) both behave as micronutrients when present at low concentrations, and lead to oxidative stress and membrane damage when in high concentrations. The carbon-based nanomaterials (carbon nanotubes, graphene oxide, fullerenes) are typically less soluble; however, they are able to infiltrate cell walls, disrupting the structural integrity and water homeostasis (Kharlamova & Kramberger, 2023). Each type has intrinsic toxicity, which is not just determined by the chemical constituents; or, it is also related to its combination with biological molecules and environmental matrices.

14.4.2 Concentration and Dosage Effects

There is a dose-dependent relationship between phytotoxicity. Low concentrations of nanoparticles are capable of inducing growth by increasing nutrient absorption, enzyme activity and stress resiliency known as hormesis in certain cases (Răcuciu et al., 2025). But, nanoparticles overload cellular defense, forming reactive oxygen species (ROS), producing genotoxic and inhibiting photosynthesis when the nanoparticles accumulate to high concentrations. Indicatively, ZnONPs were found to enhance soybean germination at 25 mg /L and suppress seedling growth at 200mg/L (Basit et al., 2023). The range between protective and toxic concentrations differs in the variety of nanoparticles and flora species, another factor that supports the need to pay close attention to optimized doses in agriculture. (Table 2).

Table 2. Studies showing dose-dependent phytotoxic effects of nanoparticles in plants, highlighting growth stimulation at low concentrations and toxicity at higher doses

	NP type	Plant / system	Concentrations tested (reported)	Study
1	ZnO NPs	Rice (As-stressed)	25 mg/L effective	Ahmad, P., Alyemeni, M. N., Al-Huqail, A. A., Alqahtani, M. A., Wijaya, L., Ashraf, M., ... & Bajguz, A. (2020). Zinc oxide nanoparticles application alleviates arsenic (As) toxicity in soybean plants by restricting the uptake of as and modulating key biochemical attributes, antioxidant enzymes, ascorbate-glutathione cycle and glyoxalase system. *Plants*, *9*(7), 825.
2	ZnO, others	Various crops	µg–mg/L ranges	Donia, D. T., & Carbone, M. (2023). Seed priming with zinc oxide nanoparticles to enhance crop tolerance to environmental stresses. *International Journal of Molecular Sciences*, *24*(24), 17612.
3	Multiple (Ag, ZnO, TiO2, etc.)	Multiple species	Wide ranges summarized	Iavicoli, I., Leso, V., Fontana, L., & Calabrese, E. J. (2018). Nanoparticle exposure and hormetic dose–responses: An update. *International Journal of Molecular Sciences*, *19*(3), 805.)
4	ZnO NPs & MWCNTs	*Arabidopsis thaliana* seedlings	µg–mg/L	Yang, S., Yin, R., Wang, C., Yang, Y., & Wang, J. (2023). Phytotoxicity of zinc oxide nanoparticles and multi-walled carbon nanotubes, alone or in combination, on Arabidopsis thaliana and their mutual effects on oxidative homeostasis. *PLoS One*, *18*(2), e0281756.
5	ZnO NPs	Pakchoi *(Brassica chinensis)*	mg/kg in soil	ShShen, M., Liu, W., Zeb, A., Lian, J., Wu, J., & Lin, M. (2022). Bioaccumulation and phytotoxicity of ZnO nanoparticles in soil-grown Brassica chinensis L. and potential risks. *Journal of Environmental Management*, *306*, 114454.
6	ZnO NPs	Soybean	Study-specific concentrations	Salac, E. L. O., Alvarez, M. R., Gaurana, R. S., Grijaldo, S. J. B., Serrano, L. M., Juan, F. D., ... & Nacario, R. C. (2022). Biological assay-guided fractionation and mass spectrometry-based metabolite profiling of Annona muricata L. cytotoxic compounds against lung cancer A549 cell line. *Plants*, *11*(18), 2380.
7	ZnO, TiO_2	Soybean (in vitro)	100–200 nm particles, conc. Varied	Leopold, L. F., Coman, C., Clapa, D., Oprea, I., Toma, A., Iancu, . D., ... & Coman, V. (2022). The effect of 100–200 nm ZnO and TiO2 nanoparticles on the in vitro-grown soybean plants. *Colloids and Surfaces B: Biointerfaces*, *216*, 112536.
8	ZnO nano-coatings	Soybean, peanut seeds	1000 mg/L soak	Montanha, G. S., Rodrigues, E. S., Marques, J. P. R., de Almeida, E., Colzato, M., & Pereira de Carvalho, H. W. (2020). Zinc nanocoated seeds: an alternative to boost soybean seed germination and seedling development. *SN Applied Sciences*, *2*(5), 857.
9	Ag, Au NPs	*Allium cepa* (root assay)	5–40 mg/L	Debnath, P., et al. (2018). Cytogenetic effects of silver and gold nanoparticles on *Allium cepa* root cells. *Journal of Genetic Engineering and Biotechnology*, 16(2), 519–526. https://doi.org/10.1016/j.jgeb.2018.07.007.
10	AgNPs	Various crops	<0.01–100 mg/L	Yan, A., & Chen, Z. (2019). Impacts of silver nanoparticles on plants: a focus on phytotoxicity and underlying mechanism. *International Journal of Molecular Sciences*, 20(5), 1003. https://doi.org/10.3390/ijms20051003.
11	PVP-AgNP (Argovit™)	*Allium cepa*	5–10 µg/mL positive; higher toxic	Casillas-Figueroa, G. O., et al. (2020). Argovit™ PVP-stabilized silver nanoparticles effects on *Allium cepa*: plant-growth promotion without cyto/genotoxic damage. *Nanomaterials*, 10(7), 1386. https://doi.org/10.3390/nano10071386
12	AgNPs	Lettuce	0.01 mg/L Ag+ vs 25–100 mg/L AgNPs	Hasan, M., Mehmood, K., Mustafa, G., Zafar, A., Tariq, T., Hassan, S. G., Loomba, S., Zia, M., Mazher, A., Mahmood, N., & Shu, X. (2021). Phytotoxic evaluation of phytosynthesized silver nanoparticles on lettuce. *Coatings*, 11(2), 225. https://doi.org/10.3390/coatings11020225

continued on following page

Table 2. Continued

	NP type	Plant / system	Concentrations tested (reported)	Study
13	CuO NPs	Rice	≥62.5 mg/L	Yang, Z., Xiao, Y., Jiao, T., Zhang, Y., Chen, J., & Gao, Y. (2020). Effects of copper oxide nanoparticles on the growth of rice (Oryza sativa L.) seedlings and the relevant physiological responses. *International journal of environmental research and public health, 17*(4), 1260.
14	CuO NPs	Wheat *(Triticum aestivum)*	EC50 ≈0.94 mg/L	Ibrahim, A. S., Ali, G. A., Hassanein, A., Attia, A. M., & Marzouk, E. R. (2022). Toxicity and uptake of CuO nanoparticles: evaluation of an emerging nanofertilizer on wheat (Triticum aestivum L.) plant. *Sustainability, 14*(9), 4914.
15	TiO_2 NPs	Lettuce *(Lactuca sativa)*	50–400 mg/L	Hu, J., Wu, X., Wu, F., Chen, W., Zhang, X., White, J. C., ... & Wang, X. (2020). TiO 2 nanoparticle exposure on lettuce (Lactuca sativa L.): Dose-dependent deterioration of nutritional quality. *Environmental Science: Nano*, 7(2), 501-513.
16	TiO_2 NPs	Multiple species	0–2500 mg/L	Albaladejo-Marico, L., Thameur, A., Garcia-Martinez, A., Carvajal, M., & Yepes-Molina, L. (2025). Nanoparticles as modulators of stress tolerance: Physiological and molecular insights into TiO_2 and ZnO effects in Cucumis melo L. subjected to salt shock. *Environmental Technology & Innovation, 38*, 104101.
17	Ag, ZnO, TiO_2, CNTs	Terrestrial plants	Multiple	Allará, C., Ciccone, G., Ciocca, M., Vasquez, S., Ibba, P., Maver, M., ... & Petti, L. (2025). Electronic Nanomaterials for Plants: A Review on Current Advances and Future Prospects. *Advanced Electronic Materials*, 2500080.
18	Multiple NPs	Crops (review)	Broad ranges	Muzammil, S., Ashraf, A., Siddique, M. H., Aslam, B., Rasul, I., Abbas, R., ... & Hayat, S. (2023). A review on toxicity of nanomaterials in agriculture: Current scenario and future prospects. *Science Progress, 106*(4), 00368504231221672.
19	Multiple NPs	Rice, others	500–1000 mg/L cause adverse effects	Liu, J., Dhungana, B., & Cobb, G. P. (2018). Environmental behavior, potential phytotoxicity, and accumulation of copper oxide nanoparticles and arsenic in rice plants. *Environmental Toxicology and Chemistry, 37*(1), 11-20.

14.4.3 Size, Shape, and Surface Chemistry of Nanoparticles

One of the very important predictors of toxicity is the nanoparticle size. The smaller particles possess a greater surface-area-to-volume which makes them more reactive and more able to penetrate plant tissues. As an example, nano-sized silver particles (10 nm) cause an increased inhibitor of growth in rice than larger (50-100 nm) particles (Pan et al., 2024). Shape is also important: the rod and plate-like particles may behave in a different way in contact with membrane and vascular tissues compared to the spherical ones. Surface chemistry, such as charge, functional groups/surface coats determine biomolecule interaction and influence on soil components. Nanoparticles, in other cases, or which have been functionalized or even coated, tend to have lesser toxicity through reduced exposure to tissue or slowing the rate of dissolution (Khan, Saeed et al., 2019).

14.4.4 Plant Species and Growth Stage Dependency

The sensitivity of different plant species to the same nanoparticles is different. Monocots like wheat and rice might be better adapted to some nanoparticles than dicots such as the soybean or tomato, which could be related to their different root architecture, cell wall thickness, or cell wall defense system (Matras et al., 2022). Susceptibility is also heavily dependent upon development stage, since germinating plants and young seedlings are less resistant to nanoparticle exposure than mature plants because their developing tissues and meristems more easily disordinate. This heterogeneity makes the evaluation of risks hard, and the assessment needs to be species and stage-specific.

14.4.5 Environmental Conditions (pH, Soil Type, Temperature)

The surrounding environment plays a principal role in the fate of nanoparticles and bioavailability. The pH of soils affects solubility and release of ions; ZnONPs therefore dissolute more in acidic soils, thereby causing higher toxicity (Cyprichová et al., 2025). Nanoparticles can be bound by soil type and content of organic elements and cause a decrease in free concentration, but a change in long-term accumulation may occur. Plant metabolic activity and reactivity of nanoparticles like photoactivation of TiO_2NPs using UV light to increase ROS generation can be modulated by temperature and light intensity. In addition, pollution will be synergistically or antagonistically correlated with other co-existing pollutants (heavy metals, pesticides), making their ecological impacts more complicated.

14.5 ENVIRONMENTAL IMPACT OF NANOPARTICLES IN AGRICULTURE

The increasing use of nanoparticles (NPs) in agriculture as fertilizers, pesticides, growth enhancers, and soil conditioners has raised serious concerns regarding their environmental safety. While nanotechnology offers solutions to challenges in modern farming, the release and accumulation of NPs in the environment can disrupt ecological balance. Their high surface area, reactivity, and persistence make them capable of interacting with soil, water, and living organisms in complex ways.

14.5.1 Soil Ecosystem Interactions

Soil is the primary sink for nanoparticles introduced through agricultural practices. Once released, nanoparticles can undergo transformations such as dissolution, aggregation, or binding to organic matter, influencing their bioavailability and toxicity. Their interactions with soil biota and processes play a key role in determining long-term environmental outcomes.

14.5.1.1 Effects on Soil Microbiota and Microbial Diversity

The microorganisms of soil mediate the processes of nutrient cycling, decomposition of organic matter and interaction between plants and microorganisms (Chen et al., 2024). Nanoparticles, including metal and metal oxide NPs like AgNPs, CuONPs, and ZnONPs, may affect microbial communities by generating reactive oxygen species (ROS), releasing ions that are harmful or providing a mechanical action on cell membranes (Ma et al., 2022). Studies have indicated that silver nanoparticles decrease considerably the abundance of helpful microorganisms such as Rhizobium, Pseudomonas, and arbuscular mycorrhizal fungi (AMF) needed to fix nitrogen and stimulate the growth of plants. Persistent selective pressure may alter the structure of the microbial community, and prefer persistent resistant and, not necessarily beneficial, strains of microorganisms to beneficial ones. Less diversity of microbes not only impacts on soil fertility, but it also compromises the soil ecosystem's resistance to environmental stressors (Table 3).

Table 3. Reported effects of different types and concentrations of nanoparticles (NPs) on soil and plant systems. Studies demonstrate that NP impacts are concentration- and context-dependent, influencing microbial community structure, enzyme activity, nutrient cycling (C, N, P), and plant–microbe interactions, with outcomes ranging from stimulation at low doses to inhibition at high doses

NP Type & Concentration	Soil/Plant System	Reported Effects	References
CuO NPs; soil microcosms (63 days)	Agricultural soil	Reduced bacterial diversity; stronger impact than Cu^{2+} ions	Qian, Y., Jiang, W., Li, R., Liu, F., Li, S., & Wang, F. (2025). Nanospecific effect rather than dissolved Cu2+ dominates the impact of CuO nanoparticles on soil microbiota. *Journal of Hazardous Materials, 490*, 137817.
AgNPs; 100 mg/kg soil	Cucumber rhizosphere & bulk soil	Altered bacterial composition; impaired C, N, P cycling	Zhang, H., Huang, M., Zhang, W., Gardea-Torresdey, J. L., White, J. C., Ji, R., & Zhao, L. (2020). Silver nanoparticles alter soil microbial community compositions and metabolite profiles in unplanted and cucumber-planted soils. *Environmental Science & Technology, 54*(6), 3334-3342.
ZnO NPs ≥10 mg/kg in Pb-contaminated soil	Pb-contaminated agricultural soil	Reduced enzyme activity & microbial respiration; tolerant taxa increased	Huang, H., Chen, J., Liu, S., & Pu, S. (2022). Impact of ZnO nanoparticles on soil lead bioavailability and microbial properties. *Science of the Total Environment, 806*, 150299.
ZnO NPs up to ~250 mg/kg soil	Soil–plant microcosm	Low concs stimulated rhizobacteria; high concs reduced diversity	Strekalovskaya, E. I., Perfileva, A. I., & Krutovsky, K. V. (2024). Zinc oxide nanoparticles in the "Soil–Bacterial Community–Plant" system: Impact on the stability of soil ecosystems. *Agronomy, 14*(7), 1588.
ZnO NPs at 400 & 800 mg/kg	Maize rhizosphere soil with AMF	High conc inhibited AMF colonization; AMF reduced NP stress	Wang, F., Liu, X., Shi, Z., Tong, R., Adams, C. A., & Shi, X. (2016). Arbuscular mycorrhizae alleviate negative effects of zinc oxide nanoparticle and zinc accumulation in maize plants–a soil microcosm experiment. *Chemosphere, 147*, 88-97.
AgNPs (10–60 nm, varied charges)	Wheat rhizosphere soil	Size/surface charge determined effect; some AgNPs suppressed rhizobacteria	Przemieniecki, S. W., Ruraż, K., Kosewska, O., Oćwieja, M., & Gorczyca, A. (2024). The impact of various forms of silver nanoparticles on the rhizosphere of wheat (Triticum aestivum L.)–Shifts in microbiome structure and predicted microbial metabolic functions. *Science of the Total Environment, 914*, 169824.
CuO NPs; 50 mg/kg soil	Wheat rhizosphere (calcareous soil)	Reduced N-fixing bacteria; altered nitrate dynamics	Guan, X., Gao, X., Avellan, A., Spielman-Sun, E., Xu, J., Laughton, S., ... & Lowry, G. V. (2020). CuO nanoparticles alter the rhizospheric bacterial community and local nitrogen cycling for wheat grown in a calcareous soil. *Environmental science & technology, 54*(14), 8699-8709.
CuO NPs; 0.1–100 mg/kg soil	Agricultural soils with plants	High dose reduced denitrification, nitrification, respiration	Simonin, M., Cantarel, A. A., Crouzet, A., Gervaix, J., Martins, J. M., & Richaume, A. (2018). Negative effects of copper oxide nanoparticles on carbon and nitrogen cycle microbial activities in contrasting agricultural soils and in presence of plants. *Frontiers in microbiology, 9*, 3102.

continued on following page

Table 3. Continued

NP Type & Concentration	Soil/Plant System	Reported Effects	References
CuO NPs; varied sizes & conc.	Agricultural vs unmanaged soils	Land use history shaped response; high doses suppressed respiration	Rippner, D. A., Margenot, A. J., Fakra, S. C., Aguilera, L. A., Li, C., Sohng, J., ... & Parikh, S. J. (2021). Microbial response to copper oxide nanoparticles in soils is controlled by land use rather than copper fate. *Environmental Science: Nano*, *8*(12), 3560-3576.
CuO, ZnO NPs; 1000 mg/kg soil	Soil with wheat & cucumber	NPs more toxic than MPs; suppressed enzymes & biomass	Kim, S., Sin, H., Lee, S., & Lee, I. (2013). Influence of metal oxide particles on soil enzyme activity and bioaccumulation of two plants. *Journal of microbiology and biotechnology*, *23*(9), 1279-1286.
TiO_2 & ZnO NPs; varied doses	Agricultural soil	N-fixing taxa decreased; decomposers increased	Ge, Y., Schimel, J. P., & Holden, P. A. (2012). Identification of soil bacteria susceptible to TiO2 and ZnO nanoparticles. *Applied and environmental microbiology*, *78*(18), 6749-6758.
ZnO NPs; moderate–high doses	Soil with wheat straw	High concs reduced fungal complexity & decomposition	Jia, Y., Gu, D., Du, X., Yang, W., Yin, X., Li, Q., ... & Liu, T. (2024). ZnO Nanoparticles and Soil Fauna Affect Nutrient Transfer via Effects on Soil Fungal Community During Returned Wheat Straw Decomposition. *International Journal of Environmental Research*, *18*(4), 63.

14.5.1.2 Impact on Nutrient Cycling and Soil Fertility

The degradation of the activity of the microbes has a downstream effect on the process of nutrient cycling. NPs are capable of blocking major microbial enzymes that include, urease, dehydrogenase, and nitrogenase, as well as, disrupting turnover of nitrogen and carbon. An example cited is how ZnONPs have been found to lower the activity of nausea in soils, resulting in nitrogen availability to plants. Equally, AgNPs can inhibit the growth of nitrifying and denitrifying bacteria, causing an imbalance in the cycle of nitrogen (Fan et al., 2024). Moreover, essential nutrients (e.g., phosphorus, iron) may be specifically targeted by nanoparticles, so that they become inaccessible to plants. Throughout history, this could erode the fertility of the soil, cause a decrease in crop outputs and dependence on artificial substances, vicious in combating the intended positive impacts of nanotechnology.

14.5.2 Aquatic Ecosystem Contamination

The agricultural fields may cause runoff, leaching, and irrigation activities, which may move nanoparticles to the water bodies. Their movement is affected by their particle size, their surface chemistry and soil characteristics. When present in aquatic systems, there is interaction of nanoparticles with the sediments, plants in water, and aquatic organisms that may cause ecotoxic effects (Kapeleka & Mwema, 2024).

14.5.2.1 Nanoparticle Leaching and Runoff

The NPs may be mobilized in groundwater and water systems through rain and irrigation. To illustrate this case, TiO_2NPs and ZnONPs that are present on soils have been found in drainage water, a fact that casts doubts on long-distance transportation (Šebesta et al., 2021). Continued solidity of nanoparticles on water implies that they are bioactive and can live a long life, with impacts on the quality of water and the aquatic ecosystems. The movement of nanoparticles is affected by criteria like soil size, pH, and organics, where sandy soils with low organic content are likely to increase the rate of leaching.

14.5.2.2 Effects on Aquatic Plants and Organisms

Nanoparticles have the potential to interfere with both primary producers and higher tropical levels once in the water. The effect of AgNPs or CuONPs on algae and other aquatic plants is that photosynthesis is more often inhibited through the formation of ROS and through the interference of the operations of chloroplasts through the disruption of chlorophyll function (Tripathi et al., 2024; Wang et al., 2024). Other rivals: the decreased algal biomass, may be transferred to the herbivores and predators in a cascade through the food web. Aquatic life, such as fish and invertebrates, is at risk also. Research has shown that AgNPs are retained in the gills and liver of fish and hinder respiration systems and the detoxification process. On the same note, TiO2NPs used in UV light can present oxidative stress in zooplankton, with carbon-based nanomaterials potentially harming the surfaces of the gills physically. These not only pose a hazard to the aquatic biodiversity, but also put food safety at risk because of contaminated fish or shellfish found in the human diet.

14.5.3 Bioaccumulation and Food Chain Transfer

One of the key issues raised by the application of nanoparticles in agriculture is their possible occurrence of bioaccumulation in plants and their further distribution in food chains. Plants may assimilate nanoparticles on their roots or leaves and may remain in the tissues or be translocated into the edible portions of grains, fruits and leaf. An example might be the discovery of ZnONPs and TiO_2 NP in wheat grains and rice shoots with exposure to soil. The nanoparticles build up in edible tissues and this increases the concern in regard to trophic transfer. Nanoparticle concentration could reach higher levels when herbivores feed on polluted vegetation, thus affecting other higher predators, including human beings. Despite the metabolism or excretion of some nanoparticles, some may accumulate, resulting in health threats in the long-term (Rana & Kalaichelvan, 2013). The chronic exposure via food chains may lead to oxidative damage, organ damage or genotoxic damage in animals and humans.

14.6 RISK ASSESSMENT AND SAFETY CONCERNS

Enhancement in crop productivity, efficiency of nutrient uptake and resistance to stress has prompted excitement to incorporate in agricultural production the nanoparticles (NPs). However, their entity body, serving as the peptidoglycan structure, presents issues such as threats to plants, ecosystems, and human beings as well because of their peculiar physicochemical properties. In comparison to traditional agrochemicals, nanoparticles exhibit novel nano-bio interface behaviors that are difficult to predict regarding their toxicological behavior. In a good manner therefore, good safety measures and correct assessment of risk should be undertaken in order to implement sustainable farming applications. In this section, the protocols, the regulatory frameworks, and the perceived challenges within nanoparticle risk assessment will be addressed.

14.6.1 Standard Protocols for Phytotoxicity Testing

The risk assessment of nanoparticles is premised on the phytotoxicity test. The investigation of the effect of diverse nanoparticles on plant germination, growth, physiology, and reproduction can be conducted through the utilization of conventional methodologies. Used predominantly as significant screening tools. Nanoparticles are cultivated, and different amounts are measured and germination rate, root length and seedling vigor are measured. OECD and ISO propose these assays as their basic toxicity indicators as an example. Nanoparticles are introduced to nutritional solutions, or soil to cultivate plants under agricultural conditions. Such endpoints are biomass, chlorophyll content, transpiration and uptake of nutrients. Extended to non-target organisms such as earthworms, soil microbes, and pollinators. These assays highlight ecosystem-level risks beyond plants. Despite progress, there is a need for standardized global protocols that account for nanoparticle diversity (metal, metal oxide, carbon-based) and agricultural variability (soil type, climate, crop species).

14.6.2 Regulatory Frameworks and Guidelines

The agricultural regulations regarding nanoparticles are not quite established yet and are different across regions. Nanomaterials, unlike pesticides and fertilizers, do not easily fit under the already established laws because of their hybrid characteristics of chemicals and new materials. European Union (EU): Nanomaterials in the European Chemicals Agency (ECHA) are regulated in the framework of the Registration (Registration) or REACH (Registration, Evaluation, Authorization and Restriction of Chemicals) regulation. Nanopesticides and nanofertilizers should be analyzed in different risk evaluations, taking into consideration the nano-specific properties. EFSA (European Food Safety Authority) also gives input regarding nanoparticles in food and feed. United States: EPA (Environmental Protection Agency) regulates nanopesticides by the Federal Insecticide, Fungicide and Rodenticide Act (FIFRA). Nanoparticles in food contact material and dietary products are regulated by the FDA, although agricultural guidelines are minimal. Asia (China, India, and Japan) is in the early stages of regulation. Other nations use, or will use, current chemical regulations to regulate nanomaterials with no particular nano-oriented consideration. India, as an example, has put draft guidelines for nanofertilizers in the Fertilizer Control Order (FCO). The OECD has come up with testing instructions for manufacturing nanomaterials. Testing standards and definitions are given by ISO (International Organization for Standardization). However, there is an uneven rate of adoption. The absence of harmonization to the extent of different regions impedes global trade and introduces uncertainty for all the parties concerned. The adoption of unified, clearer regulation of nanomaterials in agriculture is exceedingly necessary (Figure 3).

Figure 3. Global regulatory landscape of agricultural nanomaterials, showing stricter frameworks in the EU and US, early-stage or chemical-based regulations in Asia and other nations, with varying agency involvement and specificity

Nanoparticle Regulation Comparison

Characteristic	European Union (EU)	United States	Asia	Other Nations
Regulation Focus	REACH	FIFRA, FDA	Early Stages	Chemical Regulations
Specificity	Nano-specific properties considered	Minimal agricultural guidelines	No specific focus	No specific focus
Key Agencies	ECHA, EFSA	EPA, FDA	Fertilizer Control Order (India)	None

14.6.3 Challenges in Risk Assessment

Nanoparticles can cause diversity and complexity. The nanoparticles are different in terms of their sizes, shapes, surface chemistry as well as coatings. Minor alterations of these properties may drastically change toxicity and generalization of risk profiles is a challenge. Nanoparticles do not just stay on release but further dissolve, as well as, sticking together or reacting with the organic matter of soils. Such changes occur in bioavailability and toxicity, which are, however, seldom considered in laboratory analyses. Traditional toxicology is based upon concentration (mg/L), whereas in the context of nanoparticles, surface area, particle number and ion release are also valuable values. Pathways (root uptake, foliar absorption) complicate the dose-response assessment. The majority of the studies are based on acute toxicity. There are little studies about long-term effects on soils (i.e., bioaccumulation), crop genetic instability, or disruption of ecosystems. There are no uniform testing procedures between labs that are used in comparisons. With many conflicting studies documented, their use is dependent on experimental design, purity of the nanoparticles and characterization. It is hard to have both innovation and a precautionary balance. Increased regulation can disadvantage an asset like beneficial technologies, whereas loose regulations can result in environmental and food safety concerns. Social cognition, perception and willingness of the consumer are also a vital aspect, particularly when it comes to consumption of foods.

14.7 STRATEGIES TO MINIMIZE PHYTOTOXICITY AND ENVIRONMENTAL HAZARDS

14.7.1 Green Synthesis and Eco-friendly Nanoparticles

Green synthesis and environmentally friendly nanoparticles are becoming more popular since natural chemicals like plant extracts, microorganisms, or biopolymers are used instead of harsh chemical reducing, capping or stabilizing chemicals. It has been demonstrated in a vast number of studies that green-silver, metal-oxide nanoparticles (silver, zinc oxide) are usually less phytotoxic than their chemically prepared counterparts. As an example, relative work with silver nanoparticle synthesis that happened using plant extracts (in comparison to chemically synthesized AgNPs) illustrated less creation of accretive oxidizing bodies (ROS), significantly stable antioxidant reaction, and indeed encouraged chlorophyll content and photosynthetic rate instead of suppressing it. These green routes will also frequently lead to the formation of particles with natural surface coating (due to the biological material) that can either slow down ion leaching, aggregation, or surface reactivity, thus diminishing the destruction of plant cells or soil microbes (Figure 4).

Figure 4. Schematic representation of green synthesis of nanoparticles using plant extracts or microbial cultures, followed by mixing, centrifugation, drying, and sonication for final nanoparticle formation

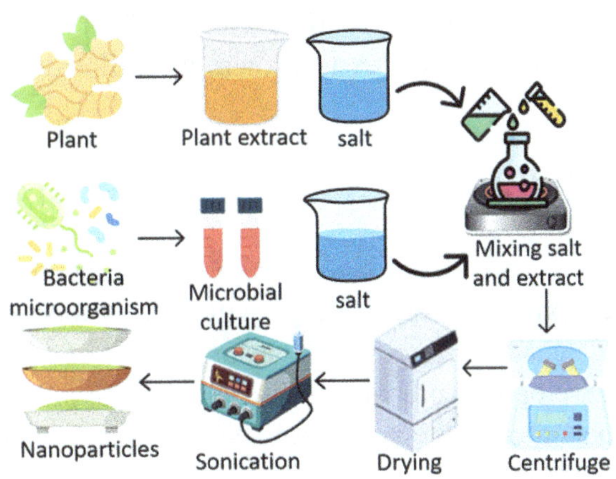

14.7.2 Controlled Release Formulations

The other major strategy is that of controlled release formulations. Rather than delivering a puffing burst of active nanoparticle or pesticide/fertilizer material that surpasses safe levels, controlled release or slow-release nanoparticle preparations provide the active ingredient over a longer duration, which is more adapted to plant needs. Polymers or nanocomposites that wrap or entrap nutrients or pesticidal actives slow up the rate of dissolution, reduce the amount of nutrient entering the environment at the start and rapidly reduce environmental leaching and non-target absorption. Nanofertilizers encapsulated in biopolymers or polymers, such as alginate, carboxymethyl cellulose or gelatin, have been shown to allow a slow release of nutrients and reduce initial exposure peaks which, in turn, decrease environmental leaching and during non-target impact. Optimal very low and high concentrations as well as methods of use are the basics. Even green or controlled-release NPs may turn dangerous with increased dosages or with the wrong mode of application. Increasing amounts of literature are reporting dose response curves exhibiting a hormetic property, with low dose concentrations being able to stimulate growth or stress tolerance, and greater concentrations adverse by influencing oxidative stress or causing DNA damage. So, setting safe threshold levels for each type of NP (metal, metal-oxide, carbon, etc.), modulating the frequency of application, selecting the best delivery mode (e.g. foliar spray rather than root or soil amendment), and timing that mode according to the specific stage of the plant cycle is all important. Experiments with lettuce, e.g. indicate that moderate doses of some NPs when used as a foliar treatment are useful, but high doses are lethal.

14.7.3 Biodegradable and Functionalized Nanoparticles

It is further proposed that biodegradable and functionalized nanoparticles lower the risk, since once they have a function to perform, they can degrade to benign components, or such particles may be functionally loosened (capped, coated, conjugated) to compromise their reactivity, control their release or deliver therapeutic benefits on-target. Herbicides or fertilizers can be encapsulated in biodegradable polymeric carriers like chitosan, poly (lactic -co-glycolic acid) (PLGA), alginate, etc.; such compositions have been found to cause fewer adverse effects in non-target organisms in trials. Surface reactivity and ion leakage have been shown to be reduced by functionalization (ADTA or phytoantioxidant capping of nanoparticles) to keep ROS generation in plant tissues down, and still maintain or even improve efficacy.

14.8 FUTURE PERSPECTIVES

The way is a sustainable, sustainable and safe strategy of nanoparticles in agriculture. Potential sustainable agriculture applications of nanotechnology will address food security challenges, nutrient exploitation effectiveness, and enhance crop tolerance against biotic and abiotic stressors at the cost of environmentally-sustainable development practices that would substantially diminish the ecological footprints. An invention of safer nanomaterials, such as biodegradable or functionalized or green-synthesis nanoparticles to reduce toxicity levels, has also been identified as one of the critical paths, to improve biocompatibility and provide targeted delivery systems of nutrients or Agrochemicals. In the meantime, a greater focus should be placed on monitoring the environment in the long term and impromptu modeling as short-term laboratory testing will not convey the important data on this matter in the long term regarding the effects that nanoparticles have on soils, water regimes, biodiversity, and food chain transfer. Predictive model establishment will then be core in the aspect of quantifying the fate of nanoparticles, their persistence, and potential threats of nanoparticles in agricultural systems. Taken together, these perceptions allude to the need to make nanoparticle applications in agriculture sustainable, safe and socially acceptable through interdisciplinary efforts in nanotechnology, plant science, soil ecology and environmental policy.

CONCLUSION

Nanoparticles reach the domain of agriculture can solve one hand and on the other, this is a revolutionary step in crop management, efficient use of nutrients and resistance to exposure to stress types, even though it is still not yet publicly identified how harmful nanoparticles can be to plants and ecosystems. It has been demonstrated that nanoparticle exposure can cause morphology, physiology and genetics plant changes, with consequences varying between the stimulation of plant growth at low doses and severely damaging occurrence at high levels. The distribution of nanoparticles in soil and water systems also makes their use even more complicated since they can cause imbalance in microbial communities, nutrient cycles, contamination of aquacosms, and bioaccumulation through food webs, threatening the non-target options and human health. Regardless of these fears, there exist sustainable solutions, including green synthesis, controlled release formulations, biodegradable carriers, and functionalize nanomaterials, offer potential ways of reducing the risks. Nonetheless, to be safe even through integration, effective risk assessment procedures, harmonized regulation, and incessant environmental surveillance are needed. The future should be on creating safer nanomaterials, defining dose limits, and predictive software on the effects when looking long-term. After all, the proper use of nanoparticles in agriculture requires an interdisciplinary strategy that will overcome the tension between technological innovation and environmental safety in assuring not only an agricultural progress but also the sustainability of the environment.

REFERENCES

Abbasi, A. J., Anas, M., Elahi, M., Khan, A., Khattak, W. A., Saleem, M. H., & Quraishi, U. M. (2025). Restoring wheat productivity and nutrient balance under cadmium stress through reducing toxicity, metal uptake, and oxidative damage using selenium nanoparticles. *Journal of Trace Elements in Medicine and Biology, 89*, 127644. PMID: 40233590

Afzal, S.. (2024). Nanostructure and plant uptake: Assessing the ecological footprint and root-to-leaf dynamics. *Plant Nano Biology, 10*, 100122. https://doi.org/https://doi.org/10.1016/j.plana.2024.100122

Ahmad, P., Alyemeni, M. N., Al-Huqail, A. A., Alqahtani, M. A., Wijaya, L., Ashraf, M., & Bajguz, A. (2020). Zinc oxide nanoparticles application alleviates arsenic (As) toxicity in soybean plants by restricting the uptake of as and modulating key biochemical attributes, antioxidant enzymes, ascorbate-glutathione cycle and glyoxalase system. *Plants, 9*(7), 825. PMID: 32630094

Akhtar, M. J.. (2012). Zinc oxide nanoparticles selectively induce apoptosis in human cancer cells through reactive oxygen species. *International Journal of Nanomedicine, 7*, 845–857. DOI: 10.2147/ijn.S29129 PMID: 22393286

Albaladejo-Marico, L., Thameur, A., Garcia-Martinez, A., Carvajal, M., & Yepes-Molina, L. (2025). Nanoparticles as modulators of stress tolerance: Physiological and molecular insights into TiO_2 and ZnO effects in Cucumis melo L. subjected to salt shock. *Environmental Technology & Innovation, 38*, 104101.

Ali, A.. (2022). Phytohormones and Transcriptome Analyses Revealed the Dynamics Involved in Spikelet Abortion and Inflorescence Development in Rice. *International Journal of Molecular Sciences, 23*(14). Advance online publication. DOI: 10.3390/ijms23147887 PMID: 35887236

Allará, C., Ciccone, G., Ciocca, M., Vasquez, S., Ibba, P., Maver, M., & Petti, L. (2025). Electronic Nanomaterials for Plants: A Review on Current Advances and Future Prospects. *Advanced Electronic Materials*, •••, 2500080.

Ameh, T.. (2022). Silver and Copper Nanoparticles Induce Oxidative Stress in Bacteria and Mammalian Cells. *Nanomaterials (Basel, Switzerland), 12*(14). Advance online publication. DOI: 10.3390/nano12142402 PMID: 35889626

Anas, M., Falak, A., Hassan, S., Khattak, W. A., Saleem, M. H., Khan, K. A., & Fahad, S. (2025). Microbial Interactions and Bacterial Responses to Metal Stress in Plants: Mechanisms, Adaptations, and Applications for Sustainable Agriculture. *Journal of Crop Health, 77*(1), 36.

Anas, M., Khattak, W. A., Fahad, S., Alrawiq, N., Alrawiq, H. S., Abdelsalam, N. R., & Quraishi, U. M. (2025). Mitigating heavy metal pollution in agriculture: A multi-omics and nanotechnology approach to safeguard global wheat production. *Journal of Hazardous Materials Advances, 17*, 100584.

Ashrafpour, S., & Ashrafpour, M. (2025). The double-edged sword of nutraceuticals: Comprehensive review of protective agents and their hidden risks. *Frontiers in Nutrition, 12*, 1524627. PMID: 40212723

Atanda, S. A., Shaibu, R. O., & Agunbiade, F. O. (2025). Nanoparticles in agriculture: Balancing food security and environmental sustainability. *Discover Agriculture, 3*(1), 26.

Avellan, A., Yun, J., Morais, B. P., Clement, E. T., Rodrigues, S. M., & Lowry, G. V. (2021). Critical review: Role of inorganic nanoparticle properties on their foliar uptake and in planta translocation. *Environmental Science & Technology*, *55*(20), 13417–13431. PMID: 33988374

Babaei, M., Tayemeh, M. B., Jo, M. S., Yu, I. J., & Johari, S. A. (2022). Trophic transfer and toxicity of silver nanoparticles along a phytoplankton-zooplankton-fish food chain. *The Science of the Total Environment*, *842*, 156807. PMID: 35750161

Bardaweel, S. K.. (2018). Reactive Oxygen Species: The Dual Role in Physiological and Pathological Conditions of the Human Body. *The Eurasian Journal of Medicine*, *50*(3), 193–201. DOI: 10.5152/eurasianjmed.2018.17397 PMID: 30515042

Basit, F.. (2023). Protective role of ZnO nanoparticles in soybean seedlings growth and stress management under Cr-enriched conditions. *Plant Growth Regulation*, *100*(3), 703–716. DOI: 10.1007/s10725-023-00965-7

Bhaduri, B., & Nath, J. (2021). Tracing of Ag-and CeO2 based engineered nanoparticles in cucumber plant system. *Journal of Environmental Chemical Engineering*, *9*(4), 105778.

Bhati, V. S., et al. Foliar uptake and translocation of nanoparticles in the plants: a review. Journal of Plant Nutrition, 1-38. https://doi.org/DOI: 10.1080/01904167.2025.2538638

Bhati, V. S., Singh, P. K., Parmar, Y. J., & Rokins, P. G. (2025). Foliar uptake and translocation of nanoparticles in the plants: A review. *Journal of Plant Nutrition*, •••, 1–38.

Brandelli, A. (2020). The interaction of nanostructured antimicrobials with biological systems: Cellular uptake, trafficking and potential toxicity. *Food Science and Human Wellness*, *9*(1), 8–20. https://doi.org/https://doi.org/10.1016/j.fshw.2019.12.003

Budhani, S.. (2019). Phytotoxic effect of silver nanoparticles on seed germination and growth of terrestrial plants. *Journal of Environmental Science and Health. Part C, Environmental Carcinogenesis & Ecotoxicology Reviews*, *37*(4), 330–355. DOI: 10.1080/10590501.2019.1676600 PMID: 31661365

Casillas-Figueroa, G. O.. (2020). Argovit™ PVP-stabilized silver nanoparticles effects on Allium cepa: Plant-growth promotion without cyto/genotoxic damage. *Nanomaterials (Basel, Switzerland)*, *10*(7), 1386. DOI: 10.3390/nano10071386 PMID: 32708646

Chen, H. (2018). Metal based nanoparticles in agricultural system: Behavior, transport, and interaction with plants. *Chemical Speciation and Bioavailability*, *30*(1), 123–134.

Chen, L., Peng, Y., Zhu, L., Huang, Y., Bie, Z., & Wu, H. (2022). CeO2 nanoparticles improved cucumber salt tolerance is associated with its induced early stimulation on antioxidant system. *Chemosphere*, *299*, 134474. PMID: 35367497

Chen, Q.. (2024). Soil Microorganisms: Their Role in Enhancing Crop Nutrition and Health. *Diversity*, *16*(12).

Chinmayi, S.. (2025). Enzymatic and nonenzymatic antioxidant defense in plants under drought stress. In Husen, A. (Ed.), *Role of Antioxidants in Mitigating Plant Stress* (pp. 75–85). Academic Press., https://doi.org/https://doi.org/10.1016/B978-0-443-26799-4.00016-5

Cyprichová, V.. (2025). Interaction of zinc oxide nanoparticles with soil colloidal suspensions. *Chemosphere, 370*, 144001. https://doi.org/https://doi.org/10.1016/j.chemosphere.2024.144001 PMID: 39708946

Das, H. S., Basak, A., & Maity, S. (2025). Materials science and nanotechnology. In *Innovations in Energy Efficient Construction Through Sustainable Materials* (pp. 175–206). IGI Global.

Debnath, P.. (2018). Cytogenetic effects of silver and gold nanoparticles on Allium cepa root cells. *Journal of Genetic Engineering and Biotechnology, 16*(2), 519–526. DOI: 10.1016/j.jgeb.2018.07.007 PMID: 30733769

Deshmukh, S. K., Kochar, M., Kaur, P., & Singh, P. P. (Eds.). (2023). *Nanotechnology in agriculture and environmental science*. CRC Press.

Ding, Y.. (2023). Recent Trends in Foliar Nanofertilizers: A Review. *Nanomaterials (Basel, Switzerland), 13*(21). PMID: 37947750

Djanaguiraman, M., Anbazhagan, V., Dhankher, O. P., & Prasad, P. V. (2024). Uptake, translocation, toxicity, and impact of nanoparticles on plant physiological processes. *Plants, 13*(22), 3137. PMID: 39599346

Donia, D. T., & Carbone, M. (2023). Seed Priming with Zinc Oxide Nanoparticles to Enhance Crop Tolerance to Environmental Stresses. *International Journal of Molecular Sciences, 24*(24). Advance online publication. DOI: 10.3390/ijms242417612 PMID: 38139445

Elahi, M.. (2025). Nitrogen Nano-fertilizers Enhance Wheat Productivity, Improve Nitrogen Use Efficiency, and Mitigate Soil Nitrogen Losses While Maintaining Grain Quality Compared to Conventional Fertilizers. *BioNanoScience, 15*(3), 477. DOI: 10.1007/s12668-025-02065-4

Etesami, H.. (2021). Interactions of nanoparticles and salinity stress at physiological, biochemical and molecular levels in plants: A review. *Ecotoxicology and Environmental Safety, 225*, 112769. https://doi.org/https://doi.org/10.1016/j.ecoenv.2021.112769 PMID: 34509968

Fais, G.. (2024). Cytotoxic Effects of ZnO and Ag Nanoparticles Synthesized in Microalgae Extracts on PC12 Cells. *Marine Drugs, 22*(12). Advance online publication. DOI: 10.3390/md22120549 PMID: 39728124

Faisal, M., Faizan, M., & Alatar, A. A. (2024). Metallic allies in drought resilience: Unveiling the influence of silver and zinc oxide nanoparticles on enhancing tomato (Solanum lycopersicum) resistance through oxidative stress regulation. *Plant Physiology and Biochemistry, 212*, 108722. PMID: 38761543

Fan, Z.. (2024). Effects of silver nanoparticles and various forms of silver on nitrogen removal by the denitrifier Pseudomonas stutzeri and their toxicity mechanisms. *Ecotoxicology and Environmental Safety, 269*, 115785. https://doi.org/https://doi.org/10.1016/j.ecoenv.2023.115785 PMID: 38056119

Fischer, D. G.. (2014). Plant genetic effects on soils under climate change. *Plant and Soil, 379*(1), 1–19. DOI: 10.1007/s11104-013-1972-x

Fu, L., Wang, Z., Dhankher, O. P., & Xing, B. (2020). Nanotechnology as a new sustainable approach for controlling crop diseases and increasing agricultural production. *Journal of Experimental Botany, 71*(2), 507–519. PMID: 31270541

Fu, Y.. (2024). Factors Influencing Seed Dormancy and Germination and Advances in Seed Priming Technology. *Plants, 13*(10). Advance online publication. DOI: 10.3390/plants13101319 PMID: 38794390

Ge, Y., Schimel, J. P., & Holden, P. A. (2012). Identification of soil bacteria susceptible to TiO2 and ZnO nanoparticles. *Applied and Environmental Microbiology, 78*(18), 6749–6758. PMID: 22798374

Geetha, A. (2019). Chapter-2 Phytotoxicity due to fungicides and herbicides and its impact in crop physiological factors. Chief Editor Dr. RK Naresh, 29.

Goncharuk, E. A., & Zagoskina, N. V. (2023). Heavy metals, their phytotoxicity, and the role of phenolic antioxidants in plant stress responses with focus on cadmium. *Molecules (Basel, Switzerland), 28*(9), 3921. PMID: 37175331

Guan, X., Gao, X., Avellan, A., Spielman-Sun, E., Xu, J., Laughton, S., & Lowry, G. V. (2020). CuO nanoparticles alter the rhizospheric bacterial community and local nitrogen cycling for wheat grown in a calcareous soil. *Environmental Science & Technology, 54*(14), 8699–8709. PMID: 32579348

Guo, S., Hu, X., Wang, Z., Yu, F., Hou, X., & Xing, B. (2024). Zinc oxide nanoparticles cooperate with the phyllosphere to promote grain yield and nutritional quality of rice under heatwave stress. *Proceedings of the National Academy of Sciences of the United States of America, 121*(46), e2414822121. PMID: 39495932

Hanif, S.. (2024). Harnessing the potential of zinc oxide nanoparticles and their derivatives as nano-fertilizers: Trends and perspectives. *Plant Nano Biology, 10*, 100110. https://doi.org/https://doi.org/10.1016/j.plana.2024.100110

Haq, I. U.. (2024). Interactions Between Nanoparticles and Tomato Plants: Influencing Host Physiology and the Tomato Leafminer's Molecular Response. *Nanomaterials (Basel, Switzerland), 14*(22). Advance online publication. DOI: 10.3390/nano14221788 PMID: 39591030

Hasan, M., Mehmood, K., Mustafa, G., Zafar, A., Tariq, T., Hassan, S. G., Loomba, S., Zia, M., Mazher, A., Mahmood, N., & Shu, X. (2021). Phytotoxic evaluation of phytosynthesized silver nanoparticles on lettuce. *Coatings, 11*(2), 225. DOI: 10.3390/coatings11020225

Hong, J., Wang, C., Wagner, D. C., Gardea-Torresdey, J. L., He, F., & Rico, C. M. (2021). Foliar application of nanoparticles: Mechanisms of absorption, transfer, and multiple impacts. *Environmental Science. Nano, 8*(5), 1196–1210.

Hu, J., Wu, X., Wu, F., Chen, W., Zhang, X., White, J. C., & Wang, X. (2020). TiO 2 nanoparticle exposure on lettuce (Lactuca sativa L.): Dose-dependent deterioration of nutritional quality. *Environmental Science. Nano, 7*(2), 501–513.

Hu, J., Wu, X., Wu, F., Chen, W., Zhang, X., White, J. C., & Wang, X. (2020). TiO 2 nanoparticle exposure on lettuce (Lactuca sativa L.): Dose-dependent deterioration of nutritional quality. *Environmental Science. Nano, 7*(2), 501–513.

Huang, Y., Bai, X., Li, C., Kang, M. E., Weng, Y., & Gong, D. (2022). Modulation mechanism of phytotoxicity on Ipomoea aquatica Forssk. by surface coating-modified copper oxide nanoparticles and its health risk assessment. *Environmental Pollution, 314*, 120288. PMID: 36180003

Iavicoli, I., Leso, V., Fontana, L., & Calabrese, E. J. (2018). Nanoparticle exposure and hormetic dose–responses: An update. *International Journal of Molecular Sciences*, *19*(3), 805. PMID: 29534471

Ibrahim, A. S., Ali, G. A., Hassanein, A., Attia, A. M., & Marzouk, E. R. (2022). Toxicity and uptake of CuO nanoparticles: Evaluation of an emerging nanofertilizer on wheat (Triticum aestivum L.) plant. *Sustainability*, *14*(9), 4914.

Ibrahim, A. S., Ali, G. A., Hassanein, A., Attia, A. M., & Marzouk, E. R. (2022). Toxicity and uptake of CuO nanoparticles: Evaluation of an emerging nanofertilizer on wheat (Triticum aestivum L.) plant. *Sustainability*, *14*(9), 4914.

Jia, Y., Gu, D., Du, X., Yang, W., Yin, X., Li, Q., & Liu, T. (2024). ZnO Nanoparticles and Soil Fauna Affect Nutrient Transfer via Effects on Soil Fungal Community During Returned Wheat Straw Decomposition. *International Journal of Environmental Research*, *18*(4), 63.

Kamal, A., Yang, J., Batool, M., Ara, U., Khattak, W. A., Touhami, D., & Elshikh, M. S. (2025). Green synthesis of zinc oxide nanoparticles, characterization and its applications in inducing disease resistance against spot blotch disease in wheat by enhancing its physiological and biochemical parameters. *Journal of Crop Health*, *77*(1), 5.

Kapeleka, J. A., & Mwema, M. F. (2024). State of nano pesticides application in smallholder agriculture production systems: Human and environmental exposure risk perspectives. *Heliyon*, *10*(20), e39225. https://doi.org/https://doi.org/10.1016/j.heliyon.2024.e39225 PMID: 39492887

Khan, A., Anas, M., Ullah, N., Khattak, W. A., Ali, M., Maaza, M., & Shinwari, Z. K. (2025). Harnessing green nanotechnology: Exploring the potential of silver, copper, and their nanocomposites for biomedical and agricultural applications. *Pakistan Journal of Botany*, *57*(6), 2197–2220.

Khan, I., Saeed, K., & Khan, I. (2019). Nanoparticles: Properties, applications and toxicities. *Arabian Journal of Chemistry*, *12*(7), 908–931.

Khan, M. R., Adam, V., Rizvi, T. F., Zhang, B., Ahamad, F., Jośko, I., & Mao, C. (2019). Nanoparticle–plant interactions: Two-way traffic. *Small*, *15*(37), 1901794. PMID: 31318142

Khan, S.. (2023). The impact of silver nanoparticles on the growth of plants: The agriculture applications. *Heliyon*, *9*(6), e16928. https://doi.org/https://doi.org/10.1016/j.heliyon.2023.e16928 PMID: 37346326

Khan, S., Zahoor, M., Khan, R. S., Ikram, M., & Islam, N. U. (2023). The impact of silver nanoparticles on the growth of plants: The agriculture applications. *Heliyon*, *9*(6). PMID: 37346326

Khanizadeh, P.. (2024). Application of Fe(2)O(3) nanoparticles improves the growth, antioxidant power, flavonoid content, and essential oil yield and composition of Dracocephalum kotschyi Boiss. *Front Plant Sci*, *15*, 1475284. DOI: 10.3389/fpls.2024.1475284 PMID: 39450081

Kharlamova, M. V., & Kramberger, C. (2023). Cytotoxicity of Carbon Nanotubes, Graphene, Fullerenes, and Dots. *Nanomaterials (Basel, Switzerland)*, *13*(9). Advance online publication. DOI: 10.3390/nano13091458 PMID: 37177003

Kim, S., Sin, H., Lee, S., & Lee, I. (2013). Influence of metal oxide particles on soil enzyme activity and bioaccumulation of two plants. *Journal of Microbiology and Biotechnology, 23*(9), 1279–1286. PMID: 23751560

Kohay, H., Wielinski, J., Reiser, J., Perkins, L. A., Ristroph, K., Giraldo, J. P., & Lowry, G. V. (2025). Nanocarrier foliar uptake pathways affect delivery of active agents and plant physiological response. *Environmental Science. Nano, 12*(1), 660–674. PMID: 39450293

Kumari, R., Suman, K., Karmakar, S., Mishra, V., Lakra, S. G., Saurav, G. K., & Mahto, B. K. (2023). Regulation and safety measures for nanotechnology-based agri-products. *Frontiers in Genome Editing, 5*, 1200987. PMID: 37415849

Leopold, L. F., Coman, C., Clapa, D., Oprea, I., Toma, A., Iancu, . D., & Coman, V. (2022). The effect of 100–200 nm ZnO and TiO2 nanoparticles on the in vitro-grown soybean plants. *Colloids and Surfaces. B, Biointerfaces, 216*, 112536. PMID: 35567806

Li, P.. (2024). The Impact of Nanomaterials on Photosynthesis and Antioxidant Mechanisms in Gramineae Plants: Research Progress and Future Prospects. *Plants, 13*(7). Advance online publication. DOI: 10.3390/plants13070984 PMID: 38611512

Li, R., Zhang, R., Li, Y., Liu, C., Wang, P., Sun, H., & Wang, L. (2024). Foliar uptake and distribution of metallic oxide nanoparticles in maize (zea mays L.) leaf. *Environmental Science & Technology, 58*(38), 16994–17003. PMID: 39258394

Liu, Y., Li, Y., Pan, B., Zhang, X., Zhang, H., Steinberg, C. E., & Peijnenburg, W. J. (2021). Application of low dosage of copper oxide and zinc oxide nanoparticles boosts bacterial and fungal communities in soil. *The Science of the Total Environment, 757*, 143807. PMID: 33288254

Lopez-Herrera, A., ... (2024). AgNP uptake in aquatic/wetland plants. Environments. MDPI https://www.mdpi.com/2076-3298/11/12/297

Ma, X.. (2022). Copper-containing nanoparticles: Mechanism of antimicrobial effect and application in dentistry-a narrative review. *Frontiers in Surgery, 9*, 905892. DOI: 10.3389/fsurg.2022.905892 PMID: 35990090

Ma, Y., He, X., Zhang, P., Zhang, Z., Ding, Y., Zhang, J., & Yang, K. (2017). Xylem and phloem based transport of CeO2 nanoparticles in hydroponic cucumber plants. *Environmental Science & Technology, 51*(9), 5215–5221. PMID: 28383248

Matras, E.. (2022). Phytotoxicity of Silver Nanoparticles with Different Surface Properties on Monocots and Dicots Model Plants. *Journal of Soil Science and Plant Nutrition, 22*(2), 1647–1664. DOI: 10.1007/s42729-022-00760-9

Milewska-Hendel, A., Zubko, M., Karcz, J., Stróż, D., & Kurczyńska, E. (2017). Fate of neutral-charged gold nanoparticles in the roots of the Hordeum vulgare L. cultivar Karat. *Scientific Reports, 7*(1), 3014. PMID: 28592798

Mishra, K. K., & Kumar, S. (Eds.). (2023). *Biotic stress management of crop plants using nanomaterials*. CRC Press.

Mishra, S., Keswani, C., Abhilash, P. C., Fraceto, L. F., & Singh, H. B. (2017). Integrated approach of agri-nanotechnology: Challenges and future trends. *Frontiers in Plant Science*, *8*, 471. PMID: 28421100

Montanha, G. S., Rodrigues, E. S., Marques, J. P. R., de Almeida, E., Colzato, M., & Pereira de Carvalho, H. W. (2020). Zinc nanocoated seeds: An alternative to boost soybean seed germination and seedling development. *SN Applied Sciences*, *2*(5), 857.

Montes de Oca-Vásquez, G. (2021). Impacto de concentraciones ambientalmente relevantes de nanopartículas de plata en la comunidad microbiana del suelo.

Moshirian Farahi, S. M.. (2023). The effects of titanium dioxide (TiO2) nanoparticles on physiological, biochemical, and antioxidant properties of Vitex plant (Vitex agnus - Castus L). *Heliyon*, *9*(11), e22144. https://doi.org/https://doi.org/10.1016/j.heliyon.2023.e22144 PMID: 38034643

Muzammil, S., Ashraf, A., Siddique, M. H., Aslam, B., Rasul, I., Abbas, R., & Hayat, S. (2023). A review on toxicity of nanomaterials in agriculture: Current scenario and future prospects. *Science Progress*, *106*(4), 00368504231221672. PMID: 38131108

Pan, R.. (2024). Stage-Specific Effects of Silver Nanoparticles on Physiology During the Early Growth Stages of Rice. *Plants*, *13*(23). PMID: 39683247

Pandey, K., & Dasgupta, C. N. (2025). Role of nanobionics to improve the photosynthetic productivity in plants and algae: an emerging approach. 3 Biotech, 15(4), 1-21.

Potapova, T., & Gorbsky, G. J. (2017). The Consequences of Chromosome Segregation Errors in Mitosis and Meiosis. *Biology (Basel)*, *6*(1). Advance online publication. DOI: 10.3390/biology6010012 PMID: 28208750

Preetha, J. S. Y.. (2023). Cerium oxide as a nanozyme for plant abiotic stress tolerance: An overview of the mechanisms. *Plant Nano Biology*, *6*, 100049. https://doi.org/https://doi.org/10.1016/j.plana.2023.100049

Przemieniecki, S. W., Ruraż, K., Kosewska, O., Oćwieja, M., & Gorczyca, A. (2024). The impact of various forms of silver nanoparticles on the rhizosphere of wheat (Triticum aestivum L.)–Shifts in microbiome structure and predicted microbial metabolic functions. *The Science of the Total Environment*, *914*, 169824. PMID: 38185142

Qian, Y., Jiang, W., Li, R., Liu, F., Li, S., & Wang, F. (2025). Nanospecific effect rather than dissolved Cu2+ dominates the impact of CuO nanoparticles on soil microbiota. *Journal of Hazardous Materials*, *490*, 137817. PMID: 40048778

Răcuciu, M.. (2025). Evaluation of phytotoxicity and genotoxicity of TMA-stabilized iron-oxide nanoparticle in corn (Zea mays) young plants. *Scientific Reports*, *15*(1), 18951. DOI: 10.1038/s41598-025-03872-1 PMID: 40442473

Raina, A.. (2021). Mechanisms of Genome Maintenance in Plants: Playing It Safe With Breaks and Bumps. *Frontiers in Genetics*, *12*, 675686. DOI: 10.3389/fgene.2021.675686 PMID: 34239541

Ramos-Pan, L.. (2024). Impact of gold nanoparticle exposure on genetic material. *Mutation Research/ Genetic Toxicology and Environmental Mutagenesis*, *900*, 503827. https://doi.org/https://doi.org/10.1016/j.mrgentox.2024.503827 PMID: 39617592

Rana, L., Kumar, M., Rajput, J., Kumar, N., Sow, S., Kumar, S., & Raigar, B. L. (2024). Nexus between nanotechnology and agricultural production systems: Challenges and future prospects. *Discover Applied Sciences*, *6*(11), 555.

Rana, S., & Kalaichelvan, P. T. (2013). Ecotoxicity of nanoparticles. *ISRN Toxicology*, *574648*. Advance online publication. DOI: 10.1155/2013/574648 PMID: 23724300

Rasheed, A.. (2022). The role of nanoparticles in plant biochemical, physiological, and molecular responses under drought stress: A review. *Front Plant Sci*, *13*, 976179. DOI: 10.3389/fpls.2022.976179 PMID: 36507430

Ravishankar, L. V.. (2025). ZnO Nanoparticles: Advancing Agricultural Sustainability. *Plants*, *14*(15). PMID: 40805779

Rim, K. T.. (2013). Oxidative DNA damage from nanoparticle exposure and its application to workers' health: A literature review. *Safety and Health at Work*, *4*(4), 177–186. DOI: 10.1016/j.shaw.2013.07.006 PMID: 24422173

Rippner, D. A., Margenot, A. J., Fakra, S. C., Aguilera, L. A., Li, C., Sohng, J., & Parikh, S. J. (2021). Microbial response to copper oxide nanoparticles in soils is controlled by land use rather than copper fate. *Environmental Science. Nano*, *8*(12), 3560–3576.

Rodrigues, S., ... (2024). Zinc phosphate-coated ZnO NPs: uptake & mobility in pepper. Environ Sci Technol. https://pmc.ncbi.nlm.nih.gov/articles/PMC11494269/

Rudenko, N. N.. (2023). Antioxidants of Non-Enzymatic Nature: Their Function in Higher Plant Cells and the Ways of Boosting Their Biosynthesis. *Antioxidants*, *12*(11). Advance online publication. DOI: 10.3390/antiox12112014 PMID: 38001867

Rudi, L.. (2025). Stage-Specific Effects of TiO2, ZnO, and CuO Nanoparticles on Green Microalga Haematococcus lacustris: Biomass and Astaxanthin Biosynthesis. *Marine Drugs*, *23*(5). PMID: 40422794

Salac, E. L. O., Alvarez, M. R., Gaurana, R. S., Grijaldo, S. J. B., Serrano, L. M., Juan, F. D., & Nacario, R. C. (2022). Biological assay-guided fractionation and mass spectrometry-based metabolite profiling of Annona muricata L. cytotoxic compounds against lung cancer A549 cell line. *Plants*, *11*(18), 2380. PMID: 36145779

Saleem, M. H., Noreen, S., Ishaq, I., Saleem, A., Khan, K. A., Ercisli, S., & Fahad, S. (2025). Omics technologies: Unraveling abiotic stress tolerance mechanisms for sustainable crop improvement. *Journal of Plant Growth Regulation*, •••, 1–23.

Saptarshi, S. R.. (2013). Interaction of nanoparticles with proteins: Relation to bio-reactivity of the nanoparticle. *Journal of Nanobiotechnology*, *11*(1), 26. DOI: 10.1186/1477-3155-11-26 PMID: 23870291

Šebesta, M., Kolenčík, M., Sunil, B. R., Illa, R., Mosnáček, J., Ingle, A. P., & Urík, M. (2021). Field application of ZnO and TiO2 nanoparticles on agricultural plants. *Agronomy (Basel)*, *11*(11), 2281.

Sembada, A. A., & Lenggoro, I. W. (2024). Transport of Nanoparticles into Plants and Their Detection Methods. *Nanomaterials (Basel, Switzerland)*, *14*(2). Advance online publication. DOI: 10.3390/nano14020131 PMID: 38251096

Shen, M., Liu, W., Zeb, A., Lian, J., Wu, J., & Lin, M. (2022). Bioaccumulation and phytotoxicity of ZnO nanoparticles in soil-grown Brassica chinensis L. and potential risks. *Journal of Environmental Management, 306*, 114454. PMID: 35007793

Shen, X., Li, S., Zhang, H., Chen, W., Yang, Y., Li, J., & Wang, X. (2018). Effect of multiwalled carbon nanotubes on uptake of pyrene by cucumber (Cucumis sativus L.): Mechanistic perspectives. *NanoImpact, 10*, 168–176.

Sidhu, A. K.. (2024). Nanomaterial strategies for enhancing plant resilience in the face of temperature stress. *CABI Agriculture and Bioscience, 5*(1), 60. DOI: 10.1186/s43170-024-00255-w

Silva, S., Dias, M. C., & Silva, A. M. (2022). Titanium and zinc based nanomaterials in agriculture: A promising approach to deal with (a) biotic stresses? *Toxics, 10*(4), 172. PMID: 35448432

Simonin, M., Cantarel, A. A., Crouzet, A., Gervaix, J., Martins, J. M., & Richaume, A. (2018). Negative effects of copper oxide nanoparticles on carbon and nitrogen cycle microbial activities in contrasting agricultural soils and in presence of plants. *Frontiers in Microbiology, 9*, 3102. PMID: 30619181

Singh, A.. (2024). In-depth exploration of nanoparticles for enhanced nutrient use efficiency and abiotic stresses management: Present insights and future horizons. *Plant Stress, 14*, 100576. https://doi.org/https://doi.org/10.1016/j.stress.2024.100576

Singh, A.. (2024). Nanoparticles in revolutionizing crop production and agriculture to address salinity stress challenges for a sustainable future. *Discover Applied Sciences, 6*(6), 317. DOI: 10.1007/s42452-024-06009-7

Strekalovskaya, E. I., Perfileva, A. I., & Krutovsky, K. V. (2024). Zinc oxide nanoparticles in the "Soil–Bacterial Community–Plant" system: Impact on the stability of soil ecosystems. *Agronomy (Basel), 14*(7), 1588.

Szczyglewska, P., Feliczak-Guzik, A., & Nowak, I. (2023). Nanotechnology–general aspects: A chemical reduction approach to the synthesis of nanoparticles. *Molecules (Basel, Switzerland), 28*(13), 4932. PMID: 37446593

Szőllősi, R.. (2020). Dual Effect of Nanomaterials on Germination and Seedling Growth: Stimulation vs. *Phytotoxicity. Plants (Basel), 9*(12). Advance online publication. DOI: 10.3390/plants9121745 PMID: 33321844

Tiwari, D. K., Dasgupta-Schubert, N., Villaseñor Cendejas, L. M., Villegas, J., Carreto Montoya, L., & Borjas García, S. E. (2014). Interfacing carbon nanotubes (CNT) with plants: Enhancement of growth, water and ionic nutrient uptake in maize (Zea mays) and implications for nanoagriculture. *Applied Nanoscience, 4*(5), 577–591.

Tripathi, D. K., Tripathi, A., Shweta, Singh, S., Singh, Y., Vishwakarma, K., ... & Chauhan, D. K. (2017). Uptake, accumulation and toxicity of silver nanoparticle in autotrophic plants, and heterotrophic microbes: a concentric review. Frontiers in microbiology, 8, 07.

Tripathi, S.. (2024). Interaction of silver nanoparticles with plants: A focus on the phytotoxicity, underlying mechanism, and alleviation strategies. *Plant Nano Biology*, *9*, 100082. https://doi.org/https://doi.org/10.1016/j.plana.2024.100082

Waani, S. P. T.. (2021). TiO2 nanoparticles dose, application method and phosphorous levels influence genotoxicity in Rice (Oryza sativa L.), soil enzymatic activities and plant growth. *Ecotoxicology and Environmental Safety*, *213*, 111977. https://doi.org/https://doi.org/10.1016/j.ecoenv.2021.111977 PMID: 33578101

Wang, F., Liu, X., Shi, Z., Tong, R., Adams, C. A., & Shi, X. (2016). Arbuscular mycorrhizae alleviate negative effects of zinc oxide nanoparticle and zinc accumulation in maize plants–a soil microcosm experiment. *Chemosphere*, *147*, 88–97. PMID: 26761602

Wang, J.. (2014). ZnO nanoparticle-induced oxidative stress triggers apoptosis by activating JNK signaling pathway in cultured primary astrocytes. *Nanoscale Research Letters*, *9*(1), 117. DOI: 10.1186/1556-276x-9-117 PMID: 24624962

Wang, N.. (2024). The Influence of Cuprous Oxide Nanoparticles on Photosynthetic Efficiency, Antioxidant Responses and Grain Quality throughout the Soybean Life Cycle. *Agronomy (Basel)*, *14*(8).

Wang, R., Mi, K., Yuan, X., Chen, J., Pu, J., Shi, X., & Zhang, H. (2023). Zinc oxide nanoparticles foliar application effectively enhanced zinc and aroma content in rice (Oryza sativa L.) grains. *Rice (New York, N.Y.)*, *16*(1), 36. PMID: 37599294

Wang, S., Fang, R., Yuan, X., Chen, J., Mi, K., Wang, R., & Zhang, H. (2023). Foliar spraying of ZnO nanoparticles enhanced the yield, quality, and zinc enrichment of rice grains. *Foods*, *12*(19), 3677. PMID: 37835330

Wang, X., ... (2016). Zinc Oxide Nanoparticles Affect Biomass Accumulation and Photosynthesis in Arabidopsis [Original Research]. Frontiers in Plant Science, Volume 6 - 2015. https://www.frontiersin.org/journals/plant-science/articles/10.3389/fpls.2015.01243

Wang, X.. (2023). Nanoparticles in Plants: Uptake, Transport and Physiological Activity in Leaf and Root. *Materials (Basel)*, *16*(8). Advance online publication. DOI: 10.3390/ma16083097 PMID: 37109933

Wohlmuth, J.. (2022). Interaction of the Nanoparticles and Plants in Selective Growth Stages—Usual Effects and Resulting Impact on Usage Perspectives. *Plants*, *11*(18). PMID: 36145807

Wu, H., & Li, Z. (2022). Nano-enabled agriculture: How do nanoparticles cross barriers in plants? *Plant Communications*, *3*(6). PMID: 35689377

Wu, P., Wang, Z., Adusei-Fosu, K., Wang, Y., Wang, H., & Li, X. (2024). Integrative chemical, physiological, and metabolomics analyses reveal nanospecific phytotoxicity of metal nanoparticles. *Journal of Environmental Management*, *354*, 120338. PMID: 38401494

Yan, A., & Chen, Z. (2019). Impacts of silver nanoparticles on plants: A focus on phytotoxicity and underlying mechanism. *International Journal of Molecular Sciences*, *20*(5), 1003. DOI: 10.3390/ijms20051003 PMID: 30813508

Yang, J., ... (2022). Effects of rice root exudates on aggregation, dissolution and bioaccumulation of differently-charged Ag nanoparticles††Electronic supplementary information (ESI) available. See DOI: . RSC Advances, 12(15), 9435-9444. https://doi.org/https://doi.org/10.1039/d2ra00229aDOI: 10.1039/d2ra00229a

Yang, J., Cao, W., & Rui, Y. (2017). Interactions between nanoparticles and plants: Phytotoxicity and defense mechanisms. *Journal of Plant Interactions*, *12*(1), 158–169.

Yang, S., Yin, R., Wang, C., Yang, Y., & Wang, J. (2023). Phytotoxicity of zinc oxide nanoparticles and multi-walled carbon nanotubes, alone or in combination, on Arabidopsis thaliana and their mutual effects on oxidative homeostasis. *PLoS One*, *18*(2), e0281756. PMID: 36791126

Yang, Z., Xiao, Y., Jiao, T., Zhang, Y., Chen, J., & Gao, Y. (2020). Effects of copper oxide nanoparticles on the growth of rice (Oryza sativa L.) seedlings and the relevant physiological responses. *International Journal of Environmental Research and Public Health*, *17*(4), 1260. PMID: 32075321

YingYing. J., et al. (2025). Green nanoparticles in agriculture: Enhancing crop growth and stress tolerance. Plant Stress, 18, 101017. https://doi.org/https://doi.org/10.1016/j.stress.2025.101017

Zafar, H., ... (2016). Effect of ZnO Nanoparticles on Brassica nigra Seedlings and Stem Explants: Growth Dynamics and Antioxidative Response [Technology Report]. Frontiers in Plant Science, Volume 7 - 2016. https://doi.org/https://www.frontiersin.org/journals/plant-science/articles/10.3389/fpls.2016.00535

Zaman, W.. (2025). Nanomaterials in Agriculture: A Pathway to Enhanced Plant Growth and Abiotic Stress Resistance. *Plants*, *14*(5). PMID: 40094659

Zhang, H.. (2024). Multifaceted impacts of nanoparticles on plant nutrient absorption and soil microbial communities. *Front Plant Sci*, *15*, 1497006. DOI: 10.3389/fpls.2024.1497006 PMID: 39606675

Zhang, H., Huang, M., Zhang, W., Gardea-Torresdey, J. L., White, J. C., Ji, R., & Zhao, L. (2020). Silver nanoparticles alter soil microbial community compositions and metabolite profiles in unplanted and cucumber-planted soils. *Environmental Science & Technology*, *54*(6), 3334–3342. PMID: 32088952

Zhang, P., Ma, Y., Zhang, Z., He, X., Zhang, J., Guo, Z., & Chai, Z. (2012). Biotransformation of ceria nanoparticles in cucumber plants. *ACS Nano*, *6*(11), 9943–9950. PMID: 23098040

Zulfiqar, F. (2021). Effect of seed priming on horticultural crops. *Scientia Horticulturae*, *286*, 110197. https://doi.org/https://doi.org/10.1016/j.scienta.2021.110197

Chapter 14
Cytotoxic and Genotoxic Effects of Silver and Gold Nanoparticles in Human Cells:
Implications for Safe Applications in Plant Health Nanotechnology

Ayush Madan
https://orcid.org/0009-0007-6861-0760
Department of Biotechnology, Shobhit Institute of Engineering and Technology, Meerut, India

Rajiv Dutta
https://orcid.org/0000-0003-3411-6530
Department of Biotechnology, School of Biological Engineering and Life Sciences, Shobhit Institute of Engineering and Technology, Meerut, India

ABSTRACT

The expanding use of silver (AgNPs) and gold nanoparticles (AuNPs) in plant health nanotechnology necessitates a clear understanding of their safety profiles in human cells. This chapter examines the cytotoxic and genotoxic consequences of AgNPs and AuNPs by integrating evidence on oxidative stress, mitochondrial impairment, DNA damage, and long-term genomic instability. Emphasis is placed on how nanoparticle size, surface chemistry, and biological interactions govern toxicity, with comparative insights into AgNP-induced ionic effects and the relative stability of AuNPs. The chapter further links human cell toxicity to broader agricultural safety, highlighting exposure risks and the need for safe-by-design nanoformulations. These insights aim to guide the responsible and sustainable application of nanotechnology in plant health.

1. INTRODUCTION

Nanotechnology has emerged as a revolution in various fields of science, which comprise medicine, agriculture, environmental science, and material engineering. Silver nanoparticles (AgNPs) and gold nanoparticles (AuNPs) are some of the most peculiar nanomaterials because they possess physicochem-

DOI: 10.4018/979-8-3373-4913-8.ch014

ical characteristics, high stability, adjustable surface functions, and even a general range of biological activities (Rokunuzzaman, 2024). Their application in plant health as nano-enabled fertilizers, pesticides and also in plant disease diagnoses, and in the system of delivering nutrients has gained pace. With the increase in the role of these materials in the agricultural process, a corresponding increase in the concerns of their possible toxicological implications on human health has also been raised. The biological mechanisms of nanoparticles are very complicated, and they must be comprehended to develop safe and sustainable nanotechnological devices in the field of plant science (Mittal et al., 2020).

The growing use of nanotechnology in the agricultural sector is bound to heighten the chances of human exposure to it either directly or indirectly. Exposure of human cells to AgNPs and AuNPs can be in the work place, in the environment, food chain biomagnification, or in the farm produce. Silver nanoparticles and especially silver nanoparticles are reported to possess a high level of antimicrobial activity, which makes them popular in the management of plant diseases (Tariq et al., 2022). Nevertheless, this antimicrobial efficacy is also strictly associated with their potential to produce reactive oxygen species (ROS) and release silver ions, as two significant sources of their cytotoxic and genotoxic actions in human cells. In turn, the general consensus on gold nanoparticles is that they are biologically inert and biocompatible, although the surface coverings, dosage, and aggregation state of nanoparticles can lead to the emergence of cellular stressor or genomic changes (Kadhim et al., 2021). These intricacies support the importance of an evidence-based assessment of their safety.

Although great literature on the use of nanoparticles in the biomedical arena has been noted, there is a major knowledge gap in toxicity profiles of nanoparticles when applied in plant health nanotechnology. A majority of the literature concentrates on either the benefits of agriculture or the medical aspect with seldom attempts to connect the two schools of thought. However, it is at this crossroad that risk assessment becomes the most applicable one. The materials used in agricultural systems eventually come into contact with human beings either by occupational exposure routes or consumption routes. Therefore, the knowledge of the effects of AgNPs and AuNPs on human cells contributes to the forecast of the risks in the long run and undoubtedly sustainable technological development. This research gap will be filled by this chapter through the provision of an integrated summary of the cytotoxic and genotoxic reactions that are a result of such nanoparticles within human cellular contexts, and putting these findings into a workable context in regards to their application in agriculture.

One of the difficulties of nanoparticle toxicity is their variability. What is even more astonishing is that even particles with the same core material can exhibit varying biological behaviors based on their size, shape, surface charge, crystallinity and the availability of stabilizing materials. Indicatively, smaller nanoparticles have a higher reactivity because they have a higher surface-to-volume ratio and can cause a more pronounced oxidative stress response in human cells (Sarkar et al., 2014). Likewise surface coating, be it polymer, organic or inorganic may attenuate or enhance toxic effects through alteration of cellular uptake, biodistribution or dissolution rates. Knowing these parameters is important not only to interpret the results of toxicity but also to develop safer nanomaterials that can maintain agricultural activity and at the same time reduce adverse effects.

Human cells react to exposure to nanoparticle via various biological pathways. Cytotoxic phenomena usually include dysfunction of the mitochondria, membrane destruction, oxidative stress, disturbed metabolism, and the stimulation of apoptosis or necrosis (Saifi et al., 2018). Genotoxic ones, in their turn, are characterized by DNA strand breakages, chromosomal abnormalities, oxidative lesions to DNA, and epigenetic alterations. Such effects have long term consequences such as potential genomic disturbance or carcinogenicity especially where the exposure is chronic or occupational. The chapter offers a basis of evaluating the impact of cellular damage caused by nanoparticles by dissecting these molecular processes, which could be relevant to long-term health effects.

The usefulness of findings is not only limited to individual toxicity mechanisms. Since plant health nanotechnology is now on the verge of being used in the real world, the aspect of safety should be integrated during the design stage. The terms safe-by-design nanomaterials and sustainable nano-agriculture are based on the concepts in which toxicological information is used to inform innovation. To take just a few examples, the size of nanoparticles, surface chemistry, or the use of green synthesis methods can be adjusted to make nanoparticles overall less toxic, without affecting their functional activity (Malik et al., 2023). There should be more transparent regulatory standards and risk evaluation models to facilitate the responsible commercialization and field implementation. The chapter is useful to researchers, policy makers, and practitioners by tying in toxicologically with their applications in agriculture to come up with both effective and safe solutions of nanotechnology. While these nano-enabled applications enhance plant health and crop productivity, they also introduce new occupational, environmental, and consumer exposure routes, making it essential to evaluate the potential human health risks associated with AgNPs and AuNPs.

The environmental fate of AgNPs and AuNPs is another complexity that arises. After being utilized in the agricultural environment, the nanoparticles experience numerous changes depending on the dynamics of the soil, microbial activities, pH, and soil organic matter. Such changes not only may influence their effectiveness in vegetal regimes but also may pose a threat to human and environmental health. As an example, the exposure risks can be changed by aggregation or dissolution processes or toxicity profiles can be altered. Thus, it is also essential to comprehend the interactions of nanoparticles in the environmental matrices, to forecast human outcomes.

The chapter is located at the crossroads between nanotoxicology, biotechnology of plants, and environmental safety. It will meet the essential safety concerns in the application of AgNPs and AuNPs in sustainable plant health nanotechnology by listing and critically analyzing evidence on cytotoxic and genotoxic effects of the two nanoparticles on human cells. The chapter strengthens the importance of informed decision-making that ensures that technological innovation is balanced with human and environmental well-being through an integrated view of the chapter.

2. PHYSICOCHEMICAL PROPERTIES OF SILVER AND GOLD NANOPARTICLES RELEVANT TO TOXICITY

Physicochemical properties of silver nanoparticles (AgNPs) and gold nanoparticles (AuNPs) are intrinsic factors of their biological behaviour and toxicological outcomes in human cells. These properties determine the behavior of nanoparticles at the cell membrane, organelles, protein, and nucleic acids, which eventually determine their cytotoxic and genotoxic effects, as depicted in Figure 1. The knowledge of these characteristics is critical in the interpretation of the toxicity information and the development of safe nanoformulations that can be used in the treatment of plants.

Figure 1. Key physicochemical properties governing toxicity of silver and gold nanoparticles

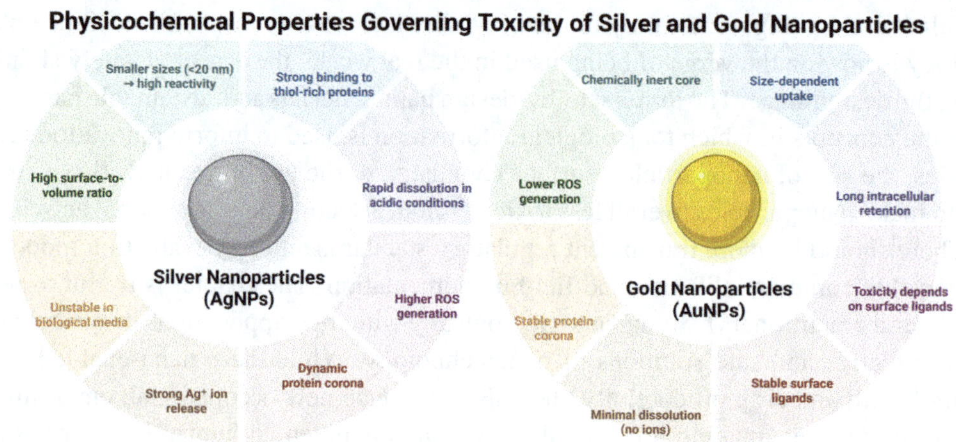

2.1 Particle Size, Shape, and Surface Area

Particle size is also one of the most impactful physicochemical parameters that determine the toxicity of nanoparticles. Smaller nanoparticles have a larger surface-volume ratio, making them reactive and allowing them to interact more with the pieces of a cell (Bilardo et al., 2022). AgNPs smaller than 20 nm can efficiently penetrate cells, reach mitochondria or the nucleus, and trigger marked oxidative stress. Although AuNPs are less toxic, small-sized AuNPs (≤10 nm) have a high internalization potential in cells and can interfere with cellular processes under high concentrations (Georgeous et al., 2024). Studies commonly report enhanced cytotoxicity for AgNPs below 10–20 nm, with mitochondrial disruption observed even at ≤ 5–10 µg/mL in human cell models.

The shape of the particles also changes biological interactions. Nanoparticles in a form of a sphere tend to follow predictable uptake behaviors whereas triangular, rod-shaped or cubic nanoparticles can cause unpredictable membrane interactions and intracellular localization. With a high aspect ratio, e.g. nanorods, penetration into cells is stronger and longer-lasting and increases the chances of responses to stress (Abbasi et al., 2023). Such shape-dependent behaviours should be noted in formulation of nanoparticles to be used in agriculture because the nano-enabled products often need to be stable, mobile, and have a uniform biological interactions.

2.2 Surface Charge and Colloidal Stability

Surface charge, which is typically measured using a zeta potential, controls the stability of nanoparticles in the suspension and it controls the contact between the nanoparticles and the biological membranes. The positive charged nanoparticles are easily bound to negatively charged phospholipid bilayers, thereby promoting internalization but escalating disruption of membrane, protein denaturation and consequently causing cytotoxicity. Those nanoparticles with negative charges, but slower in terms of internalization, tend to be more stable in the biological media and less prone to acute cellular damage (Yusuf et al., 2023).

AgNPs exhibit more problems with surface charge variation because it is easier to oxidize them, as well as react to surface species, particularly in the presence of electrolytes or biomolecules. AuNPs exist with relatively stable surface charges which depend largely on their surface ligands. These particles are of great importance in determining their behaviour in human tissues and in the environment because their stability is paramount. In AgNPs, aggregation is more prevalent in the absence of sufficient coating, and it decreases the dispersibility, although can still cause toxicity due to mechanical or inflammatory effects.

2.3 Dissolution and Ion Release

AgNP toxicity is characterized by dissolution. The release of silver ions (Ag^{+}) occurs when AgNPs enter biological fluids and plays a major role in cytotoxic and genotoxic effects. Ag + ions easily react with thiol groups, proteins, and DNA, causing oxidative stress, mitochondrial dysfunction, and possible genotoxicity via direct or indirect mechanisms (Morozova, 2021). Particle size, pH, temperature and organic compounds affect the rate of ion release. Uncoated AgNPs with a smaller size emit ions faster as compared to larger or surface-stabilized particles.

The dissolution of AuNPs is insignificant because of the inert noble-metal core. The fact that they are toxic can be greatly explained by their size, shape, protein corona interactions, or surface-functionalizing ligands and not by the ion-mediated mechanisms (Park, 2020). This nanoparticle difference between AgNPs and AuNPs underscores the reason why AgNPs can be more reactive and have greater biological effect although both nanoparticles have numerous applications in agriculture and biomedicine.

2.4 Surface Functionalization and Coatings

The stability, reactivity and cellular interaction of nanoparticles is a critical aspect that is influenced by surface coating. Aggregation may be reduced by functionalization with polymers of polyethylene glycol (PEG), citrate, dextran, polyvinylpyrrolidone (PVP) or biomolecules, decrease colloidal stability and alter cellular uptake. PEGylation can decrease protein adsorption and enhance biocompatibility and PVP stabilizes nanoparticles with no significant changes in surface reactivity (Schubert and Chanana, 2018). Surface functionalization in plant health nanotechnology has the benefit of increasing the dispersibility of nanoparticles in soil or foliar systems and controlling the active component release. For instance, positively charged AgNPs (+15 to +30 mV) have shown up to 2–3-fold greater cellular uptake than neutral or negatively charged particles at comparable exposure levels.

Green-synthesized nanoparticles produced using plant extracts or biomolecules are more biocompatible and generally exhibit lower toxicity. Such coatings can minimise ion release of AgNPs and inherently antioxidant their effects against ROS-induced damage. Nanoparticle–cell interactions are strongly influenced by the type and thickness of the surface coating. The explanation of coating-dependent behaviour is therefore important in the determination of safety profiles.

2.5 Interaction with Biomolecules and Protein Corona Formation

When nanoparticles are introduced into biological fluids, they are immediately covered by proteins, lipids and other macromolecules, and provide a dynamic protein corona. The effect of such a corona is that it is effectively redefining what the nanoparticle is, affecting how it is absorbed into cells, biodistribution, recognition by the immune system, and toxicity. The nanoparticle charge, surface chemistry, dissolution behaviour and biological milieu around the nanoparticle affect the composition of the corona.

Continuous release of ions and surface modification causes AgNPs to have a more dynamic corona, which results in reversible interactions with proteins. This can cause some erratic intracellular behavior or even change in nanoparticle fate. One of the reasons why AuNPs exhibit a comparatively high level of biocompatibility is the tendency of the nanoparticles to create a more stable corpus (Levak et al., 2017). Coronas of nanoparticles can also be generated in plant sap, soil leachates or environmental waters in agricultural settings and may further alter the behaviour of nanoparticles during human exposure.

2.6 Aggregation, Stability, and Environmental Behavior

The transport of nanoparticles, their bioavailability, and their toxicity depends on their stability in various media such as the biological, the environmental and the agricultural. Aggregation not only reduces effective surface area, but may also lead to localized toxicity due to the formation of high concentrations in the cellular compartments. Unless stabilized or coated, AgNPs have a higher tendency to aggregate in high-ionic-strength conditions, but tend to stay dispersed because of stronger ligand-binding affinities with AuNPs.

Nanoparticles are subject to environmental influences including pH, salinity, organic matter and the presence of natural colloids which modify the structure and behaviour. These changes should be taken into consideration in the assessment of the risk of human exposure due to the utilization of nanoparticles in agriculture where nanoparticles can be altered prior to contacting human cells (Čarapar et al., 2022).

3. CELLULAR UPTAKE AND INTRACELLULAR FATE OF NANOPARTICLES

Intracellular fate and cellular uptake of silver nanoparticles (AgNPs) and gold nanoparticles (AuNPs) are the focus of the cytotoxicity and genotoxicity of these substances on human cells. Their interactions in the cell environment are determined by the physicochemical characteristics of the cellular environment including size, charge, and surface chemistry, which affect internalization efficiency, organelle association, and biological reactivity. These processes define the effect of nanoparticles as harmless carriers or harmful agents of oxidative injury, inflammation, and DNA destabilization.

3.1 Cellular Entry Mechanisms

There are various endocytic pathways that nanoparticles are absorbed by human cells but the most prominent one is the clathrin- mediated endocytosis. The process includes the creation of coated vesicles which involve the uptake of nanoparticles and their delivery to early endosomes. Smaller and well dispersed AgNPs and AuNPs can tend to follow this path. Another significant route is caveolae-mediated endocytosis, which is particularly involved with nanoparticles bearing target surface ligands or those that bind with cholesterol-rich membrane domains (Varma et al., 2022). This path of uptake tends to produce alternative intracellular destinations to clathrin pathways, which affect the extent of cellular stress induced.

Bigger nanoparticles or extraneous-treated nanoparticles can be subjected to phagocytosis, especially by macrophages and dendritic cells. It leads to the rapid accumulation and may start powerful inflammatory reactions. Less selective is pinocytosis, which is a pathway that helps to background uptake in most cell types (de Almeida et al., 2025). The polarity of the particles on the surface greatly influences such processes: positively charged nanoparticles are more attracted by the negatively charged membranes and internalization is faster and results in a greater likelihood of membrane disruption. AgNPs tend to exhibit greater uptake rates because their surface reactivity is dynamic whereas AuNPs tend to depend greatly on ligand chemistry to dictate uptake trends.

3.2 Intracellular Trafficking and Vesicular Transport

After their internalization, nanoparticles are transported via endosomal lysosomal system. The exposure of nanoparticles to weakly acidic conditions by early endosomes results in alteration of surface coatings or, in the case of AgNPs, in part dissolution. When the vesicles transform to late endosomes and then merge with lysosomes the environment becomes progressively more acidic and enzyme-rich. It is a crucial point of AgNPs because acidification increases the release of Ag + ions, which boosts the potential of oxidative stress. Though structurally stable, AuNPs can be rearranged or displaced in vesicles and change their cellular interactions (Morais et al., 2024). Typical uptake has been documented at concentrations ranging from 1–50 μg/mL, depending on nanoparticle size and surface functionalization.

The lysosomal build-up over time is linked with severe toxicological effects. Lysosomes dissatisfaction with AgNP may cause a rise in intracellular Ag + levels, disrupting the lysosome membrane and releasing hydrolytic enzymes into the cytoplasm (Liu et al., 2023). AuNPs can stay behind and cause chronic cellular loading. Constant lysosomal storage has the potential to disrupt vesicular trafficking and activate inflammatory signalling.

3.3 Mitochondrial Association and Energy Disruption

Mitochondria are the major victims of nanoparticle-induced stress. Nanoparticles can be transported to mitochondria by vesicle damage, dissolved ions, or direct transport. Ag + interacts with mitochondrial proteins especially electron transport chain proteins interfering with ATP production and increasing the production of reactive oxygen species (ROS). Overproduction of ROS changes the mitochondrial membrane potential, induces changes in permeability, and activates intrinsic apoptotic pathways. The mechanisms justify the work of AgNPs that commonly results in severe cytotoxic effects despite moderate doses (Liu et al., 2023).

Although it does not emit ions, AuNPs may still affect the activity of the mitochondrion when they accumulate close to the mitochondrial membrane or when its surface ligands become reactive in the cytosolic space. Impaired mitochondrial membrane integrity is a contributor to downstream genotoxicity, which enhances oxidative damage to DNA and disables DNA repair mechanisms.

3.4 Nuclear Interactions and Genotoxic Potential

Direct entry of nanoparticles into the nucleus is size-dependent and is generally restricted to particles with diameter less than 5-10nm. Cell division leads to a temporal breakdown of the nuclear envelope, thereby enabling some nanoparticles to gain access to the nucleus. ANPs with the capacity to enter the nucleus can either react directly with DNA or release ions that produce strand breaks or disrupt replication. Nonetheless, indirect genotoxicity is more widespread and is triggered by the buildup of ROS or lipid peroxidation by-products, or the impaired work of mitochondria (Encinas-Gimenez et al., 2024).

AuNPs can be introduced into the nucleus when functionalized with nuclear-targeting ligands, and, in itself, they are inert, yet their introduction can disrupt the transcription of genes or the chromatin structure (Bonaccorsi, 2021). AgNPs and AuNPs can both alter the gene expression and stimulate stress-responsive pathways. Other epigenetic changes such as changes in DNA methylation or histone changes also enhance genomic instability caused by nanoparticles.

3.5 Autophagy, Exocytosis, and Cellular Clearance

To protect against cell exposure to nanoparticles, cells trigger autophagy. Autophagosomes surround nanoparticles and merge with lysosomes, although high readings or poor lysosomal activity can disrupt the autophagosomes. Ion-induced oxidative stress can cause autophagic flux to be inhibited by AgNPs, which cause the accumulation of damaged organelles and proteins (Chen et al., 2022). AuNPs have a tendency to remain in autophagic bodies because of low degradability. In case of dysfunction of autophagy, it adds to apoptosis, inflammation and metabolic imbalance.

Cells can get rid of nanoparticles by exocytosis, but not all cells are equally effective at clearing the nanoparticles. Retarded exocytosis causes permanent intracellular retention, particularly of AuNPs, and can cause chronic cellular stress. AgNPs dissolve during intracellular retention and enhances long-term Ag + ions exposure beyond the first uptake event, thereby augmenting chronic cytotoxic and genotoxic outcomes.

3.6 Implications for Human Toxicity

The toxicological consequences of nanoparticles depend on the uptake and intracellular fate of the nanoparticles. AgNPs which have a rapid uptake and ion-mediated reactivity pose more risks of cytotoxicity and genotoxicity. AuNPs may become more stable but with time, they can build up in a cell and disrupt the cellular functioning. Such processes explain the significance of considering nanoparticle behaviour when determining the risks of exposure to human beings because of nanotechnology in plants (Islam et al., 2023).

4. CYTOTOXIC EFFECTS OF AGNPS AND AUNPS IN HUMAN CELLS

The cytotoxicity of silver nanoparticles (AgNPs) and gold nanoparticles (AuNPs) on human cells is the product of a complicated interaction between physicochemical properties, the dynamics of cellular uptake, and intracellular bio chemistry. Although both the nanoparticles have been widely used in biomedical and agricultural nanotechnology, they exhibit different toxicity profiles because they have different modes of action. The main mechanisms of AgNPs toxicity are ion release and oxidative stress, but the cytotoxicity of the AuNPs is lower, though, it still may cause adverse effects depending on size, coating, and concentration. Table 1 provides a brief comparison between these nanoparticles and outlines the significant cytotoxic differences between them.

4.1 Oxidative Stress and Reactive Oxygen Species Generation

One of the most common mechanisms of nanoparticle-induced cytotoxicity is oxidative stress that is one of the most well-documented. Being surface-reactive and able to release Ag + ions, AgNPs are strong sources of reactive oxygen species (ROS). High levels of ROS cause an imbalance in the redox in the cell, which results in lipid peroxidation, protein denaturation, and oxidative destruction of the DNA. The events disrupt critical cell processes and cause cell death mechanisms (Choudhary et al., 2022). The ROS generation catalyzed by AgNPs is usually dose-dependent with smaller nanoparticles causing more oxidative stress as they have a larger surface area.

Generally, AuNPs have reduced ROS-inducing properties due to their chemical inertness, but surface-functionalized AuNPs or ultrasmall particles are capable of inducing oxidative stress. This happens especially when the AuNPs pile up in the mitochondria or interfere with the antioxidant systems like the glutathione or superoxide dismutase. No matter which type of nanoparticle was used, the overproduction of ROS is a key determinant of cytotoxic reaction (McLoughlin et al., 2024).

4.2 Mitochondrial Dysfunction and ATP Depletion

One of the most sensitive locations of nanoparticle action is the mitochondria, and mitochondrial dysfunction is one of the major cytotoxic pathways. AgNPs and Ag + ions interfere with mitochondrial electron transport and diminish the production of ATP and augment the leakage of ROS. This in turn results in mitochondrial membrane potential reduction and release of pro-apoptotic factors including cytochrome c. This cascade prompts intrinsic pathways of apoptosis and induces programmed cell death (He et al., 2024).

AuNPs are not ion releasing, but their contact with mitochondrial membranes can cause changes in membrane permeability, or cause energy disruption particularly in high dosages or in cases of lost ligand stability. Both types of nanoparticles can cause cellular energy imbalance, metabolic stress and eventual cell death (Nunes et al., 2022).

4.3 Loss of Membrane Integrity and Cellular Damage

Interactions of nanoparticles with membrane also play a significant role in cytotoxicity. Due to the positive surface interaction and reduced sizes, AgNPs can easily penetrate lipid bilayers, causing an increase in membrane permeability, structural deformation and leakage of intracellular contents. Ag+ ions also destabilize membrane bound proteins and ion channels. Physical and chemical membrane damage together promotes the cell lysis and necrosis, especially with increased nanoparticle concentration (Kariuki et al., 2025).

AuNPs do not normally result in extreme damage of membranes, although cationic AuNPs or high aspect ratio particles can alter membrane architecture. Membrane stability can also be indirectly impaired by repeated exposure or extracellular high concentration levels of AuNPs via the oxidative stress or lipid peroxidation (Canepa et al., 2021).

4.4 Apoptosis, Necrosis, and Autophagic Cell Death

AgNPs and AuNPs have the ability to induce different cell death depending on their concentration, exposure time, and physicochemical characteristics. The most prevalent response after moderate exposure to nanoparticles is apoptosis. Apoptosis induced by AgNP is closely connected to the impairment of mitochondria and the presence of oxidative DNA damage, whereas the effect of AuNP-induced apoptosis may be directly caused by the failure of lysosomes or excess autophagic cell death (Mihailovic et al., 2021).

With an increased concentration, AgNPs are prone to cause necrosis as it results in excessive membrane damage and metabolic breakdown. Inflammation which cannot be controlled is linked to necrosis, which is especially applicable to occupational exposure scenarios. Other causes of autophagic cell death associated with nanoparticles are accumulation in the lysosomes or autophagosomes and impairment of degradation pathways. AgNPs often disrupt autophagic fermentation, resulting in the accumulation of damaged organelles, whereas AuNPs can remain in autophagosomes because of their low degradability (Sun et al., 2021).

4.5 Dose-Response Relationships and Threshold Levels

The cytotoxic reactions to nanoparticles are associated with typical dose-response relationships. The steepness of dose-response curves with AgNPs is a characteristic typical of these nanoparticles due to the fact that even minor concentrations can greatly increase the release of ions and the production of ROS. They also have a higher toxicity with a lower size and lesser stability of surface coating. AuNPs are slower to onset dose-response, and can also have steep toxicity peaks at the ultrapure or ultrasmall particle once accumulated intracellularly.

It is important to identify the threshold concentrations of safe exposure to be used in applications in agriculture where humans may be exposed to the residues or subject to occupational exposure. Although the standardized exposure limits are not yet developed, available evidence indicates that AgNPs should be controlled at higher concentration levels than AuNPs (BalaKumaran et al., 2022).

4.6 Comparative Toxicity: AgNPs vs AuNPs

AgNPs are very cytotoxic and active at the moderate concentration through ion release, oxidative stress, and membrane disruption. AuNPs are more biocompatible, yet not completely non-toxic, especially when they are highly functionalized or when they are internalized in great numbers. The comparative lessons highlighted in Table 1 demonstrate the need to be cautious in designing nanoparticles and risk evaluation in their application to plant health technologies.

Table 1. Comparative cytotoxic characteristics of silver and gold nanoparticles

Cytotoxic Feature	Silver Nanoparticles (AgNPs)	Gold Nanoparticles (AuNPs)	Sources
Primary Toxicity Driver	Ag^+ ion release and ROS generation	Surface chemistry, accumulation, ligand effects	(Choudhary et al., 2022; McLoughlin et al., 2024)
ROS Generation	High; strong oxidative stress	Low to moderate depending on coating	(Nunes et al., 2022; He et al., 2024)
Mitochondrial Impact	Severe mitochondrial disruption; ATP loss	Moderate disruption; dose- and ligand-dependent	(Nunes et al., 2022; He et al., 2024)
Membrane Damage	High membrane permeability and structural disruption	Mild to moderate; stronger with cationic surfaces	(Canepa et al., 2021; Kariuki et al., 2025)
Apoptosis/Necrosis	Strong induction of apoptosis and necrosis at higher doses	Mainly apoptosis; necrosis rare	(Mihailovic et al., 2021; Sun et al., 2021).
Overall Cytotoxicity	High, size- and dose-dependent	Relatively low but not negligible	(BalaKumaran et al., 2022)

5. GENOTOXIC EFFECTS OF SILVER AND GOLD NANOPARTICLES

Genotoxicity is the property of a substance to cause direct or indirect damage to genetic material, resulting in the cleavage of DNA strands, chromosomal changes, mutations or a long-lasting genomic instability. It is necessary to determine the genotoxic potential of silver nanoparticles (AgNPs) and gold nanoparticles (AuNPs), as they are becoming more widely used in agriculture and biomedical, and the human body may be exposed to these nanoparticles by inhalation, ingestion, dermal contact, or handling. Although the two nanoparticles are capable of causing genotoxic damage, how this damage occurs and the extent is quite different as it is influenced by the differences in physicochemical behaviour, cellular interactions, and intracellular changes.

5.1 DNA Strand Breaks and Chromosomal Aberrations

DNA strand break is one of the most reported genotoxic effects that happen when exposed to AgNPs. The breaks can be the result of direct contact of the nanoparticles or silver ions (Ag^+) with the DNA, or indirectly, via oxidative stress. Ag+ ions are highly attracted to the nucleophilic sites in DNA, which allows them to interact with DNA and interfere with the base pairing and destabilise the double helix. It can lead to single strand breaks (SSBs) or even at higher doses, it causes double strand breaks (DSBs) which are more dangerous since they are harder to repair. DSBs may result in the loss of genetic material, translocations, or fragmentation of chromosomes (Naguib et al., 2022).

AuNPs are also chemically inert but can cause breaks to DNA strands, especially when the particles are extremely small (under 5 nm) or when the particles are made to be more nuclear localized by surface ligands. Nonetheless, the strand breaks caused by AuNP are relatively less common and are most likely to form by indirect oxidative processes but not by direct interactions with DNA. Micronucleus formation, lagging chromosomes, spindle disruption and chromosomal aberration have also been reported following exposure to either of the two nanoparticles, although these are more extreme and dose-dependent in cells that have been exposed to AgNP, as demonstrated by Figure 2.

Figure 2. Genotoxic effects of gold and silver nanoparticles

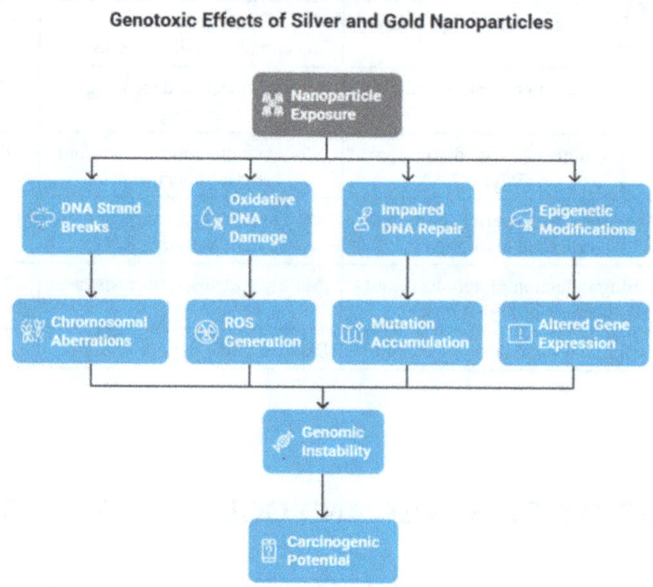

5.2 Oxidative DNA Damage and ROS-Mediated Genotoxicity

Oxidative stress is a key factor in the genotoxicity caused by nanoparticles especially AgNPs. Superoxide anions produced by nanoparticles or their ions may be too many and may cause DNA base damage to form oxidation products like 8-hydroxy-2'-deoxyguanosine (8-OHdG). It is a well-characterized oxidative damage biomarker lesion that has been extensively observed in AgNP-exposed cells. Unrepaired, the oxidative lesions may lead to mutations during DNA replication, which contributes to genomic instability in the long run (Omari Shekaftik & Nasirzadeh, 2021).

In some situations, too, AuNPs cause oxidative DNA damage and this is particularly when they accumulate in mitochondria or interfere with antioxidant systems. Even though they produce lower amount of ROS compared to AgNPs, AuNPs can still cause oxidative alterations in DNA, and this is mostly dose and ligand chemistry dependent. Both nanoparticles are capable of overloading cell antioxidant systems, such as glutathione, catalase, and superoxide dismutase result in continued oxidative stress of nuclear and mitochondrial DNA.

5.3 Impact on DNA Repair Pathways and Cell Cycle Regulation

Their genotoxic nature can be worsened by the effects of nanoparticles on the DNA repair mechanism. Particularly AgNPs have been demonstrated to disrupt base excision repair (BER) and nucleotide excision repair (NER) pathways by inhibiting repair enzymes or by modifying redox-sensitive repair proteins. Ag^+ ions can interact with zinc finger motifs found on DNA repair proteins and interfere with their shape. There is the accumulation of DNA lesions due to inhibited repair activity, which leads to an increase in mutation frequency and cell cycle arrest (Zhao et al., 2025).

The activation of the p53 pathway as a response to DNA damage is common in cells to stop the progression of the cell cycle to either repair the damage or induce apoptosis. The p53 activation triggered by AgNP is well established and usually connected with the G2M phase arrest through repair of the DSB occurred. AuNPs can also cause cell cycle arrest at the same checkpoints but at a higher concentration, or when functionalized to have better nuclear interactions. Nanoparticles lead to dysregulation of the cell cycle, which is a factor of genomic instability and can enhance carcinogenic risk in the long-term in case of chronic exposure (Almanghadim et al., 2021).

5.4 Epigenetic Modifications and Altered Gene Expression

In addition to structural DNA damage, nanoparticles also have the ability to cause epigenetic effects that are not connected to a change in the underlying DNA sequence, but cause a change in gene expression. AgNPs have been proven to cause changes in the methylation of the DNA, alter histone alterations, and change the expression levels of microRNA. These alterations may impact the pathways of oxidative stress response, inflammation, and apoptosis as well as cell differentiation. The existence of epigenetic abnormalities is especially worrisome, since it can last through several generations of cells, and can have long-term biological consequences (González-Palomo et al., 2021).

Epigenetic changes, particularly in the case of AuNPs, can also be induced but more subtle and reliant on surface chemistry (especially with functional ligands on the surface of the particle). Methylation or histone acetylation changes that occurred after exposure to AuNP have been attributed to changes in genes involved in cell cycle control and antioxidant response. Although these alterations are not as significant as the ones caused by AgNPs, they indicate that inert nanoparticles may have an effect on the gene control systems, as well.

5.5 Long-Term Genomic Instability and Carcinogenic Potential

These factors together with oxidative stress, DNA strand breaks, chromosomal abnormalities, impaired repair and epigenetic changes result in long term genomic instability after exposure to nanoparticle. The risk of AgNPs is greater because of its ability to continuously release its ions and its high reactivity that produce repeated insults to DNA and sustained oxidative stress. Occupational or chronic exposure consequently can increase the risk of mutation buildup and even carcinogenic consequences, though clinical evidence is under investigation (Dastgheib et al., 2024).

Although comparative biocompatibility would imply that auNPs would not be able to induce low-grade genomic instability, their accumulation in cells over prolonged exposure would or their interference to the pathways of DNA repair. Their prolonged intracellular retention period particularly in lysosomes is a matter of concern in terms of slow, cumulative genetic effects not necessarily evident in the short term.

6. FACTORS INFLUENCING TOXICITY IN HUMAN CELLS

Silver nanoparticles (AgNPs) and gold nanoparticles (AuNPs) are not characterized by permanent cytotoxic and genotoxic effects on human cells: these effects are formed under the influence of a broad set of physicochemical and biological conditions. These factors determine the behaviour of nanoparticles as early as they are exposed to their eventual destiny of being incorporated into the cell, which ultimately dictates the scale and nature of toxic response. These variables are the basis of predicting the human risks of agricultural nanomaterials as well as creating safer nano-enabled products. Table 2 summarises the major factors influencing it.

6.1 Concentration, Exposure Duration, and Dose Metrics

A major cause of toxicity is the concentration of nanoparticles. The increased levels of AgNPs cause a rapid release of Ag^+ and excess formation of reactive oxygen species (ROS), which causes severe oxidative and mitochondrial damage. The dose-response curve of AuNPs is more gradual, but can have spikes of sudden toxicity at high concentrations above the clearance threshold of the cell. The duration of exposure, on another hand, also adjusts toxicity, with the long-term contact leading to the accumulation of the toxin in the intracellular space and lysosomal overload, as well as a long-term oxidative stress (Niżnik et al., 2024).

Mass concentration, surface area, and particle number are dose metrics that give different information about the interactions of nanoparticles. In case the nanoparticles have the same mass, smaller particles offer more surface area and particles, which tends to increase reactivity. This holds especially with AgNPs whereby surface based oxidative and ion-release reactions become stronger with smaller size.

6.2 Influence of Protein Corona and Biological Media

When nanoparticles enter biological fluids they quickly attach themselves to proteins, lipids and other biomolecules forming a dynamic protein corona. Corona defines biological identity, which affects cellular uptake, immune recognition and toxicity. In the case of AgNPs, the corona can be unstable since there is a constant release of ions and a transformation of the surfaces, which results in variable interactions with cells (Tomak et al., 2022). Conversely, AuNPs have a tendency to form sturdier coronas leading to uniform cellular uptake patterns.

Nanoparticle behaviour is further modulated by the composition of the surrounding media, e.g., the serum content, ionic strength and the presence of antioxidants. As an example, AgNP dissolution to form AgCl complex is quickened in chloride-rich environments and this affects toxicity mechanisms. Higher protein concentration can suppress nanoparticles reactivity by covering their surfaces and low pH value improves AgNP dissolution and lysosomal degradation.

6.3 Role of pH, Ionic Strength, and Environmental Conditions

The reactivity of nanoparticles is strongly affected by biological and environmental pH. AgNPs are more easily soluble in acidic compartments, which accelerates the release of Ag + ions and heightens toxicity. This is particularly applicable within lysosomes where the pH can go as low as 4.55. Neutral or alkaline PH retards dissolution but can enhance aggregation, based on surface charge.

The ionic strength is also a significant factor. Agglutination of AgNPs in high ionic aqueous conditions, e.g., physiological saline, or soil extracts, facilitates the neutralization of repulsive forces. Aggregation modulates nanoparticle size distribution, cellular routes of uptake and toxicity effects. AuNPs, which have higher ligand stability, are not as susceptible to ionic destabilization however, they can aggregate in case of a degradation of their coating (Kyziol-Komosinska et al., 2024).

Conditions of the environment that are of concern to agricultural utilization like an organic matter, plant metabolites, and soil microbiota alter the nanoparticle surfaces prior to human exposure. This eco-corona has an influence on the transport of nanoparticles, their reactivity, and biological effects.

6.4 Cell Type Sensitivity and Biological Variability

The response of different human cell types to the exposure of nanoparticles differs because of the difference in metabolic activity, membrane structure, endocytosis rate, and antioxidant capacity. Oxidative and mitochondrial stress is especially susceptible to high metabolic cells like the hepatocyte and renal epithelial cells. Immune cells, which can phagocytose, tend to increase their loads of nanoparticle, leading to strong inflammatory and cytotoxic reactions (Shafiq et al., 2025).

Genotoxic damage is of special concern to stem cells and progenitor cells since differentiation and tissue repair require intact DNA. The fact that cell types vary highlights the significance of different models of cells to evaluate the safety of nanoparticles, especially in cases where exposure can be in various routes.

6.5 Synergistic or Antagonistic Effects with Other Environmental Agents

Combinations of Nanoparticle toxicity with other environmental stressors may either increase or decrease their toxicity. As an example, AgNPs are shown to have synergistic toxicity in cases of co-exposure to pesticides, heavy metals or UV radiation because of compounded oxidative stress. On the other hand, some antioxidants or phytochemicals, or serum proteins can lessen toxicity by inhibiting the formation of ROS or stabilizing the surfaces of nanoparticles (Yohannis et al., 2025).

Nanoparticles can react with plant hormones, chemical sprays, fertilizers, and soil pollutants in the world of agriculture. These interactions may alter the reactivity of nanoparticles, and this may produce heightened human exposure risks.

Table 2. Major factors influencing the toxicity of silver and gold nanoparticles in human cells

Influencing Factor	AgNPs (Silver Nanoparticles)	AuNPs (Gold Nanoparticles)	Sources
Dose & Exposure Time	Strong, rapid toxic response; high sensitivity	Moderate, gradual toxicity; higher thresholds	Niżnik et al., 2024)
Protein Corona Effects	Dynamic; influenced by ion release	Stable; predictable interactions	(Tomak etal., 2022)
pH & Ionic Strength	Highly sensitive; rapid dissolution in acidic media	Relatively stable; ligand-dependent effects	(Kyziol-Komosinska et al., 2024)
Cell Type Sensitivity	High toxicity in metabolic and immune cells	Moderate toxicity; ligand-modulated	(Shafiq et al., 2025)
Environmental Modifiers	Strong interactions with contaminants, UV, soil ions	Moderate influence; stable core	(Yohannis et al., 2025)

7. IN VITRO AND IN VIVO ASSESSMENT TECHNIQUES

To assess the cytotoxicity and genotoxicity of silver nanoparticles (AgNPs) and gold nanoparticles (AuNPs) it is necessary to have sound and reliable methods of assessment that reflect the complicated biological interactions. As the application of these nanoparticles in plant health nanotechnology is increasingly being used, with resultant possible human exposure, it is necessary to utilize in vitro and in vivo studies to achieve a complete toxicology picture. The methods can be used to determine the toxicity mechanisms, predict the risks of human health, and develop the safer nanoformulations.

7.1 In Vitro Cytotoxicity Assays

In vitro models are the initial tools of research in nanoparticle toxicity because of their controlled conditions, reproducibility and ethical viability. Cytotoxicity assays give a quantitative analysis of cell viability, cell metabolic activity, cell membrane integrity and cell stress response.

The MTT assay that is used to measure the mitochondrial metabolic activity is one of the most popular ones. The metabolically active cells convert the MTT reagent to formazan crystals, and the activity of impaired mitochondria, which is usually the initial victim of AgNP-induced oxidative stress produces lower signals. In line with this, WST-1 and XTT assays are also advantageous in terms of solubility and are most commonly used to assess the toxicity of AuNP, particularly when surface ligands have the potential to disrupt the formation of crystals (Kumar et al., 2024).

Lactate dehydrogenase (LDH) release assay assesses the integrity of the membrane by determining the amount of enzyme released into the medium by the cells which are damaged. Since AgNPs often act by disrupting the membranes via ion-mediated oxidative stress, LDH release is a good way of getting a clear picture of necrotic and late-stage cytotoxicity. Trypan blue exclusion assays are complementary to these results in that they used cellular membrane permeability to identify living and dead cells (Zaib et al., 2022).

These tests allow the determination of the nanoparticle concentration levels that invalidate cell viability. Nonetheless, they have to be supplemented with other methods in order to achieve accuracy since sometimes nanoparticle optical properties may interfere with colorimetric or fluorometric values.

7.2 Genotoxicity Assays

Genotoxicity tests determine the level of DNA damage, aberrant chromosomes and long-term genome stability which are major end points in determining the safety of nanoparticles that gain access to human body.

One of the most sensitive methods of detecting breaks in strands of DNA plasmids is the comet assay (single-cell gel electrophoresis). Fragmented DNA under electrophoretic conditions separates to create a comet-shaped tail and the intensity and length of the tail depends on the severity of the genotoxin. The comet assay results are frequently significant because AgNPs cause oxidative damage and strands breakages based on ion actions. AuNPs typically cause less serious effects, though in rare cases, ultrasmall or surface-reactive nanoparticles can also cause measurable genotoxicity (Dunkenberger et al., 2021).

The other method that is commonly employed is the micronucleus assay which identifies the chromosomal fragments or entire chromosomes that cannot segregate during cell division. The formation of micronucleus is closely related to AgNP exposure, which is based on their ability to distort the functioning of mitotic spindles and chromosome stability. Higher doses and with functionalization of AuNPs to increase the nuclear uptake also result in micronuclei (Kwasniewska & Bara, 2022).

The γ-H2AX foci assay possesses a high sensitivity of the detection of the double-strand breaks (DSBs). Upon the development of DSBs, the phosphorylated histone H2AX is localized at sites of damage to create discrete foci in the nucleus that can be measured with fluorescence microscopy. This is a very convenient test that is applicable in the differentiation between primary DNA breaks and oxidative by-products.

Although the Ames test was developed on chemical mutagens, it can be used to determine mutagenicity caused by nanoparticles in bacteria. It has a low applicability to nanoparticles because of differences in the uptake mechanisms, although alterations such as metabolic activation can be used to offer supportive evidence of mutagenic risk.

7.3 Oxidative Stress and Biomarker Evaluation

Because oxidative stress plays a crucial role in the nanoparticle toxicity, the oxidative biomarkers are important to measure. The real-time movement of oxidative imbalance can be determined by quantification of reactive oxygen species (ROS) with the help of fluorescent probes such as DCFH-DA. The second evidence of membrane oxidation is provided by Lipid peroxidation assays which are based on malondialdehyde (MDA) production. Carbonylation of proteins experiments and glutathione depletion analysis can be used to describe the redox imbalance caused by nanoparticles (Mohideen et al., 2023).

AgNPs are generally characterized by intense oxidative biomarker modifications as a result of Ag ions creation whereas AuNPs are more subtle based on their surface coats and aggregation. This allows the distinction of direct nanoparticle interaction and secondary oxidative effects with these biomarkers to generate a richer mechanistic insight.

7.4 Advanced Imaging and Localization Techniques

Imaging of nanoparticles on the basis of high-resolution technologies is crucial to observe internalization, distribution, and interaction of nanoparticles with intracellular structures. The gold standard in terms of observing the localization of nanoparticles in endosomes, lysosomes, mitochondria, and in selected cases, nuclei is transmission electron microscopy (TEM). TEM offers first hand data on structural disruption, including mitochondrial swelling or lysosomal lumps (Akhtar & Zuhair, 2025).

Intracellular trafficking can be dynamically observed using confocal laser scanning microscopy that is frequently used in conjunction with fluorescent labeled nanoparticles or organelle markers. Surface-enhanced Raman spectroscopy (SERS), which is made possible by the optical properties of AuNPs, has provided sensitive detection of the molecular interactions at the interface of a nanoparticle.

Such imaging methods confirm uptake pathways and offer the spatial context of cytotoxic and genotoxic responses.

7.5 In Vivo Toxicity Assessments

In vivo investigations are used to supplement in vitro results, that is, to assess systemic effect, biodistribution, clearance, and long-term effect. Animal models, which are usually rodents, give information on organ specific accumulation, effects on immunological reasons and protracted genotoxicity.

The most frequent sites of accumulation of AgNPs are the liver, the kidneys, the spleen and the lungs, and the ion release in these areas causes inflammation, oxidative stress, and histopathological changes. AuNPs are typically more predictable in their biodistribution and slower to be eliminated, however, their retention in tissues is a cause of concern when it comes to chronic exposure (Salama et al., 2023).

In vivo tests are also useful to determine the reproductive, neurotoxicity and hematological alterations, providing a complex picture of possible health hazards of repeated or professional re-exposure.

7.6 Emerging High-Throughput and Organoid-Based Models

The recent developments have also brought about high-throughput screening (HTS) systems alongside organoid models to nanoparticle toxicology. HTS allows the testing of various nanoparticle formulations in rapid mode and speeding up the process of safe-by-design. Organoids are 3D models of tissues that are obtained via stem cells and more resembling of human physiological conditions compared to traditional cultures (Madan et al., 2025). Liver, lung and intestinal organoids have also been useful in assessing nanoparticle accumulation, metabolism, and genotoxic reactions of ingestion or inhalation exposure.

8. IMPLICATIONS FOR SAFE USE IN PLANT HEALTH NANOTECHNOLOGY

8.1 Human Exposure Pathways in Agricultural Settings

Various human exposures avenues exist with the use of AgNPs and AuNPs in the health of plants. The greatest risk is on the agricultural workers when there is a nanoparticle synthesis, preparation, spraying and post-application intervention. Nanoparticles may be aerosolized, and through contact with the skin during manual procedures. Combined with their larger ion-release capacity, AgNPs are more dangerous due to the development of respiratory distress, oxidative stress, and membrane damage (Sati et al., 2025). AuNPs are less reactive; however, with repetitive exposure, they can be accumulated in the tissues. Engineering controls like enclosed mixing systems, proper training, and the use of PPEs are needed to protect the workers against the direct contact.

8.2 Consumer Exposure Through Crop Residues

When nanoparticles are used on crops, they may also remain on the surface of the leaves or they may even cross to plant tissues which is a cause of concern when it comes to human consumption. AgNPs applied as antimicrobial may be left behind as residues, which might result in gastrointestinal absorption and dissemination. The fact that they have the ability to release Ag + ions indicates that they may cause oxidative or genetic toxicity after chronic dietary intake. AuNPs although generally biocompatible might be retained in the edible tissues in case of nano-formulations intended to deliver nutrients. The quantification of the residue levels, examination of nanoparticles in plant systems and the determination of safe amounts of consumption to protect consumers therefore becomes important (Aswani et al., 2025).

8.3 Environmental Fate and Indirect Human Risks

Nanoparticles find their way into the soil, water, and plant-associated mico-bial environments after use in the field. Their associations with the soil minerals, organic matter and pH status can change their dissolution, aggregation or surface chemistry. AgNPs may become ionic or sulfidized which could impact soil microbiota and may leech into groundwater. More stable, AuNPs can survive and build up in sediments, and there are concerns about the exposure of the environment to them over a long time (Arienzo & Ferrara, 2022). These environmental changes may have an indirect impact on human health by exposing it to contaminated water, food or environmental reservoirs. Indirect exposure risks are important to be considered by monitoring the persistence and mobility of nanoparticles.

8.4 Safe-by-Design Approaches for Minimizing Toxicity

The use of nanotechnology in the area of plant health can only be safely deployed when nanoparticles are designed with lower human toxicity. These can involve the optimization of particle size such that deep tissue penetration is minimized, biocompatible polymeric or biomolecular coatings such that oxidative reactivity is limited and development of slow release or encapsulated preparations so that free nanoparticles are not exposed. AgNPs synthesized with the help of plant extracts are green, which implies that they could slow down the release of $Ag+$ ions and increase biocompatibility. Inert ligands can be attached to AuNPs so as to restrict undesirable cellular uptake. In such safe-by-design, it is possible to achieve good agricultural performance with reduced toxicity to humans and the environment.

8.5 Need for Standardized Guidelines and Regulatory Oversight

The laws regarding nanomaterials in agriculture are emerging and there is no globally harmonised laws. Some of the recommendations that are available in OECD, EPA, and EFSA have focused on toxicity testing, analysis of residues and the assessment of environmental risks. To be used safely in the field of plant health, regulatory policies should include behaviors associated with nanoparticles, such as dissolution rates, environmental transformations and bioaccumulation patterns. The standardization of any assessment protocols and allowed levels of exposure will make the evaluation of health risks consistent (Oladele, 2024). Even more responsible commercialization will be supported by increased regulatory control and open reporting by manufacturers.

9. RISK ASSESSMENT, GUIDELINES, AND REGULATORY FRAMEWORK

9.1 Importance of Comprehensive Risk Assessment

The high rate of incorporation of AgNPs and AuNPs in plant health demands the use of a sound risk assessment system that ensures the protection of human and environmental health. Risk assessment needs to consider the entire life cycle of nanoparticles, that is, the synthesis and formulation, as well as field application, environmental transformation, and possible human exposures. In the case of AgNPs, the ion release and intense oxidative potential need to be scrutinized, whereas in the case of AuNPs, long-term profiling is needed because of its continuity and an ability to accumulate in tissues. Learning assessment systems need to encompass multi-tiered information, such as in vitro toxicity, in vivo reactions, environment behaviour, and exposure modelling (Almayahi, 2024).

9.2 Hazard Identification and Characterization

The process of hazard identification can be defined as a process involving the identification of whether AgNPs or AuNPs are inherently risky due to their physicochemical characteristics and biological interactions. Generally, AgNPs are categorized as more hazardous materials because they cause the generation of oxidative stress, membrane disruption and genotoxic potential. AuNPs are also thought to be less risky and non-inert, particularly in cases of modification using reactive ligands. It is vital to characterize dose-response relationships, determine key toxic endpoints as well as analyzing organ-specific vulnerabilities. The use of omics-based instruments including transcriptomics and metabolomics can be included in mechanistic insights and facilitate predictive modelling.

9.3 Exposure Assessment Across Agricultural Pathways

The exposure assessment is used to determine the risk and extent of human exposure to nanoparticles in case they are used in agricultural practices. The exposure of the workers is by inhalation, dermal contact and handling of nano-enabled formulations. Consumers can be subjected to traces of nanoparticles in food, whereas communities can be indirectly exposed to nanoparticles in contaminated water or soil. The main parameters are the mode of application, frequency, environmental conditions, and persistence of nanoparticles. AgNPs need further consideration because they dissolve, move, and enter the soil chemistry, being able to spread further (Huang et al., 2022). Probabilistic risk frameworks are exposure modelling tools that can be used to estimate real-world exposure situations.

9.4 Risk Characterization and Threshold Determination

Risk characterization involves combining hazard and exposure data to ascertain whether the risks that are being observed are in excess of acceptable limits. It is difficult to establish safe levels of AgNPs and AuNPs because of the different formulations, turturings, and dynamics of nanoparticles. Nevertheless, the existing evidence indicates that AgNPs need much lower permissible exposure limits because they are reactively ion mediated. AuNPs are relatively safer, but nevertheless, threshold assessment is necessary in case of chronic build-up. Regulatory bodies note that there must be safety margins that facture the vulnerable groups such as agricultural workers, children, and immunocompromised people.

9.5 Existing Regulatory Frameworks and International Guidelines

The regulation of nanomaterials in agriculture is still under development in different parts of the world. Other agencies like OECD, EPA, EFSA, and BIS of India have also proposed initial provisions in the characterization of nanoparticles, toxicity studies, and environmental safety. OECD offers guidelines of testing of physicochemical analysis, toxicity evaluation of genotoxicity and aquatic toxicity. EFSA emphasises on the importance of stringent residue assessment and nano-specific risk profiling of food safety. EPA requires the environmental impact assessment of nano-enabled pesticides or plant treatments. Even with these, regulatory policy loopholes still exist, namely, in the context of chronic exposure, nano-bio interactions, and long-term environmental outcomes (EFSA Scientific Committee et al., 2021).

10. FUTURE DIRECTIONS AND RESEARCH GAPS

10.1 Enhancing Mechanistic Understanding

Despite the great achievements, much is still unknown about the nano- bio interactions. Future studies are needed to define the way in which AgNPs and AuNPs are converted within cells, and how their protein corona varies with time, and affects cellular functions, signaling pathways, and long-term genomic stability. This development will be based on high-resolution imaging and real-time tracking facilities.

10.2 Integrating Multi-Omics Approaches

Traditional toxicology is limited in the number of endpoints. The various omics tools i.e. transcriptomics, proteomics and epigenomics can give a detailed picture of cellular reactions to nanoparticles. By combining omics data with nanoparticle characterization, one will be able to use early toxicity biomarkers and make predictive models of safer nanomaterial design (Ahmad et al., 2021).

10.3 Advancing Safe-by-Design Nanoparticles

One of the priorities is to develop nanoformulations which can increase the benefits of agriculture but reduce the toxicity of human beings. In the case of AgNPs, stabilizing surfaces to slow down the rate of ion release is required. In the case of AuNPs, the long term deposition can be restricted by designing biodegradable or faster clearing formulations. Agricultural nanoparticles should have safe-by-design engineering as a natural part.

10.4 Improving Experimental Models

The models that are more life-like are required to forecast the toxicity in the real world. However, more realistic human tissue architecture and exposure dynamics can be recreated in advanced 3D cultures, organoids, and organ-on-chip platforms than in conventional cultures. The uniformity of the conditions of testing will enhance the reproducibility and acceptability of regulation.

10.5 Studying Long-Term and Low-Dose Exposure

The number of studies on acute toxicity is more than that of chronic, low-dose exposure, as is the case in agriculture. There is an urgent need to conduct long-term studies on the cumulative effects with a special focus on vulnerable populations like farm workers in order to comprehend the faint genomic or epigenetic effects.

11. CONCLUSION

Nanoparticles such as silver and gold have been found promising in plant health nanotechnology, which has led to better disease management, delivery systems as well as increased agricultural efficiency. But their growing application requires a clear picture of their possible effects on the human health. In this chapter, the cytotoxic and genotoxic actions of AgNPs and AuNPs are evaluated together with a focus on how biological reactions are determined by the physicochemical characteristics, cellular uptake mechanisms, and intracellular modifications. AgNPs are always more reactive, which is caused by the release of $Ag+$ ions and the formation of oxidative stress, which ultimately results in mitochondrial dysfunction, membrane damage, and DNA instability. AuNPs are less toxic, but their long-term retention, ligand-dependent reactivity, and ability to accumulate in cells indicate the necessity of further caution. The chapter identifies the determinant factors that affect toxicity such as dose, exposure time, biological media and changes in the environment and the significance of a strong in vitro and in vivo assessment methods. The strategies that must be developed in the future to guarantee responsible innovation are safe-by-design strategies, multi-omics strategies, and advanced testing platforms. The harmonized regulatory control, the overall risk evaluation, and the interdisciplinary research are the conditions of sustainable use of AgNPs and AuNPs in the field of plant health. Nanotechnology can still help us have resilient and safe agricultural systems by incorporating the aspect of safety in all development processes.

REFERENCES

Abbasi, R., Shineh, G., Mobaraki, M., Doughty, S., & Tayebi, L. (2023). Structural parameters of nanoparticles affecting their toxicity for biomedical applications: A review. *Journal of Nanoparticle Research*, *25*(3), 43. PMID: 36875184

Ahmad, F., Mahmood, A., & Muhmood, T. (2021). Machine learning-integrated omics for the risk and safety assessment of nanomaterials. *Biomaterials Science*, *9*(5), 1598–1608. PMID: 33443512

Akhtar, S., & Zuhair, F. (2025). Advancing Nanomedicine Through Electron Microscopy: Insights Into Nanoparticle Cellular Interactions and Biomedical Applications. *International Journal of Nanomedicine*, 2847–2878. PMID: 40078651

Almanghadim, H. G., Nourollahzadeh, Z., Khademi, N. S., Tezerjani, M. D., Sehrig, F. Z., Estelami, N., & Sargazi, S. (2021). Application of nanoparticles in cancer therapy with an emphasis on cell cycle. *Cell Biology International*, *45*(10), 1989–1998. PMID: 34233087

ALMAYAHI, B. (2024). AgNPs: An Adaptive Methodology in Malignant Growth Treatment, Disclosing Upgraded Radiotherapy and Beyond.

Arienzo, M., & Ferrara, L. (2022). Environmental fate of metal nanoparticles in estuarine environments. *Water (Basel)*, *14*(8), 1297.

Aswani, R., Radhakrishnan, E. K., & Visakh, P. M. (2025). Nanoformulations for Agricultural Applications: State-of-the-Art, New Challenges, and Opportunities. *Nanoformulations for Agricultural Applications*, 1-34.

BalaKumaran, M. D., Ramachandran, R., Balashanmugam, P., Jagadeeswari, S., & Kalaichelvan, P. T. (2022). Comparative analysis of antifungal, antioxidant and cytotoxic activities of mycosynthesized silver nanoparticles and gold nanoparticles. *Materials Technology*, *37*(6), 411–421.

Bilardo, R., Traldi, F., Vdovchenko, A., & Resmini, M. (2022). Influence of surface chemistry and morphology of nanoparticles on protein corona formation. *Wiley Interdisciplinary Reviews. Nanomedicine and Nanobiotechnology*, *14*(4), e1788. PMID: 35257495

Bonaccorsi, S. (2021). *Tracking Nanoparticles going into the nucleus of the cell* (Doctoral dissertation, UNSW Sydney).

Canepa, E., Salassi, S., Simonelli, F., Ferrando, R., Rolandi, R., Lambruschini, C., & Rossi, G. (2021). Non-disruptive uptake of anionic and cationic gold nanoparticles in neutral zwitterionic membranes. *Scientific Reports*, *11*(1), 1256. PMID: 33441958

Čarapar, I., Jurković, L., Pavičić-Hamer, D., Hamer, B., & Lyons, D. M. (2022). Simultaneous influence of gradients in natural organic matter and abiotic parameters on the behavior of silver nanoparticles in the transition zone from freshwater to saltwater environments. *Nanomaterials (Basel, Switzerland)*, *12*(2), 296. PMID: 35055313

Chen, Y. Y., Lee, Y. H., Wang, B. J., Chen, R. J., & Wang, Y. J. (2022). Skin damage induced by zinc oxide nanoparticles combined with UVB is mediated by activating cell pyroptosis via the NLRP3 inflammasome–autophagy–exosomal pathway. *Particle and Fibre Toxicology*, *19*(1), 2. PMID: 34983566

Choudhary, A., Singh, S., & Ravichandiran, V. (2022). Toxicity, preparation methods and applications of silver nanoparticles: An update. *Toxicology Mechanisms and Methods*, *32*(9), 650–661. PMID: 35403559

Dastgheib, Z. S., Abolmaali, S. S., Farahavar, G., Salmanpour, M., & Tamaddon, A. M. (2024). Gold nanostructures in melanoma: Advances in treatment, diagnosis, and theranostic applications. *Heliyon*, *10*(15). PMID: 39170173

de Almeida, J. V. F. C., Faccioli-Martins, P. Y., Ferrante, M., Monteiro, J. P., Pinheiro, R. R., Lima, A. M. C., & de Mello Brandão, H. (2025). In vitro infection model in primary macrophages and in vivo evaluation of benzathine cloxacillin nanoparticles for the treatment of Corynebacterium pseudotuberculosis in goats. *Small Ruminant Research*, •••, 107561.

Dunkenberger, L., Reiss, K., & Del Valle, L. (2021). Comet assay for the detection of single and double-strand DNA breaks. In *Immunohistochemistry and Immunocytochemistry: Methods and Protocols* (pp. 263–269). Springer US.

EFSA Scientific Committee, More, S., Bampidis, V., Benford, D., Bragard, C., Halldorsson, T., ... & Schoonjans, R. (2021). Guidance on risk assessment of nanomaterials to be applied in the food and feed chain: Human and animal health. *EFSA Journal*, *19*(8), e06768. PMID: 34377190

Encinas-Gimenez, M., Martin-Duque, P., & Martín-Pardillos, A. (2024). Cellular alterations due to direct and indirect interaction of nanomaterials with nucleic acids. *International Journal of Molecular Sciences*, *25*(4), 1983. PMID: 38396662

Georgeous, J., AlSawaftah, N., Abuwatfa, W. H., & Husseini, G. A. (2024). Review of gold nanoparticles: Synthesis, properties, shapes, cellular uptake, targeting, release mechanisms and applications in drug delivery and therapy. *Pharmaceutics*, *16*(10), 1332. PMID: 39458661

González-Palomo, A. K., Saldaña-Villanueva, K., Cortés-García, J. D., Fernández-Macias, J. C., Méndez-Rodríguez, K. B., & Maldonado, I. P. (2021). Effect of silver nanoparticles (AgNPs) exposure on microRNA expression and global DNA methylation in endothelial cells EA. hy926. *Environmental Toxicology and Pharmacology*, *81*, 103543. PMID: 33166681

He, J., Ma, Y., Niu, X., Pei, J., Yan, R., Xu, F., & Ma, W. (2024). Silver nanoparticles induce endothelial cytotoxicity through ROS-mediated mitochondria-lysosome damage and autophagy perturbation: The protective role of N-acetylcysteine. *Toxicology*, *502*, 153734. PMID: 38290605

Huang, D., Dang, F., Huang, Y., Chen, N., & Zhou, D. (2022). Uptake, translocation, and transformation of silver nanoparticles in plants. *Environmental Science. Nano*, *9*(1), 12–39.

Islam, M. A., Riaz, T. A., Hossain, M. A., Lee, I., Akter, J., Shrestha, S., & Hahn, J. R. (2023). Controlled release of Ag+ ions to human cancer cells selectively neutralized with silver nanoparticles of different sizes produced by a green synthesis method. *Journal of Molecular Structure*, *1294*, 136384.

Kadhim, R. J., Karsh, E. H., Taqi, Z. J., & Jabir, M. S. (2021). Biocompatibility of gold nanoparticles: In-vitro and In-vivo study. *Materials Today: Proceedings*, *42*, 3041–3045.

Kariuki, R., Mirihana, K. A., Penman, R., Hung, A., Meftahi, N., Bryant, G., & Elbourne, A. (2025). Interactions of nanoparticles with living and synthetic bio-membrane. *Chemical Society Reviews*. PMID: 41025227

Kumar, A., Rai, Y., & Bhatt, A. N. (2024). Anti-cancer drug-mediated increase in mitochondrial mass limits the application of metabolic viability-based MTT assay in cytotoxicity screening. *Cytotechnology*, *76*(3), 301–311. PMID: 38736730

Kwasniewska, J., & Bara, A. W. (2022). Plant cytogenetics in the micronuclei investigation—The past, current status, and perspectives. *International Journal of Molecular Sciences*, *23*(3), 1306. PMID: 35163228

Kyziol-Komosinska, J., Dzieniszewska, A., & Czupioł, J. (2024). Behavior of Silver Species in Soil: Ag Nanoparticles vs. Ionic Ag. *Molecules (Basel, Switzerland)*, *29*(23), 5531. PMID: 39683691

Levak, M., Burić, P., Dutour Sikirić, M., Domazet Jurašin, D., Mikac, N., Bačić, N., & Lyons, D. M. (2017). Effect of protein corona on silver nanoparticle stabilization and ion release kinetics in artificial seawater. *Environmental Science & Technology*, *51*(3), 1259–1266. PMID: 28075572

Liu, J., Huang, Z., Yin, S., Zhou, X., Jiang, Y., & Shao, L. (2023). The lysosome-mitochondrion crosstalk engaged in silver nanoparticles-disturbed mitochondrial homeostasis. *The Science of the Total Environment*, *889*, 164078. PMID: 37209729

Madan, A., Saini, R., Dhiman, N., Juan, S.-H., & Satapathy, M. K. (2025). Organoids as Next-Generation Models for Tumor Heterogeneity, Personalized Therapy, and Cancer Research: Advancements, Applications, and Future Directions. *Organoids*, *4*(4), 23. DOI: 10.3390/organoids4040023

Malik, A. Q., Mir, T. U. G., Kumar, D., Mir, I. A., Rashid, A., Ayoub, M., & Shukla, S. (2023). A review on the green synthesis of nanoparticles, their biological applications, and photocatalytic efficiency against environmental toxins. *Environmental Science and Pollution Research International*, *30*(27), 69796–69823. PMID: 37171732

McLoughlin, C. D., Nevins, S., Stein, J. B., Khakbiz, M., & Lee, K. B. (2024). Overcoming the blood–brain barrier: Multifunctional nanomaterial-based strategies for targeted drug delivery in neurological disorders. *Small Science*, *4*(12), 2400232. PMID: 40213484

Mihailovic, V., Katanic Stankovic, J. S., Selakovic, D., & Rosic, G. (2021). An Overview of the Beneficial Role of Antioxidants in the Treatment of Nanoparticle-Induced Toxicities. *Oxidative Medicine and Cellular Longevity*, *2021*(1), 7244677. PMID: 34820054

Mittal, D., Kaur, G., Singh, P., Yadav, K., & Ali, S. A. (2020). Nanoparticle-based sustainable agriculture and food science: Recent advances and future outlook. *Frontiers in Nanotechnology*, *2*, 579954.

Mohideen, K., Chandrasekar, K., Ramsridhar, S., Rajkumar, C., Ghosh, S., & Dhungel, S. (2023). Assessment of oxidative stress by the estimation of lipid peroxidation marker malondialdehyde (MDA) in patients with chronic periodontitis: A systematic review and meta-analysis. *International Journal of Dentistry*, *2023*(1), 6014706. PMID: 37288387

Morais, M., Dias, F., Figueiredo, P., Tavares, I., Escudeiro, C., Teixeira, M. R., & Medeiros, R. (2024). Silver nanoparticles (AgNPs) uptake by caveolae-dependent endocytosis is responsible for their selective effect towards castration resistant prostate cancer. *International Journal of Nanomedicine*, •••, 9091–9107. PMID: 39258003

Morozova, O. V. (2021). Silver nanostructures: Limited sensitivity of detection, toxicity and anti-inflammation effects. *International Journal of Molecular Sciences*, *22*(18), 9928. PMID: 34576088

Naguib, M., Mekkawy, I. A., Mahmoud, U. M., & Sayed, A. E. D. H. (2022). Genotoxic evaluation of silver nanoparticles in catfish Clarias gariepinus erythrocytes; DNA strand breakage using comet assay. *Scientific African*, *16*, e01260.

Niżnik, Ł., Noga, M., Kobylarz, D., Frydrych, A., Krośniak, A., Kapka-Skrzypczak, L., & Jurowski, K. (2024). Gold nanoparticles (AuNPs)—toxicity, safety and green synthesis: A critical review. *International Journal of Molecular Sciences*, *25*(7), 4057. PMID: 38612865

Nunes, Á. M., da Silva Filho, R. C., da Silva, K. R., Bezerra, S. M., de Figueiredo, R. C., Saraiva, K. L., & Meneghetti, M. R. (2022). Gold nanoparticles with different shapes can cause distinct effect on mitochondria bioenergetics. *Journal of Nanoparticle Research*, *24*(2), 31.

Oladele, S. (2024). Risk Assessment in Nanoparticle Technology Adoption for Healthcare.

Omari Shekaftik, S., & Nasirzadeh, N. (2021). 8-Hydroxy-2′-deoxyguanosine (8-OHdG) as a biomarker of oxidative DNA damage induced by occupational exposure to nanomaterials: A systematic review. *Nanotoxicology*, *15*(6), 850–864. PMID: 34171202

Park, S. J. (2020). Protein-Nanoparticle Interaction: Corona Formation and Conformational Changes in Proteins on Nanoparticles. *International Journal of Nanomedicine*, *15*, 5783–5802. DOI: 10.2147/IJN.S254808 PMID: 32821101

Rokunuzzaman, M. K. (2024). The Nanotech Revolution: Advancements in Materials and Medical Science. *Journal of Advancements in Material Engineering*, *9*(2), 1–10.

Saifi, M. A., Khan, W., & Godugu, C. (2018). Cytotoxicity of nanomaterials: Using nanotoxicology to address the safety concerns of nanoparticles. *Pharmaceutical Nanotechnology*, *6*(1), 3–16. PMID: 29065848

Salama, B., Alzahrani, K. J., Alghamdi, K. S., Al-Amer, O., Hassan, K. E., Elhefny, M. A., & Fathalla, A. S. (2023). Silver nanoparticles enhance oxidative stress, inflammation, and apoptosis in liver and kidney tissues: Potential protective role of thymoquinone. *Biological Trace Element Research*, *201*(6), 2942–2954. PMID: 36018545

Sarkar, A., Ghosh, M., & Sil, P. C. (2014). Nanotoxicity: Oxidative stress mediated toxicity of metal and metal oxide nanoparticles. *Journal of Nanoscience and Nanotechnology*, *14*(1), 730–743. PMID: 24730293

Sati, A., Ranade, T. N., Mali, S. N., Ahmad Yasin, H. K., & Pratap, A. (2025). Silver nanoparticles (AgNPs): Comprehensive insights into bio/synthesis, key influencing factors, multifaceted applications, and toxicity— a 2024 update. *ACS Omega*, *10*(8), 7549–7582. PMID: 40060826

Schubert, J., & Chanana, M. (2018). Coating matters: Review on colloidal stability of nanoparticles with biocompatible coatings in biological media, living cells and organisms. *Current Medicinal Chemistry, 25*(35), 4553–4586. PMID: 29852857

Shafiq, M. S., Saleem, M., Ishaque, I., Zahra, T., Sarfraz, A., Sohail, M., & Rafique, S. (2025). Nanoparticle-Induced Oxidative Stress in Human Cell Lines: Enzymatic Biomarkers and Gene Expression Disturbances. *Scholars Academic Journal of Biosciences, 7*, 926–959.

Sun, J., Wan, J., Zhai, X., Wang, J., Liu, Z., Tian, H., & Xin, L. (2021). Silver nanoparticles: Correlating particle size and ionic Ag release with cytotoxicity, genotoxicity, and inflammatory responses in human cell lines. *Toxicology and Industrial Health, 37*(4), 198–209. PMID: 33625315

Tariq, M., Mohammad, K. N., Ahmed, B., Siddiqui, M. A., & Lee, J. (2022). Biological synthesis of silver nanoparticles and prospects in plant disease management. *Molecules (Basel, Switzerland), 27*(15), 4754. PMID: 35897928

Tomak, A., Yilancioglu, B., Winkler, D., & Karakus, C. O. (2022). Protein corona formation on silver nanoparticles under different conditions. *Colloids and Surfaces. A, Physicochemical and Engineering Aspects, 651*, 129666.

Varma, S., Dey, S., & Palanisamy, D. (2022). Cellular uptake pathways of nanoparticles: Process of endocytosis and factors affecting their fate. *Current Pharmaceutical Biotechnology, 23*(5), 679–706. PMID: 34264182

Yohannis, E., Urugo, M. M., Teka, T. A., Getachew, P., Tola, Y. B., Forsido, S. F., & Teferra, T. F. (2025). Aflatoxin Contamination in Agri-Food Systems: A Comprehensive Review of Toxicity, Food Security, Economic Impacts, and Sustainable Mitigation Across the Value Chain. *Food Science & Nutrition, 13*(10), e71104. PMID: 41111890

Yusuf, A., Almotairy, A. R. Z., Henidi, H., Alshehri, O. Y., & Aldughaim, M. S. (2023). Nanoparticles as drug delivery systems: A review of the implication of nanoparticles' physicochemical properties on responses in biological systems. *Polymers, 15*(7), 1596. PMID: 37050210

Zaib, S., Hayyat, A., Ali, N., Gul, A., Naveed, M., & Khan, I. (2022). Role of mitochondrial membrane potential and lactate dehydrogenase A in apoptosis. *Anti-Cancer Agents in Medicinal Chemistry-Anti-Cancer Agents, 22*(11), 2048–2062. PMID: 34825878

Zhao, Y., Tan, F., Zhao, J., Zhou, S., Luo, Y., & Gong, C. (2025). Targeting the Enhanced Sensitivity of Radiotherapy in Cancer: Mechanisms, Applications, and Challenges. *MedComm, 6*(6), e70202. PMID: 40384989

Chapter 15
Global Regulatory Guidelines for Nanoparticles Use in Plant Health Management

Ghaida Mustafa
https://orcid.org/0009-0003-1577-4439
University of Bradford, UK

Zakir Ullah
https://orcid.org/0000-0002-1819-5753
University of Bradford, UK

Anant Paradkar
https://orcid.org/0000-0003-1704-9858
University of Bradford, UK

ABSTRACT

Currently, there is an extensive demand for the application of nanotechnology in sustainable agricultural systems, nutrient delivery, pest control and addressing resistance of plant diseases. This quick implementation of nanoparticles-based agricultural products has outperformed the development of comprehensible regulatory frameworks. In this chapter the current global rules for the nanoparticles-based products used in plant health management are comprehensively reviewed. Scientific assessment as well as the harmonization efforts led by international frames such as FAO/WHO are also highlighted in this chapter. Reviewing the current guidelines reveals the absence of standardized definitions, limitation of risk assessment protocols and contradictory labelling requirements. Addressing these issues and adoption of rational harmonized evidence-based regulatory strategy will ensure environmental safety and biological systems health while fostering innovation in nano-based agricultural technologies.

DOI: 10.4018/979-8-3373-4913-8.ch015

1. INTRODUCTION

1.1. Development of nanotechnology as a transformative tool in plant health management.

Presently agricultural systems are faced by high demand due to growing world population which accompanied by other challenges such as climate change, biotic and abiotic stresses and decline water resources (Kumawat et al., 2023). To address this urgent global issues, potentially transformative strategies such as nanotechnology have emerged in the field of agricultural industries to boost food production sustainably (Palit, 2020).

Nevertheless, nanoparticles (NPs) have occurred naturally in ecosystems for thousands of years for instance as byproducts from living microorganisms such as bacteria, however engineered NPs have received remarkable consideration recently (Silva et al., 2021). Nanotechnology involves engineering, synthesis and application of NPs with submicron size from 1-100 nanometres (Khan & Upadhyaya, 2019). At this nanoscale, materials exhibit unique physicochemical properties such as large surface area to volume ratio, enhanced efficacy, and high absorptivity in comparison to larger particles equivalents. These features can be utilized for various applications in agricultural sector (Tarafdar et al., 2013). Therefore, NPs based products have been introduced into protocols for managing plant diseases as pesticides, bactericides and fungicides to enhance the proficiency, accuracy and sustainability of plant health management procedures. In addition, NPs have also been used as fertilizers for plants to relieve nutrients deficiency conditions and complement essential elements (Tarafder et al., 2020).

Nonetheless, considering their great advantages to agricultural sector, recently there are raised concerns about the potential health and environmental risks of NPs (Dikshit et al., 2021). Therefore, it is necessary to understand the importance of the rational use of NPs and development of comprehensive regulation frameworks.

The next subsection provides insights into scope of application of NPs in agricultural systems.

1.2. Rationale for Regulatory requirements of nanoparticles in agroecosystems.

Although the application of nanotechnology in agricultural sector especially for plant health management practices brings several beneficial advantages, consumers safety, ecological impacts and regulatory monitoring remain crucial. The United Kingdom (UK), European countries, United States of America (USA), Canada, India, China and Australia are among countries that have established solid regulatory frameworks for nano based agrochemicals. These policies aim to outline safety guidelines and legislation for rational application of nanomaterials in agricultural systems on an international level.

1.3. Chapter objectives and overview of structure.

This chapter provides overview of NPs in agroecosystems, illustrates global regulatory frameworks, highlights challenges and gaps in the current regulatory practices, and discusses risk assessment protocols and global harmonization efforts.

2. NANOPARTICLES IN AGROECOSYSTEMS

2.1. Types of nanoparticles commonly used in plant health management

Generally, NPs used in plant health management can be categorized according to their structure and function. Each category has characteristic properties that affect their interactions with plant systems and mechanisms of action. Metallic and metal oxide NPs e.g. silver NPs and Zinc oxide NPs are used for their antimicrobial, immunostimulant and delivery systems for nutrients (Wassie & Yemata, 2025; Seleiman et al., 2023). Carbon based NPs such as nanotubes have reported to enhance nutrient uptake, and increase plant tolerance mechanisms against abiotic stresses such as salinity (Zhu et al., 2022). In addition, this category of NPs has the ability to detect stress signalling molecules such as reactive oxygen species providing precise control of agricultural systems and early detection of diseases (Zhu et al., 2022). Polymeric class of NPs are applied for encapsulation of agrochemicals to provide controlled and targeted release patterns (Vemula & Reddy, 2023). Furthermore, fabricated natural polymeric NPs are green and biocompatible with intrinsic antimicrobial properties (Vemula & Reddy, 2023). Lipidic and nano emulsion types of NPs are used to enhance aqueous solubility and bioavailability of hydrophobic pesticides reducing the required dose and environmental impacts (Hosseinpour Jajarm et al., 2021; Fincheira et al., 2023). Moreover, they are applied to achieve stimuli and controlled release of pesticides (Campos et al., 2015).

2.2. Mechanisms of interaction of nanoparticles with plant health systems

The interactions between NPs and plant systems can occur through various mechanisms depending on NPs structure, concentration, size, physicochemical properties, route of application, exposure time and plant's species (Mohamed et al., 2017). One of the main mechanisms for interactions of NPs with plants is the enhancement of nutrients uptake. NPs can easily penetrate plants cellular membranes and cell wall structures facilitating the delivery of nutrients to the intended tissues (Liu et al., 2009). In addition, nanocarriers can interact with plant systems in a controlled release pattern that enables sustained and retarded release of agrochemicals with minimization of leaching and volatilization from plant tissues (Chen et al., 2021). Certain NPs such as metal oxides and chitosan NPs can activate specific signalling pathways that boost plant defence mechanisms against pathogens (Joshi, 2025). Moreover, metallic NPs possess a remarkable antimicrobial activity through disruption of vital metabolic pathways within the pathogens and activation of reactive oxygen species generation signalling in plant cells (Wassie & Yemata, 2025; Seleiman et al., 2023).

2.3. Advantages of nano-based agrochemicals over conventional counterparts

Compared to classical agrochemicals, NPs formulations have several advantages. In the arena of pesticides delivery, NPs provide unique advantage of improved delivery effectiveness through encapsulation of pesticides compared to traditional pesticides products (Adisa et al., 2019; Jiang et al., 2022). This strategy permits targeted delivery of pesticides, minimizing their unwanted effects and environmental pollution. In addition, nanotechnology formulation approach enhances the solubility of unsolvable active ingredients, permits adjusted release and improves stability.

Owing to their high surface to volume ratio, silica NPs are commonly utilized in NPs based pesticides and nano fertilizers (Naaz et al., 2023). For instance, nano fertilizers have been reported to have significant enhancement in efficiency over the typical ones (Li et al., 2020). Furthermore, mesoporous silica NPs are fabricated to allow controlled dispersion of insecticides in crops like rice (Zhu et al., 2018).

Also, nano formulations exhibit enhanced bioactivity with plant biological systems. It has been recorded that botanical pesticides inclusion into star polycation nanocarrier resulted in improvement of their bioactivity both in vitro and in vivo in terms of toxicity and persistence owing to particle size reduction (Yan et al., 2019). In another investigation, star polycation nano systems exhibited increased defending reactions against potato late blight via boosting of plant endocytosis pathways (Wang et al., 2021).

Among the valuable benefits of NPs especially polymeric one is their environmentally friendly composition. Polymeric NPs are green, biodegradable that permit precise delivery at minimal dosing therefore they are ideal matrices for pesticides (Pascoli et al., 2018).

Moreover, one of the remarkable advantages of nanotechnology is the ability to design stimuli-responsive systems usually by application of biodegradable polymeric carriers (Camara et al., 2019). These smart systems react with changes in the surrounding environmental parameters such as temperature, pH, and light allowing optimization of agrochemicals performance under filed conditions.

Additionally, NPs based formulations can be integrated with advanced technologies such as artificial intelligence offering a platform for design of smart products and precise control of agricultural systems (Ashique et al., 2025).

For more exploration of the advantages and applications of nano agrochemicals the reader can refer to the cited literature (Wahab et al., 2024).

It can be clearly recognized that, nano-based agrochemicals possess several advantages over traditional products; however, their use must be applied cautiously.

2.4. Alarming concerns and missing piece.

Currently there is a high demand for application NPs in plant health management approaches to boost crop growth, improve soil quality and nutrients uptake whilst facilitating targeted pesticides delivery and reduced utilization of agrochemicals (Khot et al., 2012). However, there are raising concerns related to risks of NPs to human health and ecological systems including accumulation in biological tissues, toxicity to other organisms and disruption of soil microflora (He et al., 2019). Therefore, regulatory bodies have founded safety principles, accentuating evaluation of toxicity and environmental influences assessments (Kookana et al., 2014).

The next section provides outline of international regulatory protocols.

3. OVERVIEW OF GLOBAL REGULATORY FRAMEWORKS

3.1. Safety measures and regulatory guidelines in the UK and European Union.

The UK's regulatory framework is in accordance with broader European union framework to affirm harmonization of safety procedures and regulatory adherence. There are multiple governmental organizations in the UK which are in charge for regulations of nano-based agrochemicals, remarkably Health and Safety Executive (HES), Food Standards Agency (FSA), and Department for Environment Food and Rural Affairs (DEFRA). The HSE offers wide-ranging safety guideline for applications of NPs at industrial scale by highlighting risk control measures and workplace exposure thresholds (Health and Safety Executive, n.d.). FSA guidance focuses on food safety though supporting rigorous risk measurements and traceable labelling of NPs based products (Food Standards Agency, n.d.). The DEFRA prioritises health and ecological impacts of nano based preparations and it has been updated in 2018 to support rational scientific evidence approach (Department for Environment, Food & Rural Affairs, 2025). Largely, the regulation of NPs based agricultural products in UK is continuously modifying according to the data generated by scientific investigations related to potential risks and benefits of NPs utilization and application.

For European countries, there are restricted regulations for the use of NPs based pesticides, and fertilizers and forage additives to protect human health and sustainability of environmental systems. Regulatory bodies such as European Food Safety Authority (EFSA) and European Chemicals Agency (ECHA) are responsible for providing guidelines related to risk assessment protocols and offering recommendations regarding safety, toxicity and requirements for regulation compliance.

In European countries there are several laws to control the use of NPs in biological, food production environmental and agricultural sectors. Table 1 provides illustration of selected major legislations and regulations for nano based agrochemicals (European Food Safety Authority, n.d.; European Chemicals Agency, n.d.).

Table 1. European regulatory framework for NPs based agrochemicals

Regulation	Year	Scope	Focus
Regulation (EC) No 1935/2004	2004	Materials in contact with food.	Indicates safety measures for food-contact materials to limit the potential of transfer of NPs into food.
Regulation (EC) No 396/2005	2005	Pesticides residue in food and animal feed products.	Determines the maximum allowed residual level for classical and nano pesticides.
Regulation (EC) No 1107/2009	2009	Plant health management products.	Emphasizes on the detailed analysis regarding safety and toxicity of NPs based pesticides to biological and ecological systems before approval for marketing.
Regulation (EC) No 767/2009	2009	NPs based animal feed additives	Requires detailed overview of NPs properties to protect healthiness of animal, humans, and environment.
Regulation (EU) No 2019/1009	2019	CE-marketed fertilizers	Specifies safety and quality references for NPs fertilizers to ensure their safety to soil, plant and aquatic systems.

Several European agencies have been established to support the existing regulatory frameworks in addressing the challenges encountered by NPs in agricultural and other food related practices. There are two essential institutions known as the European Food Safety Authority (EFSA) and the European Chemicals Agency (ECHA) which have significantly participated in the efforts applied for adaption of risk assessment principles for NPs and clarifying the regulatory compliance requirements within European continent.

EFSA has provided a detailed guidance on NPs risk assessment for their applications in food and animal feed supply as additives, plant health management products and materials in direct or indirect contact with food. The last updated one was in 2021 which highlights comprehensive multi-level method for safety assessment indicating the complications and diversity of fabricated NPs (European Food Safety Authority Scientific Committee et al., 2021). EFSA guideline underlines the importance of comprehensive physicochemical analysis of NPs in terms of their size and size distribution, morphology, surface energy, solubility, and accumulation property because these attributes significantly affect their bioavailability and toxicity data. Additionally, EFSA guidance supports the adoption of specific considerations when conducting toxicological assessment of NPs. The guideline suggests conduction of degradation studies using in vitro models, toxicant biokinetics evaluations, and mutagenicity assessment. These assessments are recommended to be followed by advanced studies on long-term toxicity, immune system toxicity, and impacts on the gut microbiota if required according to NPs mode of action and exposure hazards.

EFSA control protocol also emphasizes the importance of combined testing approaches and the utilization of innovative techniques to reduce animal testing whereas confirming rigorous hazard evaluation (European Food Safety Authority Scientific Committee et al., 2021).

Accompanying EFSA's effort, ECHA has established detailed protocol under the Registration, Evaluation, Authorisation, and Restriction of Chemicals (REACH) rule to certify the harmless usage of NPs across manufacturing and farming areas (European Chemicals Agency, n.d.). In January 2020, ECHA makes it mandatory for companies to provide specific information in their REACH profiles for completion of product registration process. The data must include detailed analysis of NPs, their dissolution profiles on potential hazards to both human health and ecological systems (European Chemicals Agency, n.d.).

On its appendix R7-1, ECHA provides endpoint-focused references for NPs physicochemical analysis, ecological providence, and toxicity assessments, taking not consideration attributes for example aqueous solubility, aerosolization potential, and sorption/ desorption dynamics. ECHA protocol also tackles practical obstacles, for instance those involved during sample preparation and dispersal stability. These limitations are essential for producing replicable and systematically authentic data. Moreover, ECHA advocates the use of non-animal assessment models in scientific alignment with European commission obligation to the 3Rs attitude (Replacement, Reduction, Refinement) in animal experimentation (European Commission, n.d.).

Together EFSA and ECHA emphasize the necessity for context-dependant assessments, assumed the miscellany of NPs and their condition-specific interactions within living and ecological systems. These rules jointly intend to complement nano specific risk evaluation conductions across the Europe, guaranteeing that invention in nanotechnology is applied with precise safety principles. Nonetheless, although these progresses, argues remain in executing consistent analysis procedures and tackling data breaks for chronic ecotoxicological impacts. It is mandatory for the refinement of NPs based products regulatory frameworks and rational practice in agroecosystems a presence of constant association amongst regulatory authorities, manufacturing investors, and research foundations.

To summarize, the regulatory frameworks in UK and European countries for NPs based agrochemicals are established based on precise safety measures and reliable scientific evidence. Nevertheless, their efficiency centres on multidisciplinary collaboration and flexibility to adapt the evolving hazards in agricultural systems and environmental control.

3.2. Safety measures and regulatory guidelines in the United States of America (USA)

NPs-based agrochemicals in the USA are regulated by various authorization bodies including the Food and Drug Administration (FDA), Environmental Protection Agency (EPA) and US Department of Agriculture (USDA) to certify their safety to consumers, ecological sustainability and fulfilment of regulations requirements (Hannon et al., 2015). The FDA assesses and ensure the safety of NPs used in food and nutritional supplement preparations before they can be utilized by the consumers (Grobe & Rissanen, 2012). The EPA evaluates the degree of ecological risks of NPs applied in pest control products and other agrochemicals (Kookana et al., 2014). The USDA regulates the application of NPs in sustainable and common agricultural practices by following certain programs such as the National Organic Program (NOP) (Dasgupta et al., 2015).

There are numerous NPs based products which have been authorized for use in numbers of food and farming applications. For instance, preparations such as sweeteners, chewing gum commonly include titanium dioxide as colour-removing agent products containing silica NPs as free-flowing agent such as coffee creamer have been approved by FDA (U.S. Food and Drug Administration, n.d.). Ingredients like zinc oxide (colorant) and iron oxide (pigmentation agent) are utilized in USA upon passing the regulations standards by FDA (U.S. Food and Drug Administration, n.d.). Despite their consideration as safe materials, there is a mandatory requirement for assessing their long-term impacts on biological tissues and ecosystems.

There are several essential laws and regulatory guidance for the preparation, approval and control of nano agrochemicals in the USA, Table 2 provides summary of these legislations.

Table 2. Summary of the USA key laws and regulation frameworks indicating scope, requirements and references

Law/ Regulation	Regulatory Body	Scope	Requirement / Focus	References
The Toxic Substances Control Act (TSCA)	EPA	Chemical materials including NPs	Requires the manufacturer to provide safety data regarding potential human and ecological health influences prior marketing.	(U.S. Environmental Protection Agency, n.d.)
The Federal Insecticide, Fungicide, and Rodenticide Act (FIFRA)	EPA	Pesticides based on NPs	Requires the manufacturer to demonstrate safety and efficacy profiles of NPs products before registration and application.	(U.S. Environmental Protection Agency, n.d.)
The Food, Drug, and Cosmetic Act (FD&C Act)	FDA	Nanotechnology based food, drugs and cosmetic products	Obligates disclosure of NPs characteristics and properties to confirm consumer and environment safety.	(U.S. Food and Drug Administratio, n.d.)
The National Organic Program (NOP)	USDA	Organic agricultural production	Emphasize NPs producer to fulfil with federal regulatory frameworks applied for agricultural production.	(U.S. Department of Agriculture, n.d.)
The Nanotechnology Research and Development Act (NRDA)	Miscellaneous	NPs	Encourages collaborations between regulatory agencies on nanotechnology research and innovation to ensure safety and implementation of regulatory practices.	(Congress.gov, n.d.)

In addition to laws and regulations, several safety measures exist in USA for manufacturers to guarantee the sensible application of NPs in agroecosystems. Summary of these safety practices is provided in Table 3.

Table 3. Key safety measures for NPs, focusing on toxicity assessment, labelling, environmental evaluation and occupational risk control

Safety Practice	Objective	Key requirements/ Actions	Reference
Rigours safety testing	Evaluate human and environmental healthiness hazards of NPs.	Perform massive toxicity study and risk assessment for NPs	(Kookana et al., 2014)
Labelling requirements	Provide consumers with product's information related to composition, possible risks, safe application and enable risk assessment.	Reveal NPs properties (e.g. size, shape, reactivity…etc) on the product label to enable the risk considerations measures.	(U.S. Food and Drug Administration, n.d.)
Assessment of environmental impact	Protect natural ecosystems from disruption by encouraging the manufacturing companies to perform risk assessment studies.	Assess the interactions of NPs with water, soil and biota to minimize environmental risks.	(Hansen et al., 2016)
Health and safety training and risk control	Reduce exposure at workplace and contamination by providing training programs for employees related to safe handling, storage and disposal of NPs based products.	Fulfil safety guidelines, apply protective equipments and follow waste management protocols.	(Scott & Chen, 2012)

The regulation of NPs based agricultural products in USA is an incessantly progressing field, necessitating the authorities to adopt the emerging research on efficiency and safety of nanomaterials. This can be achieved by collaborations with scientific institutions, industrial firms, policymakers, stakeholders and regulatory agencies.

3.3. Safety measures and regulatory guidelines in Canada

The Canadian regulatory framework for application of nano-based agricultural products including those used in plant health management is based on multiple laws and regulatory principles. These multi-agency legislations are designed to ensure the safety of consumers and ecological systems. The Canadian regulatory rules are applied to a broad range of NPs based products including nano pesticides, nano fertilizers and nano based food additives.

Canadian federal agencies in charge for NPs based agrochemicals regulations are Health Canada, Environment and Climate Change Canada (ECCC), and the Canadian Food Inspection Agency (CFIA) (Institute for Science, Society and Policy, 2022). Each of these agencies works with its command to ensure that application of NPs based agricultural formulations comply with comprehensive safety and quality measures prior to commercialization.

According to these regulations, agricultural NPs including nano pesticides, nano fertilizers, and nano engineered food additives are exposed to comprehensive pre-approval assessments. These involve series of characterisation including risk evaluations, toxicologic reporting, and amenability with labelling and safety values, before authorization for market introduction and distribution. These evaluations are crucial for assessing the possible risks introduced by NPs based agrochemicals to consumers, workers and environment. The process is supported by subsequent regulatory tools cooperatively define the legal framework governing NPs based agrochemicals in Canada.

One of the key laws is the Canadian Environmental Protection Act, 1999 (CEPA) which indicates the effects of materials including NPs on biological and ecological systems (Government of Canada, n.d.a). Under the umbrella of this rule, there is ''New Substances Notification Regulations''. These legislations mandate the manufacturers to provide rigorous information on the chemical identification, particle size surface properties and ecological fate of NPs. This allows the regulation authorities to perform risk assessments and introduce control standards where required.

The Food and Drugs Act (FDA) has an essential role in regulation of NPs based food, drug and cosmetic formulations (U.S. Food and Drug Administration, n.d.). It obligates the disclosure of NPs properties such as size, composition, and other functional characteristics to ensure safety of consumers. This is especially related to NPS based agrochemicals which may drop residues on crops or be applied in food manufacture approaches. The FDA rules ensure that such formulations are assessed for possible toxicity, other interactions with biological systems beforehand approval.

The regulation of both conventional and NPs based pesticides is ruled by Pest Control Products Act (PCPA) (Government of Canada, n.d.b). It indicates the requirements for registration, safety evaluation and labelling of NPs based pesticides. The law affirms that these products are assessed for their effectiveness, ecological residue and probable risks to non-target creatures such as aquatic organisms and pollinators. The PCPA regulation also monitoring of long-term impacts and compliance of NPs formulations by emphasizes of post-market surveillance.

The Canada Agricultural Products Act (CAPA) pairs other Canadian regulations by supervising the commercialization and quality control of agrochemicals including those based on NPs (Government of Canada, n.d.c). It ensures that NPs based agricultural products comply with founded measures for purity, stability, and performance whilst being accurately labelled with information to consumers about composition and intended use.

All together, these rules create multilayered regulatory framework that incorporates product safety, ecological system protection and customer transparency. Canadian's regulation system for NPs based agrochemicals has a precautionary feature involving risk assessment prior to market entry and providing comprehensive guidelines throughout the product lifecycle. The participation of several agencies affirms that NPs based formulations are analysed from various aspects of health, environment and commercialization which provides a rigorous protection against possible risks.

Table 4 provides summary of some of key laws and regulation guidelines in Canada.

Table 4. Selected examples of Canadian regulatory framework of NPs application in agrochemicals

Law/ guideline	Scope	Focus	References
Canadian Environmental Protection Act, 1999 (CEPA)	Impacts of NPs on human and environmental health	Indicates the protocol of risk assessment and control under ''New Substances Notification Regulations''.	(Government of Canada, n.d.a)
Food and Drugs Act (FDA)	Nanotechnology based food, drugs and cosmetic products	Obligates disclosure of NPs characteristics and properties to confirm consumer safety.	(U.S. Food and Drug Administration, n.d.)
Pest Control Products Act (PCPA)	Regulation of classical and NPs based pesticides	Indicates requirements of registration, safety, and labeling of NPs based pesticides.	(Government of Canada, n.d.b)
Canada Agricultural Products Act (CAPA)	Agricultural products	Regulates marketing and assessment of quality control standards of agrochemicals.	(Government of Canada, n.d.c)

3.4. Safety measures and regulatory guidelines in Australia

There are several structured regulatory guidelines for application of NPs in agricultural systems in Australia. These rules have developed to ensure safe utilization of NPs based agrochemicals, integrate risk assessment measures, guarantee workplace safety and implement global collaboration. Australian regulatory framework is mainly governed by two key systems: Australian Pesticides and Veterinary Medicines Authority (APVMA) and regulations for workplace safety under Work Health and Safety (WHS) laws & The National Industrial Chemicals Notification and Assessment Scheme (NICNAS).

The APVMA is the chief authorization responsible for indication of guidelines for registration and regulation of pesticides and veterinary medicines, comprising NPs-based agrochemicals (Australian Pesticides and Veterinary Medicines Authority, n.d.). This legislation covers aspects of pre-market evaluation, risk assessment, and compliance control to ensure that products comply with rigorous safety and effectiveness standards prior to commercialization. Regarding NPs based products, APVMA requires comprehensive data on physicochemical characteristics, ecological persistence, toxicity data and exposure routes taking into consideration the effect of NPs properties on interaction with biological systems and toxicity. Significantly, APVMA cooperates with global regulatory bodies such as ECHA and US EPA to complement risk assessment protocols and align with international best procedure. This global collaboration enables consistency in assessing specific hazards related to NPs and support mutual identification of safety measures and reducing replication of regulatory efforts.

Moreover, Australian authorities accentuate occupational health and safety via WHS rules and NICNAS (Safe Work Australia, n.d.). These legislations require the employers to create rigorous workplace standards to reduce risks associated with handling, labelling and disposal of NPs based agrochemicals. Key specifications include identification of risks, strategies to control exposure, application of personal protecting equipment and cautious waste control protocols. In addition, NICNAS requires notification and evaluation of synthetic chemicals including NPs before their introduction into the workplaces. These measures ensure that NPs are assessed for their possible hazards to workforce and environment and that suitable risk moderation protocols are in place.

Concurrently, these Australian regulatory frameworks reflect a dual core; protection of human health and ecological integrity whilst permitting technical innovation in agriculture. The Australian legislation approaches address hazards and risks related to both NPs products and their processing. This is achieved by requiring rigorous risk evaluations and enforcing workplace safety measures. Owing to translational nature of agricultural trade chains and nanotechnology research, it is crucial to implement global collaboration for ensuring robustness of regulatory bodies.

Table 5 provides summary of selected key laws, safety measures and regulations for NPs based agrochemicals in Australia.

Table 5. Selected examples of Australian regulatory framework of NPs application in agrochemicals

Law/ guideline	Scope	Focus	References
Australian Pesticides and Veterinary Medicines Authority (APVMA)	Pesticides & veterinary medicine.	• Indicates the required guidelines for registration and risk evaluation of NPs agrochemicals. • Collaborates with other international regulatory bodies such as ECHA and EPA	(Australian Pesticides and Veterinary Medicines Authority, n.d.)
Work Health and Safety (WHS) laws & The National Industrial Chemicals Notification and Assessment Scheme (NICNAS)	Workplace exposure to NPs	Requires the employers to establish workplace measures to ensure employees safety and adherence to minimizing risks protocols during handling, labelling and disposal of NPs.	(Safe Work Australia, n.d.)

In summation, Australian regulatory guidelines for NPs based agrochemicals are science based, precautionary and internationally aligned. APVMA makes certain that NPs based pesticides and other related products undertake precise assessment before release into the market. While WHS and NICNAS laws offer a shielding layer for workforce and environment during manufacturing and application of those NPs based agrochemicals. Continual enhancement of these legislations along with engagement of stakeholder and research facilities on risk factors associated with nano formulations is crucial to maintain safety and sustainability. That is because there is a massive involvement of nanotechnology in the modern agricultural systems.

3.5. Safety measures and regulatory guidelines in China

The regulation of NPs based agrochemicals in China is based primarily on horizontal laws and guidance which apply to existing chemicals and classical pesticides rather than through an individual nano-specific regulation. Management is distributed across several governmental authorities such as the Ministry of Agriculture and Rural Affairs (MARA), the State Administration for Market Regulation (SAMR), the National Health Commission (NHC) and the Ministry of Ecology and Environment (MEE). Cooperatively, these authorities organize series of requirements for before market assessment, product standardization, labelling, traceability, and post-market evaluation. These measures are to control potential human and environmental risks related to NPs based agrochemicals such as nano pesticides, and nano fertilizers.

Within this framework, NHC's Safety Requirements for Food and Food Additives Containing Nanomaterials (NHC No. 13) — 2011 offers guidelines for manufacturers of NPs that may enter the food chain (Kumari et al., 2023). Even though these legislations are designed for food and food additives, the protocol is directly applicable to agricultural products. The requirements specify expectations of toxicology data, physicochemical properties evaluation (e.g. size, agglomeration and surface charge), exposure limits and labelling. In addition, these provisions link primary production with food safety management as they determine the standard of risk assessment in harmonization with labelling transparency. This is important for regulation of agrochemicals which may present as traces in eatable commodities.

Moreover, MARA's Technical Guidelines for Safety Assessment of Nano-Scale Agricultural Products (MARA No. 198)—2014 provides a layered, class-specific framework for nano pesticides, nano fertilizers, nano-coated seeds (Kumari et al., 2023). The rules underline assessment related to stability (e.g. alteration in soil and water bases), risk (acute and chronic toxicity to biological, ecological to non-target organisms such as beneficial arthropods) and exposure (release, consequence and transfer). Significantly, these legislations mention nano-specific analysis methods under consumption conditions (e.g. the particle size distribution in the product and following application). In addition, the guidelines recognize that the properties of NPs and accordingly their risks are circumstances dependent. The (MARA NO 198 – 2014) standards anchor pre-commercialization data requirements and advise on labelling and management conditions for agricultural application of NPs.

Another key regulation in China is Administrative Measures for Safety Evaluation of New Varieties of Agricultural Genetically Modified Organisms (SAMR Order No. 8) — 2020 which is under control of SAMR (Kumari et al., 2023). Although, these rules centre on genetically modified organisms, they can be applied for nano-based products used in agricultural settings. Practically, SAMR's measures strengthen the prospect that rigorous, pre-approval safety verification is essential when novel technologies such as nanotechnology are introduced into viable agricultural settings. This provides an extra regulatory scope that merges with MARA's specialized risk assessment for nano-based products.

From an ecological viewpoint, the MEE's Measures for the Administration of Environmental Safety Assessment of Agricultural Genetically Modified Organisms (MEE Order No. 12)—2021 offers a road-map for assessment of NPs risks on ecosystems (Kumari et al., 2023). These risks include persistence in ecosystems, bioaccumulation possibility, and effects on non-target organisms and performance of agricultural systems. MEE's rules are framed around genetically modified organisms, however, they can be adopted for evaluation of NPs-based agricultural products on ecosystems. The guidelines include protocols focused on recognizing the scope of risk assessment and identifying the possible environmental risks, detection of exposure pathways and stepwise testing approaches for potential risks.

The data presented in Table 6 provides overview of selected key acts, safety measures and regulations related to application of NPs products in China (Kumari et al., 2023).

Table 6. Key regulations of application of nanotechnology in agriculture in China

Law/ Regulation	Authority	Focus
Safety Requirements for Food and Food Additives Containing Nanomaterials (NHC No. 13)—2011	NHC	Indicate safety standards and labelling requirements for NPs used in food products.
Technical Guidelines for Safety Assessment of Nano-Scale Agricultural Products (MARA No. 198)—2014	MARA	Provide guidelines for assessment of stability, toxicity and ecological impacts of nano pesticides, nano fertilizers, nano-coated seeds.
Administrative Measures for Safety Evaluation of New Varieties of Agricultural Genetically Modified Organisms (SAMR Order No. 8)—2020	SAMR	Emphasize restrict safety measures before approval of NPs based products into the market.
Measures for the Administration of Environmental Safety Assessment of Agricultural Genetically Modified Organisms (MEE Order No. 12)—2021.	MEE	Provides layout of requirements and procedures for assessment of environmental safety of NPs based products.

China's regulatory framework of NPs demonstrates deliberate alignment of biological systems safety, ecological sustainability along with technological advancement and it is expected to accommodate the evolving challenges in the field of NPs based agrochemicals.

3.6. Safety measures and regulatory guidelines in India.

India has established multidisciplinary approach for regulation of NPs agrochemicals. This inter-ministerial approach of regulation is to ensure safe development, manufacturing and commercialization.

The Indian framework of regulation integrates ecological, agricultural, food and pharmaceuticals laws echoing the complexity of applications of nanotechnology aspects in agriculture settings.

The Environment (Protection) Act, 1986, under the control of Ministry of Environment, Forest and Climate Change (MoEFCC) acts as a principal law for managing hazardous materials compromising NPs (Government of India, 1986). Under this legislation, manufacturing, commercialization and application of products based on NPs are regulated with consideration of the commitment to environmental safety standards. In addition to the previous rule, another legislations known as the Hazardous Waste (Management, Handling, and Transboundary Movement) Rules, 2016 comes also under MoEFCC (Central Pollution Control Board, n.d.). They focus on the management of disposal and cross-border movement of hazardous substances. These laws require registration and approval of operational units that produce, gather and dispose of hazardous materials including NPs therefore, reducing environmental risks.

Furthermore, there is the Food Safety and Standards Act, 2006 which addresses food safety and enforced by Food Safety and Standards Authority of India (FSSAI) (Kumari et al., 2023). This legislation regulates the inclusion of NPs in agrochemicals and food additives to ensure the protection of consumers and compliance with healthiness standards. Relatedly, the Insecticides Act, 1968 governed by Central Insecticides Board & Registration Committee (CIBRC), regulates the registration, marketing and application of NPs based insecticides. This ensures that NPs based pesticides comply with efficiency and safety specifications before getting into the market.

MoEFCC additionally issues special statements, such as Manufacture, Storage and Import of Hazardous Chemicals Rules (1989), which framework safety and ecological measures for NPs manufacturing and import (Ministry of Environment, Forest and Climate Change, 1989). These laws underline risk assessment and protective policies to moderate hazards during production and storage of NPs.

In the research arena, the Indian Council of Agricultural Research (ICAR) established Guidelines on Nanotechnology Research in Agriculture (2010) to raise in authority invention (Kumari et al., 2023). Thess regulations provide a constituted methodology for safe application of NPs in agricultural research and development with accentuation on safety of biological as well as environmental systems. Moreover, the Indian Pharmacopoeia Commission (IPC) issued Guidelines on Nanoparticle Characterization (2019), providing standardized procedure for analysis of NPs properties in agrochemical formulations (Kumari et al., 2023). This rule guarantees reliability in quality control and enables regulatory conformity.

It can be clearly observed that several regulatory bodies exist in India. Table 7 provides summary of selected examples including their scope, focus and responsible authorities.

Table 7. Summary of selected Indian regulatory measures of application of nanotechnology in agro-chemicals

Law/ guideline	Authorization body	Scope	Focus	References
The Environment (Protection) Act, 1986	Ministry of Environment, Forest and Climate Change (MoEFCC)	Hazardous materials including NPs	Regulates manufacturing, global trade and application of NPs.	(Government of India, 1986)
The Hazardous Waste (Management, Handling, and Transboundary Movement) Rules, 2016	MoEFCC	Management of hazardous substances disposal	Involves registration and approval of operational units that produce, gather and dispose of hazardous materials including NPs.	(Central Pollution Control Board, n.d.)
The Food Safety and Standards Act, 2006	Food Safety and Standards Authority of India (FSSAI)	Food additives including NPs	Regulates the application of NPs in agrochemicals.	(Kumari et al., 2023)
The Insecticides Act, 1968	Central Insecticides Board & Registration Committee (CIBRC)	Insecticides	Regulates registration, marketing and application of NPs based insecticides.	(Government of India, 1968)
MoEFCC Notification on Manufacture, Storage and Import of Hazardous Chemicals Rules (1989)	(MoEFCC)	Management of hazardous substances	States the requirements for safety and environmental measures of manufacturing and import of NPs products.	(Ministry of Environment, Forest and Climate Change, 1989)
ICAR Guidelines on Nanotechnology Research in Agriculture (2010)	Indian Council of Agricultural Research (ICAR)	Research in agricultural nanotechnology	Provides structure for safe and liable utilization of NPs for research and development purposes.	(Kumari et al., 2023)
Indian Pharmacopoeia Commission (IPC) Guidelines on Nanoparticle Characterization (2019)	Indian Pharmacopoeia Commission (IPC)	Analysis of NPs	Provides protocols for characterization of NPs included in agrochemicals.	(Kumari et al., 2023)

It is worth noting that, Indian framework of regulation of NPs based agrochemicals demonstrates scientific and risk-based approach. However, it is still developing necessitating the requirement for more inclusive and combined strategy to grantee safety and effectiveness. The present regulation polices are modifying according to constant assessments and may be accompanied by extra specific protocols and provincial appropriate principles.

4. CHALLENGES AND GAPS IN THE CURRENTLY APPLIED REGULATORY FRAMEWORKS

There is a substantial variation in the safety data for control and assessment of NPs based agro-chemicals due to the absence of standardised protocols in the current regulatory frameworks. The main regulatory hurdle is the lack of global agreement on terminologies, evaluation procedures and analysis methods. There is an absence of an internationally accepted definition of the term ''nanomaterials'' or ''nanoparticles'' among the different regulatory authorizations that complicates product registration and global marketing (Ijaz et al., 2024). Moreover, regulatory bodies are currently using risk assess-ment measures commonly adapted from general chemicals protocols. This approach fails to sufficiently detect the inimitable properties of NPs such as accumulations in biological and eco systems along with possibility of interactions with such systems resulting in doubt in regulatory approval or considerations. Furthermore, a critical gap exists in regulatory guidelines of NPs based agrochemicals owing to the absence of authenticated methods tailored specifically for assessment of their toxicity and fate in agricultural systems (Karahalil, 2018).

In addition to the above-mentioned challenges, other limitations arise from restricted transparency of data, regulatory incoherence, variable labelling and documentation requirements. Also, for the developing communities there might be some social, ethical and legal concerns for application of NPs agrochemicals due to limited regulatory capacities.

One of the suggested solutions to overcome the challenges associated with regulations of NPs use in agroecosystems is the adoption of inclusive regulatory platform in which polices, and guidelines harmonized to ensure safe and sustainable practice of such products (Karahalil, 2018).

5. RISK ASSESSMENT PROTOCOLS FOR NANOPARTICLES IN AGROECOSYSTEMS

Risk assessment is a systematic approach for identification, evaluation and analysis of the possible hazards related to application of a substance or technique, and to assess the probability and severity of adverse impacts under certain exposure conditions. In arena of nanotechnology, estimation of possible risks involves characterizing physicochemical properties of NPs, their ecological impacts, exposure routes and evaluation of their toxicity to environmental and biological systems (European Food Safety Authority Scientific Committee et al., 2021).

Development of a robust risk assessment protocol for NPs based agrochemicals is crucial to estimate their hazards, backbone implementation of regulatory guidelines, permits rational safety-based design of developing methodologies. Despite the absence of specific risk evaluation practice for NPs, some global authorities such as EFSA and Organisation for Economic Co-operation and Development (OECD) indorse adaptation of classical risk assessment measures to consider the unique properties of NPs such as size, charge and aggregation behaviour (Usmani et al., 2024).

It is worth noting that, there are cutting-edge methods that can be applied for robust anticipation of toxicity and environmental effects of NPs while reducing the reliance on animal models. These approaches include Integrated Approaches to Testing and Assessment (IATAs) and New Approach Methodologies (NAMs) which merge in silico modelling, in vitro tests and omics-based technologies (Usmani et al., 2024). These procedures provide understanding of the potential interactions between NPs with plants, soil and other biological systems through determination of certain parameters such as dissolution kinetics, antioxidant profile, and bioaccumulation. OECD has established an internal agreement known as Mutual Acceptance of Data Framework to permit harmonization in NPs risk assessment methods. According to this approach safety data generated and documented based on OECD guidelines in one member country must be allowed by other contributing countries for entire regulatory purposes (Rasmussen et al., 2019).

Regardless of progresses in NPs risk assessment measures, remarkable limitations exist in evaluating their long-term impacts in agroecosystems. The exposure modelling and hazard estimation methods are complicated for assessment of influences of NPs owing to their tiny size and other distinctive attributes (Islam, 2025). In addition, some recent literature suggested the importance of toxicity assessment protocols to consider interactions between NPs and ecological and biological systems and to gather real-field samples for analysis.

Therefore, to ensure green application of NPs based products in agroecosystems, future risk assessment protocols should incorporate scientific based innovative approaches that allow continuous monitoring, regulatory harmonization and safe-by-design application (Usmani et al., 2024).

6. GLOBAL REGULATORY HARMONIZATION APPROACH

One of the essential solutions to overcome challenges in innovation, safety and international trade of NPs based agrochemicals is the application of global regulatory harmonization strategy. Core priorities involve establishment of a globally standardized definition of NPs depending on their size, morphology and functional characteristics. Additionally, design integrated NPs focused risk assessment principles based on international research collaboration and regulatory authorities' agreements as stated in the previous section for OECD guidelines.

International frameworks such as Food and agriculture organization (FAO) and World Health Organization (WHO) in partnership with OECD and International organization for Standardization (ISO) to initiate comprehensible international standards for regulation of NPs based agricultural products (Rasmussen et al., 2019).

Moreover, creation of universal open-access register office for NPs based agrochemicals would improve regulation transparency and implementation and support life cycle safety monitoring. This registry should contain information related to product composition profile, application, safety records, and approval status in various countries. It can be accommodated by unbiassed international authority such as OECD and FAO. Furthermore, the global registry approach will promote mutual identification agreements to avoid regulatory replication. It is important to highlight the importance of establishment of regulatory offices in low- and middle-income countries by providing scientific training and creation of the required infrastructures. Finally, collaboration and transparent dialogue at international level among stakeholders, society, regulatory bodies, research institutions are mandatory to support safe, effective and sustainable application of NPs based products in agroecosystems.

7. CONCLUSION

There is no doubt that nanotechnology is reshaping plant health management sector through offering an innovative platform to fabricate precise preparations for delivering nutrients, controlling acquired diseases and withstanding environmental stresses such as salinity and dryness. These NPs based products possess superior properties compared to their conventional counterparts in terms of efficiency, sustainability, and adjusted delivery control and much more advantageous characteristics that enable smart agricultural practices.

Despite the above-mentioned benefits of NPs, their hasty incorporation into agricultural ecosystems creates crucial challenges for regulatory authorities to ensure their efficacy, safety and sustainability to biological and ecological systems. Considering their unique physiochemical properties of submicron size that permits high interactions with surrounding environment including biological and other ecological systems leading to safety and toxicity concerns. These complications battle classical risk assessment methodologies as they were originally planned for bulk chemical substances. Therefore, there is a pressing need to tailor those protocols to address the inimitable characteristics of NPs and their short- and long-term impacts on living organisms' health and ecological safety.

Major gaps exist in nano-tailored risk assessment protocols, limitation of long-term toxicological data and absence of international harmonization in regulation guidelines. Tackling these issues requires adoption of innovative evaluation approaches such as IATAs and NAMs that merge advanced tools. These approaches are crucial for apprehending the complex nature of NPs systems across wide diverse of ecological and biotic frameworks.

The coordination between the international regulatory authorizations is of paramount importance. The fragmentation in regulatory frameworks can hinder innovation, create global trade boundaries, and weaken consumers trust. Global regulatory harmonization efforts led by organizations such as FAO and WHO in association with other agencies such as OECD and ISO should be highlighted to guarantee reasonable and green practices of NPs based products in agroecosystems. These collaborations are aimed to establish internationally accepted comprehensive description of NPs, analysis methods and risk evaluation measures for agricultural applications.

Establishment of regulatory capacity in low- and middle-income countries should be prioritized. In many of the developing areas, there is an absence of logistics infrastructures and authorization guidelines to assess NPs based agrochemicals aggregating the risk of unsupervised market or elimination from technical benefits. Therefore, global partnership should focus on information and technology transfer, training and design of cost-effective analytical tools to allow fair access and rational application.

Eventually, the reliable implementation of nanotechnology in agroecosystems require scientific based regulation frameworks that progress with scientific innovations. Through integration of NPs tailored risk assessment protocols, boosting international regulatory harmonization and reinforcing institutional infrastructures, investors can ensure that nanotechnology participates to green agriculture whereas maintain preservation of ecological systems.

REFERENCES

Adisa, I. O., Pullagurala, V. L. R., Peralta-Videa, J. R., Dimkpa, C. O., Elmer, W. H., Gardea-Torresdey, J. L., & White, J. C. (2019). Recent advances in nano-enabled fertilizers and pesticides: A critical review of mechanisms of action. *Environmental Science. Nano*, *6*(7), 2002–2030. DOI: 10.1039/C9EN00265K

Ashique, S., Raikar, A., Jamil, S., Lakshminarayana, L., Gajbhiye, S. A., De, S., & Afzal, O. (2025). Artificial intelligence integration with nanotechnology: A new frontier for sustainable and precision agriculture. *Current Nanoscience*, *21*(2), 242–273. DOI: 10.2174/1573413720666230911111809

Australian Pesticides and Veterinary Medicines Authority. (n.d.). *Home page*. https://www.apvma.gov.au/

Camara, M. C., Campos, E. V. R., Monteiro, R. A., do Espírito Santo Pereira, A., de Freitas Proença, P. L., & Fraceto, L. F. (2019). Development of stimuli-responsive nano-based pesticides: Emerging opportunities for agriculture. *Journal of Nanobiotechnology*, *17*(1), 100. Advance online publication. DOI: 10.1186/s12951-019-0533-7 PMID: 31542052

Campos, E. V. R., de Oliveira, J. L., da Silva, C. M. G., Pascoli, M., Pasquoto, T., Lima, R., & Fraceto, L. F. (2015). Polymeric and solid lipid nanoparticles for sustained release of carbendazim and tebuconazole in agricultural applications. *Scientific Reports*, *5*(1), 13809. Advance online publication. DOI: 10.1038/srep13809 PMID: 26346969

Central Pollution Control Board. (n.d.). *Rules*. https://cpcb.nic.in/rules/

Chen, H., Zhi, H., Liang, J., Yu, M., Cui, B., Zhao, X., & Sun, C. (2021). Development of leaf-adhesive pesticide nanocapsules with pH-responsive release to enhance retention time on crop leaves and improve utilization efficiency. *Journal of Materials Chemistry. B*, *9*(3), 783–792. DOI: 10.1039/D0TB02468E PMID: 33333547

Congress.gov. (n.d.). *Senate Report 108-147*. https://www.congress.gov/committee-report/108th-congress/senate-report/147/1

Dasgupta, N., Ranjan, S., Mundekkad, D., Ramalingam, C., Shanker, R., & Kumar, A. (2015). Nanotechnology in agro-food: From field to plate. *Food Research International*, *69*, 381–400. DOI: 10.1016/j.foodres.2015.01.005

Department for Environment. Food & Rural Affairs. (2025). *Search: Guidelines on the use handling and disposal of nanomaterials in agriculture*. https://www.gov.uk/search/all?keywords=guidelines+on+the+use+handling+and+disposal+of+nanomaterials+in+agriculture

Dikshit, P. K., Kumar, J., Das, A. K., Sadhu, S., Sharma, S., Singh, S., & Billing, B. K. (2021). Green synthesis of metallic nanoparticles: Applications and limitations. *Catalysts*, *11*(8), 902. Advance online publication. DOI: 10.3390/catal11080902

European Chemicals Agency. (n.d.). *Nanomaterials*. https://echa.europa.eu/regulations/nanomaterials

European Commission. (n.d.). *Animals – Science*. https://environment.ec.europa.eu/topics/chemicals/animals-science_en

European Food Safety Authority. (n.d.). *EFSA journal*. https://www.efsa.europa.eu/en/efsajournal/pub/2140

European Food Safety Authority Scientific Committee, More, S., Bampidis, V., Benford, D., Bragard, C., Halldorsson, T., Younes, M. (2021). Guidance on technical requirements for regulated food and feed product applications to establish the presence of small particles including nanoparticles. *EFSA Journal*, *19*(8), e06769. Advance online publication. DOI: 10.2903/j.efsa.2021.6769 PMID: 34377191

Fincheira, P., Hoffmann, N., Tortella, G., Ruiz, A., Cornejo, P., Diez, M. C., & Rubilar, O. (2023). Eco-efficient systems based on nanocarriers for the controlled release of fertilizers and pesticides: Toward smart agriculture. *Nanomaterials (Basel, Switzerland)*, *13*(13), 1978. Advance online publication. DOI: 10.3390/nano13131978 PMID: 37446494

Food and Agriculture Organization of the United Nations & World Health Organization. (2013). *State of food and agriculture 2013: Food systems for better nutrition*. https://www.fao.org/4/i3281e/i3281e.pdf

Food Standards Agency. (n.d.). *Home page*. https://www.food.gov.uk

Government of Canada. (n.d.). *Canadian Environmental Protection Act, 1999 (CEPA)*. https://laws-lois.justice.gc.ca/eng/acts/C-15.31/

Government of Canada. (n.d.). *Pest Control Products Act (PCPA)*. https://laws-lois.justice.gc.ca/eng/acts/p-9.01/

Government of Canada. (n.d.). *Canada Consumer Product Safety Act*. https://laws-lois.justice.gc.ca/eng/acts/c-0.4/

Government of India. (1968). *The Insecticides Act, 1968*. https://www.indiacode.nic.in/bitstream/123456789/1551/1/A1968-46.pdf

Government of India. (1986). *The Environment (Protection) Act, 1986*. https://www.indiacode.nic.in/bitstream/123456789/4316/1/ep_act_1986.pdf

Grobe, A., & Rissanen, E. M. (2012). Nanotechnologies in agriculture and food – An overview of different fields of application, risk assessment and public perception. *Recent Patents on Food, Nutrition & Agriculture*, *4*(3), 176–186. DOI: 10.2174/2212798411204030176 PMID: 23013411

Hannon, J. C., Kerry, J., Cruz-Romero, M., Morris, M., & Cummins, E. (2015). Advances and challenges for the use of engineered nanoparticles in food contact materials. *Trends in Food Science & Technology*, *43*(1), 43–62. DOI: 10.1016/j.tifs.2015.01.008

Hansen, S. F., Heggelund, L. R., Besora, P. R., Mackevica, A., Boldrin, A., & Baun, A. (2016). Nanoproducts—What is actually available to European consumers? *Environmental Science. Nano*, *3*(1), 169–180. DOI: 10.1039/C5EN00182J

He, X., Deng, H., & Hwang, H.-M. (2019). The current application of nanotechnology in food and agriculture. *Yao Wu Shi Pin Fen Xi*, *27*(1), 1–21. DOI: 10.1016/j.jfda.2018.12.002 PMID: 30648562

Health and Safety Executive. (n.d.). *Nanotechnology*. https://www.hse.gov.uk/nanotechnology/

Hosseinpour Jajarm, F., Moravvej, G., Modarres Awal, M., & Golmohammadzadeh, S. (2021). Insecticidal activity of solid lipid nanoparticle loaded by *Ziziphora clinopodioides* Lam. against *Tribolium castaneum* (Herbst, 1797) (Coleoptera: Tenebrionidae). *International Journal of Pest Management*, *67*(2), 147–154. DOI: 10.1080/09670874.2020.1737339

Ijaz, S., Iqbal, J., Abbasi, B. A., Ullah, Z., Ijaz, N., Yaseen, T., & Yaseen, T. (2024). Regulatory and ethical concerns of nanotechnology in agriculture. In Husen, A. (Ed.), *Revolutionizing agriculture: A comprehensive exploration of agri-nanotechnology* (pp. 395–427). Springer.

Institute for Science, Society and Policy. (2022). *Primer on nanomaterial regulation*. University of Ottawa. https://www.uottawa.ca/research-innovation/sites/g/files/bhrskd326/files/2022-07/primer_on_nanomaterial_regulation.pdf

International Organization for Standardization. (n.d.). *ISO/TC 229 Nanotechnologies*. https://www.iso.org/committee/381983.html

Islam, S. (2025). Toxicity and transport of nanoparticles in agriculture: Effects of size, coating, and aging. *Frontiers in Nanotechnology*, *7*, 1622228. Advance online publication. DOI: 10.3389/fnano.2025.1622228

Jiang, Q., Xie, Y., Peng, M., Wang, Z., Li, T., Yin, M., & Zhao, P. (2022). A nanocarrier pesticide delivery system with promising benefits in the case of dinotefuran: Strikingly enhanced bioactivity and reduced pesticide residue. *Environmental Science. Nano*, *9*(3), 988–999. DOI: 10.1039/D1EN01056B

Joshi, H. (2025). Exploring the efficacy of green nanoparticles in enhancing plant defense: A mechanistic investigation into immune response activation. *Journal of Nanoparticle Research*, *27*(2), 29. Advance online publication. DOI: 10.1007/s11051-025-05678-9

Karahalil, B. (2018). Challenges, recommendations, and strategies for nanotoxicology evaluation and its management. In Dasgupta, R., & Kumar, A. (Eds.), *Nanotoxicology* (pp. 649–655). CRC Press.

Khan, Z., & Upadhyaya, H. (2019). Impact of nanoparticles on abiotic stress responses in plants: An overview. In Singh, V. P., Singh, S., Tripathi, D. K., Prasad, S. M., & Chauhan, D. K. (Eds.), *Nanomaterials in plants, algae and microorganisms* (pp. 305–322). Academic Press., DOI: 10.1016/B978-0-12-811488-9.00015-8

Khot, L. R., Sankaran, S., Maja, J. M., Ehsani, R., & Schuster, E. W. (2012). Applications of nanomaterials in agricultural production and crop protection: A review. *Crop Protection (Guildford, Surrey)*, *35*, 64–70. DOI: 10.1016/j.cropro.2012.01.014

Kookana, R. S., Boxall, A. B. A., Reeves, P. T., Ashauer, R., Beulke, S., Chaudhry, Q., & van der Voet, H. (2014). Nanopesticides: Guiding principles for regulatory evaluation of environmental risks. *Journal of Agricultural and Food Chemistry*, *62*(19), 4227–4240. DOI: 10.1021/jf500232f PMID: 24754346

Kumari, R., Suman, K., Karmakar, S., Mishra, V., Lakra, S. G., Saurav, G. K., & Maurya, A. (2023). Regulation and safety measures for nanotechnology-based agri-products. *Frontiers in Genome Editing*, *5*, 1200987. Advance online publication. DOI: 10.3389/fgeed.2023.1200987 PMID: 37415849

Kumawat, K. C., Sharma, B., Nagpal, S., Kumar, A., Tiwari, S., & Nair, R. M. (2023). Plant growth-promoting rhizobacteria: Salt stress alleviators to improve crop productivity for sustainable agriculture development. *Frontiers in Plant Science*, *13*, 1101862. Advance online publication. DOI: 10.3389/fpls.2022.1101862 PMID: 36714780

Li, W., Fan, R., Zhou, H., Zhu, Y., Zheng, X., Tang, M., & Li, Y. (2020). Improving the utilization rate of foliar nitrogen fertilizers by surface roughness engineering of silica spheres. *Environmental Science. Nano*, *7*(11), 3526–3535. DOI: 10.1039/D0EN00687A

Liu, Q., Chen, B., Wang, Q., Shi, X., Xiao, Z., Lin, J., & Fang, X. (2009). Carbon nanotubes as molecular transporters for walled plant cells. *Nano Letters*, *9*(3), 1007–1010. DOI: 10.1021/nl803083u PMID: 19191500

Ministry of Environment. Forest and Climate Change. (1989). *Manufacture, Storage and Import of Hazardous Chemical Rules, 1989*. https://ifbgoa.goa.gov.in/sites/default/files/Manufacture_Storage%26Import_of_Hazardous_Chemical_Rules%201989.pdf

Mohamed, A. K. S., Qayyum, M. F., Abdel-Hadi, A. M., Rehman, R. A., Ali, S., & Rizwan, M. (2017). Interactive effect of salinity and silver nanoparticles on photosynthetic and biochemical parameters of wheat. *Archives of Agronomy and Soil Science*, *63*(12), 1736–1747. DOI: 10.1080/03650340.2017.1308475

Naaz, H., Rawat, K., Saffeullah, P., & Umar, S. (2023). Silica nanoparticles synthesis and applications in agriculture for plant fertilization and protection: A review. *Environmental Chemistry Letters*, *21*(1), 539–559. DOI: 10.1007/s10311-022-01510-6

Palit, S. (2020). Recent advances in the application of nanotechnology in food industry and the vast vision for the future. In *Nanoengineering in the beverage industry* (pp. 1–34). Academic Press.

Pascoli, M., Lopes-Oliveira, P. J., Fraceto, L. F., Seabra, A. B., & Oliveira, H. C. (2018). State of the art of polymeric nanoparticles as carrier systems with agricultural applications: A minireview. *Energy, Ecology & Environment*, *3*(3), 137–148. DOI: 10.1007/s40974-018-0092-6

Rasmussen, K., Rauscher, H., Kearns, P., González, M., & Sintes, J. R. (2019). Developing OECD test guidelines for regulatory testing of nanomaterials to ensure mutual acceptance of test data. *Regulatory Toxicology and Pharmacology*, *104*, 74–83. DOI: 10.1016/j.yrtph.2019.02.008 PMID: 30831158

Safe Work Australia. (n.d.). *Nanotechnology*. https://www.safeworkaustralia.gov.au/topic/nanotechnology

Scott, N., & Chen, H. (2012). Nanoscale science and engineering for agriculture and food systems. *Industrial Biotechnology (New Rochelle, N.Y.)*, *8*(6), 340–343. DOI: 10.1089/ind.2012.1543

Seleiman, M. F., Ahmad, A., Alhammad, B. A., & Tola, E. (2023). Exogenous application of zinc oxide nanoparticles improved antioxidants, photosynthetic, and yield traits in salt-stressed maize. *Agronomy (Basel)*, *13*(10), 2645. Advance online publication. DOI: 10.3390/agronomy13102645

Silva, L. F., Santosh, M., Schindler, M., Gasparotto, J., Dotto, G. L., Oliveira, M. L., & Hochella, M. F.Jr. (2021). Nanoparticles in fossil and mineral fuel sectors and their impact on environment and human health: A review and perspective. *Gondwana Research*, *92*, 184–201. DOI: 10.1016/j.gr.2021.01.003

Tarafdar, J. C., Sharma, S., & Raliya, R. (2013). Nanotechnology: Interdisciplinary science of applications. *African Journal of Biotechnology*, *12*(3), 219–226. DOI: 10.5897/AJB12.2481

Tarafder, C., Daizy, M., Alam, M. M., Ali, M. R., Islam, M. J., Islam, R., & Rahman, M. M. (2020). Formulation of a hybrid nanofertilizer for slow and sustainable release of micronutrients. *ACS Omega*, *5*(37), 23960–23966. DOI: 10.1021/acsomega.0c03399 PMID: 32984716

U.S. Department of Agriculture. (n.d.). *National Organic Program*. https://www.ams.usda.gov/about-ams/programs-offices/national-organic-program

U.S. Environmental Protection Agency. (n.d.). *Summary of the Toxic Substances Control Act*. https://www.epa.gov/laws-regulations/summary-toxic-substances-control-act

U.S. Environmental Protection Agency. (n.d.). *Federal Insecticide, Fungicide, and Rodenticide Act (FIFRA) and federal facilities*. https://www.epa.gov/enforcement/federal-insecticide-fungicide-and-rodenticide-act-fifra-and-federal-facilities

U.S. Food and Drug Administration. (n.d.). *Types of food ingredients*. https://www.fda.gov/food/food-additives-and-gras-ingredients-information-consumers/types-food-ingredients

U.S. Food and Drug Administration. (n.d.). *Federal Food, Drug, and Cosmetic Act (FD&C Act)*. https://www.fda.gov/regulatory-information/laws-enforced-fda/federal-food-drug-and-cosmetic-act-fdc-act

Usmani, S. M., Bremer-Hoffmann, S., Cheyns, K., Cubadda, F., Dumit, V. I., & Escher, S. E.European Food Safety Authority. (2024). Review of new approach methodologies for application in risk assessment of nanoparticles in the food and feed sector: Status and challenges. *EFSA Supporting Publications, 21*(9). *Article, 8826E*. Advance online publication. DOI: 10.2903/sp.efsa.2024.EN-8826

Vemula, M., & Reddy, A. V. B. (2023). Polymeric nanoparticles as effective delivery systems in agriculture sustainability. *Nanotechnology for Environmental Engineering*, *8*(3), 805–814. DOI: 10.1007/s41204-023-00357-5

Wahab, A., Muhammad, M., Ullah, S., Abdi, G., Shah, G. M., Zaman, W., & Alamri, S. (2024). Agriculture and environmental management through nanotechnology: Eco-friendly nanomaterial synthesis for soil-plant systems, food safety, and sustainability. *The Science of the Total Environment*, *926*, 171862. Advance online publication. DOI: 10.1016/j.scitotenv.2024.171862 PMID: 38527538

Wang, X., Zheng, K., Cheng, W., Li, J., Liang, X., Shen, J., & Zhao, X. (2021). Field application of star polymer-delivered chitosan to amplify plant defense against potato late blight. *Chemical Engineering Journal*, *417*, 129327. Advance online publication. DOI: 10.1016/j.cej.2021.129327

Wassie, K. B., & Yemata, G. (2025). Metal oxide nanoparticles as promising agents for triggering defense mechanisms in plants against bacterial diseases. *Discover Nano*, *20*(1), 1. Advance online publication. DOI: 10.1186/s11671-024-04134-8 PMID: 40762749

Yan, S., Hu, Q., Li, J., Chao, Z., Cai, C., Yin, M., & Du, X. (2019). A star polycation acts as a drug nanocarrier to improve the toxicity and persistence of botanical pesticides. *ACS Sustainable Chemistry & Engineering*, *7*(20), 17406–17413. DOI: 10.1021/acssuschemeng.9b04455

Zhu, F., Liu, X., Cao, L., Cao, C., Li, F., Chen, C., & Cai, D. (2018). Uptake and distribution of fenoxanil-loaded mesoporous silica nanoparticles in rice plants. *International Journal of Molecular Sciences*, *19*(10), 2854. Advance online publication. DOI: 10.3390/ijms19102854 PMID: 30241357

Zhu, L., Chen, L., Gu, J., Ma, H., & Wu, H. (2022). Carbon-based nanomaterials for sustainable agriculture: Their application as light converters, nanosensors, and delivery tools. *Plants*, *11*(4), 511. Advance online publication. DOI: 10.3390/plants11040511 PMID: 35214844

Chapter 16
Future Directions in Nano-Virology:
From Lab to Field Deployment

Shumaila Ijaz
https://orcid.org/0009-0003-9614-9802
Shenzhen University, China

Naila Ijaz
University of Agriculture Faisalabad, Pakistan

Navid Hussain Shah
Minzu University of China, China

Javed Iqbal
https://orcid.org/0000-0002-9032-2622
Department of Botany, Bacha Khan University, Pakistan

Banzeer Ahsan Abbasi
https://orcid.org/0000-0003-3036-4481
Department of Botany, Rawalpindi Women University, Pakistan

Rahmat Ali
https://orcid.org/0009-0007-8777-1148
Texas Tech University, USA

Syeda Anber Zahra
https://orcid.org/0009-0009-2933-9001
Institute of Biomaterials and Biomedical, School of Food and Pharmacy, Shanghai Zhongqiao Vocational and Technical University, Shanghai, China

Ayush Madan
https://orcid.org/0000-0001-5089-1794
School of Research and Technology (SORT), People's University, Bhopal, India

Ghulam Murtaza
School of Agriculture, Yunnan University, China

ABSTRACT

Nano-virology provides new ways of diagnosing, treating, and preventing viral infections, marking a new frontier of infectious disease management. Lab-on-a-chip and nanosensors enhance diagnostic speed and precision to facilitate rapid outbreak response. Nano-tech-based delivery systems maximize antiviral efficacy and safety to minimize adverse reactions and resistance. Next-generation vaccine technologies utilize nanomaterials and bio-engineered nanostructures to improve understanding of viruses. The chapter includes nanomaterial preparation for vaccine delivery, nanosensors to enhance rapid detection, and nano-based delivery of drugs to provide targeted therapy. New solutions being developed are presented

DOI: 10.4018/979-8-3373-4913-8.ch016

along with considerations of scalability, lack of biocompatibility, regulation, and environmental effects. The chapter finishes with prospects that identify translating laboratory findings into operational virus surveillance, pandemic preparation, and global health enhancement.

INTRODUCTION

In the context of virology, nanotechnology represents a truly revolutionary force that promises unparalleled opportunities to understand, identify, and control infectious viruses on the nanoscale. The devices and materials engineered through nanoscience, with dimensions between 1 to 100 nanometers, can act directly and selectively on viruses and molecular building blocks, owing to the inherent nanoscale character of viruses, typically with dimensions between 20 to 300 nanometers (Pirzada and Altintas, 2022; Singh et al., 2006). Such integration, or nano-virology, forms an interdisciplinary field in which the tenets of nanoscience and virology combine to improve conventional diagnostic and treatment methods. Nano-virology enables a superior ability to study the attributes of viruses such as attachment, entry, replication, and mechanisms of immune evasion with a degree of selectivity not possible through conventional approaches (Saini et al., 2006; Ray et al., 2025). Such developments hold unique relevance owing to the persistence of outbreaks of viruses that represent a significant danger to public health, exemplified by the recent catastrophic pandemics of the coronavirus known as COVID-19 (Khan et al., 2020). The necessity to create rapid, specific, and selective systems of virus detection appropriate to deployment on a battlefield underscores the significant research efforts that can translate laboratory tenets of nano-virology into potential, large-scale solutions deployable to real-time situations.

The exploitation of nanomaterials' unique physicochemical characteristics that differ significantly from their bulk counterparts by virtue of their small size and large surface-to-volume ratio is a cornerstone of nano-virology. New novel virus detection platform technologies increasingly employ nanoparticles of various sorts, including silver, quantum dots, and magnetic nanostructures, as key building blocks. For instance, the coating of viruses with viral antigens or nucleic acids attached to the surfaces of silver or gold nanoparticles (AuNPs) can trigger dramatic color shifts through their unique optical characteristics, such as localized surface plasmon resonance (LSPR). Such an effect leads to rapid and simple colorimetric detection technologies that enjoy excellent specificity and sensitivity (Fog & Yung et al., 2013; Draz & Shafiee, 2018). In addition to their activity as probes, silver nanoparticles have inherent antiviral activity that damages virus membranes and prevents replication, and thus find useful application as diagnostic and therapeutical tools. Semiconductor nanocrystal-based quantum dots with size-tunable fluorescence emission allow multiplexed detection by allowing simultaneous identification of multiple viruses with excellent photostability and signal brightness, and thus greatly improve diagnostic outputs. Additionally, magnetic nanoparticles help enhance assay purification and sensitivity through the magnetic concentration and purification of viruses out of rich biological matrices and consequently decrease background noise. Such functionalized nanomaterials play a key role in nano-sensors that provide results significantly faster and with higher accuracy than traditional diagnostic probes by converting biological interactions with viruses into quantitative electrical, optical, or magnetic outputs (Luceri et al., 2023; Ratan et al., 2021; Galdiero et al., 2011).

In addition, integration of nano-sensors into transportable diagnostic devices has enabled real-time, point-of-care monitoring, which is of crucial importance to outbreak control. Such devices with nanoscale sensors that can quickly identify very low viral loads, often within a minute or so, enable health care workers and public health officials to rapidly and effectively respond to newly emerging infectious diseases. Devices such as lateral flow assays with amplification by gold nanoparticles are extremely helpful by providing readily interpretable visual outputs and thus extending diagnostic capacity to limited resource settings (Thwala et al., 2023; Lakshmanan and Liu, 2025). Moreover, microfluidic devices with these nanoparticles enable automated sample preparation and multiplexed detection, with the results including abbreviated assay turn times and human error minimized. Electrochemical sensors, through the utilization of nanomaterials on the surfaces of electrodes, convert the virus adsorption-based events into electrical impulses conveniently measured by portable devices and thus enable cost-effective, real-time sensing within health care and field settings (Pirzada and Altintas, 2022; Monteil et al., 2021; Gimondi et al., 2023).

The science of nano-virology holds enormous therapeutic potential that transcends mere diagnostic uses. Nanoscience enables the construction of extremely targeted delivery systems of medication that can deliver antiviral drugs directly to infected cells while causing minimal adverse effects systemically. The therapeutically effective efficacy can be achieved through the design of several nanocarriers, such as liposomes, polymeric nanoparticles, and dendrimers, that strive to enhance the solubility of drugs, stability, and cellular internalization (Curulli, 2023; Palmerston Mendes et al., 2017; Saad et al., 2008). Furthermore, the design of particle-based vaccine technology has been of substantial interest. A perfect example consists of the novel COVID-19 mRNA vaccine that made use of lipid nanoparticles to effectively encapsulate and deliver genetic information that encodes viral antigens to cells of the host to trigger vigorous immunogenic reactions. There is furthermore growing research into antiviral nanomaterials that represent a new class of therapeutics designed to augment conventional antiviral agents that can bind directly to and neutralize viruses or can suppress their infective potential to host cells (Jain et al., 2021; Fang et al., 2022; Vasu et al., 2024).

In spite of these important developments, several issues require addressing before nano-virology technologies can gain wider application across clinical and public health scenarios. Nanomaterials must be designed on a scalable and repeatable basis to offer safety and consistency, with stringent quality control procedures. Comprehensive regulatory evaluation of the pharmacokinetics of the medicine, its environmental implications, toxicity, and biocompatibility should be conducted before approval can be obtained across clinics. Issues of accessibility and cost-effectiveness remain significant handicaps, with particular relevance to integration within low-resource settings in which infectious diseases often impact populations to a biased degree. Moreover, procedures already existing within the clinical realm, data management systems, and surveillance systems should be compatible with nano-virology devices to make their integration into health systems effective (Sharmile et al., 2025; Wang et al., 2023).

Under the umbrella of antiviral delivery systems, the integration of nanocarriers can be regarded as a revolutionary move toward enhancing the efficacy of treatments alongside their safety profiles. The reasoning is derived based on the novel capability of nanocarriers to target infected cells specifically and, therefore, significantly enhance local access to antiviral agents within the infection site while simultaneously reducing systemic exposure that inevitably induces toxicity and off-target implications. The consequence of such targeting allows sustained and regulated delivery of pharmacologic agents such that concentrations therapeutically can prevail over prolonged periods without repeated dosing (Lembo et al., 2018; Zigrayova et al., 2023; Chen et al., 2021). One notable demonstration of this success is the clinical success evidenced by lipid nanoparticles utilized in mRNA vaccine candidates against SARS-CoV-2. The lipid-based nanocarriers triggered vigorous immunological responses by safeguarding prone molecules of mRNA against degradation while promoting cellular internalization and efficient delivery within host cells. The rapid evolution and rapid advances of these mRNA vaccines significantly compressed developing timelines of conventional vaccines, therefore revolutionizing the vaccine science landscape and serving as a demonstration of how nanomedicine can emerge rapidly to respond to emerging global health pandemics through efficient, scalable, and safe platforms (Lozano et al., 2023; Wilson & Geetha, 2022). Beyond reconfiguring vaccine approaches, these advances have exemplified the potential of nano-carrying therapeutics through delivery to protect against various viruses of different kinds beyond vaccine-induced protection boundaries.

Aside from lipid nanoparticles, numerous polymeric and inorganic nanocarriers have emerged with bright prospects of antiviral treatments through delivering various therapeutic agents, including CRISPR-Cas elements, small molecule therapeutics, and small interfering RNAs (siRNAs) (Jan et al., 2025). Because of their tunable characteristics, biocompatibility, and degradation rates, polymeric nanocarriers provide pinpoint control over drug delivery kinetics and target-specific delivery (Abasian et al., 2021). Functionalization of these nanocarriers with ligands that identify specific markers of virus-infected cells allows selective internalization and increased therapeutic indices (Cardoso et al., 2012). Inorganic nano-carriers made of silica and metal oxide nanoparticles have large surfaces and multifunctional abilities that can be engineered to provide synergistic therapies that combine medicinal and diagnostic methods (theragnostic) (Nguyen et al., 2015; Zhao, Cui et al., 2020). The imposition of these nanocarriers to carry siRNAs and CRISPR elements enables the direct targeting of the genetic material of viruses, consequently breaking the infection replication life cycle of viruses at the molecular level through mechanisms of genome editing or gene suppression. Such novel gene-based therapeutics provide an important addition to the treatment of viruses that cannot be treated by conventional treatments of both emerging and chronic cases (Motamedi et al., 2024; Tong et al., 2023).

Besides the direct transfer of antiviral agents, nano-virology employs nanoscale engineered structures and virus-like nanoparticles (VLPs) that mimic viral capsids to craft vaccines and regulate the immune response. Such VLPs, by definition nontoxic, can trigger robust and specific immune responses through virtue of their structural similarity to true viruses while lacking the genome necessary to induce replication (Ray et al., 2025). Such characterizations of the particles and repeated presentation of antigens on the surfaces of VLPs enhance immunological recognition by favoring efficient internalization and presentation of antigens by dendritic cells and other antigen-presenting cells. A vigorous humoral and cellular immunity thus follows, with none of the hazards of live-attenuated or killed-virus vaccines. The recent advances in the design of viral-like nanostructures that permit modular design have enabled the inclusion of multiple antigens, immunostimulants, or immunomodulating agents within single platforms. Such mutability allows custom-tailored immune responses maximized to different viral challenges and promises to enhance vaccine efficacy across many populations (Zepeda-Cervantes et al., 2020; Beyer et al., 2001).

The combination of nanotechnology and artificial intelligence (AI) forms the key to augmenting the discipline of nano-virology by allowing nanomaterials and biomolecular interactions to be optimized through high-throughput and predictive modeling. Computationally-based frameworks driven by AI scrutinize vast datasets generated through both experimental and clinical studies to uncover the important attributes that govern the stability, safety, and efficacy of virus-like particles (VLPs) and nanocarriers. Data-driven approaches of this kind allow rational design and dramatically lower the costs and timescales of the process of developing nanomedicine by predicting the nanomedicine's behavior in realistic biological settings before actual synthesis and testing. Beyond this, machine learning algorithms hold promise of ushering into being a new generation of custom-tailorable antiviral therapy that permits nanomedicine formulations to be specifically tuned to patient-specific variants of viruses, to patients' immune system profiles, and to their genetic backgrounds. To tackle viruses of high mutation rate or of different strains evading conventional treatments, a custom-tailored approach that boosts contemplated effectiveness while dwindling adverse effects is depotentiated (Nooraei et al., 2021; Jekhmane et al., 2017; Mohsen et al., 2017).

However, there are several challenges in translating laboratory-based nano-virological developments into field-deployable solutions. Public acceptance, moral considerations, scalable manufacturing, official approvals, and comprehensive safety and toxicity evaluations are a few of these. Addressing environmental impacts and ensuring equitable access globally are also necessary for sustainable deployment. Standardization of tests, manufacturing processes, and clinical trial frameworks is necessary to support clinical acceptance and regulatory compliance. Future research should concentrate on incorporating multifunctional nano-platforms that can modify the immune system, deliver targeted antivirals, and detect viruses concurrently in compact, user-friendly devices that may be utilized in a range of field settings. Utilizing advancements in data analytics, wearable technologies, biosensing, and microfluidics will be essential. Additionally, strengthening interdisciplinary collaborations across nanotechnology, clinical medicine, virology, material science, and policy-making would accelerate the translation of nano-virology results into practical public health tools (Ehidiamen and Oladapo, 2024; Kabir et al., 2024).

This chapter has provided a thorough examination of the most recent developments and prospects in nano-virology, emphasizing the vital role that nanotechnology will play in transforming the identification, management, and avoidance of viral illnesses. We also looked at personalized nanomedicine; namely how precise drug delivery and tailored vaccination formulations can be made possible by engineering nanoscale platforms to accept patient-specific virus variations and immune profiles. By addressing the difficulties caused by host heterogeneity and virus mutation, this individualized strategy pushes the limits of traditional antiviral treatments. In order to expedite the conversion of nano-virological advances into clinically approved, reasonably priced, and extensively available medicinal solutions, the chapter further underlined the significance of international cooperation and strong policy frameworks. In order to build public trust and adoption, we looked at ways to address ethical issues, create regulatory harmonization, and promote interdisciplinary and multinational collaborations. We talked about the proactive use of nanotechnology to counter new viral dangers. This includes the creation of environmental nano-biosensors, broad-spectrum antiviral nanomaterials, and modular nano-vaccine platforms that facilitate effective viral epidemic containment, improved surveillance, and quick reaction.

1. CURRENT STATUS AND ADVANCES OF NANO-SYSTEMS FOR VIRAL TREATMENT

Nano-systems that exploit nanotechnology hold potential roles as novel agents within the rapidly developing field of viral disease control by enabling advances in vaccine preparation, therapy, and diagnostic approaches (TSAKIRI, 2020). The frequency of viral epidemic outbreaks, including the current SARS-CoV-2 pandemic, coupled with influenza, Ebola, Zika, and HIV/AIDS, has emphasized the necessity of effective targeted therapies and rapid, sensitive detection methodologies (Bukasov et al., 2021; Kalai et al., 2022). Although they provide foundational strategies, traditional diagnostic modalities and antiviral treatments suffer shortcomings, including poor sensitivity, late identification, systemic toxicity, poor absorption, and vaccine difficulties against rapidly evolving viral strains. Through the exploitation of unique physicochemical characteristics within the nanoscale, nano-systems offer innovative solutions to achieving increased specificity of drug delivery, rapid preparation of vaccines, and increased detection of viruses (Bukasov et al., 2021; Kalai et al., 2022; Kumar et al., 2025).

1.1 Nanodiagnostics for Viral Detection

The nano-diagnostics area has essentially revolutionized the detection of viruses by enabling ultrasensitive, rapid, and point-of-care testing strategies that are very important for efficient outbreak control and patient care. The nanomaterials carbon nanotubes (CNTs), magnetic nanoparticles, gold nanoparticles (AuNPs), and quantum dots (QDs) have large surface-to-volume ratios and unique optical, electronic, and magnetic characteristics that make them ideal for multiplexed detection, biofunctionalization with viral probes, and amplification of the signal (Pei et al., 2020; Wang et al., 2022). For instance, plasmonic AuNPs improve diagnostic speed and sensitivity over conventional PCR or ELISA-based strategies by allowing rapid, ambient temperature-assisted, and visual readout-based colorimetric tests of SARS-CoV-2 antigens within minutes (Yakoubi and Dhafer, 2023; Mauriz, 2020). Quantum dots provide durable, tunable fluorophores for multiplexed viral genotyping within small volumes of the reaction. The isolation and concentration of the RNA of viruses are effectively made possible by magnetically responsive nano-systems, which improve detection thresholds. The combination of nano-diagnostics with portable devices and microfluidics ensures decentralized testing that aids epidemiological monitoring and early intervention within resource-poor environments. Additionally, the development of CRISPR-Cas-based nano-detection systems allows immediate discrimination of the variants of viruses and thus improves programmability and specificity (Cui et al., 2023; Zhao, Yu et al., 2020; Ali et al., 2015).

1.2 Nanomedicine and Targeted Drug Delivery

Nanomedicine overcomes the pharmacokinetic and pharmacodynamic limitations of conventional formulations by employing nanoscale nanoparticles to enhance the delivery of antiviral medications (Lembo et al., 2018). Liposomes, polymeric nanoparticles, dendrimers, and solid lipid nanoparticles are examples of nanocarriers that reduce systemic toxicity and enhance therapeutic indices by offering tailored dispersion, controlled release, and improved cellular absorption. Encapsulation stops labile drugs from breaking down and facilitates the co-administration of combination therapies that prevent secondary infections and virus growth. Lipid nanoparticles (LNPs), for instance, have been used to deliver siRNAs and mRNA treatments that precisely and efficiently target viral genes. Nanocarriers designed with ligands that target viral entry receptors or infected cell markers enable selective accumulation in infected tissues, increasing antiviral effectiveness and decreasing off-target effects (Chaudhary et al., 2021; Wakaskar, 2018; Yetisgin et al., 2020).

1.3. Nonvaccine Development

Capitalizing on the specific qualities of nanomaterials, nanovaccines constitute a revolutionary step within the realm of immunology and efficiently target enduring obstacles within vaccine manufacturing, specifically relating to highly variant viruses. Among the best researched and therapeutically promising platforms can be found virus-like particles (VLPs) and self-assembling nanoparticles that structurally mimic natural viral capsids without any infectious genetic material. Utilizing this principle of biomimicry, VLPs can effectively induce the host's immune response with the aid of an exceptionally high level of safety, resulting in vigorous humoral and cellular immune responses. Recent literature concludes that vaccine candidates based on VLPs induce extensive immunogenicity through the presentation of antigens in repetitive forms that augment B cell recognition and access to dendritic cells and efficiently activate both branches of adaptive immunity (Alsaiari et al., 2024). Due to such qualities, nanovaccines are best suited to target viruses that are marked by significant antigen variability, including coronavirus and influenza, wherein antigen shift and drift frequently hamper conventional vaccine efforts.

Clinical confirmation of lipid nanoparticle (LNP) platforms is a historical milestone in the evolution of developing nano-vaccines, and this was achieved specifically during the SARS-CoV-2 pandemic. The chloroquine potential mechanisms of action against coronaviruses are illustrated in Figure 1. The Pfizer-BioNTech and Moderna mRNA-based vaccines encapsulated into LNPs attained unprecedented success against the prevention of COVID-19 by substantially reducing the timescale of vaccine production from a period of years to a period of months. This success signifies the promise of nanocarrier delivery systems to encapsulate and stabilize delicate mRNA molecules against enzymatic cleavage, to engineer cell-specific delivery, and to elicit effective antigen expression within cells (Milane and Amiji, 2021; Mohamed et al., 2022). The rapid emergency approval and large-scale deployment of such vaccines signify a proof of nanotechnology's potential to revolutionize pandemic preparation by offering effective, flexible, and scalable vaccine production processes during emergencies.

Figure 1. Possible ways that chloroquine (CQ) works to combat coronaviruses. Viral entry may be decreased by CQ's inhibition of clathrin-mediated endocytosis and clathrin coat formation; endosomal acidification and viral genome release may be disrupted; MAPK signaling and viral transcription may be changed; and membrane trafficking and endoplasmic reticulum function may influence post-translational protein processing and virion assembly. The suggested interference of CQ at several stages of the coronavirus life cycle, including entry, replication, and virion maturation, is depicted in the schematic

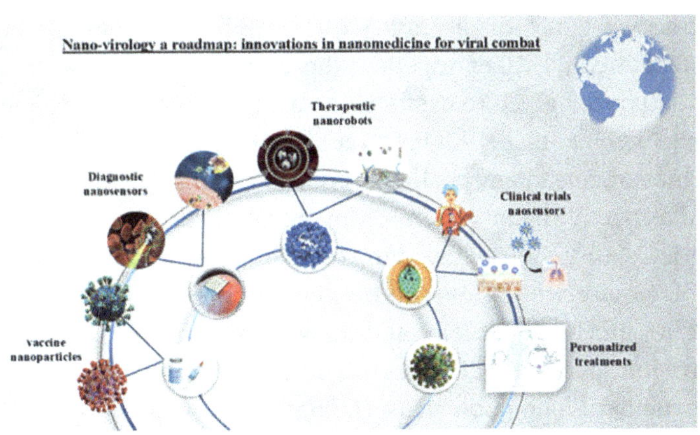

Nanoparticulate structures can not only serve as efficient delivery systems but can potentially function as potent immune adjuvants. Adjuvant formulations that employ nanoparticles can enhance the presentation of antigens, induce the maturation of dendritic cells, and activate innate immune processes, enhancing overall vaccine efficacy. For example, nanoadjuvant formulations that consist of aluminum hydroxide or new-generation polymeric materials exhibit a better ability to induce the production of co-stimulatory molecules and the secretion of cytokines and subsequently enhance adaptive immune responses. In developing vaccine designs that target various diverse viral pathogens with unique immunological demands, careful design of these adjuvants is important to determine the type and extent of immune activation (Akagi et al., 2011; Filipic et al., 2023). The construction of multifunction nanovaccines that combine antigen delivery with agents that possess immunomodulatory activity permits even finer-tuning of immune responses, necessary to achieve balanced Th1/Th2 outcomes or to induce long-term memory populations of T cells, each of which is necessary to provide sustained protection (Kidd, 2003; Moss et al., 2004).

Modularity of nano-vaccine design pays dividends with ease of loading of antigens, modification of the surface, and adjustment of physicochemical properties such as size, charge, and hydrophobicity. The modularity enables rapid iteration and enhancement of vaccine candidates and significantly shortens timelines of preparation. The latter capability is of the essence, particularly during pandemics when timelines are of the essence. By way of illustration, nanoparticles can be engineered to co-deliver multiple antigens or be co-administered with immunostimulants of the immune system, thereby developing polyvalent vaccines that confer broad-spectrum protection against rapidly evolving variants of viruses (Tan et al., 2022; Ren et al., 2023). Beyond that, developments of protein engineering and of synthetic biology contribute to such flexibility and potentially operate with nanotechnology to design new presentations of antigens to maximize immune responses.

Nanotechnology contributes significantly to the control of viral illness through diagnostic and therapeutic strategies, and to the delivery of vaccinations. Nanodiagnostics provide rapid and very sensitive detection strategies that can be taken into the field or to the point of care; these qualities were instrumental to the rapid suppression of outbreaks during the COVID-19 pandemic. Through the use of nanoparticles, such as magnetic or gold particles, these technologies improve monitoring capacity that is of key importance to efficient pandemic management, allowing the identification of viral nucleic acids or proteins with very considerable speed and specificity relative to standard testing technologies. Beyond this, nanomedicine platforms fine-tune the delivery of antiviral therapeutics through increased targeted specificity and diminished systemic toxicity, and with the ability to achieve control of delivery, to achieve maximal efficacy therapeutically. Taken together, these technologies have transformed diagnostic and therapeutic strategies and have helped to establish a vast variety of nanotechnology-based resources to control viral infection (Singh et al., 2017; Qasim et al., 2014; Koteswara, 2021).

In spite of these achievements, many challenges need to be resolved to achieve the global potential of nanovaccines and their respective nano-systems. Major challenges include the scaling up of nanomaterial manufacturing with strict quality control, receiving regulatory clearance that guarantees long-term safety, and ensuring accessible equity within low- and middle-income nations. To prevent worsening global health inequalities, addressing the intricacies and costs of manufacturing is crucial. Moreover, thorough studies of the long-term immunologic safety and stability of these innovative platforms are necessary to conform to regulation standards and public trust (Quinn and Kumar, 2014; Guasch-Girbau et al., 2021).

2. CHALLENGES IN FIELD DEPLOYMENT

Nanotechnology has greatly increased global capacity to control viral infections with unparalleled precision and effectiveness through revolutionary advances in viral diagnostic tools, treatments, and vaccine production. Though impressive results were witnessed in research laboratories and clinical practice of nano-virology hold promise, the shift from contained research facilities to large-scale general application is a significant challenge, particularly in the context of low-resource or high-needs situations that emerge during epidemic outbreaks. To maximize the public health advantages of nanotechnology-based future treatments of viruses, public health should address questions of scalability, manufacturing technology, biocompatibility, protection against adverse reactions, regulatory challenges, morality, and ecological effects (Mohite et al., 2024; Weerarathna et al., 2024).

2.1 Scalability and Manufacturing

The extension of laboratory-scale synthesis of nanomaterials into efficient, cost-effective, and throughput-effective manufacturing processes that embody stringent specifications of quality and reproducibility is a challenge to the effective integration of nanotechnology into viral therapeutics and diagnostic agents. To attain enduring biological efficacy, nanoparticles deployed within these therapeutics require stringent control of important physicochemical parameters such as size, morphology, surface chemistry, and functionalization. Attainment of such specificity and consistency on a large-scale manufacturing basis is a Herculean challenge (Ahmad et al., 2022; Yusuf et al., 2023). The origin of this limitation is primarily owing to variability between production batches, inherent complexities associated with multi-step synthesis processes, and the necessity to hold stringent conditions of asepsis. Moreover, to attain safety and clinical efficacy, of equivalently supreme significance, of paramount necessity exists to uphold the structural integrity of the highly elaborate nanomedicine assemblies that consist of multiple interacting constituents assembled into precise three-dimensional configurations, throughout the entire manufacturing process. Novel manufacturing processes, such as microfluidics-based processes of continuous synthesis, hold potential avenues to circumvent these bottlenecks by enhancing control of reaction kinetic processes and of mixing efficiencies to attain repeatable large-scale fabrication of closely distributed nanoparticles. Moreover, fabrication processes through self-assembly offer useful avenues by consolidating the fabrication process into simpler steps. However, regulation compliance, costs, and quality control continue to present enormous hurdles to large-scale manufacturing and to act to impede smooth integration of these new platforms into developed frameworks of pharmaceutical manufacturing and global supply chains (Riehemann et al., 2009; Ahmed et al., 2022; Pirzada and Altintas, 2022).

The pharma manufacturing industry faces specialized issues while switching over from traditional production models to those of nano-based products. Unlike classical small-molecule therapeutics, nanomedicine consists of intricate multicomponent delivery systems that require thorough functional and physicochemical studies to determine key quality attributes and identify crucial process parameters necessary to achieve reproducibility. The usage of lipid nanoparticles (LNPs) to deliver mRNA vaccines represents the fine balance necessary to control particle size, stability, and surface properties, all of which have significant roles to play in controlling drug delivery, biodistribution, and immune activation (Riehemann et al., 2009; Cheng et al., 2023; Liu et al., 2022). The stability and effectiveness of these products directly depend on technologies involving microfluidizers that exert significant and repeated shear forces within specialized interacting chambers to accurately decrease and homogenize particle size on a large production scale. Additionally, to achieve the goal of reproducibility and pick up any deviation early during the process, production models should employ in-process analytical testing using rapid and sensitive methods to monitor critical parameters in real-time. In spite of these technologies, issues of scaling of "in situ" nanomedicine formulations synthesized within the situation of treatment or within the bedside of the patient cause concern with respect to quality control and reproducibility outside the traditional manufacturing settings. Such complexities create the need for stringent, scalable, and standardized production procedures, of which the industry is continuously refining.

2.2 Biocompatibility and Safety

For the productive application of nanomaterials into practice, their safety and biocompatibility are critically important to ensure. The rationale lies in their danger of having unknown biological interactions that can induce immunogenicity, bioaccumulation, or off-target toxicity. Unlike classical toxicological paradigms, the increased surface reactivity and nanoscale size of nanoparticles allow intricate interactions with proteins, cells, and tissue. Thus, rigorous in vitro and in vivo testing with inclusion of parameters such as the following: cytotoxicity, genotoxicity, immunotoxicity, pharmacokinetics, and biodistribution, should be undertaken to capture such interactions that trigger inflammatory reactions or cytotoxic responses. A vivid and prudent example is lipid nanoparticles (LNPs), applied to a COVID-19 mRNA vaccine (Verbeke et al., 2021; Wilson and Geetha, 2022; Tenchov et al., 2021). Although these nanoparticles are mostly safe and effective, a limited set of rare adverse cases relating to complement activation or localized inflammatory reactions were experienced, and these instances intensify the necessity of continuous safety evaluations and post-market scrutiny. Moreover, the presence of occupational health hazards to both end-user and production staff stresses the significant requirement of strict safety measures and preventive action. The design and application of biodegradable or excretable nanomaterials exemplify optimistic strategies to prevent long-term toxicity risks and prevent their buildup within biological systems, and consequently allow wider and safer usage.

2.3 Regulatory and Ethical Considerations

Even with its slow evolution, however, the legal landscape that controls viral therapies made possible by nanotechnology is complex, disjointed, and daunting. To regulate nanomedicines and diagnostic devices, agencies like the European Medicines Agency (EMA) and the U.S. Food and Drug Administration (FDA) require detailed physicochemical characterization, reproducibility of manufacturing processes, stringent safety testing, and demonstration of clinical effectiveness. The lack of articulated, standardized measures that specifically account for nanomaterials' unique characteristics, like their size-dependent behaviors, their surface alterations, and their multifunctional potentialities, adds to the difficulty of approval processes of these products. Such difficulties are further exacerbated by ethical considerations of fair distributive justice of innovative nanotechnology-based therapeutics and vaccine access to all population's world over, informed consent by virtue of the novelty of methodologies adopted and uncertainty regarding prolonged safety, and the requirement of transparent risk explanation to ensure public trust (Singh et al., 2017; Chakravarty & Vora, 2021; Cetintas et al., 2021). A further setback to accessibility is posed by intellectual property and patent limitations that very frequently present obstacles to the adoption of such late-breaking technologies by low- and middle-income countries. To achieve fair access and to encourage future innovation, open collaboration and open models of innovation have increasingly been advocated.

2.4 Environmental Impact

Mass production, consumption, and dumping of nanomaterials hold significant environmental and ecological consequences that demand exhaustive research. Nanoparticles can potentially bioaccumulate into water, plant, and animal life and induce toxicity or microbial ecosystem process disruption that is important to environmental integrity. Hence, this necessitates sustainable design of nanomaterials and effective nano waste management approaches to minimize environmental persistence and avoid unauthorized ecological consequences (Kabir et al., 2018; Pucciarelli, 2023; Nizam et al., 2021). Life cycle analyses (LCAs) can serve as valuable tools to comprehensively evaluate and minimize the environmental footprint of nanotechnology-based viral therapeutics through each step of production, consumption, and disposal. To assist in enhancing compliance among industrially based nanomaterials manufacturers, National Regulatory Agencies may require broadening their monitoring to guidelines relating to the environmental release of engineered nanomaterials (Pucciarelli, 2023; Nizam et al., 2021). Long-term success of nanotechnology-based virus technologies can be solidified by capitalizing on green chemistry pillars of tenets within the synthesis and fabrication processes of nanomaterials. The green chemistry principle includes employing renewable starting materials, reducing harmful chemicals, and promoting material biodegradation (Islam and Islam, 2025; Khosravi et al., 2024).

3. EMERGING STRATEGIES FOR TRANSITION

The fields of nanomedicine and nano-virology stand at a turning point, offering revolutionary technologies to treat, diagnose, and vaccine against virus infections. Despite considerable laboratory advances, their transfer into real-world, practical, scalable, and cost-efficient and clinically acceptable forms of application proves daunting. The emerging strategies place significant focus on the design of novel nano-designs specifically suited to practical deployment, upgrading manufacturing processes to quality and affordability, and rigorous real-world evaluations and validations necessary to gain regulatory approval and clinical acceptance to fill this crucial gap.

3.1 Innovative Nano-Designs for Practical Use

Nanotechnology holds specific advantages within the realm of biomedicine through its capability to precisely manipulate molecules and combine various functions. The new advances within design nanomaterials are aimed at creating entities that not only show increased antiviral activity but also align with clinical demands, such as biocompatibility, stability, manufacturing capability, and usage ease (Figure 2) (Rai et al., 2025; Malik et al., 2023).

At the frontier of research are hybrid nanomaterials that combine inorganic or metallic components with organic polymers. The nanosystems have diverse roles, including the delivery of targeted medicine and genes, the regulation of immune responses, and simultaneous viral targeting. For example, polymer-polymer hybrids of nanoparticles improve biodistribution and offer custom delivery to virus-implicated tissue with the alteration of their surfaces by specific ligands, such as aptamers or antibodies (Sanchez et al., 2011; Maghsoudnia et al., 2020). The process of targeting boosts adherence among patients and safety profiles by diminishing off-target toxicity and frequencies of dosing. Another breakthrough that enables the release of medicine in response to precise external stimuli, including pH, redox gradients, or diseased tissue enzymatic activity, is the use of stimuli-responsive nanocarriers (Mura et al., 2013). Such "smart" designs restrict systemic exposure while maximizing therapeutic indices. In addition, virus-like particles (VLPs) that resemble the morphology of the source virus but lack infectious agents are utilized in biomimetic nanovaccines to induce vigorous immunogenic responses and cross-protective immunity

(Mura et al., 2013; Zeltins, 2013). The idea of multifunctionality transcends the realm of therapeutic payloads; co-delivery systems enable theragnostic functionality that enables personalized, real-time adjustments of dosages and the early detection of treatment responses through the integration of active antiviral agents with immunostimulatory adjuvants and diagnostic image compounds. Additionally, structural alterations like PEGylation improve circulation times by evading immune clearance, thus ensuring prolonged bioavailability and efficient delivery to target sites (Brai et al., 2023; Hussain et al., 2019).

Figure 2. A roadmap to nanomedicine against viruses, with emphasis on important areas and uses: diagnostic nanosensors, vaccine nanoparticles, therapeutic nanorobots, nanosensors for clinical trials, individualized treatments, and wider integration into the healthcare system. The schematic diagram represents how nanomedicine innovations come together to speed up detection, prevention, and treatment of viruses over the coming ten years

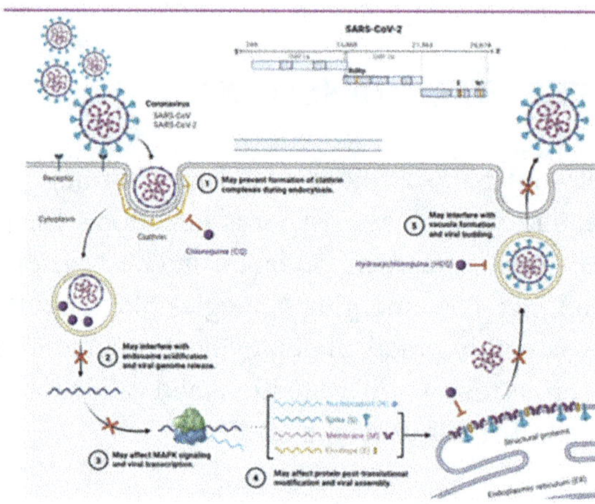

3.2 Cost-Effective Manufacturing Techniques

To achieve success with clinical and global health deployment, scalable manufacturing processes that control cost, quality, and reproducibility effectively should be implemented. Traditional nanomaterials production typically relies on batch processes that cannot be applied to large-scale production owing to their inherent variability and poor scalability. New continuous flow microreactor-based synthesis enables accurate control of nanoparticle size, shape, and composition and ensures high-throughput manufacturing with consistent quality. Bottom-up strategies, including chemical vapor deposition and self-assembly, and top-down strategies, including laser ablation and milling, have been developed for industrial production (Charitidis et al., 2014; Shegokar and Nakach, 2020; Ran et al., 2023). The deployment of Process Analytical Technology (PAT) frameworks controls batch-to-batch variations by including quality control steps and real-time measurement, and ensures compliance with rules of regulation (Floyd et al., 2025).

Each of these scalable nanomanufacturing technology advances is exemplified by lipid nanoparticle (LNP) vaccine platforms. Contrary to previous prolonged timescales, mass production of mRNA vaccine shots of COVID-19 surged to respond to global demand within months using fully automated microfluidic mix technologies (Schmidt et al., 2021; Fardoost et al., 2024). The approaches recruit modular vaccine manufacturing facilities that can rapidly shift to respond to new emerging infectious diseases. The facilities are supported by nanomaterial collections and standardized processes that significantly shorten development timescales. As part of an effort to decrease costs, green synthesis processes that minimize environmental footprints and purification demands using green solvents and biodegradable nanomaterials are sought by scientists (Pralong et al., 2014; Virkutyte and Varma, 2011). The sustainable process of manufacturing fulfills the global challenge of having accessible and affordable nanomedicines.

3.3 Real-World Testing and Validation

For proper recreation of clinical scenarios, including fluctuating viral loads, co-infections, comorbid patients, and immune competence, preclinical in vitro and animal research must advance. Richer predictive understanding of pharmacodynamics and toxicity can be achieved using advanced 3D tissue and organ-on-chip technologies, which allow fine-tuning (Georgakopoulou, 2024; Gotting et al., 2025). Phase I–III trials should adopt adaptive designs that allow adjustments to doses and biomarker-based inclusion criteria of participants. The rich data set relating to immunogenicity, durability, and safety is exemplified by research into nanovaccines, which strengthens approval processes by regulators (Antoniou et al., 2016; Ha et al., 2024). Through post-authorization, real-world evidence (RWE) assists decision-making on manufacturing and distribution adjustments by monitoring efficacy and infrequent adverse outcomes across heterogeneous populations. The inclusion of digital health solutions allows continuation of remote monitoring, thus improving patient compliance and data richness (Shastry and Shastry, 2023). Generalizability is assured through addressing accessibility by proving clinical validity across multiple sociocultural and economic settings. Networking with global approval agencies streamlines approval processes and converges standards to enhance broader approval (Abraham and Reed, 2002; Shastry and Shastry, 2023).

4. Nanoparticles in Viral Therapy

Nanoparticles (NPs) represent a key focus of recent antiviral research, profoundly transforming the practice of viral therapy through their remarkable ability to address key problems associated with viral detection, inhibition, and clearance. The physico-chemical properties of nanoparticles enable selective interactions with viral entities and host cells at comparable dimensional scales (Chakravarty & Vora, 2021). Properties of large size ratios of surface area to volume, tunable surface functionalization, and controllable optical and electronic structures enable nanoparticles to interfere with viral life cycles through a variety of mechanisms and enhance the specificity and efficacy of antiviral methods (Milovanovic et al., 2017). The present section provides an exhaustive account of how nanoparticles tackle viral dimensions and mechanisms of invasion and an account of the latest therapeutic methods and clinical applications that exploit these nanoscale phenomena.

4.1 Addressing Viral Size and Invasion Mechanisms

Viruses, owing to their nanoscale size, utilize characteristic cell mechanisms and molecular interactions to infect target cells and induce infection. The similarity of derived and virus nanoparticles' size forms significant potential grounds to design antiviral nanomaterials that mimic, intercept, or interfere with mechanisms of viruses (Figure 3) (Jackman et al., 2021; Singh et al., 2017).

Figure 3. Step-by-step overview of the antiviral mechanism of nanoparticles. An illustration of the various steps that nanotechnology can take to affect antiviral results. Arrows depict the continuum from nanoscale interactions to clinical and regulatory preparation by showing the flow and links between stages

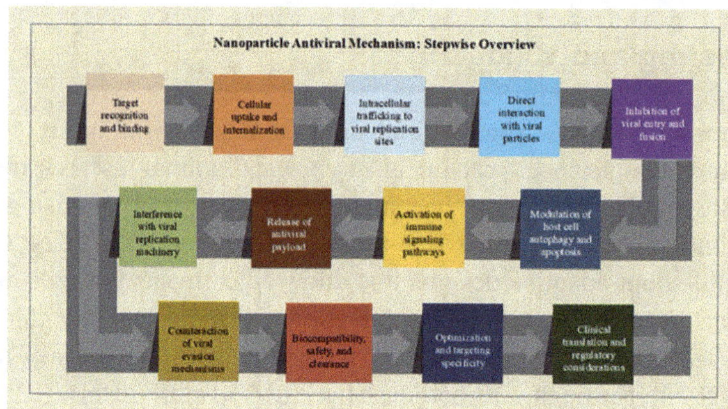

4.2. Competitive Binding and Steric Blocking

One of the notable strategies is nanoparticles designed to interfere physically with viral attachment to host cellular receptors. Gold nanoparticles (AuNPs), among others, have been functionalized with ligands that target viral surface proteins like HIV gp120 or influenza hemagglutinin to mask viral docking sites and hinder entry that follows (Gulati et al., 2020; Draz & Shafiee, 2018). Such AuNPs, with optimal particle size (~20 nm) and surface chemistry, sterically prevent viral entry through receptor site saturation, effectively neutralizing viral infectivity in vitro. The biomimetic intervention takes advantage of the size and multivalent binding capability of the nanoparticle to compete with viral particles by interacting with the receptor (Jampílek & Kralova, 2019; Draz & Shafiee, 2018).

4.3. Membrane Fusion Inhibition

The enveloped viruses that rely on lipid envelope fusion with host membrane cells to gain entry into cells offer other mechanistic targets to interfere with using nanoscale delivery. Specifically designed lipid-based nanoparticles (LNPs) that fuse or alter membrane fluidity can destabilize the viral envelope and make the virus unable to infect. The capacity of LNPs integrated with fusion inhibitors to mimic viral lipid components enables the disruption of viral fusion in a competitive process. The nanoparticles hinder necessary conformational changes that membrane fusion necessitates during entry, essentially preventing infection early on (Lee et al., 2025; Gulati et al., 2020; Draz & Shafiee, 2018).

4.4. Intracellular Trafficking Disruption

Upon entry, viruses utilize intracellular trafficking and endocytic pathways to unpack their genomes for replication. Small interfering RNAs (siRNAs) or CRISPR-Cas elements encapsulated by polymeric nanoparticles target specifically the RNA or DNA of viruses, inhibiting replication and assembly (Mercer et al., 2010; Lin et al., 2022). Polymeric nanoparticles that load siRNAs and localize within compartments of the endo-lysosome effectively knock down hepatitis-C virus genes within liver cell models infected with hepatitis-C, thereby ceasing intracellular propagation of viruses. The architecture of nanocarriers of this sort allows for surface modification to enhance cellular uptake, escape from the endosome, and controlled delivery of the payload (Dyawanapelly et al., 2014; Ma et al., 2007).

4.5. Immune System Modulation via Size Mimicry

The size of nanoparticles also determines their interactions with immune cells. Nanoparticles of virus-sized dimensions increase the uptake by antigen-presenting cells (APCs) such as dendritic cells and macrophages and trigger innate and adaptive immune responses (Moon et al., 2012; Kumar et al., 2015). The characteristic was made use of during the preparation of virus-like particles (VLPs), which mimic the morphology of viruses without having infectious genetic material, eventually leading to vigorous immune activation and the production of immunological memory. The precise control of size and architecture of the surface of VLPs maximizes their perception by pattern recognition receptors (PRRs), consequently amplifying immunogenicity while being safe (Zepeda-Cervantes et al., 2020; Tariq et al., 2022).

4.6. New Therapeutic Strategies and Clinical Applications

The nanoscale benefits listed above have been utilized to see significant expansion of therapeutical methods using nanoparticles to regulate infectious viruses over the last several years. The strategies not only include the direct antiviral mechanisms but also vaccine preparations, dual treatment regimes, and theranostic devices (Chakravarty & Vora, 2021).

4.7. Targeted Antiviral Drug Delivery

Nanocarriers function as delivery systems to improve the pharmacological characterization of antiviral compounds. By encapsulating within nanoparticles, the solubility of the drug, stability, and bioavailability are enhanced. Targeting by attaching virus- or tissue-specific ligands (e.g., antibodies, aptamers) to the particle's surface allows selective targeting of infected cells, obtaining higher local concentrations of the drug and lower systemic toxicity. The delivery of Remdesivir and of siRNAs by lipid nanoparticles targets SARS-CoV-2 infected pulmonary cells, facilitating higher intracellular concentration of the drug and better efficacy (Lembo and Cavalli, 2010; Gurunathan et al., 2020).

4.7. Multifunctional Nanoplatforms

The nanoparticles that can be co-loaded with antiviral compounds and immunomodulatory compounds, or with intrinsic antiviral activity, provide broader synergistic activity. Silver nanoparticles (AgNPs) possess inherent antiviral capability through the production of reactive oxygen species (ROS) and their reactions with viral proteins, and serve to transport traditional antiviral compounds, thus enhancing their antiviral capability. Moreover, photothermal therapy with gold nanorods or graphene oxide allows localized destruction of viruses with near-infrared light and holds potential against enveloped viruses, such as the viruses of influenza and coronavirus (Elbehiry and Abalkhail, 2025; Naghib et al., 2024).

4.8. Nanovaccines and Immunotherapy

Virus-like particles (VLPs) and nanoscale particle vaccines are among the preeminent approaches to the induction of robust and lasting immune responses. Notable examples include mRNA vaccines encapsulated within lipid nanoparticles, such as Pfizer/BioNTech's BNT162b2 and Moderna's mRNA-1273, which have been remarkably effective at inducing adaptive immune responses during the COVID-19 pandemic. Nanovaccines offer rapid reconfigurability with respect to viral mutations and permit individualized immunostimulation through simultaneous delivery of adjuvants (Gupta et al., 2023; Parvin et al., 2024).

4.9. Gene Editing and RNA Therapeutics

Nanovectors transport CRISPR/Cas systems and RNA interference devices to target the precise genome of viruses. CRISPR-containing nanoparticles effectively eliminate dormant pools of viruses of HIV models, a capability unrealizable by conventional antivirals. Gene-editing treatments hold potential to revolutionize treatment strategies by virtue of their ability to induce excision or functional gene suppression of the genome of viruses (Lin et al., 2022; Chen et al., 2020).

4.10. Clinical Progress and Trials

Some of these nanoparticle-based antivirals and vaccine candidates have moved into clinical development or approval by the regulators. The first RNAi therapeutic to be approved by LNP delivery, patisiran, represents a seminal moment of gene therapy with clinical relevance to antiviral RNA therapeutics. Inhalable formulations of nanoparticles against respiratory viruses are tested in early-phase clinical trials with potential to provide non-invasive effective delivery to the lungs (Welgan and Doyle, 2022; Baig & Chien, 2025). Point-of-care nanodiagnostics with lateral flow technology using gold nanoparticles provide rapid detection of viruses important to control pandemic outbreaks (Motamedi et al., 2024; Wen et al., 2016; Pirzada and Altintas, 2022). Many of the advances notwithstanding, issues of scalable and reproducible manufacturing, overall safety evaluation of immunogenicity and toxicity, adaptability of the regulators to new technologies, environmental effects, and equitable access remain. Future advances will concentrate on stimulus-responsive "smart" nanocarriers, artificial intelligence-assisted design optimalization, and nanocarrier integration with diagnostic platforms to achieve theranostic functionality to maximize efficacy and reduce adverse effects.

5. FUTURE DIRECTIONS IN NANO-VIROLOGY: INTEGRATION, PERSONALIZATION, COLLABORATION, AND EMERGING THREATS

This comprehensive introduction enters into future directions within the field of nano-virology, with a focus on four key areas central to its future: integration with information technologies, the rise of personalized nanomedicine, the role of global cooperation and regulation, and the need to respond to new emerging viral threats through nanotechnology. The area of nano-virology is rapidly moving beyond its root technological advances into a future of integrative innovation, personalized medicine, global cooperation, and anticipatory responses to emerging viral challenges. As the failures of traditional viral treatments and diagnoses come into clearer focus, nanotechnology holds a transformative promise to revolutionize the practice of addressing viral infections through precision, flexibility, and integrated functionality.

5.1 Integrating Nano-Virology with Digital Technologies

Digital technologies and nano-virology can combine to revolutionize the therapy and diagnosis of viral infections by achieving unprecedented levels of scale, agility, and specificity. To create new diagnostic and therapeutical systems that combine real-time sensing and dynamic tailoring of responses, digital technologies, including wearable biosensors, the Internet of Things (IoT), large data analytics, artificial intelligence (AI), and machine learning (ML), intersect with nanotechnology. By bringing nanosensors together with portables or wearables, repeated sensing of real-time physiological and biochemical markers of viral infection can be achieved. For example, nanosensors designed to detect viral antigens, nucleic acids, or host immune responses can upload data to medical platforms without wires. This enables early detection and tracking of viral kinetics (Pirzada and Altintas, 2022; Younis et al., 2020). By informing individualized treatment planning and rapid outbreak identification, these technologies eliminate the lag times of traditional diagnostic approaches.

Artificial intelligence platforms analyze massive and dynamic data streams gleaned from health records, epidemiologic databases, and nanosensor networks. Such technologies can discern fine patterns, predict the progression of diseases, and accurately customize treatment protocols. For instance, AI-based nanomedicine platforms can dynamically shift medication doses based on fluctuations of the viral load or fluctuations of the stage of immune activation, thereby optimizing the efficacy of therapy and minimizing toxicity. The use of blockchain technology obviates ethical and legal hurdles to digital health by maintaining data integrity and safeguarding patients' privacy. Moreover, digital twin models of computational avatars of patient-specific biological systems enable rapid formulations of medications and vaccines by connecting nano-virological data to simulate interactions between viruses and their hosts and to forecast results of therapy (Saylan and Denizli, 2020; Pirzada and Altintas, 2022; Younis et al., 2020). The deployment of telemedicine enables distant management of patients, which is valuable within resource-poor locations and during pandemics.

5.2 Personalized Nanomedicine

Personalized medicine is transforming the health care industry by adapting therapy to the individual's specific genetic, biochemical, and environmental profiles. Nanomedicine is perfectly suited to achieve this practice within the realm of virology because of its tunable dimensions, surface chemistry, and ability to deliver payloads. By targeting specific strains and variants of viruses unique to each patient, personalized nanotherapeutics essentially escape the endemic mutability of viruses that typically defies traditional antiviral therapy. Nanosystems achieve pinpoint delivery of nucleic acid-based therapeutics, including antisense oligonucleotides, small interfering RNAs (siRNAs), and CRISPR-Cas gene-editing tools that target either evolutionarily-conserved or patient-specific regions of viruses and ultimately achieve sustained viral suppression (Safo et al., 2025; Wen and Steinmetz, 2016).

Individual immunogenetic profiles can be applied to engineer nanovaccines that customize optimal combinations of antigens and adjuvants and therefore engender optimal immunoresponses. The technology exploits insights into immune checkpoint dynamics, epitope maps, and human leukocyte antigens (HLA) variability (Acurcio et al., 2024; Khatua et al., 2024). Such individualization alleviates adverse reactions and poor responsiveness that accompany broad-based vaccine initiatives. Personalized nanomedicine also thwarts host determinants of therapeutic outcomes, that is, pharmacogenomics and microbiota. To ensure maximal safety and patients' compliance, nanoparticles can be engineered to be released by stimuli-responsive mechanisms that are commensurate with each individual's specific immunological and metabolic profile (Rahman, 2024; Ishaque, 2025). Before their therapeutic application, treatment protocols can be refined by preclinical toxicity and efficacy testing with patient-source organoids and microfluidic "organ-on-a-chip" systems perfused with nanoformulations.

5.3 Global Collaboration and Policy Frameworks

To achieve rapid innovation and fair technology distribution, scientists, physicians, regulators, and legislators need to harmonize their international efforts, owing to the highly globalized nature of viral attacks. The success of developing the vaccine against COVID-19 illustrates the effectiveness of such global initiatives that combine interdisciplinary knowledge and resource mobilization to expedite the deployment of nanomedicine (Prasad et al., 2025; Ishaque, 2025). To streamline and improve post-market monitoring of clinical trials, global consortiums encourage open data disclosure, align regulation guidelines, and standardize the methods of nano-virology. The consortiums also discuss questions of access and pricing, to ensure that low- and middle-income countries gain access to nanotherapeutic and diagnostic nanoproducts. The policy guidelines that propel responsible innovation define safety, ethical, and environmental standards of research and deployment of nanomedicine. To combat misinformation causing vaccine hesitation and skepticism about digital health, public outreach and educational programs help to earn societal trust and rationalize informed consent (Wu et al., 2025; Ogugua et al., 2024). The commercialization of antiviral therapies based on nano-materials is accelerated by funding models that encourage partnerships between academia and industry, and thus help to facilitate translational research. Moreover, global dialogues of regulators on mutual and harmonized recognition help to decrease obstacles to the global accessibility of new nanotherapeutics.

5.4 Addressing Emerging Viral Threats with Nanotechnology

Urbanization, zoonotic spillovers, climate change, and expanding global travel represent a myriad of factors leading to the rapid dissemination of viruses and associated pandemic risk, causing emerging viral diseases to be ever-increasing global health issues. The global preparedness and response mechanisms against infectious diseases are bolstered by nanotechnology through access to rugged, modular platforms that can promptly respond to such novel and dynamic viral attacks. The integration of CRISPR gene-editing technologies and nanoparticle-based hybrid platforms leads to the creation of rapid nano-enabled diagnostic sets that represent new-age diagnostic tools for the rapid identification and characterization of novel viral pathogens. Such platforms enable highly accurate and sensitive identification of viral nucleic acids rapidly, without the need to install large-scale laboratory facilities. Such functionality is of benefit to the prompt utilization of containment measures and public health policies, particularly during the initial phases of an epidemic. Contrary to conventional RT-PCR protocols, CRISPR-Cas-based detection technologies, combined with colorimetric-based detection with the help of gold nanoparticles, enable point-of-care detections with detection levels reaching down to a few copies per milliliter, greatly reducing turnaround times (Baig & Chien, 2025; Van Dongen, 2022). Such diagnostic responsiveness greatly amplifies the ability to achieve real-time identification of novel or mutated strain viruses and facilitates rapid epidemiological response.

Along with diagnostics, modular nanovaccine platforms have transformed vaccine research through the encouragement of scalable mass manufacturing that can be adjusted to fit new emerging strains of viruses and can be rapidly switched to enable modification of antigens. Representative examples of nanoparticle platforms, including lipid nanoparticles, virus-like particles, and self-assembling nanostructures, enable flexible vaccine design through the capability to rapidly integrate new antigenic sequences while offering optimal transport and immunogenicity properties. Modularity greatly reduces vaccine development timelines relative to traditional approaches, while offering key functionalities to augment pandemic preparation and response. For instance, mRNA-based vaccines against SARS-CoV-2 variants encapsulated within lipid nanoparticles were quickly adapted and deployed globally, and represent efficient development with rapid deployment of future vaccines against emerging viral threats (Chalkias et al., 2023).

By focusing on conserved virus structures or host-pathogen interaction pathways with diminished mutational evasive capacity, broad-spectrum antiviral nanotherapeutics yield increased therapeutic success and decreased resistance formation likelihood. Multi-modal nanoplatforms combine direct viral inactivation mechanisms, such as disruption of viral envelopes or capsids, with immunomodulatory activities, including interferon response enhancement and antiviral macrophage activity augmentation. The designs of some of these nanoplatforms include regenerative cues that induce tissue repair and return following infection, to provide optimal therapeutic breadth and long-term longevity. Of note, nanoparticles that were functionalized with broad-spectrum antiviral agents or immune-stimulatory ligands can be capable of concurrent killing of a diverse set of viral strains while enhancing the host's antiviral defense, an important feature while targeting unknown or evolving pathogens (Christ et al., 2025).

In addition to its short-term application to therapy, nanotechnology enables the development of very sensitive environmental monitoring devices that can detect viral particles on surfaces, sewer water, or aerosols. Such environmental nanobiosensors, conventionally employing electrochemical or optical detection procedures facilitated by nanoparticles, yield important data streams to facilitate forecasting of outbreaks and enhancing epidemiologic models. Measurement of viral load within sewer or aerosol samples offers population-based, non-intrusive infection prevalence trend insights and therefore enables timely containment measures before the symptomatic diagnoses peak. In instances of the onset of viral epidemic outbreaks, integration of these monitoring technologies with higher-order data analytics with spatial mapping capability serves to enhance public health decision-making velocity.

6. CONCLUSION

The highly complex opportunities that exist with respect to controlling viral infections, an atmosphere of integration, custom design, collaboration, and innovation is depicted through an overview of emerging approaches. The combination of nano-virology and higher-order digital technologies, specifically artificial intelligence (AI) and machine learning (ML), is key to this future progress because, through the enhancement of therapeutic delivery, diagnostic accuracy, and predictive modeling capabilities, real-time, individualized, and adaptive intervention is made possible. Through biomimetic platforms and the programmability of nanoparticles, personalized nanomedicine brings specific therapy to each individual's unique immunological background and strain of virus to enhance effectiveness while reducing off-target toxicity. The current SARS-CoV-2 pandemic has highlighted the relevance of developing nanovaccines that can be rapidly reconfigured to respond to new emerging variants of viruses, delving beyond simple therapeutic remedies. International cooperation has been proven to be a crucial component of stimulating innovation, harmonizing regulation, and promoting equal access to viral interventions made possible by nanotechnology. Collective global consortia and public-private collaborations provide rapid translational routes while maintaining stringent standards of environmental care, protection of public safety, and scientific ethics. In order to make these technologies consistent and scalable worldwide, standardized manufacturing and quality control processes should be implemented. Nanotechnology offers versatile and rapid-response platforms of vaccines, therapies, and diagnostics against the relentless discovery of new viral threats. The unique physicochemical attributes of nanoparticles provide not only target-specific delivery of viruses but also broad-spectrum antiviral activity, and both of these play significant roles in epidemic control and lowering morbidity rates. Sophisticated nano-biosensors that combine environmental monitoring with nanotechnology improve real-time epidemiological monitoring that is significant to public health preparation. To take full advantage of the promise of nano-virology, significant technical obstacles associated with scalable manufacturing, safety verification, social approval, and clarity on regulation should be resolved. Ongoing interfacing research is key to developing nanotherapeutics that have wider safety profiles, efficiency, and economies, aided by design strategies of artificial intelligence and superior characterization methods. The pandemic of COVID-19 emphasized the importance of nanotechnology, particularly the delivery of mRNA vaccines by lipid nanoparticles. There seems to be a promise of a future world where nanomedicine can be a key protection against pandemics of viruses.

REFERENCES

Abasian, P., Shakibi, S., Maniati, M. S., Nouri Khorasani, S., & Khalili, S. (2021). Targeted delivery, drug release strategies, and toxicity study of polymeric drug nanocarriers. *Polymers for Advanced Technologies*, *32*(3), 931–944.

Abraham, J., & Reed, T. (2002). Progress, innovation and regulatory science in drug development: The politics of international standard-setting. *Social Studies of Science*, *32*(3), 337–369. PMID: 12269340

Acurcio, R. C., Kleiner, R., Vaskovich-Koubi, D., Carreira, B., Liubomirski, Y., Palma, C., & Satchi-Fainaro, R. (2024). Intranasal Multiepitope PD-L1-siRNA-Based Nanovaccine: The Next-Gen COVID-19 Immunotherapy. *Advancement of Science*, *11*(40), 2404159. PMID: 39116324

Ahmad, F., Salem-Bekhit, M. M., Khan, F., Alshehri, S., Khan, A., Ghoneim, M. M., & Elbagory, I. (2022). Unique properties of surface-functionalized nanoparticles for bio-application: Functionalization mechanisms and importance in application. *Nanomaterials (Basel, Switzerland)*, *12*(8), 1333. PMID: 35458041

Akagi, T., Baba, M., & Akashi, M. (2011). Biodegradable nanoparticles as vaccine adjuvants and delivery systems: regulation of immune responses by nanoparticle-based vaccine. In *Polymers in nanomedicine* (pp. 31–64). Springer Berlin Heidelberg.

Ali, Z., Liang, W., Jin, L., Tang, Y., Mou, X., Shah, M. A. A., & Li, Z. (2015). Development of magnetic nanoparticles based nucleic acid extraction method and application in hepatitis c virus chemiluminescent detection. *Science of Advanced Materials*, *7*(7), 1233–1240.

Antoniou, M., Jorgensen, A. L., & Kolamunnage-Dona, R. (2016). Biomarker-guided adaptive trial designs in phase II and phase III: A methodological review. *PLoS One*, *11*(2), e0149803. PMID: 26910238

Baig, M. M. F. A., & Chien, W. T. (2025). Nanotechnological approaches for the targeted delivery of CRISPR-Cas systems for genomic modifications, biomolecular sensing, and precision medicine. *Biomaterials Science*. PMID: 40704512

Beyer, T., Herrmann, M., Reiser, C., Bertling, W., & Hess, J. (2001). Bacterial carriers and virus-like-particles as antigen delivery devices: Role of dendritic cells in antigen presentation. *Current Drug Targets. Infectious Disorders*, *1*(3), 287–302. PMID: 12455402

Brai, A., Poggialini, F., Pasqualini, C., Trivisani, C. I., Vagaggini, C., & Dreassi, E. (2023). Progress towards adjuvant development: Focus on antiviral therapy. *International Journal of Molecular Sciences*, *24*(11), 9225. PMID: 37298177

Bukasov, R., Dossym, D., & Filchakova, O. (2021). Detection of RNA viruses from influenza and HIV to Ebola and SARS-CoV-2: A review. *Analytical Methods*, *13*(1), 34–55. PMID: 33283798

Bukasov, R., Dossym, D., & Filchakova, O. (2021). Detection of RNA viruses from influenza and HIV to Ebola and SARS-CoV-2: A review. *Analytical Methods*, *13*(1), 34–55. PMID: 33283798

Cardoso, M., & Peca, N, I., & CA Roque, A. (2012). Antibody-conjugated nanoparticles for therapeutic applications. *Current Medicinal Chemistry*, *19*(19), 3103–3127. PMID: 22612698

. Cetintas, H. C., Tonbul, H., Şahin, A., & Çapan, Y. (2021). Regulatory guidelines of the US food and drug administration and the european medicines agency for actively targeted nanomedicines. *Drug Delivery with Targeted Nanoparticles*, 725-741.

Chakravarty, M., & Vora, A. (2021). Nanotechnology-based antiviral therapeutics. *Drug Delivery and Translational Research*, *11*(3), 748–787. PMID: 32748035

Chakravarty, M., & Vora, A. (2021). Nanotechnology-based antiviral therapeutics. *Drug Delivery and Translational Research*, *11*(3), 748–787. PMID: 32748035

Chakravarty, M., & Vora, A. (2021). Nanotechnology-based antiviral therapeutics. *Drug Delivery and Translational Research*, *11*(3), 748–787. PMID: 32748035

Chalkias, S., Whatley, J. L., Eder, F., Essink, B., Khetan, S., Bradley, P., & Das, R. (2023). Original SARS-CoV-2 monovalent and Omicron BA. 4/BA. 5 bivalent COVID-19 mRNA vaccines: Phase 2/3 trial interim results. *Nature Medicine*, *29*(9), 2325–2333. PMID: 37653342

Charitidis, C. A., Georgiou, P., Koklioti, M. A., Trompeta, A. F., & Markakis, V. (2014). Manufacturing nanomaterials: From research to industry. *Manufacturing Review*, *1*, 11.

Chaudhary, K., Parihar, S., & Sharma, D. (2021). A critical review on nanoscience advancement: In treatment of viral infection. *Journal of Drug Delivery and Therapeutics*, *11*(6), 225–237.

Chen, F., Alphonse, M., & Liu, Q. (2020). Strategies for nonviral nanoparticle-based delivery of CRISPR/Cas9 therapeutics. *Wiley Interdisciplinary Reviews. Nanomedicine and Nanobiotechnology*, *12*(3), e1609. PMID: 31797562

Chen, R., Wang, T., Song, J., Pu, D., He, D., Li, J., & Zhang, J. (2021). Antiviral drug delivery system for enhanced bioactivity, better metabolism and pharmacokinetic characteristics. *International Journal of Nanomedicine*, •••, 4959–4984. PMID: 34326637

Cheng, F., Wang, Y., Bai, Y., Liang, Z., Mao, Q., Liu, D., & Xu, M. (2023). Research advances on the stability of mRNA vaccines. *Viruses*, *15*(3), 668. PMID: 36992377

Christ, C., Waldl, A., Liu, Y., Johnson, L., Auer, V., Cardozo, O., & Li, Y. (2025). Nano-scaled advanced materials for antimicrobial applications–mechanistic insight, functional performance measures, and potentials towards sustainability and circularity. *Environmental Science. Nano*.

Cui, H., Song, W., Ru, X., Fu, W., Ji, L., Zhou, W., & Jiang, G. (2023). A simplified viral RNA extraction method based on magnetic nanoparticles for fast and high-throughput detection of SARS-CoV-2. *Talanta*, *258*, 124479. PMID: 36966663

Curulli, A. (2023). Functional nanomaterials enhancing electrochemical biosensors as smart tools for detecting infectious viral diseases. *Molecules (Basel, Switzerland)*, *28*(9), 3777. PMID: 37175186

Draz, M. S., & Shafiee, H. (2018). Applications of gold nanoparticles in virus detection. *Theranostics*, *8*(7), 1985. PMID: 29556369

Dyawanapelly, S., Ghodke, S. B., Vishwanathan, R., Dandekar, P., & Jain, R. (2014). RNA interference-based therapeutics: Molecular platforms for infectious diseases. *Journal of Biomedical Nanotechnology*, *10*(9), 1998–2037. PMID: 25992447

Ehidiamen, A. J., & Oladapo, O. O. (2024). Enhancing ethical standards in clinical trials: A deep dive into regulatory compliance, informed consent, and participant rights protection frameworks. *World Journal of Biology Pharmacy and Health Sciences*, *20*(1), 309–320.

Elbehiry, A., & Abalkhail, A. (2025). Antimicrobial Nanoparticles Against Superbugs: Mechanistic Insights, Biomedical Applications, and Translational Frontiers. *Pharmaceuticals*, *18*(8), 1195. PMID: 40872586

Fang, E., Liu, X., Li, M., Zhang, Z., Song, L., Zhu, B., & Li, Y. (2022). Advances in COVID-19 mRNA vaccine development. *Signal Transduction and Targeted Therapy*, *7*(1), 94. PMID: 35322018

Fardoost, A., Karimi, K., Govindaraju, H., Jamali, P., & Javanmard, M. (2024). Applications of microfluidics in mRNA vaccine development: A review. *Biomicrofluidics*, *18*(6). PMID: 39553921

Filipic, B., Pantelić, I., Nikolić, I., Majhen, D., Stojić-Vukanić, Z., Savić, S., & Krajišnik, D. (2023). Nanoparticle-based adjuvants and delivery systems for modern vaccines. *Vaccines*, *11*(7), 1172. PMID: 37514991

Floyd, T. G., Gurnani, P., & Rho, J. Y. (2025). Characterisation of polymeric nanoparticles for drug delivery. *Nanoscale*, *17*(13), 7738–7752. PMID: 40018862

Galdiero, S., Falanga, A., Vitiello, M., Cantisani, M., Marra, V., & Galdiero, M. (2011). Silver nanoparticles as potential antiviral agents. *Molecules (Basel, Switzerland)*, *16*(10), 8894–8918. PMID: 22024958

Georgakopoulou, V. E. (2024). Insights from respiratory virus co-infections. *World Journal of Virology*, *13*(4), 98600. PMID: 39722753

Gimondi, S., Ferreira, H., Reis, R. L., & Neves, N. M. (2023). Microfluidic devices: A tool for nanoparticle synthesis and performance evaluation. *ACS Nano*, *17*(15), 14205–14228. PMID: 37498731

Gotting, J., Medeiros, P., Sanders, J. G., Li, N., Phan, L., Elabd, K., . . . Donoughe, S. (2025). Virology capabilities test (VCT): a multimodal virology Q&A benchmark. *arXiv [preprint]. arXiv: 2504.16137*, 2-31.

Guasch-Girbau, A., & Fernàndez-Busquets, X. (2021). Review of the current landscape of the potential of nanotechnology for future malaria diagnosis, treatment, and vaccination strategies. *Pharmaceutics*, *13*(12), 2189. PMID: 34959470

Gulati, S., Singh, P., Diwan, A., Mongia, A., & Kumar, S. (2020). Functionalized gold nanoparticles: Promising and efficient diagnostic and therapeutic tools for HIV/AIDS. *RSC Medicinal Chemistry*, *11*(11), 1252–1266. PMID: 34095839

Gupta, R., Arora, K., Roy, S. S., Joseph, A., Rastogi, R., Arora, N. M., & Kundu, P. K. (2023). Platforms, advances, and technical challenges in virus-like particles-based vaccines. *Frontiers in Immunology*, *14*, 1123805. PMID: 36845125

Gurunathan, S., Qasim, M., Choi, Y., Do, J. T., Park, C., Hong, K., & Song, H. (2020). Antiviral potential of nanoparticles—Can nanoparticles fight against coronaviruses? *Nanomaterials (Basel, Switzerland)*, *10*(9), 1645. PMID: 32825737

Ha, H., Lee, H. Y., Kim, J. H., Kim, D. Y., An, H. J., Bae, S., & Kang, J. H. (2024). Precision oncology clinical trials: A systematic review of phase II clinical trials with biomarker-driven, adaptive design. *Cancer Research and Treatment: Official Journal of Korean Cancer Association*, *56*(4), 991–1013. PMID: 38726510

Hussain, Z., Khan, S., Imran, M., Sohail, M., Shah, S. W. A., & de Matas, M. (2019). PEGylation: a promising strategy to overcome challenges to cancer-targeted nanomedicines: a review of challenges to clinical transition and promising resolution. *Drug Delivery and Translational Research*, *9*(3), 721–734. PMID: 30895453

Ishaque, P. I. (2025). Convergence at the Nanoscale: Transformative Advances in Drug Delivery, Vaccinology, and Biomedical Diagnostics. *Sch Acad J Pharm*, *6*, 128–162.

Islam, F. S., & Islam, M. N. (2025). AI-driven integration of nanotechnology and green nanotechnology for sustainable energy and environmental remediation. *Journal of Engineering Research and Reports*, *27*(7), 260–311.

Jackman, J. A., Yoon, B. K., Ouyang, L., Wang, N., Ferhan, A. R., Kim, J., & Cho, N. J. (2021). Biomimetic nanomaterial strategies for virus targeting: Antiviral therapies and vaccines. *Advanced Functional Materials*, *31*(12), 2008352.

Jain, S., Venkataraman, A., Wechsler, M. E., & Peppas, N. A. (2021). Messenger RNA-based vaccines: Past, present, and future directions in the context of the COVID-19 pandemic. *Advanced Drug Delivery Reviews*, *179*, 114000. PMID: 34637846

. Jampílek, J., & Kráľová, K. (2019). in the Therapy of Viral Diseases Attacking. *Nanotheranostics: Applications and Limitations*, 137.

Jan, S. A., Debnath, A., Singh, R. K., Tyagi, P. K., Singh, S., & Singh, A. K. (2025). Targeting Undruggable Proteins: The siRNA Revolution Beyond Small Molecules-Advances, Challenges, and Future Prospects in Therapeutic Innovation. *Current Gene Therapy*. PMID: 39917924

Jekhmane, S., De Haas, R., Paulino da Silva Filho, O., van Asbeck, A. H., Favretto, M. E., Hernandez Garcia, A., & De Vries, R. (2017). Virus-like particles of mRNA with artificial minimal coat proteins: Particle formation, stability, and transfection efficiency. *Nucleic Acid Therapeutics*, *27*(3), 159–167. PMID: 28437166

Kabir, E., Kumar, V., Kim, K. H., Yip, A. C., & Sohn, J. R. (2018). Environmental impacts of nanomaterials. *Journal of Environmental Management*, *225*, 261–271. PMID: 30096714

Kabir, M., Rana, M. R. H., & Debnath, A. (2024). The Role of Quality Assurance in Accelerating Pharmaceutical Research and Development: Strategies for Ensuring Regulatory Compliance and Product Integrity. *Integrative Biomedical Research*, *8*(12), 1–11.

Kalai, T., Pongrácz, J. E., & Mátyus, P. (2022). Recent advances in influenza, HIV and SARS-CoV-2 infection prevention and drug treatment—The need for precision medicine. *Chemistry (Basel, Switzerland)*, *4*(2), 216–258.

Kálai, T., Pongrácz, J. E., & Mátyus, P. (2022). Recent advances in influenza, HIV and SARS-CoV-2 infection prevention and drug treatment—The need for precision medicine. *Chemistry (Basel, Switzerland)*, *4*(2), 216–258.

Khan, M., Adil, S. F., Alkhathlan, H. Z., Tahir, M. N., Saif, S., Khan, M., & Khan, S. T. (2020). COVID-19: A global challenge with old history, epidemiology and progress so far. *Molecules (Basel, Switzerland)*, *26*(1), 39. PMID: 33374759

Khatua, R., Bhar, B., Dey, S., Jaiswal, C., & Mandal, B. B. (2024). Advances in engineered nanosystems: Immunomodulatory interactions for therapeutic applications. *Nanoscale*, *16*(27), 12820–12856. PMID: 38888201

Khosravi, A., Zarepour, A., Iravani, S., Varma, R. S., & Zarrabi, A. (2024). Sustainable synthesis: Natural processes shaping the nanocircular economy. *Environmental Science. Nano*, *11*(3), 688–707.

Kidd, P. (2003). Th1/Th2 balance: The hypothesis, its limitations, and implications for health and disease. *Alternative Medicine Review*, *8*(3), 223–246. PMID: 12946237

Koteswara Rao, V. (2021). Point of care diagnostic devices for rapid detection of novel coronavirus (SARS-nCoV19) pandemic: A review. *Frontiers in Nanotechnology*, *2*, 593619.

Kumar, A., Saini, S., Anvikar, A., Mishra, N., & Misra, G. (2025). Evolving Landscape of Emerging Virus Diagnosis: Challenges and Innovations. *Molecular Biotechnology*, •••, 1–23. PMID: 40042766

Kumar, S., Anselmo, A. C., Banerjee, A., Zakrewsky, M., & Mitragotri, S. (2015). Shape and size-dependent immune response to antigen-carrying nanoparticles. *Journal of Controlled Release*, *220*, 141–148. PMID: 26437263

Lakshmanan, K., & Liu, B. M. (2025). Impact of point-of-care testing on diagnosis, treatment, and surveillance of vaccine-preventable viral infections. *Diagnostics (Basel)*, *15*(2), 123. PMID: 39857007

Lee, S., Nyamzaya, K., Han, J., Song, Y., Lee, J., Jo, J. H., & Kim, E. (2025). Membrane Fusion-Inspired Nanomaterials: Emerging Strategies for Infectious Disease and Cancer Diagnostics. *Advanced Healthcare Materials*, •••, 2500485. PMID: 40599079

Lembo, D., & Cavalli, R. (2010). Nanoparticulate delivery systems for antiviral drugs. *Antiviral Chemistry & Chemotherapy*, *21*(2), 53–70. PMID: 21107015

Lembo, D., Donalisio, M., Civra, A., Argenziano, M., & Cavalli, R. (2018). Nanomedicine formulations for the delivery of antiviral drugs: A promising solution for the treatment of viral infections. *Expert Opinion on Drug Delivery*, *15*(1), 93–114. PMID: 28749739

Lin, Y., Wagner, E., & Lächelt, U. (2022). Non-viral delivery of the CRISPR/Cas system: DNA versus RNA versus RNP. *Biomaterials Science*, *10*(5), 1166–1192. PMID: 35103261

Lin, Y., Wagner, E., & Lächelt, U. (2022). Non-viral delivery of the CRISPR/Cas system: DNA versus RNA versus RNP. *Biomaterials Science*, *10*(5), 1166–1192. PMID: 35103261

Liu, T., Tian, Y., Zheng, A., & Cui, C. (2022). Design strategies for and stability of mRNA–lipid nanoparticle COVID-19 vaccines. *Polymers*, *14*(19), 4195. PMID: 36236141

Luceri, A., Francese, R., Lembo, D., Ferraris, M., & Balagna, C. (2023). Silver nanoparticles: Review of antiviral properties, mechanism of action and applications. *Microorganisms*, *11*(3), 629. PMID: 36985203

Ma, Y., Chan, C. Y., & He, M. L. (2007). RNA interference and antiviral therapy. *World Journal of Gastroenterology: WJG*, *13*(39), 5169. PMID: 17876887

Maghsoudnia, N., Eftekhari, R. B., Sohi, A. N., Zamzami, A., & Dorkoosh, F. A. (2020). Application of nano-based systems for drug delivery and targeting: A review. *Journal of Nanoparticle Research*, *22*(8), 245.

Malik, S., Muhammad, K., & Waheed, Y. (2023). Emerging applications of nanotechnology in healthcare and medicine. *Molecules (Basel, Switzerland)*, *28*(18), 6624. PMID: 37764400

Mauriz, E. (2020). Recent progress in plasmonic biosensing schemes for virus detection. *Sensors (Basel)*, *20*(17), 4745. PMID: 32842601

Mercer, J., Schelhaas, M., & Helenius, A. (2010). Virus entry by endocytosis. *Annual Review of Biochemistry*, *79*(1), 803–833. PMID: 20196649

Milane, L., & Amiji, M. (2021). Clinical approval of nanotechnology-based SARS-CoV-2 mRNA vaccines: Impact on translational nanomedicine. *Drug Delivery and Translational Research*, *11*(4), 1309–1315. PMID: 33512669

Milovanovic, M., Arsenijevic, A., Milovanovic, J., Kanjevac, T., & Arsenijevic, N. (2017). Nanoparticles in antiviral therapy. In *Antimicrobial nanoarchitectonics* (pp. 383–410). Elsevier.

Mohamed, N. A., Abou-Saleh, H., Mohamed, H. A., Al-Ghouti, M. A., Crovella, S., & Zupin, L. (2022). Think like a virus: Toward improving Nanovaccine Development against SARS-CoV-2. *Viruses*, *14*(7), 1553. PMID: 35891532

Mohite, P., Yadav, V., Pandhare, R., Maitra, S., Saleh, F. M., Saleem, R. M., & Uti, D. E. (2024). Revolutionizing cancer treatment: Unleashing the power of viral vaccines, monoclonal antibodies, and proteolysis-targeting chimeras in the new era of immunotherapy. *ACS Omega*, *9*(7), 7277–7295. PMID: 38405458

Mohsen, M. O., Zha, L., Cabral-Miranda, G., & Bachmann, M. F. (2017, December). Major findings and recent advances in virus–like particle (VLP)-based vaccines. [). Academic Press.]. *Seminars in Immunology*, *34*, 123–132. PMID: 28887001

Monteil, S., Casson, A. J., & Jones, S. T. (2021). Electronic and electrochemical viral detection for point-of-care use: A systematic review. *PLoS One*, *16*(9), e0258002. PMID: 34591907

Moon, J. J., Huang, B., & Irvine, D. J. (2012). Engineering nano-and microparticles to tune immunity. *Advanced Materials*, *24*(28), 3724–3746. PMID: 22641380

Moss, R. B., Moll, T., El-Kalay, M., Kohne, C., Soo Hoo, W., Encinas, J., & Carlo, D. J. (2004). Th1/Th2 cells in inflammatory disease states: Therapeutic implications. *Expert Opinion on Biological Therapy*, *4*(12), 1887–1896. PMID: 15571451

Motamedi, H., Ari, M. M., Alvandi, A., & Abiri, R. (2024). Principle, application and challenges of development siRNA-based therapeutics against bacterial and viral infections: A comprehensive review. *Frontiers in Microbiology*, *15*, 1393646. PMID: 38939184

Mura, S., Nicolas, J., & Couvreur, P. (2013). Stimuli-responsive nanocarriers for drug delivery. *Nature Materials*, *12*(11), 991–1003. PMID: 24150417

Naghib, S. M., Bakhshi, M., Ahmadi, B., & Bakhshi, A. (2024). Gold nanoparticles based antibacterial and antiviral functional materials. In *Antibacterial and Antiviral Functional Materials* (Vol. 2, pp. 221–256). American Chemical Society.

Nguyen, K. T., & Zhao, Y. (2015). Engineered hybrid nanoparticles for on-demand diagnostics and therapeutics. *Accounts of Chemical Research*, *48*(12), 3016–3025. PMID: 26605438

Nizam, N. U. M., Hanafiah, M. M., & Woon, K. S. (2021). A content review of life cycle assessment of nanomaterials: Current practices, challenges, and future prospects. *Nanomaterials (Basel, Switzerland)*, *11*(12), 3324. PMID: 34947673

Nooraei, S., Bahrulolum, H., Hoseini, Z. S., Katalani, C., Hajizade, A., Easton, A. J., & Ahmadian, G. (2021). Virus-like particles: Preparation, immunogenicity and their roles as nanovaccines and drug nanocarriers. *Journal of Nanobiotechnology*, *19*(1), 59. PMID: 33632278

Ogugua, J. O., Anyanwu, E. C., Olorunsogo, T., Maduka, C. P., & Ayo-Farai, O. (2024). Ethics and strategy in vaccination: A review of public health policies and practices. *International Journal of Science and Research Archive*, *11*(1), 883–895.

Palmerston Mendes, L., Pan, J., & Torchilin, V. P. (2017). Dendrimers as nanocarriers for nucleic acid and drug delivery in cancer therapy. *Molecules (Basel, Switzerland)*, *22*(9), 1401. PMID: 28832535

Parvin, N., Joo, S. W., & Mandal, T. K. (2024). Enhancing vaccine efficacy and stability: A review of the utilization of nanoparticles in mRNA vaccines. *Biomolecules*, *14*(8), 1036. PMID: 39199422

Pei, X., Tao, G., Wu, X., Ma, Y., Li, R., & Li, N. (2020). Nanomaterial-based multiplex optical sensors. *Analyst*, *145*(12), 4111–4123. PMID: 32490466

Pirzada, M., & Altintas, Z. (2022). Nanomaterials for virus sensing and tracking. *Chemical Society Reviews*, *51*(14), 5805–5841. PMID: 35735186

Pralong, A., Levine, H. L., Lilja, J., Gaasvik, Å. S. A., & Hummel, H. (2014). Paradigm shift for vaccine manufacturing facilities: The next generation of flexible, modular facilities. *Engineering in Life Sciences*, *14*(3), 244–253.

Prasad, R., Ghosh, A., Patel, V., Peng, B., Mendes, B. B., Win, E. H. A., & Conde, J. (2025). Voices of Nanomedicine: Blueprint Guidelines for Collaboration in Addressing Global Unmet Medical Needs. *ACS Nano*, *19*(3), 2979–2991. PMID: 39792961

Pucciarelli, M. (2023). *Life Cycle Thinking for the sustainability assessment of nanoparticles' manufacturing and applications* (Doctoral dissertation, UCL (University College London)).

Qasim, M., Lim, D. J., Park, H., & Na, D. (2014). Nanotechnology for diagnosis and treatment of infectious diseases. *Journal of Nanoscience and Nanotechnology*, *14*(10), 7374–7387. PMID: 25942798

Quinn, S. C., & Kumar, S. (2014). Health inequalities and infectious disease epidemics: A challenge for global health security. *Biosecurity and Bioterrorism*, *12*(5), 263–273. PMID: 25254915

Rahman, M. (2024). *Metabolomics for Personalized Vaccinology*. Elsevier.

Rai, B., Srivastava, J., & Saxena, P. (2025). Challenges in translating laboratory findings to drug development. In *Protein Misfolding in Neurodegenerative Diseases* (pp. 319–354). Academic Press.

Ran, J., Wang, X., Liu, Y., Yin, S., Li, S., & Zhang, L. (2023). Microreactor-based micro/nanomaterials: Fabrication, advances, and outlook. *Materials Horizons*, *10*(7), 2343–2372. PMID: 37139613

Ratan, Z. A., Mashrur, F. R., Chhoan, A. P., Shahriar, S. M., Haidere, M. F., Runa, N. J., & Cho, J. Y. (2021). Silver nanoparticles as potential antiviral agents. *Pharmaceutics*, *13*(12), 2034. PMID: 34959320

Ray, A., Simpson, J. D., Demir, I., Gisbert, V. G., Gomes, D. B., Amadei, F., & Alsteens, D. (2025). From viral assembly to host interaction: AFM's contributions to virology. *Journal of Virology*, *99*(1), e00873–e24. PMID: 39655953

Ren, H., Jia, W., Xie, Y., Yu, M., & Chen, Y. (2023). Adjuvant physiochemistry and advanced nanotechnology for vaccine development. *Chemical Society Reviews*, *52*(15), 5172–5254. PMID: 37462107

Riehemann, K., Schneider, S. W., Luger, T. A., Godin, B., Ferrari, M., & Fuchs, H. (2009). Nanomedicine—Challenge and perspectives. *Angewandte Chemie International Edition*, *48*(5), 872–897. PMID: 19142939

Saad, M., Garbuzenko, O. B., Ber, E., Chandna, P., Khandare, J. J., Pozharov, V. P., & Minko, T. (2008). Receptor targeted polymers, dendrimers, liposomes: Which nanocarrier is the most efficient for tumor-specific treatment and imaging? *Journal of Controlled Release*, *130*(2), 107–114. PMID: 18582982

Safo, K. K., Ogoe, M., George, N., Gupta, A., Janiyani, K., Singh, H., & Pandya, S. R. (2025). Nanotechnology for RNA and DNA Delivery in Cancer Therapeutics. In *Nanoparticle Delivery in Cancer Care* (pp. 199–234). Springer Nature Singapore.

Saini, V., Zharov, V. P., Brazel, C. S., Nikles, D. E., Johnson, D. T., & Everts, M. (2006). Combination of viral biology and nanotechnology: New applications in nanomedicine. *Nanomedicine; Nanotechnology, Biology, and Medicine*, *2*(3), 200–206. PMID: 17292143

Sanchez, C., Belleville, P., Popall, M., & Nicole, L. (2011). Applications of advanced hybrid organic–inorganic nanomaterials: From laboratory to market. *Chemical Society Reviews*, *40*(2), 696–753. PMID: 21229132

Saylan, Y., & Denizli, A. (2020). Virus detection using nanosensors. In *Nanosensors for smart cities* (pp. 501–511). Elsevier.

Schmidt, A., Helgers, H., Vetter, F. L., Juckers, A., & Strube, J. (2021). Fast and flexible mrna vaccine manufacturing as a solution to pandemic situations by adopting chemical engineering good practice—Continuous autonomous operation in stainless steel equipment concepts. *Processes (Basel, Switzerland), 9*(11), 1874.

Sharmile, N., Chowdhury, R. R., & Desai, S. (2025). A comprehensive review of quality control and reliability research in micro–nano technology. *Technologies, 13*(3), 94.

. Shastry, K. A., & Shastry, A. (2023). An integrated deep learning and natural language processing approach for continuous remote monitoring in digital health. *Decision analytics journal, 8*, 100301.

Shegokar, R., & Nakach, M. (2020). Large-scale manufacturing of nanoparticles—An industrial outlook. In *Drug delivery aspects* (pp. 57–77). Elsevier.

Singh, L., Kruger, H. G., Maguire, G. E., Govender, T., & Parboosing, R. (2017). The role of nanotechnology in the treatment of viral infections. *Therapeutic Advances in Infectious Disease, 4*(4), 105–131. PMID: 28748089

Singh, L., Kruger, H. G., Maguire, G. E., Govender, T., & Parboosing, R. (2017). The role of nanotechnology in the treatment of viral infections. *Therapeutic Advances in Infectious Disease, 4*(4), 105–131. PMID: 28748089

Singh, L., Kruger, H. G., Maguire, G. E., Govender, T., & Parboosing, R. (2017). The role of nanotechnology in the treatment of viral infections. *Therapeutic Advances in Infectious Disease, 4*(4), 105–131. PMID: 28748089

Singh, P., Gonzalez, M. J., & Manchester, M. (2006). Viruses and their uses in nanotechnology. *Drug Development Research, 67*(1), 23–41.

Tan, J., Ding, B., Teng, B., Ma, P. A., & Lin, J. (2022). Understanding structure–function relationships of nanoadjuvants for enhanced cancer vaccine efficacy. *Advanced Functional Materials, 32*(16), 2111670.

Tariq, H., Batool, S., Asif, S., Ali, M., & Abbasi, B. H. (2022). Virus-like particles: Revolutionary platforms for developing vaccines against emerging infectious diseases. *Frontiers in Microbiology, 12*, 790121. PMID: 35046918

Tenchov, R., Bird, R., Curtze, A. E., & Zhou, Q. (2021). Lipid nanoparticles— From liposomes to mRNA vaccine delivery, a landscape of research diversity and advancement. *ACS Nano, 15*(11), 16982–17015. PMID: 34181394

Thwala, L. N., Ndlovu, S. C., Mpofu, K. T., Lugongolo, M. Y., & Mthunzi-Kufa, P. (2023). Nanotechnology-based diagnostics for diseases prevalent in developing countries: Current advances in point-of-care tests. *Nanomaterials (Basel, Switzerland), 13*(7), 1247. PMID: 37049340

Tong, L., Liu, D., Cao, Z., Zheng, N., Mao, C., Liu, S., & Liu, S. (2023). Research status and prospect of non-viral vectors based on siRNA: A review. *International Journal of Molecular Sciences, 24*(4), 3375. PMID: 36834783

TSAKIRI. M. A. (2020). *EVALUATION OF NANO-SYSTEMS AS VACCINE PLATFORMS AGAINST INFECTIOUS DISEASES* (Doctoral dissertation, National and Kapodistrian University of Athens).

TSAKIRI. M. A. (2020). *EVALUATION OF NANO-SYSTEMS AS VACCINE PLATFORMS AGAINST INFECTIOUS DISEASES* (Doctoral dissertation, National and Kapodistrian University of Athens).

. Van Dongen, J. E. (2022). Novel Biosensors for Early Cancer Detection: Development of Gold Nanoparticle and CRISPR/Cas Assays for the Optical Detection of Cancer Biomarkers in Urine.

Vasu, M., Yadav, P., Satyanarayana, Y., Ahlawat, J., Sandeep, D., & Kumar, K. N. (2024). Golden Threads of Lipid Nanoparticles (LNPs) contributing to COVID-19 mRNA Vaccine: A Review. *BioNanoScience*, *14*(3), 3430–3439.

Verbeke, R., Lentacker, I., De Smedt, S. C., & Dewitte, H. (2021). The dawn of mRNA vaccines: The COVID-19 case. *Journal of Controlled Release*, *333*, 511–520. PMID: 33798667

Virkutyte, J., & Varma, R. S. (2011). Green synthesis of metal nanoparticles: Biodegradable polymers and enzymes in stabilization and surface functionalization. *Chemical Science (Cambridge)*, *2*(5), 837–846.

Wakaskar, R. R. (2018). General overview of lipid–polymer hybrid nanoparticles, dendrimers, micelles, liposomes, spongosomes and cubosomes. *Journal of Drug Targeting*, *26*(4), 311–318. PMID: 28797169

Wang, H., Chen, Y., Wang, L., Liu, Q., Yang, S., & Wang, C. (2023). Advancing herbal medicine: Enhancing product quality and safety through robust quality control practices. *Frontiers in Pharmacology*, *14*, 1265178. PMID: 37818188

Wang, Y., Gao, Y., Yin, Y., Pan, Y., Wang, Y., & Song, Y. (2022). Nanomaterial-assisted microfluidics for multiplex assays. *Microchimica Acta*, *189*(4), 139. PMID: 35275267

Weerarathna, I. N., Doelakeh, E. S., Kiwanuka, L., Kumar, P., & Arora, S. (2024). Prophylactic and therapeutic vaccine development: Advancements and challenges. *Molecular Biomedicine*, *5*(1), 57. PMID: 39527305

Wen, A. M., & Steinmetz, N. F. (2016). Design of virus-based nanomaterials for medicine, biotechnology, and energy. *Chemical Society Reviews*, *45*(15), 4074–4126. PMID: 27152673

Wilson, B., & Geetha, K. M. (2022). Lipid nanoparticles in the development of mRNA vaccines for COVID-19. *Journal of Drug Delivery Science and Technology*, *74*, 103553. PMID: 35783677

Wu, Z., Huang, J., & Zuo, K. (2025). Oscillation of doubt: A historical odyssey of vaccine hesitancy and the evolution of public trust. *Hist Philos Med*, *7*(1), 4.

Yakoubi, A., & Dhafer, C. E. B. (2023). Advanced plasmonic nanoparticle-based techniques for the prevention, detection, and treatment of current COVID-19. *Plasmonics*, *18*(1), 311–347. PMID: 36588744

Yetisgin, A. A., Cetinel, S., Zuvin, M., Kosar, A., & Kutlu, O. (2020). Therapeutic nanoparticles and their targeted delivery applications. *Molecules (Basel, Switzerland)*, *25*(9), 2193. PMID: 32397080

Younis, S., Taj, A., Zia, R., Hayat, H., Shaheen, A., Awan, F. R., & Bajwa, S. Z. (2020). Nanosensors for the detection of viruses. In *Nanosensors for smart cities* (pp. 327–338). Elsevier.

Yusuf, A., Almotairy, A. R. Z., Henidi, H., Alshehri, O. Y., & Aldughaim, M. S. (2023). Nanoparticles as drug delivery systems: A review of the implication of nanoparticles' physicochemical properties on responses in biological systems. *Polymers*, *15*(7), 1596. PMID: 37050210

Zeltins, A. (2013). Construction and characterization of virus-like particles: A review. *Molecular Biotechnology*, *53*(1), 92–107. PMID: 23001867

Zepeda-Cervantes, J., Ramírez-Jarquín, J. O., & Vaca, L. (2020). Interaction between virus-like particles (VLPs) and pattern recognition receptors (PRRs) from dendritic cells (DCs): Toward better engineering of VLPs. *Frontiers in Immunology*, *11*, 1100. PMID: 32582186

Zepeda-Cervantes, J., Ramírez-Jarquín, J. O., & Vaca, L. (2020). Interaction between virus-like particles (VLPs) and pattern recognition receptors (PRRs) from dendritic cells (DCs): Toward better engineering of VLPs. *Frontiers in Immunology*, *11*, 1100. PMID: 32582186

Zhao, S., Yu, X., Qian, Y., Chen, W., & Shen, J. (2020). Multifunctional magnetic iron oxide nanoparticles: An advanced platform for cancer theranostics. *Theranostics*, *10*(14), 6278. PMID: 32483453

Zhao, Z., Cui, H., Song, W., Ru, X., Zhou, W., & Yu, X. (2020). A simple magnetic nanoparticles-based viral RNA extraction method for efficient detection of SARS-CoV-2. BioRxiv, 2020-02.

Zigrayova, D., Mikušová, V., & Mikuš, P. (2023). Advances in antiviral delivery systems and chitosan-based polymeric and nanoparticulate antivirals and antiviral carriers. *Viruses*, *15*(3), 647. PMID: 36992356

About the Contributors

Javed Iqbal earned his Ph.D. in Plant Biochemistry and Molecular Biology from Quaid-i-Azam University, Islamabad, Pakistan. He is currently serving as an Assistant Professor (TTS) in the Department of Botany at Bacha Khan University, Charsadda, Pakistan. A double gold medalist and HEC-approved supervisor, Dr. Iqbal has made significant academic contributions in the fields of green nanotechnology, plant biochemistry, and medicinal plant-based bioassays. His research focuses on the green synthesis of metal oxide nanoparticles and their biological applications. He completed a six-month research fellowship at the University of California, Los Angeles, and has published over 30 research articles in reputed international journals. He is an active reviewer for numerous SCI-indexed journals and participates in academic leadership roles within his department.

Muhammad Anas is a plant physiologist specializing in abiotic stress biology, nanotechnology-based interventions, and multi-omics approaches in crop plants. He earned his PhD in Plant Sciences from Quaid-i-Azam University, Islamabad, where his research focused on mitigating heavy-metal toxicity in bread wheat through biogenic quantum dots and nano-enabled physiological, ionomic, and antioxidant regulation. He completed a TÜBİTAK–TWAS Research Fellowship at Selçuk University, Türkiye, working on miRNAomics of wheat under cadmium and chromium stress using small-RNA sequencing and bioinformatics. Dr. Anas has authored more than 40 Scopus-indexed peer-reviewed research articles and over 30 scholarly book chapters with leading international publishers. He has also served as a visiting lecturer and teaching assistant at Quaid-i-Azam University, with academic work focused on strengthening crop resilience and sustainable agricultural systems.

Ayush Madan earned his Ph.D. in Biotechnology from Shobhit University, India. He is currently affiliated as an Assistant Professor at Department of Biotechnology, People's University, Bhopal, India. Previously, He worked as a Graduate Research Assistant at Universiti Malaysia Terengganu, Malaysia, where he contributes to projects on nanotechnology-driven bioremediation and sustainable agriculture. He completed his M.Tech. in Biotechnology from Graphic Era University, India. Dr. Madan's academic background includes research at Biocon Research Limited and ONGC, with expertise in microbiology, bioinformatics, and nanobiotechnology. He has published several book chapters with prominent international publishers, focusing on topics such as environmental sustainability, cancer immunotherapy, and AI-driven recycling technologies. His research emphasizes the intersection of biotechnology and environmental health.

Ghulam Murtaza received his Ph.D. in Environmental Science and is currently a researcher at the School of Ecology and Environmental Sciences, Yunnan University, China. His research centers on the application of modified and aged biochars for environmental remediation, particularly for pollutant removal in soil and water systems. Dr. Murtaza has authored numerous peer-reviewed articles in journals such as Scientific Reports, Chemosphere, and BMC Plant Biology. His academic work has contributed to advancing sustainable solutions in agriculture and environmental management. With a focus on biochar-soil-plant interactions, he plays an active role in promoting ecological resilience and soil restoration through innovative biotechnological approaches.

Banzeer Ahsan Abbasi received her graduate degree in Botany from Quaid-i-Azam University, Islamabad, Pakistan. Dr. Banzeer is currently working as an Assistant Professor at Department of Botany, Rawalpindi Women University, Rawalpindi, Pakistan. She has published more than 60 research articles in various international journals and 10 chapters in reputed publishers (ELSIEVER, RSC and SPRINGER). Currently, she is supervising 2 Ph.D, 6 MPhil and 6 BS students. She has expertise in the synthesis of novel nanomaterials from medicinal plants and their multiple biological and envirnemental applications. Her research is also focusing on phytochemistry, natural products of medicinal plants and Phylogenetics. She has a very decent research background with research published in well-reputed journals with H-index 28, i-10 index 44 and overall citations 3317.

Rajiv Dutta earned M.Tech. in Biotech. and Engg. IIT, Kharagpur and Ph.D. in Bioelectronics from BITS, Pilani. Prof. Dutta is Professor in Bioengineering at School of Engg. & Tech., Sharda University, India. He was also Faculty at the Department of Botany at Oklahoma State University, Stillwater, OK and Department of Biological Sciences at Purdue University, West Lafayette, IN, and carried out advanced research in the area of Ion channels electrophysiology and Membrane Nanopore. He is Fellow of American Association of Science & Technology (FAAST). He is Hon. Scientific Advisor to Serra BioLife Ltd., Melbourne, Australia. He had received various awards including JCI Outstanding Ten Young Person of India and ISBEM Dr. Ramesh Gulrajani Memorial Award 2006 and Outstanding Researcher Award-2011 at Ion Channel Retreat, Vancouver, Canada. He is Editorial Board Member of several journals including American Journal of Plant Physiology, American Journal of Science & Technology, International Journal of Engineering etc. He has published 44 publications in highly cited journals and three patents to his credits. Prof. Dutta visited USA, UK, Germany, France, Japan, Italy, and China.

Rashid Abbas Khan is a Fellow of UNESCO-UNISA-iTLABS/NRF, Africa Chair in Nano, Pretoria-Republic of South Africa. His research explores plant-microbe interactions, plant stress physiology, and the use of bio-fertilizers and nanoparticles for biocontrol. He has contributed significantly to understanding plant growth-promoting rhizobacteria (PGPR) and its role in enhancing plant stress resilience. His work also delves into the biochemical and antioxidant responses of plants under stress conditions, with a focus on improving agricultural productivity through sustainable practices.

Waseem Ahmed Khattak is a researcher and PhD scholar in Plant Sciences at Quaid-i-Azam University (QAU), Islamabad, specializing in applied mycology, plant physiology, nanotechnology, and sustainable agriculture. He earned his MPhil in Plant Sciences (Applied Mycology) from QAU, where his work focused on plant–fungus interactions under stress. His research spans nanoparticles for heavy metal stress alleviation, microbial ecology, fungal biotechnology, and functional foods, with projects exploring selenium nanoparticles, chitosan-coated nanofertilizers, and ascorbic acid-based mitigation of lead toxicity in crops. Waseem has authored research articles, reviews, and book chapters on topics such as nanosensors, bioremediation, microbial interactions, and sustainable agriculture. Proficient in plant physiology assays, nanoparticle synthesis, and scientific writing, he actively mentors students and contributes to interdisciplinary collaborations. His long-term goal is to advance eco-friendly innovations in agriculture and environmental sustainability.

Ghaida Mustafa is a Sudanese pharmacist and lecturer in pharmaceutics at University of Khartoum, Sudan. She is currently a PhD researcher in pharmaceutical technologies at University of Bradford with aim to develop innovative formulations from natural materials to tackle neglected diseases in Sudan. Her PhD project is funded by the prestigious Faculty For The Future (FFTF) program offered by Schlumberger Foundation. Beyond the lab, Ghaida is passionate about community-based, supporting women and STEM engaging activities and programs that turn science into impact and inspire future innovators.

Sonia Navvuru Deeply rooted in the rich tradition of Indian agriculture and inspired by the resilience of its farmers, I have embarked on a passionate journey towards a career as a plant breeder, driven by a commitment to contribute to global agricultural development. My academic pursuit began at the University of Agricultural Sciences, Bangalore, India, where I immersed myself in the multifaceted realms of Plant Sciences, Agronomy, Biotechnology, and more. With a thirst for deeper understanding, I endeavored to excel in my field, earning the prestigious AIR 42 rank in the ICAR examination and securing a scholarship for my master's degree at Punjab Agricultural University, Ludhiana. Here, specializing in Plant Breeding and Genetics, I delved into cutting-edge research, employing techniques like GWAS and qRTPCR to unravel the genetic intricacies of aphid resistance breeding in Brassica juncea showcased my dedication to practical solutions in agricultural challenges. Throughout my academic journey, I seized every opportunity to share my knowledge and passion, actively participating in seminars, debates, elocution, and poster presentations, earning accolades along the way. Fueled by my research ex

Anant Paradkar is a Professor and Director of the Institute of Health and Social Care (IHSC). He previously served as Director of the Centre for Pharmaceutical Engineering Science, an interdisciplinary research and knowledge transfer centre. Anant joined the University of Bradford following a distinguished 20-year academic career in India. His research focuses on the development of innovative products and processes, supported by funding from EPSRC, Innovate UK, regional agencies, and industry partners across multiple sectors. He has successfully developed and commercialised several novel technologies in collaboration with corporate partners. Anant has authored over 170 peer-reviewed research papers and holds seven patents. He has supervised 54 PhD students to completion and was shortlisted for the Times Higher Education (THE) PhD Supervisor of the Year Award in 2020.

Zakir Ullah is a multidisciplinary researcher specialising in cancer biology, nanotechnology, virology, phytochemistry, and molecular biology. He earned his Ph.D. in Biology for developing phytofabricated metal oxide nanoparticles with biomedical and agricultural applications. At the University of Bradford (UK), his current research nanoplatforms for glioblastoma therapy. His previous work explored nanoparticle-mediated antiviral defence in plants using molecular, transcriptomic, and biochemical approaches. Dr Ullah's expertise spans nanomaterial synthesis, drug delivery, gene transformation, and transgenic plant development, contributing to advancements in nanomedicine and sustainable biotechnology.

Index

I

L

M

N

O

P

Q

R

S

www.ingramcontent.com/pod-product-compliance
Lightning Source LLC
LaVergne TN
LVHW080531300926
849162LV00007B/1585